AF819735

Homogeneous Catalysis by Metal Complexes

VOLUME I

Activation of Small Inorganic Molecules

Homogeneous Catalysis by Metal Complexes

VOLUME I

Activation of Small Inorganic Molecules

M. M. TAQUI KHAN

Department of Chemistry
Nizam College
Osmania University
Hyderabad, India

ARTHUR E. MARTELL

Department of Chemistry
Texas A & M University
College Station, Texas

ACADEMIC PRESS New York and London 1974

A Subsidiary of Harcourt Brace Jovanovich, Publishers

COPYRIGHT © 1974, BY ACADEMIC PRESS, INC.
ALL RIGHTS RESERVED.
NO PART OF THIS PUBLICATION MAY BE REPRODUCED OR TRANSMITTED IN ANY FORM OR BY ANY MEANS, ELECTRONIC OR MECHANICAL, INCLUDING PHOTOCOPY, RECORDING, OR ANY INFORMATION STORAGE AND RETRIEVAL SYSTEM, WITHOUT PERMISSION IN WRITING FROM THE PUBLISHER.

ACADEMIC PRESS, INC.
111 Fifth Avenue, New York, New York 10003

United Kingdom Edition published by
ACADEMIC PRESS, INC. (LONDON) LTD.
24/28 Oval Road, London NW1

Library of Congress Cataloging in Publication Data

Taqui Khan, M M
Homogeneous catalysis by metal complexes.

Includes bibliographical references.
CONTENTS: v. 1. Activation of small inorganic molecules.–v. 2. Activation of alkenes and alkynes.
1. Catalysis. 2. Metal catalysts. 3. Complex compounds. I. Martell, Arthur Earl, Date joint author. II. Title.
QD505.T36 546 72-9982
ISBN 0-12-406101-X (v. 1)

PRINTED IN THE UNITED STATES OF AMERICA

Contents

PREFACE vii
CONTENTS OF VOLUME II xi

1 Activation of Molecular Hydrogen

I. Introduction 1
II. Structure and Reactivity in Metal Hydride Complexes 2
III. Activation of Hydrogen by Copper and Silver Ions 11
IV. Activation by Transition Metal Halides 17
V. Activation of Alkyl and Acyl Halides 23
VI. Activation of the Carbon–Hydrogen Bond by Platinum Group Metal Complexes 30
VII. Activation of Molecular Hydrogen by Pentacyanocobaltate (II) 33
VIII. Hydrogenation of π Complexes 42
References 72

2 Activation of Molecular Oxygen

I. Introduction 79
II. Reaction Types 79
III. Energy Requirements of Oxygen Reduction 82
IV. Non-Insertion Reactions Involving the Reduction of Oxygen to Hydrogen Peroxide 84
V. Non-Insertion Reactions Involving Complete Reduction of Oxygen to Water 127
VI. Non-Insertion Reactions of Hydrogen Peroxide 136
VII. Partial Insertion Reactions 143
VIII. Total Insertion Reactions 170
References 174

3 Activation of Molecular Nitrogen

I. Introduction 181
II. Activation by Microorganisms 182
III. Transition Metal Complexes of Molecular Nitrogen 200

IV. Fixation of Molecular Nitrogen by Metal Ions and Metal Complexes 266
V. Summary 285
References 285

4 Activation of Carbon Monoxide

I. Introduction 293
II. Carbon Monoxide Exchange Reactions 294
III. Carbon Monoxide Insertion Reactions 296
IV. Synthetic Applications of Carbonyl Insertion Reactions 316
V. Catalysis of Oxidation of Carbon Monoxide by Metal Ions and Metal Complexes 344
VI. Reactions of Coordinated Carbon Monoxide 351
References 353

5 Activation of Nitric Oxide

I. Introduction 357
II. Transition Metal Nitrosyls 358
III. Nitrosation of Organic Compounds 375
IV. Applications of Metal-Catalyzed Nitrosation Reactions 382
References 387

Appendix I **Formulas, Bonding, and Formal Charges** 391

Appendix II **Glossary of Terms and Abbreviations** 394

AUTHOR INDEX 397
SUBJECT INDEX 415

Preface

During the past twenty-five years the development of the field of coordination chemistry has gone through several general yet discernible phases. After a classic beginning based largely on descriptive and stereochemical studies, a more physical approach developed in which quantitative equilibrium studies led to the understanding of the thermodynamics of complex formation in solution. Subsequently the development of the ligand field theory and bonding concepts made possible successful correlations between electronic and magnetic spectra and the constitution and properties of coordination compounds. Presently, improvements in X-ray crystallographic techniques are providing a large body of structure–property correlations and producing a new level of understanding of coordination chemistry. Also, synthesis of new types of metal complexes of small molecules, many of which contain metal–carbon sigma and pi bonds, has opened new vistas in coordination chemistry, since these compounds are frequently catalysts or intermediates in the synthesis of new organic and organometallic compounds. In addition to their general applications to homogeneous catalysis, the complexes of nitrogen and oxygen in particular are of interest as models for biological oxidation and nitrogen fixation.

The purpose of this two-volume work is to review and systematize the chemistry of reaction of metal ions with small molecules, and to include chapters on hydrogen, oxygen, nitrogen, carbon monoxide, nitric oxide, and the alkenes and alkynes. Because many coordination chemists are primarily interested in the inorganic complexes, the decision was made to publish the metal ion activation of the small diatomic molecules as Volume I. Metal ion activation of alkenes and alkynes comprises Volume II. While the subject of metal ion activation of alkenes and alkynes is of vital importance to the field of coordination chemistry, it is broader in scope and would be of interest to organic and organometallic chemists as well as to coordination chemists.

As with many monographs, this work is the result of an earlier attempt to review and systematize the subject of interest for our own purposes. This effort began about seven years ago with a review of metal ion activation of molecular oxygen, a general subject closely related to previous research work on which we had collaborated. Completion of this review led over the next few years to the development of similar reviews on the activation of hydrogen, carbon monoxide, and unsaturated hydrocarbons. In the initial phase, the subject of nitrogen activation was not given sufficient weight to justify a full chapter, however the rapid growth of that field over the past few years led to the development of a nitrogen chapter that is now the most extensive in the work. Similar recent developments in nitrosation reactions have led to a chapter which is greatly expanded over its original concept; however in this case the field does not seem to have developed sufficiently to justify a treatment comparable to the other subjects.

Although the work began as a review for our own use only, we were prevailed upon by friends to prepare the work for publication. In a treatment of this nature, it is impossible to be expert in all phases of the field. Thus first-hand knowledge of metal ion activation of oxygen and nitrogen does not provide much insight into the fine points of reactions such as catalytic hydrogenation and nitrosation. In areas outside our range of expertise we have tried to present reactions and interpretations in a manner that seems reasonable to us. If our points of view do not coincide with those who are more experienced with the reactions under consideration, we hope for a charitable judgment of our treatment. We hope we have managed to present interpretations that are sufficiently original to make positive contributions to the subject at hand.

We have used our own preferred conventions for representation of covalent and coordinate bonds of organometallic compounds and metal complexes. We have related this to a consistent distribution of formal charges in the formulas used to represent these compounds. Details of the method employed have been presented in Appendix I. It is our hope that the readers will consider this approach to be both reasonable and satisfying. In any case one of its strong points is the designation of all formal charges for metal ions and ligand atoms, and where reasonably possible, their locations. While the formal charges assigned to the metal ions in many complexes seem to deviate widely from the accepted oxidation states, it is believed that the formalism employed corresponds more closely to (or reflects more satisfactorily variations in) the true charge distribution. In any case it is felt that our methods allow consistent and complete electron bookkeeping for the formulas under consideration.

We express our thanks to all those who provided valuable assistance in the preparation of the manuscript. Reviews of various chapters by the following professional friends and associates were particularly helpful: Professor Minoru Tsutsui (nitrogen), Professor Gordon Hamilton (oxygen), Dr. Elmer Wymore (hydrogen), and Dr. J. L. Herz (oxygen). Thanks and appreciation are also extended to Dr. R. Motekaitis and Dr. Badar Taqui Khan for assistance in proofreading the typescript and the galleys, to Robert M. Smith for literature searches, and to Mary Martell for editorial assistance and for typing the several versions of the manuscript that were produced over the long period of time during which this work developed.

M. M. Taqui Khan
Arthur E. Martell

Contents of Volume II

Introduction
Migration of π Bonds
The Oxo Reaction
Hydrosilation of Alkenes and Alkynes
Oxidation of Alkenes and Alkynes
Multiple Insertion Reaction of Alkenes and Alkynes
Bibliography
Appendix I: Formulas, Bonding, and Formal Charges
Appendix II: Glossary of Terms and Abbreviations
Author Index—Subject Index

Homogeneous Catalysis by Metal Complexes

VOLUME I

Activation of Small Inorganic Molecules

1

Activation of Molecular Hydrogen

I. Introduction

The activation of molecular hydrogen in homogeneous systems by metal complexes has received considerable attention because of possible commercial and synthetic applications, as well as interest in the catalytic processes and reaction intermediates in themselves as examples of new types of inorganic chemical reactions. Since many of the reactions of molecular hydrogen that are catalyzed by metal ions are carried out in common solvents and even in aqueous solution, and since the reactions frequently occur under mild conditions, there has been considerable success in interpreting the reaction mechanisms in terms of the nature of reactive intermediates and the properties of the metal catalysts.

In the extensive investigations of the activation of molecular hydrogen in aqueous solution by Halpern and co-workers [1–4], the metal ions copper(II), mercury(II), mercury(I), copper(I), and silver(I) were reported to be effective catalysts. It has been noted that metals that are good heterogeneous catalysts—ruthenium, cobalt, nickel, palladium, and platinum—have the same number of electrons in their valence shell as the catalytically active metal ions—palladium(II), copper(II), copper(I), silver(I), and mercury(II), respectively. Thus homogeneous and heterogeneous catalysis of the activation of molecular hydrogen seems to arise from similar electron characteristics, the most important of which are the *d* electron configurations and electron affinities of the catalytic species. From the results published thus far it may be concluded that only transition metal ions that possess electron configurations in the d^5–d^{10} range are catalysts in the activation of molecular hydrogen.

The electronic configuration, however, is not the only factor required for catalytic activity of a metal ion. The transition metal ions manganese(II), cobalt(II), nickel(II), and iron(III) are all inactive in spite of the fact that they have electrons in the d^5–d^{10} range. It has also been observed that metal ions that possess catalytic hydrogenation activity seem to be those that form labile hydrido complexes in solution [4]. It seems, therefore, that hydrido complexes are probably essential intermediates in metal-catalyzed homogeneous hydrogenation reactions, and in metal-catalyzed reactions in which hydrogen acts as a reducing agent.

Since metal hydrides apparently have an important role in the mechanism of activation of molecular hydrogen, it is useful at this stage to consider the nature of metal–hydrogen bonds and the factors influencing lability and stability in metal hydride complexes.

II. Structure and Reactivity in Metal Hydride Complexes

Considerable physical data are now available on the structures of stable hydrido complexes in the solid state [5]. The nature of the metal–hydrogen bond in stable hydrido complexes, such as ReH_4^-, $HCo(CN)_5$, and $HPtBr(PEt_3)_2$, has been shown to be covalent by NMR (nuclear magnetic resonance) studies of Wilkinson [6] and of Griffith and Wilkinson [7]. In these complexes the hydride ion may be considered as an anionic donor that occupies one of the normal coordination positions of the metal ion. Further examples of stable metal hydride complexes for which X-ray data show the hydride ion to occupy one of the coordination positions in the metal complex are the following: *trans*-$PtHBr(PEt_3)_2$ [8], $HMn(CO)_4$ [9], $OsHBr(CO)(PPh_3)_2$ [10], $RhH(CO)(PPh_3)_3$ [11], and $K_2(ReH_9)$ [12]. The metal–hydrogen distance estimated for all these complexes is that expected for a normal covalent bond.

The presence of a hydride ion in the coordination sphere of a metal ion is also indicated by the presence of PMR (proton magnetic resonance) resonances at high field (~20–30 τ) [13]. Spin–spin coupling with the central metal ion has been studied for the hydrido derivatives of ^{103}Rh, ^{195}Pt, and ^{183}W complexes. The coupling constant J_{M-H} increases with increasing strength of the metal–hydrogen bond and thus lends strong support for the concept of highly covalent bonds in these complexes.

The infrared spectra of the hydrido complexes show bands due to the metal–hydrogen stretching vibrations, ranging from 1726 to 2242 cm^{-1}. The presence in the *trans* position of ligands, such as cyanide and carbon monoxide that usually have large *trans* effects, reduces the strength of the metal–hydrogen bond, and the observed metal–hydrogen stretching frequencies are accordingly reduced. In the complexes $Ir(H)_2Cl(CO)(Ph_3P)_2$ (**1**) and

$Os(H)_2(CO)(PPh_3)_3$ (**2**) the stretching frequencies of the metal–hydrogen bonds *trans* to the carbonyl group are 2100 and 1882 cm^{-1}, respectively. The hydrogen *trans* to Cl^- in **1** has a metal–hydrogen stretching frequency of 2196 cm^{-1}, reflecting a change of 96 cm^{-1} reulting from the *trans* effect of the carbonyl group in this iridium complex [14]. The hydrogen *trans* to triphenylphosphine in **2** absorbs at 2051 cm^{-1}, indicating a large difference, 169 cm^{-1}, in the *trans* effects of CO and PPh_3 in this osmium complex [14]. Although not indicated by these data, the *trans* effect of a tertiary phosphine substituent is usually much greater than that of the chloride ligand. On deuteration all the metal–hydrogen bands shift to lower frequencies as expected. The infrared spectrum of the dihydrido complex cation $Ir(H)_2(Ph_2PCH_2CH_2PPh_2)_2^+$ (**3**) shows metal–hydrogen absorption bands at 2091 and 2080 cm^{-1} [15]. In this complex the two hydrogens probably occupy *cis* positions, as in the case of $Ir(H)_2Cl(CO)(Ph_3P)_2$ (**1**).

Although considerable data have accumulated on the structures of solid hydrido complexes, the structures of labile hydrido complexes in solution are not known with certainty. According to Halpern [4] the nature of the metal–hydrogen bond in solution is probably the same as that of crystalline hydrido complexes.

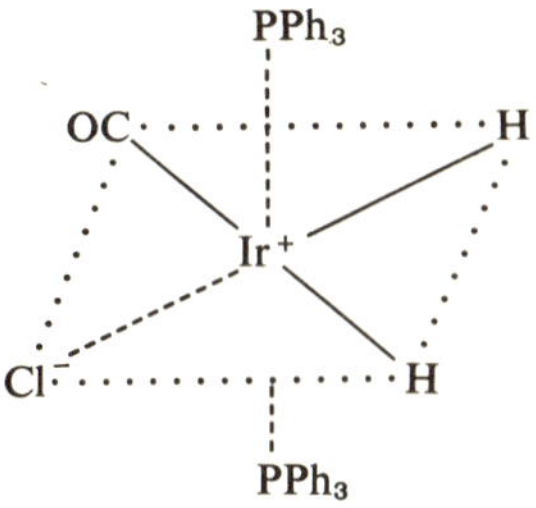

1

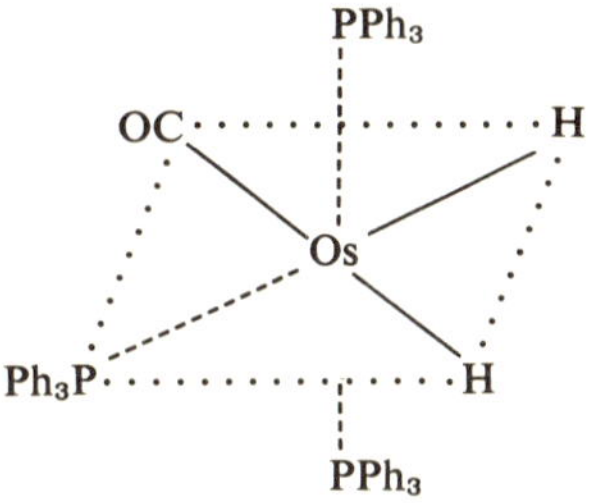

2

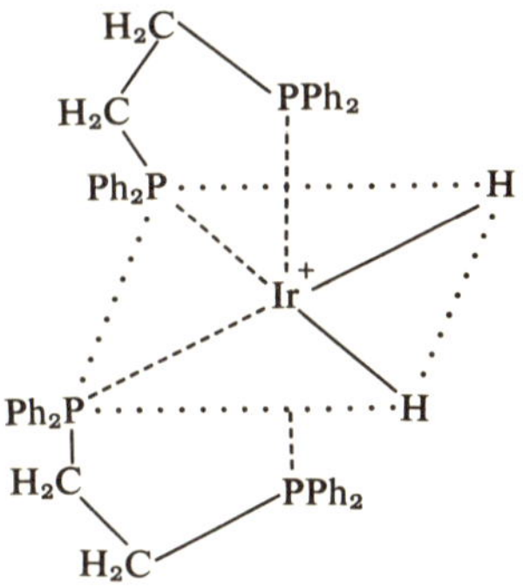

3

The reactivities of labile hydrides as intermediates in the catalytic reactions of molecular hydrogen seem to depend on their stabilities. In order that a particular transition metal ion may act as a catalyst in the activation of molecular hydrogen, its hydrido complex should have sufficient thermodynamic stability to be formed readily in solution, but should be sufficiently labile to react rapidly with the substrate [4]. If the hydrido complex is thermodynamically too unstable it is not formed at all, as may be the case for manganese(II), cobalt(II), nickel(II), and iron(III). The oxidation potential of the metal ion or complex seems to offer an important criterion for predicting the formation of thermodynamically stable hydrido intermediates.

Although the cobalt(II) ion by itself is inactive as a hydrogenation catalyst, pentacyanocobaltate(II) is a very active catalyst for the activation of molecular hydrogen. For each of the metal ions that acts as a catalyst there seems to be a rough correlation between catalytic activities and the stabilities (or oxidation potentials) of its complexes. Thus for silver(I) complexes, the catalytic activities decrease in the order: $Ag(C_2H_3O_2)_2^- > Ag(NH_3)_2^+ \gg Ag(CN)_2^-$. This is also the inverse order of stability. The oxidation potentials of these complexes vary in the order: -0.643 v, -0.373 v, and $+0.31$ v, respectively.

Although superficially the lack of catalytic activity of the most stable silver(I) complex, $Ag(CN)_2^-$, and the high catalytic activity of pentacyanocobaltate(II), $Co(CN)_5^{3-}$, seem mutually contradictory, the catalytic effects may be readily explained on the basis of stability. Silver(I) complexes function by heterolytic fission of hydrogen to form mixed complexes of silver(I) in which the hydride ion is one of the ligands. Thus the less stable silver(I) complexes will have the greatest affinity for the hydride ligand, whereas the cyanide ions are so strongly bound that they are not displaced by hydride ion. On the other hand, a pentacoordinated cobalt(II) complex functions through homolytic fission of hydrogen and simultaneous oxidation of the cobalt(II) to the very stable cobalt(III) species. Thus the cyanide ion will assist the electron transfer to hydrogen by forming the most stable monohydrido-cobalt(III) complex, $HCo(CN)_5^{3-}$.

The hydrido complex of a catalytic species must also be labile, in order that hydrogen (or hydride ion) be transferred to a substrate and regenerate the original catalytic complex. The lability of a hydrido complex depends on the total ligand-field stabilization energies of the ligands in the complexes of both the original catalyst and the hydrido complex formed after hydrogen fission [16]. According to Chatt and Shaw [17,18], the energy separation between the highest occupied bonding or nonbonding levels and the lowest vacant antibonding levels (Δ) must be greater than some critical value in order to confer stability on the hydrido complex. In the case of octahedral aqueous transition metal ions, especially those of the first transition series, the separation between the t_{2g} and $e_g(\sigma^*)$ levels (Fig. 1) is not enough to achieve stability of the hydride. This results in the formation of labile hydrido complexes in solution. In such cases, bonding electrons from the metal–hydrogen σ bond are easily promoted through the t_{2g} level into the vacant antibonding $e_g(\sigma^*)$ level, with the effective dissociation of the metal–hydrogen bond and the consequent separation of a hydride ion or hydrogen atom that reacts with substrate or solvent and is thus destroyed. In the absence of a substrate the metal ion itself is reduced, in some cases with the liberation of a proton in solution. In the presence of π-bonding ligands such as tertiary phosphines and carbon monoxide, the backbonding interaction of metal t_{2g} orbitals with the

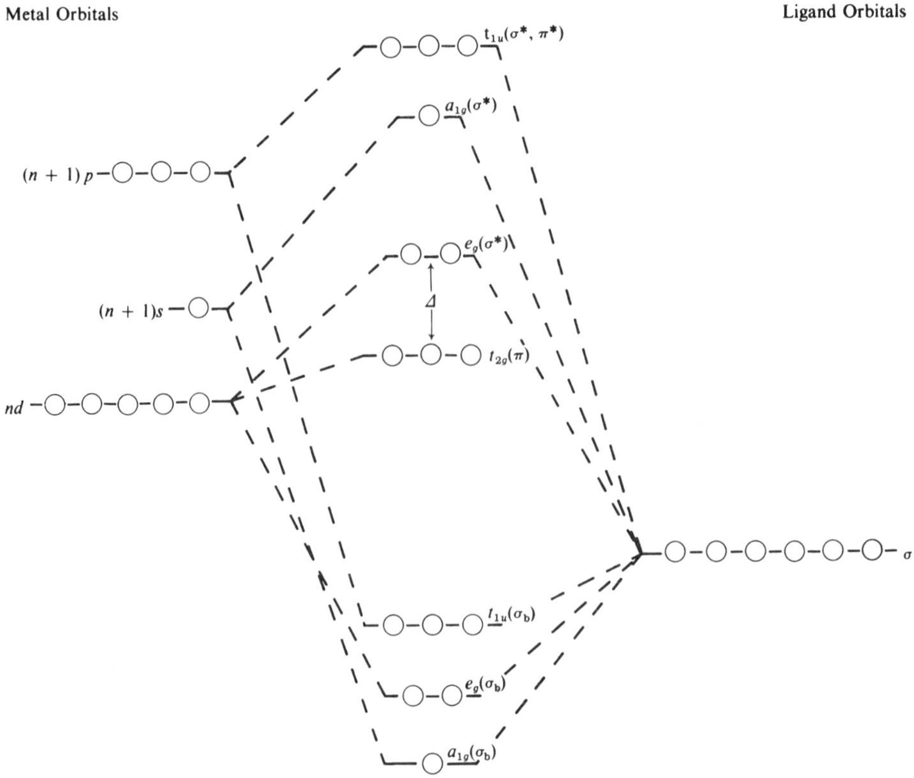

Fig. 1. Molecular orbital diagram for an octahedral complex with only σ-bond contributions from the ligands. Asterisk indicates antibonding; b, bonding.

ligand π and π^* orbitals results in a lowering of the t_{2g} energy level and in a greater separation of the t_{2g} and $e_g(\sigma^*)$ levels (Fig. 2). Thus promotion of the electrons to the antibonding level becomes more difficult and additional stability is conferred on the metal hydride.

For heavier transition metal ions of the second and third series, the hydride complexes in the presence of π-bonding donors are even more stable than the corresponding complexes of the first series because of an already greater separation of t_{2g} and $e_g(\sigma^*)$ levels. In the series of transition metal complexes with a given ligand, Δ increases by about 30% from first transition series to the second and by approximately the same amount from the second to the third series [19]. For the metal ions of the second and third transition series, ligands of moderate strength can produce enough separation in t_{2g} and e_g levels to produce stable hydrido complexes. Gillard and Wilkinson [20] have been able to prepare stable hydrido complexes of rhodium(III) with

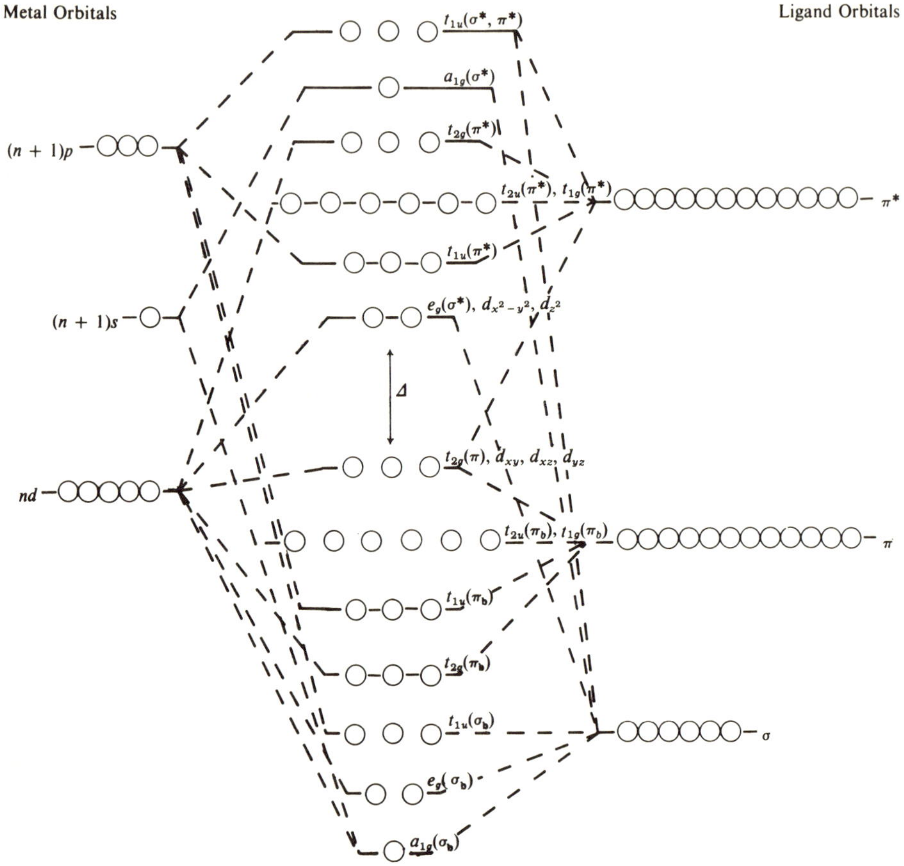

Fig. 2. Molecular orbital diagram for an octahedral complex with contributions from π-bonding and antibonding orbitals of the ligands.

ethylenediamine or triethylenetetramine (trien) as stabilizing ligands. The energy separation Δ also increases considerably with charge on the metal ion. Higher-valent metal ions of the second and third series may induce a stabilizing influence on the hydrido complex because of their ability to increase Δ. This may be the reason for an enhanced catalytic effect in hydrogenation reactions of lower-valent metal ions of the second, and third transition series, such as iridium(I), rhodium(I), and ruthenium(II), since catalytic activity requires facile dissociation of the hydride ion and hence a value of Δ that is intermediate in magnitude.

The catalytic activities of the square-planar d^8 metal ions such as iridium(I), rhodium(I), palladium(II), and platinum(II) may be explained on the basis of the separation between the highest occupied $b_{2g}(\pi^*)$ and the lowest empty

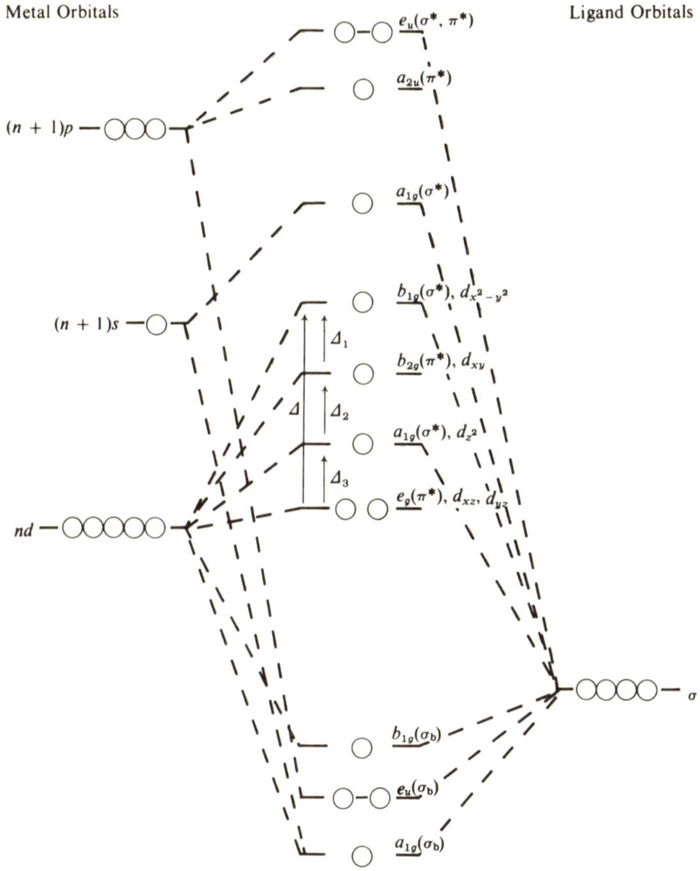

Fig. 3. Molecular orbital diagram for a square-planar complex with only σ-bond contributions from the ligands. Asterisk indicates antibonding; b, bonding.

$b_{1g}(\sigma^*)$ levels. The five degenerate levels of the uncomplexed metal ion separate into four non-bonding orbitals in a square-planar complex, as shown in Fig. 3 and 4. The relative magnitudes of the separations between the antibonding and nonbonding *d* levels is $\Delta_1 > \Delta_2 > \Delta_3$ [21]. Complexing of the metal ion with a strong π-bonding ligand such as CN^- causes a marked increase in Δ_1 and Δ_2 and a decrease in Δ_3 (Fig. 4). The catalytic activation of molecular hydrogen by the cyano complexes of the metal ions cited above is thus greatly reduced since the promotional energy for an electron to the antibonding b_{1g}^* level is increased beyond the optimal intermediate value.

For the tetrahedral complexes of d^9 copper(II) and d^{10} metal ions, silver(I), copper(I), and mercury(II), a schematic diagram for the ligand field of pure

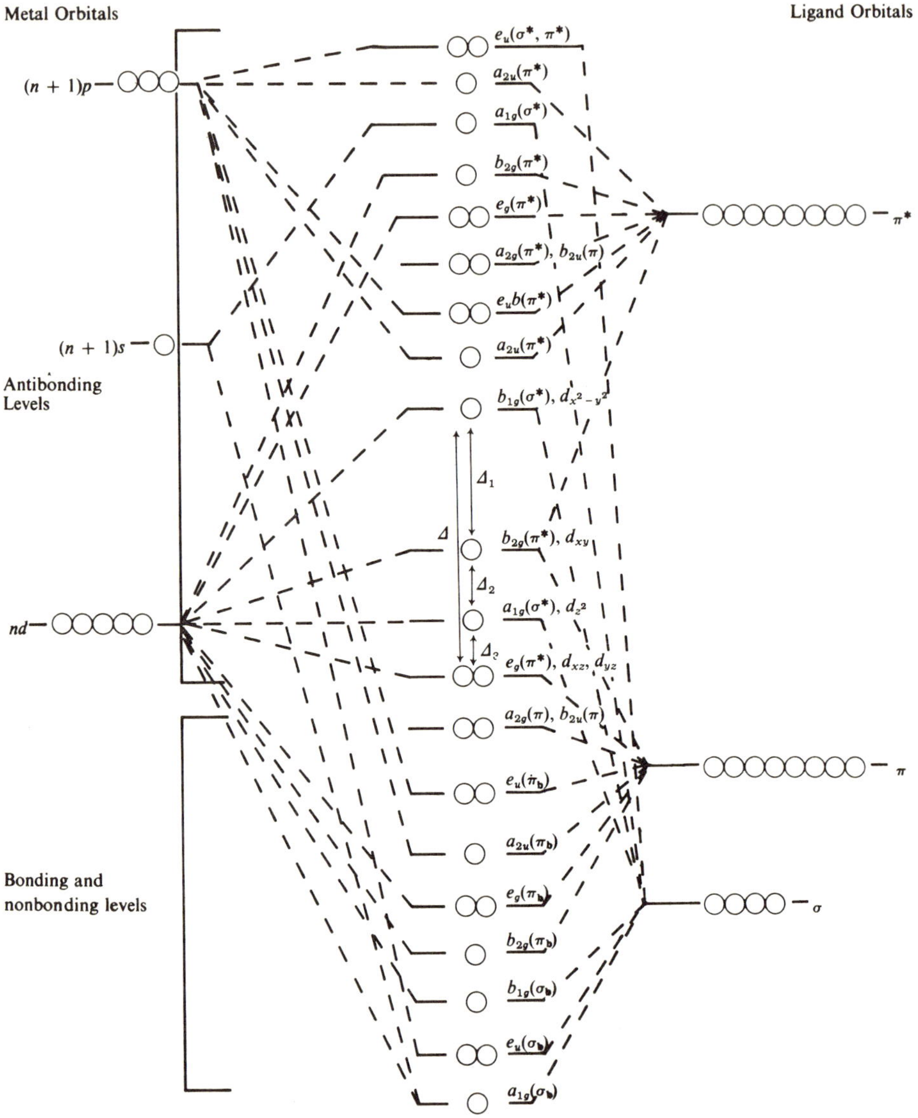

Fig. 4. Molecular orbital diagram for a square-planar complex with contributions from π-bonding and antibonding levels of the ligand. Asterisk indicates antibonding; b, bonding.

σ-bonding ligands is depicted in Fig. 5. The levels $e(\pi^*)$ and $t_2(\sigma^*, \pi^*)$ are completely occupied by metal d electrons. The catalytic activities of these metal ions may be explained on the basis of the relative separation of the occupied metal $t_2(\sigma^*, \pi^*)$ and vacant $a_1(\sigma^*)$ levels. The dissociation of the

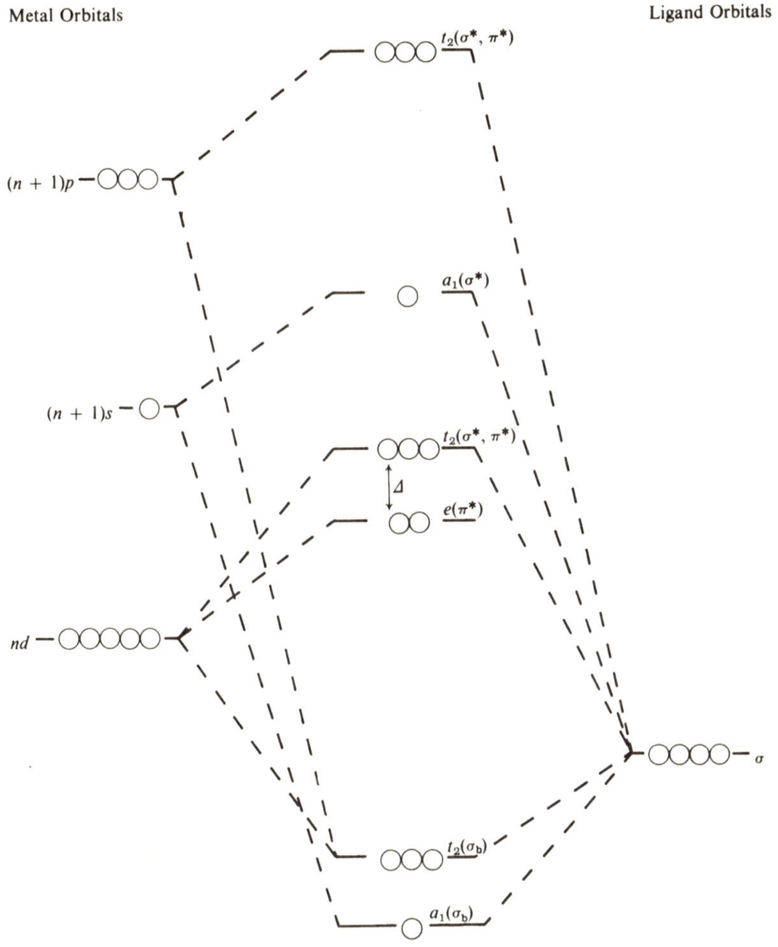

Fig. 5. Molecular orbital diagram for a tetrahedral complex with only σ-bond contributions from the ligands. Asterisk indicates antibonding; b, bonding.

labile metal hydride may be visualized as taking place by the transfer of σ-bonding electrons of the hydride ligand to the antibonding $a_1(\sigma^*)$ through the metal $e(\pi^*)$ or $t_2(\sigma^*, \pi^*)$ levels.†

In the examples cited in this section for the catalytic activation of molecular hydrogen, dissociation of the metal–hydrogen bond takes place in preference

† This tentative explanation is based on a similar explanation for octahedral complexes by Chatt and Shaw [17,18]. For the octahedral case, σ-bonding electrons of the hydride group move through the filled t_{2g} level to the empty antibonding e_g level, causing either a reduction of the metal ion and liberation of a proton in the absence of a substrate, or transfer of a hydride ion or hydrogen atom to the substrate.

to the bonds of other ligands bound to the metal ion. This has been well established experimentally in the case of the activation of molecular hydrogen by square-planar carbonyl complexes of univalent iridium and rhodium [22]. The *trans*-[MX(CO)$(PPh_3)_2$] complexes [where M = iridium(I), rhodium(I), X = Cl^-, Br^-] absorb molecular hydrogen in solution to form the dihydrides, *trans*-[$M^{III}X(CO)(PPh_3)_2H_2$]. The hydrido complex catalyzes hydrogenation of ethylene, propylene, and acetylene in homogeneous solution without undergoing any change in the *trans* configuration of PPh_3 and X(CO) groups. The configuration of both the hydrido complex and the complex left after hydrogenation has been very well established by infrared and NMR studies [22]. The complex left after hydrogenation of unsaturated hydrocarbons is identical with the starting material, *trans*-[MX(CO)$(PPh_3)_2$].

The mechanisms of hydrogen activation covered in this chapter are of two types. One type involves homolytic or heterolytic fission of molecular hydrogen to yield a labile hydride which either reduces the metal complex in the absence of a substrate or hydrogenates an appropriate substrate. The metal complexes involved in this type of activation are those of copper(I) [23–25]; copper(II), mercury(I), mercury(II), silver(I) [1–4]; ruthenium(III) [26]; rhodium(III) [27]; palladium(II) [28]; and cobalt(II) [29,30]. The other type involves formation of stable hydrido complexes by the oxidative addition of molecular hydrogen to four- or five-coordinate d^8 complexes of cobalt(I), rhodium(I), iridium(I), ruthenium(I), and osmium(0). In such cases *cis*-dihydrido metal complexes are formed with the concomitant oxidation of the metal ion by two units (see Table III). Stable metal hydrido complexes may also be formed by the displacement of coordinated negative or neutral donor groups in d^6 metal complexes of ruthenium(II), platinum(II), and palladium(II). These hydrides have been well characterized and take part in the catalytic hydrogenation of unsaturated substrates. The numerous stable transition metal hydrides [5] that are not catalysts for the hydrogenation of unsaturated substrates have not been discussed in this section.

Activation of molecular hydrogen by d^8 complexes of the transition group metal ions have a close parallel in the activation of alkyl or aryl halides [31–60], as in processes that proceed by oxidative addition to the metal complex. This process affords a convenient method for the synthesis of σ-alkyl or aryl transition metal complexes and is discussed in Section III,E. Metal–carbon σ bonds are also formed by the intramolecular cleavage of C—H bonds from a coordinated ligand (phenyl group) of certain iron(0) [61], ruthenium(0) [62], iridium(I) [63,64], platinum(II) [65,66], and palladium(II) [67] complexes. In the case of iron(0), ruthenium(0), and iridium(I) both the σ-carbon and hydrido moieties are added to the metal ion by a reversible oxidative addition process. The platinum(II) [65,66] and palladium(II) [67] complexes, however, eliminate the hydride ion and form only the σ-carbon-bonded species. These processes are very little understood at this stage, but developments in this area may become one of the important new advances in metal complex chemistry and homogeneous catalysis. Such reactions are not only important in themselves as new reaction types but also may lead to the development of new synthetic pathways, as well as to a better understanding of natural metal–hydrocarbon σ complexes such as those in reaction intermediates of vitamin B_{12} and other biocatalysts.

III. Activation of Hydrogen by Copper and Silver Ions

The first report of the homogeneous activation of molecular hydrogen by a metal ion was catalysis by cuprous ion in quinoline solution, described by Calvin [23,24], and Calvin and Wilmarth [25]. The catalysts employed were the acetate and salicylaldehyde complexes of copper(II). The rate of reduction of copper(II) and quinone substrates by the Cu(I)–H_2 system was found to be independent of the concentration of the substrate. The following mechanism was proposed [23] for the hydrogenation of the substrate:

$$\mathrm{Cu(I)} + \mathrm{Q} \rightleftharpoons \mathrm{Cu(I)Q}$$

$$2\,\mathrm{Cu(I)Q} \rightleftharpoons [\mathrm{Cu(I)Q}]_2$$

$$[\mathrm{Cu(I)Q}]_2 + \mathrm{H_2} \underset{k_1'}{\overset{k_1}{\rightleftharpoons}} [\mathrm{Cu(I)Q}]_2\mathrm{H_2}$$

$$[\mathrm{Cu(I)Q}]_2\mathrm{H_2} + 2\,\mathrm{S} \xrightarrow{k_2} [\mathrm{Cu(I)Q}]_2 + 2\,\mathrm{SH}$$

$$[\mathrm{Cu(I)Q}]_2\mathrm{H_2} \xrightarrow{k_3} 2\,\mathrm{Cu^0} + 2\,\mathrm{Q} + 2\,\mathrm{H^+}$$

where $k_2 > k_1' > k_1 > k_3$; Q = quinoline; S = substrate.

The model employed by Calvin [23,24] involved homolytic splitting of the hydrogen molecule in a Cu(I)–quinoline complex. Both homolytic and heterolytic mechanisms have now been proposed as pathways for homogeneous catalytic hydrogenation by metal ions. These two classifications are used as a basis for the following discussion.

A. Heterolytic Splitting

The general mechanism for the heterolytic splitting of hydrogen reported by Halpern *et al.* in the case of Cu^{2+} [68–70], Hg_2^{2+} [71], $RuCl_6^{3-}$ [26],

$RhCl_6^{3-}$ [27], and $PdCl_4^{2-}$ [28] is as given in Eqs. (1)–(3):

$$MX_i^{n+} + H_2 \underset{k_2}{\overset{k_1}{\rightleftharpoons}} MHX_{i-1}^{(n-1)+} + H^+ + X \qquad (1)$$

$$MHX_{i-1}^{(n-1)+} + MX_i^{n+} \xrightarrow{k_3} 2\,M^{(n-1)+} + H^+ + (2i - 1)X \qquad (2)$$

(absence of substrate)

$$MHX_{i-1}^{(n-1)+} + S + X \xrightarrow{\text{fast}} SH^- + MX_i^{n+} \qquad (3)$$

(presence of substrate)

The kinetic observations [70] indicating heterolytic fission of hydrogen are (1) the dependence of the rate of the reaction (step k_1) on hydrogen ion concentration [28], (2) isotopic tracer studies with D_2O [70], and (3) the dependence of the rate on the basicity of the ligand X [72].

Increasing the basicity of the ligand up to a certain limit seems to enhance greatly the catalytic activity of the metal complex for the heterolytic fission of molecular hydrogen. The basicity of the ligand helps heterolytic fission by stabilization of the proton released in the reaction. In the case of Cu(II), the rate was found to increase in the order of the basicity of the ligand, $H_2O < Cl^- < SO_4^{2-} < CH_3COO^- < CH_3CH_2COO^- < CH_3(CH_2)_2COO^-$. With greatly enhanced basicity of the ligand, as in the case of the polyamines, the strength of the metal–ligand bond increases to the point that displacement of the ligand by the hydride ion becomes more difficult. Thus low catalytic activity would be expected for complexes of higher stability formed from more basic ligands than those given by the above series.

Two transition states, **4** and **5**, have been proposed for cases where the ligand X or the solvent B take part in a disproportionation reaction of the hydrogen molecule.

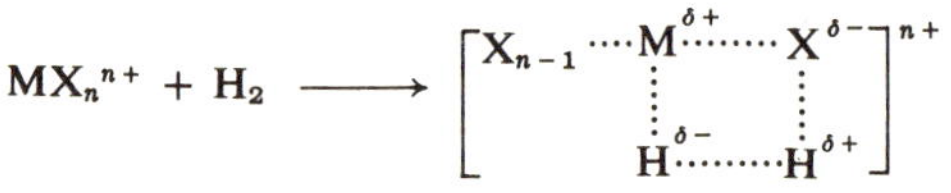

4

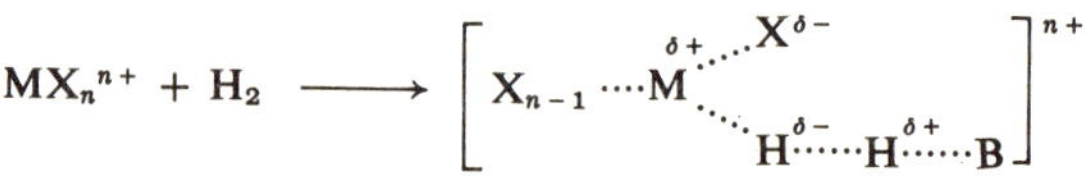

5

While it is difficult to distinguish between **4** and **5**, it may be reasonably assumed that **4** would be favored in solvents of low polarity [73] since it would have the lower dipole moment. In polar solvents and in low-polarity solvents containing an additional basic catalyst, the extended form of the intermediate, **5**, would be expected to compete more favorably with **4** and may even predominate.

For all the cases of heterolytic splitting of hydrogen the rate of the reaction was found to be first order with respect to molecular hydrogen and first order with respect to the catalyst, but independent of substrate concentration, thus supporting the proposed mechanism. The activation energies are in the range 20–24 kcal/mole [3]. In the absence of metal ions, heterolytic splitting of hydrogen with the formation of hydride ions and protons has also been observed with strong bases such as hydroxide ion [74], boranes [75–77], or alkoxide ion (RO^-) [78,79]. Such reactions, however, require much higher activation energies and extreme conditions of temperature and hydrogen pressure as compared to metal ion catalysis. Base-catalyzed hydrogenation has recently been reported by Walling and Bollyky [79]. The authors have studied the hydrogenation of benzophenone by potassium *t*-butoxide in *t*-butyl alcohol at 130°–200°C and 100 atm H_2 pressure. The following mechanism has been suggested:

$$RO^- + H_2 \rightleftharpoons ROH + H^-$$

$$H^- + R_2C{=}O \rightleftharpoons R_2CH{-}O^-$$

$$R_2CHO^- + ROH \rightleftharpoons R_2CHOH + RO^-$$

In a similar manner carbonium ions can abstract hydride ion from hydrogen since they are very strong bases. The hydrogenation of isobutene and cyclohexene by an 8–15% solution of $AlBr_3$ in cyclohexene at 150°C and 82 atm of pressure of H_2 was reported by Walling and Bollyky [79]. Although the reactions in the absence of metal ion catalyst have little practical application up to the present time, they are of interest for comparison with metal ion and metal complex catalysis.

B. Homolytic Splitting of Hydrogen

A mechanism for the homolytic splitting of molecular hydrogen by Ag^+, substantiated by the work of Webster and Halpern [80], and Nagy and Simandi [81], is given in Eqs. (4)–(6):

$$H_2 + 2\,Ag^+ \xrightarrow{k} 2\,AgH^+ \tag{4}$$

$$2\,AgH^+ \xrightarrow{\text{fast}} 2\,Ag^0 + 2\,H^+ \tag{5}$$

$$2\,AgH^+ + S \xrightarrow{\text{fast}} SH_2 + 2\,Ag^+ \tag{6}$$

The reaction follows third-order kinetics, first-order with respect to molecular hydrogen and second-order with respect to silver ion. The energy of activation was reported by Webster and Halpern [80] to be 14 kcal/mole. The enthalpy of activation thus favors the less probable termolecular mechanism over a bimolecular reaction, which would have an energy of activation of about 24 kcal/mole, as observed for Cu(II) [3]. The bimolecular mechanism, however, with associated heterolytic splitting of hydrogen, predominates at higher temperatures and in nonaqueous solutions such as pyridine [82]. Thus, the rate at a given temperature is frequently a combination of termolecular and bimolecular rates, given by Eq. (7) [k_i's refer to Eqs. (1) and (2)].

$$\frac{-d[H_2]}{dt} = k[H_2][Ag^+]^2 + \frac{k_1k_3[H_2][Ag^+]^2}{k_3[Ag^+] + k_2[H^+]} \tag{7}$$

The first term of Eq. (7) corresponds to homolytic and the second term to heterolytic splitting. The first term is pH-independent and corresponds to the formation of the hydrido species AgH^+, which then reacts rapidly with substrate. In the absence of substrate, however, AgH^+ is reduced to Ag^0. The bimolecular mechanism corresponding to an intermediate hydrido species, AgH, has been substantiated by Webster and Halpern [83,84] using D_2O as a tracer. It predominates over the termolecular reaction mechanism in nonaqueous solvents (e.g., silver(I) salts in pyridine [82]).

Nagy and Simandi [81] have followed the initial suggestion of Webster and Halpern [80,83,84] that the mechanism involves the homolytic splitting of hydrogen by silver(I) at low temperatures. They disagreed, however, with the termolecular step involving the interaction of two positively charged silver ions with molecular hydrogen. On the basis of analogous reactions of molecular hydrogen taking place on the surface of a platinum electrode, they [81] proposed the formation of a complex, $Ag_2H_2^{2+}$, by two successive bimolecular steps. Since the chemical bonding in a complex of molecular hydrogen would seem to be rather difficult to explain, the Halpern mechanism, in which hydrogen is homolytically split during complex formation to give AgH^+, would seem more logical.

Webster and Halpern [83] have suggested homolytic splitting of hydrogen in the silver ion-catalyzed oxidation of molecular hydrogen by permanganate ion. The rate-determining step of the mechanism proposed by the authors [84] is as shown in Eq. (8). Ag(I) and MnO_4^- probably form a complex,

$$Ag^+ + MnO_4^- + H_2 \longrightarrow AgH^+ + MnO_4^{2-} + H^+ \tag{8}$$

$AgMnO_4$, which splits molecular hydrogen by a homolytic path to form AgH^+, H^+, and MnO_4^{2-}. The homolytic fission of hydrogen is probable in this case because of the high electron affinity of MnO_4^-, which combines with H to give MnO_4^{2-} and H^+. The other hydrogen atom simultaneously combines with Ag(I) to give the hydride AgH^+. The AgH^+ and MnO_4^{2-} then react in subsequent fast steps to yield the final products, manganese dioxide and the silver(I) ion. The role of permanganate in the reaction is to replace one of the silver ions in the termolecular step involving the reaction of hydrogen and silver(I). The activation energy of 9 kcal/mole supports fully the possible participation of a complex, $AgMnO_4$, in the rate-determining step of the activation of molecular hydrogen, and is the lowest activation energy yet reported for a hydrogenation process.

Cuprous salts that are inactive in aqueous solutions have been reported by Chalk and Halpern [72] to activate molecular hydrogen in nonaqueous solvents such as amines, carboxylic acids, and hydrocarbons. The splitting of molecular hydrogen is homolytic, with the possible formation of CuH^+ species. This suggestion is supported by a third-order reaction observed for copper(I) in quinoline [23–25] in which the rate is second order with respect to the metal. The activity of Cu(I) ion was reported to increase with the polarity of the solvents, octadecane < biphenyl < heptanoic acid [72].

C. Ag(I) Complexes

Halpern and Milne [85] reported that the rate of reduction of various silver–amine complexes by molecular hydrogen involves second-order kinetics, first order with respect to silver amine and first order with respect to molecular hydrogen. The mechansim of Eqs. (9) and (10) was suggested for this reaction:

$Ag(NR_3)_2^+ + H_2O + H_2 \longrightarrow$

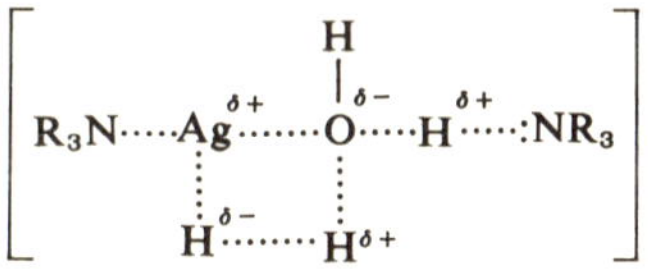

6

$\xrightarrow{k}$ $R_3N{-}Ag{-}H + HNR_3^+ + H_2O$ (9)

$$R_3N{-}Ag{-}H + Ag(NH_3)_2^+ \xrightarrow{\text{fast}} 2\,Ag^0 + HNR_3^+ + 2\,NR_3 \qquad (10)$$

A four-centered transition state **6** in which molecular hydrogen splits heterolytically was suggested. The rate-determining step involves the formation of a mixed hydrido-amine complex R_3NAgH by the displacement of one of the tertiary amine groups through protonation. The reduction of R_3NAgH to Ag^0 takes place in a subsequent fast step. The rate constant was found to vary nearly linearly with the stabilities of Ag(I)–amine complexes. The only deviation observed was for $[Ag(NH_3)_2]^+$, for which the rate was found to be much slower than that expected from the linear relationship.

Indirect support of Halpern's mechanism [85] is provided by the work of Beck [86] and Beck and Gimesi [87], who obtained a hundredfold increase in the activation of molecular hydrogen by using fluoride ion instead of an amine as a ligand for stabilizing the intermediate Ag(I)–H_2 complex. The

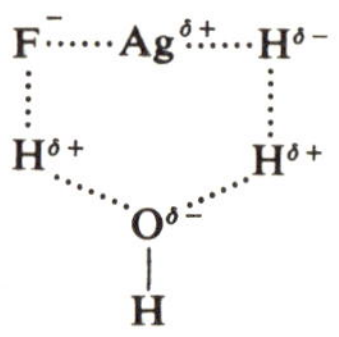

7

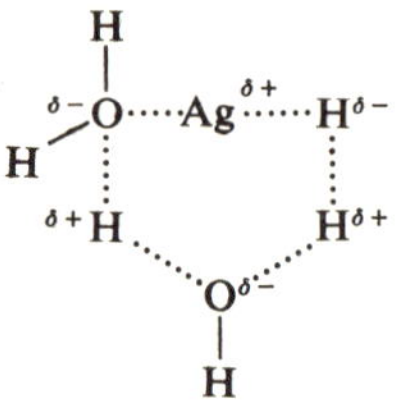

8

influence of fluoride ion was ascribed to the low stability of the monofluorosilver(I) complex [88] and the affinity of fluoride ion for a proton. The monofluorosilver(I) complex completes its coordination requirement by combining with a hydride ion to form a fluorohydridosilver(I) complex. Since fluoride ion is strongly polarizing and solvated in aqueous solution, Beck *et al.* [89] suggested the formation of a six-centered activated complex, where the polarizing effect of fluoride ion operates through a water molecule, as indicated in structure **7**. The activated complex suggested by Beck *et al.* [89] would be in conformity with large kinetic deuterium isotope effects for both hydrogen and solvent, since the transition state would involve interaction of polarized hydrogen and water molecules. According to Beck [86], the aquo silver ion forms a similar cyclic intermediate, **8**, much less readily, because the affinity of water for silver ion is relatively small (much lower than that of the fluoride ion) while its affinity for the proton is high.

It is interesting to compare the stabilities of the various silver(I) complexes

TABLE I

ACTIVATION OF MOLECULAR HYDROGEN BY Ag(I) COMPLEXES

Ligand	k^a (M^{-1} sec^{-1}) 70°C	log K_{ML}	$\Delta H^\ddagger$ (kcal/mole)	$\Delta S^\ddagger$ (eu)	References
H_2O	6.3×10^{-4}	—	23.3	−6	80
CN^-	0	7.7 (25°C)	—	—	80
$NH_2(CH_2)_2NH_2$	1.6×10^{-2}	4.7 (20°C) ($\mu = 0.1$ KNO_3)	20.8	−7	80
$C_2H_5NH_2$	3.8×10^{-3}	3.37 (25°C) ($\mu = 0.1$ KNO_3)	21.6	−7	85
$(C_2H_5)_2NH$	4.7×10^{-2}	2.98 (30°C) ($\mu = 0.5$ KNO_3)	20.5	−5	85
$(C_2H_5)_3N$	7.4×10^{-1}	2.31 (25°C) (50% C_2H_5OH)	18.8	−5	85
F^-	1.6×10^{1}	−0.17 (25°C) (0.5 $NaClO_4$)	18.3	+3	89

[a] *k* is the second-order rate constant for the reduction of Ag(I) complex by molecular hydrogen.

with the rates corresponding to the activation of molecular hydrogen and the activation parameters of the reaction. It may be readily seen from the data in Table I that as the stabilities of 1:1 complexes of Ag(I) decrease, the catalytic activities increase, in accord with the increasing tendency of complexes of low stability to form the hydrido–ligand complexes postulated as intermediates. The very stable complex, $Ag(CN)_2^-$, is thus inactive as a catalyst, since the strongly bound cyanide ion is not displaced to any appreciable extent by the weaker hydride ligand. The values reported for the fluoride complex are in accord with the discussion given above.

IV. Activation by Transition Metal Halides

Chlororuthenate(III) solutions, $RuCl_n^{(3-n)+}$ in 3 *M* hydrochloric acid, activate molecular hydrogen through a heterolytic splitting process. The rate of the reaction was found by Harrod *et al.* [26] to be first order with respect to molecular hydrogen and first order with respect to the catalyst. One of the interesting properties of the catalyst is that it is not reduced to ruthenium(II) by molecular hydrogen. The mechanism proposed [26] for the reduction of iron(III) and ruthenium(IV) by molecular hydrogen in the presence of ruthenium(III) is shown in Eqs. (11)–(13).

$$RuCl_6^{3-} + H_2 \underset{k_{-1}}{\overset{k_1}{\rightleftharpoons}} RuCl_5H^{3-} + H^+ + Cl^- \quad (11)$$

$$RuCl_5H^{3-} + 2\,RuCl_n^{(4-n)+} + [1 + 2(6 - n)]Cl^- \xrightarrow{k_3} 3\,RuCl_6^{3-} + H^+ \quad (12)$$

$$RuCl_5H^{3-} + 2\,FeCl_3 \xrightarrow{k_2} RuCl_6^{3-} + 2\,FeCl_2 + H^+ + Cl^- \quad (13)$$

The first step is irreversible in the presence of ruthenium(IV) or iron(III) but reversible in the absence of the substrate. The magnitudes of the rate constants vary in the order $k_2,k_3 > k_{-1} > k_1$, thus making the first reaction the rate-determining step. The reversibility of the step corresponding to the reaction of ruthenium(III) with hydrogen has recently been substantiated by Halpern [90] and Halpern *et al.* [155] by the study of the deuterium–H_2O exchange catalyzed by the hexachlororuthenium(III) ion, in the absence of ferric chloride, with measurement of the ratio of $HD:D_2$ in the reactions (14a)–(14d). Hexachlororhodate(III) has been reported as a catalyst that

$$RuCl_6^{3-} + D_2 \xrightarrow{k_1} RuCl_5D^{3-} + D^+ + Cl^- \quad (14a)$$

$$RuCl_5D^{3-} + H^+ + Cl^- \xrightarrow{k_{-1}} RuCl_6^{3-} + HD \quad (14b)$$

$$RuCl_5D^{3-} + H_2O \xrightarrow{k_2} RuCl_5H^{3-} + HDO \quad (14c)$$

$$RuCl_5H^{3-} + H^+ + Cl^- \xrightarrow{k_{-1}} RuCl_6^{3-} + H_2 \quad (14d)$$

activates hydrogen in a manner similar to that illustrated above for hexachlororuthenate(III). Unlike ruthenium(III), rhodium(II) is reduced to lower oxidation states in the absence of the substrate, or when the reduction of the substrate is completed.

In contrast to aquopalladium(II) [27], which involves considerable difficulty in homogeneous activation of molecular hydrogen because of the facile reduction of metallic palladium, tetrachloropalladate(II) was reported by Halpern *et al.* [28] to activate molecular hydrogen by a heterolytic splitting process similar to those described for chlororuthenate(III) [26] and chlororhodate(III) [27]. The rates and activation parameters for the homogeneous activation of molecular hydrogen by transition metal–chloro complexes are presented in Table II.

TABLE II

ACTIVATION OF MOLECULAR HYDROGEN BY CHLORO COMPLEXES OF TRANSITION METALS

Complex	k^a(80°C)	Substrate	Rate law	$\Delta H^{\dagger}$ (kcal/mole)	$\Delta S^{\dagger}$ (eu)
$RuCl_n^{(3-n)+}$	1.0	Ru(IV)Fe(III)	$k[H_2][Ru(III)]$	2.3	+6
$RhCl_6^{3-}$ or $RhCl_5(H_2O)^{2-}$	1.8×10^{-1}	Fe(III)	$k[H_2][Rh(III)]$	25.2	+9
$PdCl_4^{2-}$	3.5×10^{-1}	Fe(III)	$k[H_2][Pd(II)]$	20.0	−6.8

[a] k is the second-order rate constant for the reduction of $M^{n+}(Ru^{3+}, Rh^{3+}, Pd^{2+})$ complexes by molecular hydrogen.

A. CATALYSIS OF LIGAND EXCHANGE BY MOLECULAR HYDROGEN

Gillard *et al.* [92] reported that the reaction of trichlorotripyridinerhodium(III), **9**, with pyridine to form *trans*-dichlorotetrapyridinerhodium(III), **11**, is catalyzed by molecular hydrogen at room temperature and atmospheric pressure. The catalysis of the reaction was explained on the basis of the formation of an intermediate hydrido complex **10**, formed by heterolytic splitting of hydrogen by $[RhCl_3Py_3]$. Pyridine then displaces hydride ion from dichlorohydridotripyridinerhodium(III), **10**, to give the product, *trans*-$[RhCl_2Py_4]Cl$, as shown in Eqs. (15)–(17).

A similar explanation whereby the coordinated Cl^- ion is replaced directly by H^- ion was advanced [92] for the acceleration of the aquation of $[Rh(H_2O)_3Cl_3]$ to form $[Rh(H_2O)_4Cl_2]$ in the presence of molecular hydrogen. The concept of nucleophilic displacement of chloride ion by hydride ion was supported by the synthesis of a number of ethylenediamine- and triethylenetetraminerhodium(III) hydrido complexes directly from the corresponding

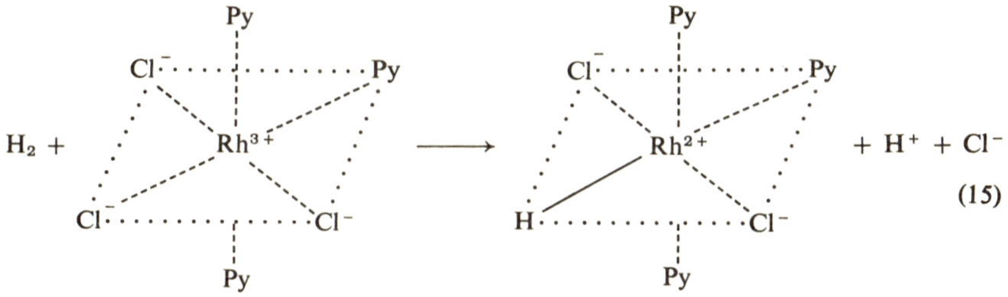

9 10 (15)

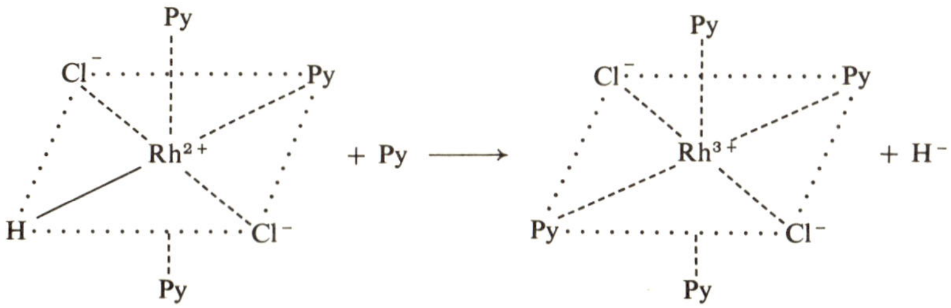

(16)

10 11

$$H^- + H^+ \longrightarrow H_2 \qquad (17)$$

rhodium(III) complexes with sodium borohydride as a source of hydride ion [20]. Hydride ion was reported [93] to be a stronger ligand than the halogens as measured by the magnitude of Δ (ligand-field splitting) and by the large *trans* effect of the hydrido ligand. Weakening of the bonds *trans* to the hydride ligand is well supported by X-ray studies of some stable hydride complexes such as *trans*-$PtHBr(PEt_3)_2$ [7] and $OsHBr(CO)(PPh_3)_2$ [9]. The Pt—Br and Os—P bond distances in these complexes *trans* to the M—H bond are 0.13 Å and 0.22 Å longer than the normal covalent distance of the Pt—Br and Os—P bonds, respectively. The general weakening of the bond *trans* to the hydride ion makes the *trans* ligand relatively more labile. The *trans* effect of the hydride donor has also been detected by kinetic measurements [94] of the more rapid substitution reactions of ligands *trans* to the hydride ion.

$$cis\text{-}[Rh(trien)Cl_2]^+ + H^- \longrightarrow cis\text{-}[Rh(trien)ClH]^+ + Cl^- \qquad (18)$$

$$trans\text{-}[Rh(en)_2Cl_2]^+ + H^- \longrightarrow trans\text{-}[Rh(en)_2ClH]^+ + Cl^- \qquad (19)$$

The preparation and properties of pentacoordinate "hydrogen" complexes of Ir(I) and hexacoordinate "hydrogen" complexes of Ir(III), Ru(II), and Os(II) have been described by Vaska [95–97], Vaska and Diluzio [98,99], Bath and Vaska [100], and Vaska and Catone [15]. Examples of the principal initial complexes and reactions studies for various metal ions are given in Table III.

TABLE III

REACTIONS OF MOLECULAR HYDROGEN WITH COMPLEXES OF GROUP VIII METALS[a]

	Reaction	Coordination no. of metal ion	References
1	$[Co(diphos)_2]Y + H_2 \rightarrow [Co(diphos)_2(H)_2]Y$	$4 \rightarrow 6$	101
2	$CoH(N_2)L_3 + H_2 \rightarrow Co(H)_3L_3 + N_2$	$5 \rightarrow 6$	102
3	$RhClL_3 + H_2 \rightarrow RhCl(H)_2L_2 + L$	$4 \rightarrow 5$	29
4	$[Rh(dmpe)_2]Cl + H_2 \rightarrow [Rh(dmpe)_2(H)_2]Cl$	$4 \rightarrow 6$	30
5	$[Rh\{P(OR)_3\}_4]BPh_4 + H_2 \rightarrow [Rh\{P(OR)_3\}_4(H)_2]BPh_4$	$4 \rightarrow 6$	97–99, 103,109
6	$RhCl(CO)(PPh_3)_2 + H_2 \rightarrow Rh(H)_2Cl(CO)(PPh_3)_2$	$4 \rightarrow 6$	15
7	$IrCl(CO)(PPh_3)_2 + H_2 \rightarrow Ir(H)_2Cl(CO)(PPh_3)_2$	$4 \rightarrow 6$	104
8	$IrH(CO)(PPh_3)_3 + D_2 \rightarrow IrD(CO)(PPh_3)_3 + HD$	$5 \rightarrow 5$	97–100
9	$[Ir(diphos)_2]X + H_2 \rightarrow [Ir(diphos)_2(H)_2]X$	$4 \rightarrow 6$	15,101
10	$[Ir(CO)_3(PPh_3)_2]BPh_4 + H_2 \rightarrow [Ir(CO)_2(H)_2(PPh_3)_2]BPh_4 + CO$	$5 \rightarrow 6$	105
11	$[Ir(CO)(PMe_2Ph)_4]BPh_4 + H_2 \rightarrow [Ir(CO)(H)_2(PMe_2Ph)_3]BPh_4 + PMe_2Ph$	$5 \rightarrow 6$	106
12	$[Ir(CO)(PMe_2Ph)_3]Cl + H_2 \rightarrow [Ir(CO)(H)_2(PMe_2Ph)_3]Cl$	$4 \rightarrow 6$	107
13	$IrCl(PPh_3)_3 + H_2 \rightarrow IrCl(H)_2(PPh_3)_3$	$4 \rightarrow 6$	108
14	$IrH(CO)(PPh_3)_2 + H_2 \rightarrow Ir(H)_3(CO)(PPh_3)_2$	$4 \rightarrow 6$	109
15	$Ru(CO)_3(PPh_3)_2 + H_2 \rightarrow Ru(H)_2(CO)_2(PPh_3)_2 + CO$	$5 \rightarrow 6$	110
16	$Os(CO)_3(PPh_3)_2 + H_2 \rightarrow Os(H)_2(CO)_3(PPh_3) + PPh_3$	$5 \rightarrow 6$	110
17	$Os(CO)_5 + H_2 \rightarrow Os(H)_2(CO)_4 + CO$	$5 \rightarrow 6$	111
18	$IrHCl_2(PPh_3)_3 + D_2 \rightarrow IrD_2Cl(PPh_3)_3 + HCl$	$6 \rightarrow 6$	95,97
19	$MHX(CO)(PPh_3)_3 + D_2 \rightarrow MDX(CO)(PPh_3)_3 + HD$	$6 \rightarrow 6$	97,98

[a] M = Ru, Os; X = Cl^-, Br^-, I^-; (S) = solvent; Y = Cl^-, OH^-, BPh_4^-; L = PPh_3 = triphenylphosphine; diphos = $Ph_2P(CH_2)_2PPh_2$; dmpe = $(CH_3)_2P(CH_2)_2P(CH_3)_2$; R = CH_3, C_6H_5.

The activation of molecular hydrogen by d^8 metal complexes results in a direct conversion of hydrogen to two hydride ion ligands with the simultaneous oxidation of the d^8 (M^{1+}, M^0) metal ion to the d^6 (M^{3+}, M^{2+}) state. In all the cases studied [15,29,30,97–111] (Table III) the addition of molecular hydrogen to the d^8 complex results in the formation of *cis*-dihydride species accompanied by increase in the coordination number of the metal ion from 4 to 6.

The ease of oxidative addition of molecular hydrogen to a four- or five-coordinate d^8 metal ion depends to a large extent on the nature of the metal ion and the coordinated ligands. For the platinum group metal ions, the tendency for oxidative addition increases from top to bottom (lower to higher atomic weight) in each group and from right to left in each triad [112]. Thus for an M^{n+} group the reactivity sequence Os(0) > Ru(0) > Fe(0) > Ir(I) > Rh(I) > Co(I) and Pt(II) > Pd(II) > Ni(II). For the triads the reactivity sequence is Fe(0) > Co(I) > Ni(II); Ru(0) > Rh(I) > Pd(II); Os(0) > Ir(I) > Pt(II). Soft σ donors favor oxidative addition by increasing the electron density on the metal ion, whereas π acids seem to reduce this tendency. Thus the activity of $IrX(CO)(PPh_3)_3$ for the activation of molecular hydrogen decreases [113] in the order $I^- > Br^- > Cl^-$. The tendency to form a dihydride is always enhanced when the coordinated CO is replaced by a tertiary phosphine. Alkyl-substituted phosphines are better than triphenylphosphine.

Subtle changes in the neutral and anionic ligands coordinated to a d^8 metal ion bring about changes in the reactivity pattern for oxidative addition. The complex $Ir(diphos)_2^+$ activates [15,101] molecular hydrogen whereas the rhodium analog, $Rh(diphos)_2^+$, is inactive [101]. The methyl-substituted phosphine complex $Rh(dmpe)_2^+$, however, is reactive [30] and activates molecular hydrogen under ambient conditions. The studies made so far of the variation of ligand on the reactivity patterns of a d^8 complex are qualitative in nature and much more remains to be done and understood before quantitative predictions about the course of a reaction can be made.

Oxidative addition to the complex $Ir(X)(CO)(PPh_3)_2$ (**12**) has been studied extensively with regard to variation of the anionic ligand and of the neutral ligand species carbon monoxide and triphenylphosphine (Table III) [97–99, 104,112,113]. The complex $IrCl(CO)(PPh_3)_2$ forms the *cis*-dihydride $Ir(H)_2Cl(CO)(PPh_3)_2$ (**13**) characterized by the appearance of two M—H stretching bands at 2190 and 2100 cm^{-1} in the infrared [97–99]. The lower-frequency band is considered to be due to the hydride *trans* to the carbonyl group. The NMR spectrum of the dihydride exhibits two proton resonances at τ 28.4 and τ 17.3 split into triplets by coupling with the ^{31}P atoms in the *cis* position. A shift of the carbonyl frequency to higher wave number correlates with the decreased electron density around the metal as a result of the oxidative addition reaction. The complex $IrX(CO)(PPh_3)_2$ (**12**), also combines with a number of other diatomic molecules [97–99], resulting in the conversion of each molecule to two ligand atoms, except in the case of molecular oxygen which retains one O—O bond. Examples of these reactions are given in Eqs. (20a)–(20e) for hydrogen, oxygen, hydrogen chloride, chlorine, and alkyl halides.

$$Ir(X)(CO)(PPh_3)_2 \;(\mathbf{12}) \xrightarrow{H_2} Ir(H)_2(X)(CO)(PPh_3)_2 \quad (20a)$$

$$\xrightarrow{HCl} Ir(H)(Cl)(X)(CO)(PPh_3)_2 \quad (20b)$$

$$\xrightarrow{Cl_2} Ir(Cl)_2(X)(CO)(PPh_3)_2 \quad (20c)$$

$$\xrightarrow{RX'} IrR(X')(X)(CO)(PPh_3)_2 \quad (20d)$$

$$\xrightarrow{O_2} Ir(O_2)X(CO)(PPh_3)_2 \quad (20e)$$

$(X = X' = Cl^-, Br^-, \text{and } I^-)$

The kinetics of addition of hydrogen and oxygen to **12** has been studied by Chock and Halpern [113]. The reactions exhibited second-order kinetics, first order with respect to hydrogen or oxygen and first order with respect to the complex. The enthalpy and entropy of activation of molecular hydrogen and oxygen were found to be in the range 10.8 to 13.1 kcal/mole and −20 eu, respectively. A polar transition state (with one hydrogen atom closer to the metal than the other) was suggested for the activation of molecular hydrogen with the concerted insertion of metal ion in the H—H bond. The higher rate of the reaction in polar solvents supports the polar transition state for

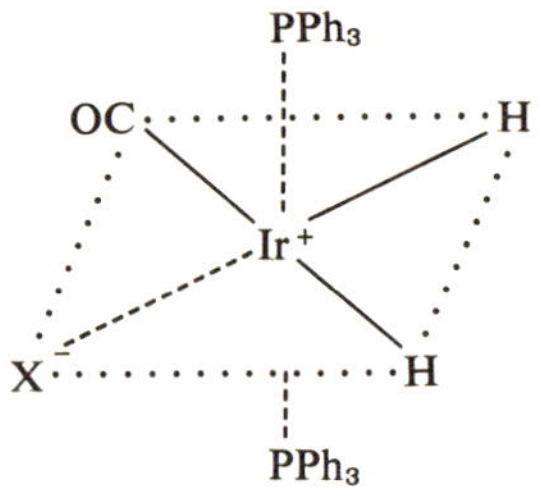

13

$(X = Cl^-, Br^-, \text{and } I^-)$

the hydrogen reaction. Dichlorohydridotris(triphenylphosphine)Ir(III) exchanges [97] with D_2 to give a compound which is analogous to that obtained by the addition of deuterium to chlorotris(triphenylphosphine)Ir(I). Accordingly, deuteration of dichlorohydridotris(triphenylphosphine)Ir(III) is believed to occur via S_N1-type dissociation to $Ir(I)Cl(PPh_3)_3$ [97], followed by addition of D_2:

$$Ir(III)HCl_2(PPh_3)_3 \rightleftharpoons Ir(I)Cl(PPh_3)_3 + HCl \quad (21)$$

$$Ir(I)Cl(PPh_3)_3 + D_2 \longrightarrow Ir(III)(D)_2Cl(PPh_3)_3 \quad (22)$$

For the hydrogen–deuterium exchange reactions of complexes 8 and 19 in Table III (coordination numbers 5 and 6, respectively), Vaska [97] has proposed formation of seven- and eight-coordinate intermediates, followed by dissociation (S_N2-type reaction). Suggestions for an elucidation of the structures of such intermediates with expanded coordination shells would be of considerable interest. Determination of the structures of the complex $RhH(CO)(PPh_3)_3$ by Laplaca and Ibers [11] and of $OsHBr(CO)(PPh_3)_3$ by Orioli and Vaska [10] by X-ray analysis provides valuable information for the interpretation of the reactivities of these substances. Changes in the carbonyl stretching frequencies are also very valuable for following and interpreting kinetic studies of the reactions of these complexes [117,115].

$$[IrH(CO)(PPh_3)_3] + D_2 \rightleftharpoons [Ir(D)_2H(CO)(PPh_3)_3] \longrightarrow [IrD(CO)(PPh_3)_3] + HD \quad (23)$$

$$MbHX(CO)(PPh_3)_3 + D_2 \rightleftharpoons MbHX(D)_2(CO)(PPh_3)_3 \longrightarrow MbDX(CO)(PPh_3)_3 + HD \quad (24)$$

V. Activation of Alkyl and Acyl Halides

The activation of alkyl and acyl halides by platinum group metal complexes is closely related to the activation of molecular hydrogen, because both processes proceed by an oxidative addition of the molecule to the d^8 metal ion. The main difference between the reaction types is that the addition of R or RCO and X moieties usually occurs in a *trans* configuration, while addition of hydrogen invariably produces two *cis*-hydride moieties. The studies on the activation of RX species have been conducted with iridium(I) [31–72,106], rhodium(I) [42–45,103], cobalt(I) [42], osmium(0) [50], platinum(II) [51,52], and palladium(II) [52] complexes (Table IV). The iridium(I) [31–42,106] complexes have been the most widely investigated. Investigations of d^{10} metal complexes have been initiated recently and thus far cover platinum(0) [53–57], palladium(0) [53,54,57], and osmium(0) [58].

The changes in the nature of the neutral and anionic ligand species that influence the oxidative addition of molecular hydrogen also influence the activation of RX in a related manner. The oxidative addition reactions of alkyl halides have been more extensively investigated than the corresponding reactions of molecular hydrogen. Information on complexes having a variety of ligands coordinated to the d^8 metal ion has helped in an assessment of the electronic and steric factors of the coordinated ligands that control the oxidative addition of these compounds. Ligands that place electron density on the metal ion (σ donors) assist the activation of alkyl halides, whereas good π acids retard the reaction. Complexes of alkyl and aralkyl phosphines are more reactive than those of aryl phosphines (Table IV). The order of reactivity for platinum group metal complexes is the same as that for the activation of molecular hydrogen, the activity increasing down a group and across a triad from right to left [112].

TABLE IV

ACTIVATION OF AKLYL HALIDES BY PLATINUM GROUP METAL COMPLEXES[a]

	Complex	Adduct	Product	References
	A. d^8 Complexes			
1	$IrCl(CO)(PPh_3)_2$	RI	$IrCl(I)(CO)R(PPh_3)_2$ R = CH_3, $CH_2{=}CH{-}CH_2$	31
2	$IrX(CO)(PEt_2Ph)_2$	$CH_2{=}CHCH_2X$	$IrX_2(\sigma\text{-}C_3H_5)(CO)(PEt_2Ph)_2$ X = Br, Cl	32
3	$IrBr(CO)(PEt_2Ph)_2$	CH_3Br	$IrBr_2(CO)(CH_3)(PEt_2Ph)_2$	33
4	$IrX(CO)(PEt_2Ph)_2$	RCOX	$IrX_2(RCO)(CO)(PEt_2Ph)_2$ X = Cl, R = *n*-Pr, iso-Pr, Et, Me X = Br, R = Me	32
5	$IrBr(CO)(PPh_3)_2$	CH_3COBr	$IrBr_2(CH_3CO)(CO)(PPh_3)_2$	32
6	$IrX(CO)(PPh_2Me)_2$	CH_3COY	$IrXY(CH_3CO)(PPh_2Me)_2$ X = Y = Cl, Br X = Cl, Y = Br X = Br, Y = Cl	33
7	$IrX(CO)(PPh_2Me)_2$	CH_3Y	$IrXY(CH_3)(CO)(PPh_2Me)_2$ X = Y = Cl, Br X = Cl, Y = Br, I X = Br, Y = Cl	33
8	$IrX(CO)L_2$	CH_3I	$IrXI(CH_3)L_2$ X = Cl, I, SCN L = PPh_3, PPh_2CH_3	34
9	$IrCl(CO)(PPh_2Me)_2$	$CH_3CHBrCOOC_2H_5$	$IrClBr(CO)(\sigma\text{-}CH(CH_3)COOC_2H_5)(PPh_2Me)_2$	34
10	$IrCl(CO)(PMe_3)_2$	*trans*-1-Bromo-2-fluorocyclohexane	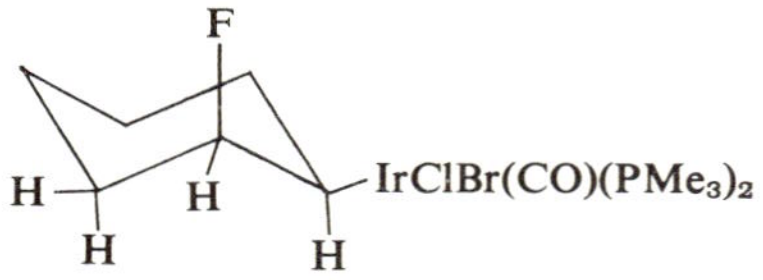	35

11	$[Ir(CO)_2(PMe_2Ph)_3]BPh_4$	CH_3X	$[IrX_2(CH_3)(CO)(PMe_2Ph)_2]$	106
12	$[Ir(CO)_2(PMe_2Ph)_3]ClO_4$	$CH_2{=}C(CH_3){-}CH_2Cl$	$[IrCl_2(\sigma\text{-}C_4H_7)(CO)(PMe_2Ph)_2]$	106
13	$[Ir(CO)_3(PMe_2Ph_3)_2]BPh_4$	$CH_2{=}C(CH_3){-}CH_2Cl$	$[IrCl(\pi\text{-}C_4H_7)(CO)(PMe_2Ph)_2]BPh_4$	36
14	$IrX(CO)L_2$	CH_3I	$[IrXI(CH_3)(CO)L_2]$ X = Cl, Br L = PBu^tMe_2, PBu^tEt_2, $PBu^tPr_2{}^n$, $PBu_3{}^t$	37
15	$IrX(CO)(PBu^tMe_2)_2$	RCl	$[IrXCl(CO)(R)(PBu^tMe_2)_2]$ R = CCl_3, $CH_2{=}CH{-}CH_2$, CH_3CO, $PhCH{=}CHCH_2$, $(CH_3)_2C{=}CHCO$, CH_3OCO	37
16	$IrCl(N_2)(PPh_3)_2$	RCOCl	$IrCl_2(RCO)(PPh_3)_2$ R = CH_3, C_2H_5, C_6F_5, $C_6H_5CH_2$, C_6H_5	38
17	$Ir(NO)(PPh_3)_2$	CH_3I	$Ir(CH_3)(I)(NO)(PPh_3)_2$	39
18	$IrCl(PPh_3)_3$	RCOCl R = $(CH_3)_2CH$, $CH_3CH_2CH(CH_3)$, $CH_3CH(C_2H_5)$, $(CH_3)(C_6H_5)CH$	$IrCl_2(RCO)(PPh_3)_2$ R = $-CH_2CH_2CH_3$, $-CH_2(CH_2)_2CH_3$, $-CH_2(CH_2)_3CH_3$, $-CH_2CH_2C_6H_5$	40
19	$A^+[Ir(CO)_2X_2]^-$	RI R = CH_3, CH_3CO	$A^+[IrX_2I(R)(CO)_2]^-$ X = Cl, Br, I A^+ = $NH_4{}^+$, $NEt_4{}^+$, $PPh_4{}^+AsPh_4{}^+$	41
20	$C_5H_5Ir(CO)(PPh_3)_3$	CH_3I	$[C_5H_5Ir(CH_3)(CO)(PPh_3)_3]I$	42
21	$C_5H_5Rh(CO)(PPh_3)_3$	CH_3I	$[C_5H_5RhI(COCH_3)(PPh_3)_3]$	42

TABLE IV *(continued)*

	Complex	Adduct	Product	References
22	$C_5H_5Rh(CO)(PPh_3)_3$	RX R = $C_6H_5CH_2$, $CH_2{=}CH{-}CH_2$	$[C_5H_5RhX(COR)(PPh_3)_3]$	43
23	$RhCl(PPh_3)_3$	CH_3I	$RhClI(CH_3)(PPh_3)_2$	44
24	$RhCl(PPh_3)_3$	$CH_2{=}CR{-}CH_2Cl$ R = H, CH_3	$RhCl_2(\sigma\text{-}C_3H_4R)(PPh_3)_2$ + $RhCl_2(\pi\text{-}C_3H_4R)(PPh_3)_2$	44
25	$RhCl(AsPh_3)_3$	$CH_2{=}C(CH_3)CH_2Cl$	$RhCl_2[\pi,\sigma\text{-}C_3H_4(CH_3)](AsPh_3)_2$	44
26	$RhCl(CO)(PR_3)_2$	CH_3I	$Rh(Cl)(I)(CH_3)(CO)(PR_3)_2$ R = C_2H_5, $n\text{-}C_4H_9$	45
27	$RhCl(CO)(MeOC_6H_4NC)_2$	CH_3I	$Rh(Cl)(I)(CO)(Me)(MeOC_6H_4NC)_2$	46
28	$Rh(NO)(PPh_3)_3$	RX	$Rh(NO)(R)(X)(PPh_3)_2$ R = alkyl, acyl	47
29	$\{Rh[cis\text{-}(Ph)_2AsCH{=}CHAs(Ph)_2]_2\}^+, X^-$	CH_3I	$\{RhI(CH_3)[cis\text{-}(Ph)_2AsCH{=}CHAs(Ph)_2]_2\}^+$ $X^- = BPh_4^-, PF_6^-$	48
30	$[Rh(dien)(PPh_3)_2]^+, X^-$	CH_3I	$[Rh(CH_3)I(dien)(PPh_3)_2]^+$ $X^- = BPh_4^-, PF_6^-$	49
31	$\{Rh[P(OCH_3)_3]_4\}BPh_4$	CH_3I	$\{Rh[P(OCH_3)_3]_4(CH_3)I\}BPh_4$	103
32	$[RhL_4]BPh_4$	$CH_2{=}CH{-}CH_2X$ X = Cl, Br	$[Rh(\pi\text{-}C_3H_5)L_4](BPh_4)_2$ L = $P(OR)_3$, R = CH_3, C_2H_5, C_3H_7 L = PMe_2Ph	103
33	$[RhL_5]BPh_4$ L = $P(OR)_3$, R = CH_3, C_2H_5	$CH_2{=}CH{-}CH_2X$ X = Cl, Br	$[RhX(\pi\text{-}C_3H_5)L_3]BPh_4$	103
34	$C_6H_5Co(CO)(PPh_3)_3$	CH_3I	$C_6H_5CoI(COCH_3)(PPh_3)_3$	42
35	$[OsH(CO)_4]^-$	CH_3I	$cis\text{-}Os(CH_3)(H)(CO)_4$ $cis\text{-}Os(CH_3)_2(CO)_4$ $cis\text{-}Os(CH_3)I(CO)_4$	50

36	$Pt(CH_3)_2(dias)$	RX R = CH_3, CH_3CO X = Cl, Br, I	*trans*-$Pt(CH_3)_2(dias)RX$	51
37	$Pt(CH_3)_2(AsMe_2Ph)_2$	RX R = CH_3, CH_3CO X = Cl, Br, I	*trans*-$Pt(CH_3)_2(AsMe_2Ph)_2RX$	51
38	$Pr(CH_3)_2(dias)$	RX R = $CH_2{=}CH{-}CH_2$, $CH_2{=}C(CH_3){-}CH_2$	*cis*-$Pt(CH_3)_2(dias)RX$	51
39	$M(CH_3)_2L_2$ L = rac-diars, meso-diars M = Pd, Pt	RX R = CH_3, CH_3CO X = Cl, Br, I,	*trans*-$M(CH_3)_2RXL_2$	52
		$CH_2{=}CH{-}CH_2Cl$	*cis*-$M(CH_3)_2(C_3H_5)ClL_2$	52
	B. d^{10} Complexes			
40	$Pd(CO)(PPh_3)_3$	RX R = CH_3, X = I R = $CH_2{=}CH{-}CH_2$, $CH_2{=}CH$ X = Cl	*trans*-$PdX(COR)(PPh_3)_2$	53
41	$M(PPh_3)_4$ M = Pd or Pt	RN_3 R = CH_3, C_2H_5	*trans*-$M(PPh_3)_2(N_3)_2$ *trans*-$M(PPh_3)_2(N_3)_2$	54
42	$Pt(PPh_3)_4$	$C(Me)(CN)_3$	$Pt(PPh_3)_2(CN)[CMe[CN]_2]$	55
43	$Pt(C_2H_2)(PPh_3)_2$	RX	$Pt(R)(X)(PPh_3)_2$ R = CH_3, CH_3CO	56
44	ML_4 L = $Ph_2PC{\equiv}CCH_3$ M = Pd, Pt	CH_2Cl_2	MCl_2L_2	57
45	$[Os(CO)_4]^{2-}$	RI R = CH_3, C_2H_5	*cis*-$Os(R)_2(CO)_4$	58

[a] dias = $(PPh_3)_2AsCH_2CH_2As(PPh_3)_2$; rac = racemic; diars = $PhMeAsCH_2CH_2AsMePh$.

The addition of alkyl and acyl halides (RX, RCOX) to d^8 metal complexes results in the formation of an oxidative addition product of the d^6 metal complex with R and X coordinated to the metal ion in a *trans* manner. For the d^8 complexes of osmium(0) [50], platinum(II) [51,52], and palladium(II) [52], the oxidative addition of RX to the d^8 complex may be *cis* or *trans* depending on the nature of the neutral ligand. Thus addition [51] of RX to $Pt(CH_3)_2$-$(AsMe_2Ph)_2$ gives the *trans* product, *trans*-$Pt(CH_3)_2R(AsMe_2Ph)_2X$, whereas $Pt(CH_3)_2$(dias) gives the *cis* adduct, *cis*-$Pt(CH_3)_2R$(dias)X (entries 37 and 38, Table IV). The stereochemistry of the addition of RX to d^{10} metal complexes, however, has not been throughly investigated.

In a few cases the oxidative addition of alkyl or acyl halides to d^8 metal complexes gives products other than the simple *trans* addition products of R and X. The oxidative addition of an alkyl group to a d^8 metal complex with a coordinated carbonyl group sometimes results in the formation of an acyl complex [42] (entries 21 and 34, Table IV). Such products are invariably formed by the rearrangement of an initial metal–alkyl complex by migration of the alkyl group to a coordinated carbonyl group, to form the coordinated acyl group [42]. The reverse reaction, cleavage of an acyl group to coordinated alkyl and carbonyl groups has sometimes been observed with iridium(I) complexes [38–40] (entries 16–18, Table IV). In some cases [40] (entry 18, Table IV) branched acyl halides give straight-chain alkyl derivatives of the metal ion. Alkyl halides add on in a *trans* manner to form both σ- and π-allyl complexes [31,32,37,43,51,103,106] (entries 1, 2, 12, 15, 21, 24, 32, 33, and 38, Table IV).

The oxidative addition of RX to d^8 metal complexes is strongly influenced by steric factors and depends on the size of the neutral ligand L. Thus iridium(I) complexes of the type *trans*-$[IrX(CO)L_2d$ ($L = PBu^tMe_2$) add a large numbers of molecules [37] such as RCOCl, ROCOCl, CCl_4, and $CH_2{=}CH{-}CH_2Cl$, whereas *trans*-$[IrX(CO)L_2]$ ($L = PBu^tEt_2, PBu^tPr^n, PBu_3{}^t$) adds CH_3I slowly and does not react with $CH_2{=}CH{-}CH_2Cl$ or RCOCl [37]. The σ-donor capacity of the anionic ligand X is also an important factor. The reactivity of the complexes increase with the σ-donor capacity of X in the order $Cl < Br < I$ [113].

Chock and Halpern [113] have studied the kinetics of addition of methyl iodide to *trans*-$[IrX(CO)(PPh_3)_2]$. The rate of the reaction is first order with respect to methyl iodide and the complex, and is sensitive to the nature of X, decreasing in the order $Cl > Br > I$. The reaction is faster in a polar solvent such as DMF than in nonpolar solvents such as benzene (by a factor of about 17 at 35°C for these two examples). The enthalpy and entropy of activation were reported as 5.6–8.8 kcal/mole and -43 to -51 eu. The large negative entropy of activation and a pronounced solvent dependence of the reaction was explained on the basis of formation of a highly polar activated complex.

The activation of RX was considered as an S_N2 displacement at the carbon–halogen bond by the nucleophilic transition metal ion.

In line with the S_N2 mechanism, Labinger *et al.* [35] have demonstrated that the reaction of *trans*-1-bromo-2-fluorocyclohexane (entry 10, Table IV), whose *trans* configuration was confirmed by its ^{19}F NMR sprectum, proceeds with inversion of configuration at the carbon center and forms a *cis* product. Pearson and Muir [34], however, on the basis of the reaction of CH_3CHBr-$COOC_2H_5$ ($[\alpha]_D{}^{25}$ $-6°$) with $IrCl(CO)(PPh_2CH_3)_2$ concluded that there is no inversion of configuration. The product $IrCl(CO)(PPh_2CH_3)_2(CH_3CH$-$COOC_2H_5)Br$ was found to be optically active, $[\alpha]_D{}^{25}$ $-20°$. On cleavage with bromine, the original ester was obtained with rotation $[\alpha]_D^{25}$ $-4°$. Since there was a slight change in the rotation of the final product it was assumed that there should be two retentions or two inversions in the process of addition of the optically active ester to the iridium complex and subsequent cleavage. Since the cleavage of alkyl–metal bonds by bromine was found to proceed with retention of configuration [59], it was argued that the addition of ester to the metal ion should also proceed with retention of configuration. In view of these findings, it was suggested [34] that the process of oxidative addition of alkyl halides to the iridium(I) complexes occurs by a one-step concerted process without the formation of intermediates and with the rentention of configuration of the alkyl carbon atom.

The two mechanisms proposed for oxidative addition of alkyl halides are illustrated by Eqs. (25) and (26), respectively.

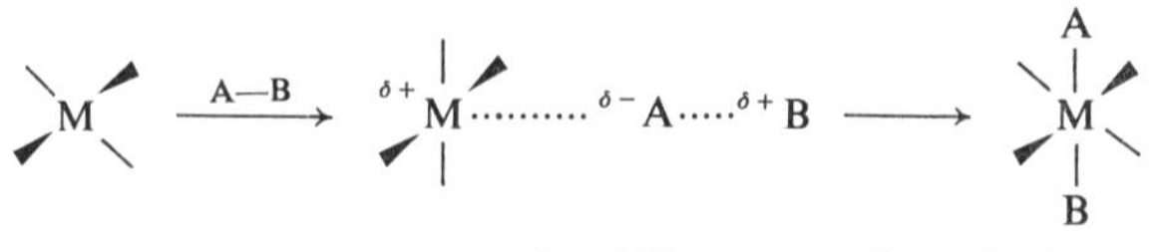

(25)

nucleophilic process (inversion)

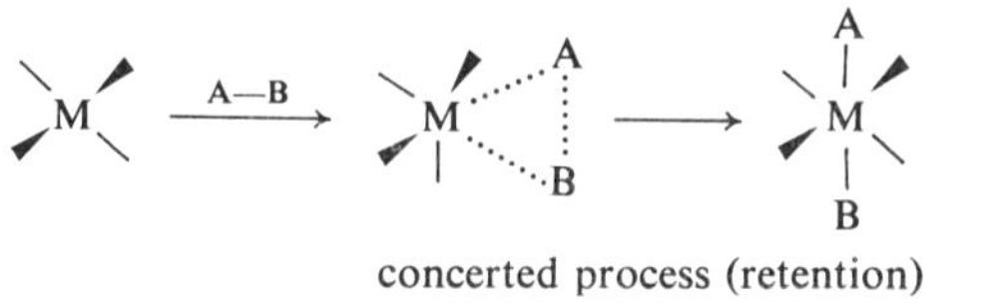

(26)

concerted process (retention)

Osborn [60] has recently studied the configuration of the compound PhCHFCHDBr and its oxidative addition product,

$$IrCl(CO)(PMe_3)_2(PhCHFCHD)Br,$$

by ^{19}F and ^{1}H NMR spectroscopy and came to the conclusion that the reaction proceeds by both inversion and retention of configuration. Osborn [60] proposed a free-radical mechanism, similar to those proposed by Schneider *et al.* [116], Marzilli *et al.* [117,118], and Halpern and Phelan [119] for the reaction of bis(dioximato)cobalt(II) with organic halides. In view of conflicting suggestions and interpretations, the mechanism of oxidative addition of RX to d^5 metal complexes may be regarded as not yet completely elucidated.

VI. Activation of the Carbon–Hydrogen Bond by Platinum Group Metal Complexes

Activation of the carbon–hydrogen bond of hydrocarbons by a process of oxidative addition to a d^8 or d^{10} metal complex similar to that observed for the activation of molecular hydrogen and alkyl halides has not yet been achieved. Intramolecular C—H bond cleavage in coordinated ligands in reactive d^8 complexes of iron(0) [6], ruthenium(0) [62] and iridium(I) [63] has been recently reported [120] (Table V). The reversible intramolecular C—H oxidative addition reactions to iron(0) and ruthenium(0) complexes are illustrated by the reactions (27) and (28), respectively. The iridium complex that is formed by the intramolecular C—H bond cleavage of a coordinated phosphite or phosphine group is very stable [63] and an X-ray structure of the compound has been reported [64].

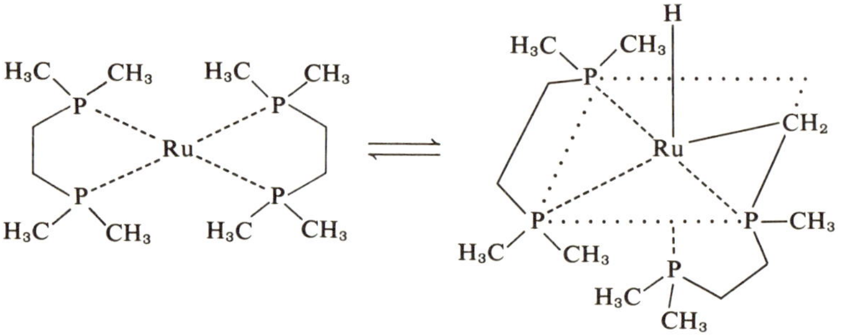

(27)

14 15

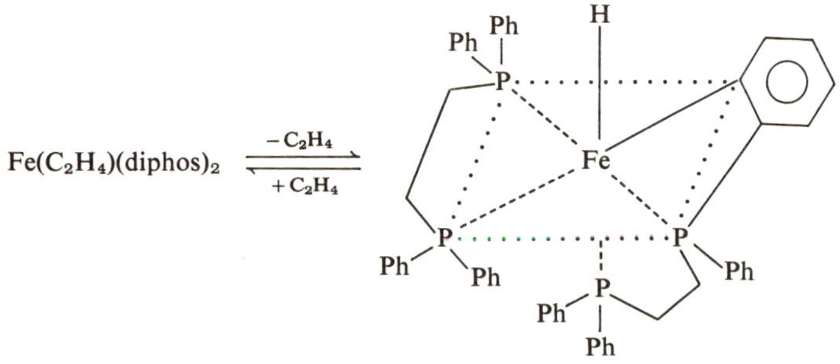

(28)

16 17

TABLE V

ACTIVATION OF THE C—H BOND BY PLATINUM GROUP METAL COMPLEXES

	Complex	Adduct	Product	References
1	$Fe(C_2H_4)(diphos)_2$	C—H cleavage of a coordinated diphos	$FeH(diphos)[\sigma\text{-}C_6H_4P(Ph)CH_2CH_2P(Ph)_2]$ (formula **17**)	61
2	$Ru(dmpe)_2$	C—H cleavage of a coordinated dmpe	$RuH(dmpe)[\sigma\text{-}CH_2P(CH_3)CH_2CH_2P(CH_3)_2]$ (formula **15**)	62
3	$IrClL_3$ L = PPh_3, $AsPh_3$, $SbPh_3$, $P(p\text{-}CH_3\text{—}C_6H_4)_3$	C—H cleavage of coordinated L	$IrHL_2Cl(\sigma\text{-}C_6H_4PPh_2)$ (analogous to formula **17**)	63
4	$IrHCl_2[P(OPh_3)_3]_3$	C—H cleavage from $P(OPh)_3$	$IrHCl[P(OPh_3)_3]_2[\sigma\text{-}C_6H_4OP(OPh)_2]$	64
5	$Pt(PPh_3)_4$	CH_3NO_2	$Pt(PPh_3)(CNO)_2$	121
6	*cis*-$PtCl_2[P(OPh)_3]_2$	C—H cleavage from $P(OPh)_3$	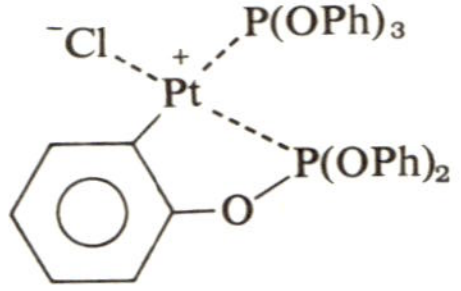	
7	$PtCl_2L_2$ L = $P(Bu^t)_2Ph$, $P(Bu^t)_2Pr^n$, $P(C_6H_4OMe)_3$	C—H cleavage from an L group	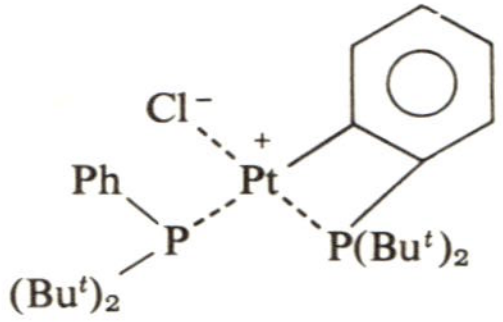 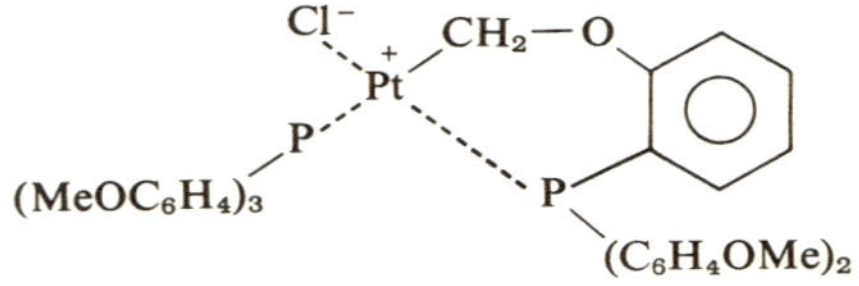	

TABLE V (*continued*)

ACTIVATION OF THE C—H BOND BY PLATINUM GROUP METAL COMPLEXES

	Complex	Adduct	Product	References
8	$PdCl_4^{2-}$	Benzo[*h*]quinoline	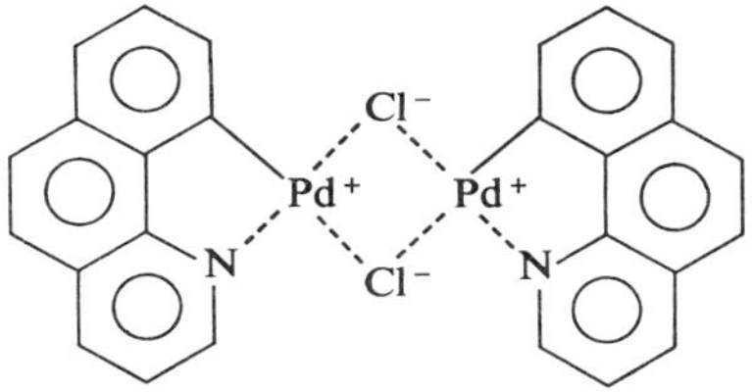	67
		8-Methylquinoline	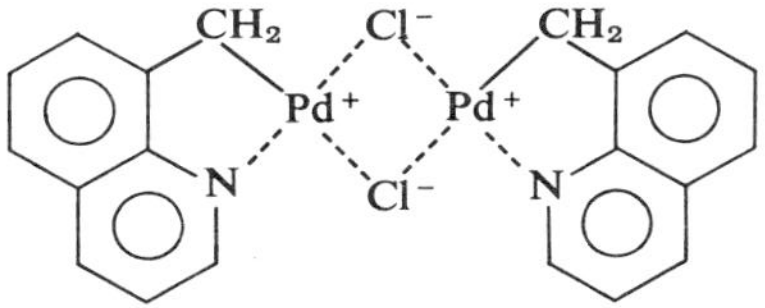	
9	$PtCl_4^{2-}$	C_nH_{2n+2} (n = 1 to 6)	Deuterated hydrocarbons	122

Platinum complexes in the zero-valent state seem to be very effective in the cleavage of the C—H bond. Tetrakis(triphenylphosphine)platinum(0) reacts [121] with nitromethane to form the bifulminato complex, $Pt(PPh_3)_2(CNO)_2$, **18** (entry 5, Table V). As suggested by Beck [121], the reaction probably proceeds through an intermediate formed by the oxidative addition of hydride and $—CH_2NO_2$ moieties to platinum(II) followed by the elimination of water and molecular hydrogen to form the product, as indicated in Eq. (29).

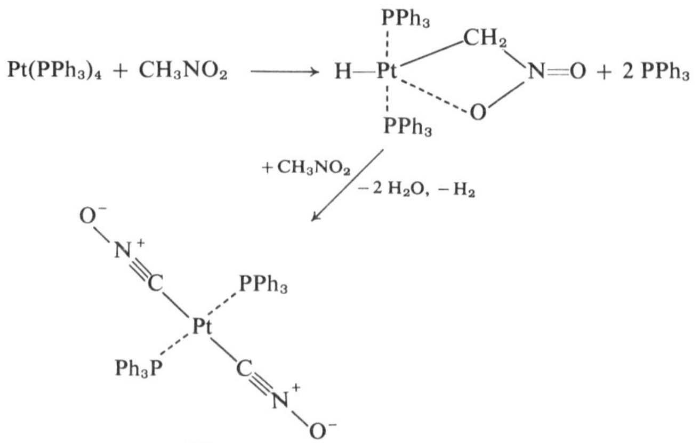

18

(29)

Intermolecular carbon–hydrogen bond cleavage by platinum(II) [65,66] and palladium(II) [67] complexes (entries 6, 7, and 8, Table V) forms σ-carbon-bonded complexes with obviously no change in the oxidation state of the metal ion. It is possible that such reactions may proceed via platinum(IV) or palladium(IV) intermediates formed by intramolecular oxidative addition of hydride and σ-carbon-bonded moieties, followed by a reductive elimination of hydrogen chloride from platinum(IV) or palladium(IV) complexes to give corresponding platinum(II) or palladium(II) complexes.

A very interesting reaction involving catalysis of hydrogen–deuterium exchange of saturated hydrocarbons, C_nH_{2n+2} (n = 1–6) with $PtCl_4^{2-}$ has been reported by Hodges *et al.* [122]. Although the mechanism of the reaction is not known, C—H bond cleavage of the hydrocarbon followed by interaction of the C_nH_{2n+1} moiety with deuterium is probably involved in the hydrogen–deuterium exchange process.

VII. Activation of Molecular Hydrogen by Pentacyanocobaltate(II)

Pentacyanocobaltate(II) is a versatile catalyst for homogeneous hydrogenation reactions. This complex has been found to bring about the hydrogenation of a wide variety of substrates [123,124] such a aliphatic dienes, conjugated aromatic olefins, α,β-unsaturated acids, aldehydes, 1,2-diketones and azoxy compounds. Hydrogenation is usually carried out in aqueous solutions at room temperature and under a pressure of 1 atm of hydrogen [123,125–132]. Before discussing the types of reactions in which $Co(CN)_5^{3-}$ activates molecular hydrogen, the nature of the reaction and the intermediates formed when pentacyanocobaltate(II) reacts with molecular hydrogen will first be considered.

The absorption of molecular hydrogen by aqueous solutions of pentacyanocobaltate(II) was first observed by Iguchi [133]. The stoichiometry by hydrogen uptake according to Iguchi [133] is 0.8 hydrogen atoms per atom of cobalt. Later investigations by Mills *et al.* [134], however, showed an absorption of nearly one atom (0.95) of hydrogen per atom of cobalt. The nature of the intermediate formed when pentacyanocobaltate(II) reacts with hydrogen was thoroughly investigated by King and Winfield [135,136], and Griffith and Wilkinson [7]. A detailed spectroscopic study by King and Winfield [136] has confirmed the formation of $[HCo(CN)_5]^{3-}$ as one of the intermediates when hydrogen is absorbed by $[Co(CN)_5]^{3-}$. The hydrido complex is characterized by an absorption band at 305 nm that is absent in other cobalt(III) complexes. The complex has also been isolated as the cesium-sodium salt, $Cs_2Na[HCo(CN)_5]$, by Banks and Pratt [137], and was characterized by its ultraviolet and infrared spectra. These investigators also identified another possible catalytically active intermediate, the dihydro complex, $H_2Co(CN)_4^{3-}$, as well as the probably inactive hexacyanocobaltate-(III) ion, $Co(CN)_6^{3-}$. Based on the Raman spectrum of a solution of pentacyanocobaltate(II) in aqueous methanol, a slightly distorted square-pyramidal geometry has been assigned [138] to this complex. The ESR (electron spin resonance) studies of Tsay *et al.* [139] seem to support this geometry and further indicate that there is no solvent molecule to complete the octahedron.

The reaction of $Co(CN)_5^{3-}$ with molecular hydrogen is reversible, the hydrido complex losing hydrogen when the solution is heated, or when hydrogen is pumped off, to reform $Co(CN)_5^{3-}$. The formation of pentacyanocobaltate(II) from $[HCo(CN)_5]^{3-}$ is, however, far from complete because of the greater stability of Co(III) complexes. A solution of pentacyanocobaltate(II) loses its paramagnetism with time. This loss of paramagnetism, which has been referred to as "aging," was originally thought to be due to dimerization [134]. It has now been shown by King and Winfield [136] that "aging" is due to the formation of the complexes of $[Co(CN)_5OH]^{3-}$ and $[Co(CN)_5H]^{3-}$ that absorb at 380 and 305 nm, respectively.

The formation of the hydrido complex $[HCo(CN)_5]^{3-}$ has been confirmed by the NMR studies of Griffith and Wilkinson [7]. On the basis of kinetic studies, King and Winfield [136] have suggested the mechanism shown in Eqs. (30)–(31) for the absorption of molecular hydrogen by $[Co(CN)_5]^{3-}$.

$$[Co(CN)_5]^{3-} + H_2 \underset{}{\overset{k_1}{\rightleftharpoons}} \underset{\mathbf{19}}{[H_2Co(CN)_5]^{3-}} \tag{30}$$

$$[H_2Co(CN)_5]^{3-} + [Co(CN)_5]^{3-} \overset{k_2}{\rightleftharpoons} 2\,\underset{\mathbf{20}}{[HCo(CN)_5]^{3-}} \tag{31}$$

The reversibility of the reaction of pentacyanocobaltate(II) with molecular hydrogen has been verified by deuterium exchange studies of Mills *et al.* [134]. As stated by Halpern [4], the driving force for the formation of hydrido complex in solution is the tendency of the coordinately unsaturated $Co(CN)_5{}^{3-}$ to expand its coordination shell by reaction with hydride ion. The higher stability of the cobalt(III) complex is an important driving force for the formation of the hydrido complex.

DeVries [140,141] has studied the equilibria involved in the reaction of pentacyanocobaltate(II) and hydrogen. The concentration of $Co(CN)_5{}^{3-}$ was measured by following the intensity of the band at 970 nm, the optical density of which is directly proportional to cobalt(II) concentration. The position of the maximum absorption of this complex anion is not changed by a change in temperature or by the presence of inert electrolyte. Since highly charged species are involved, the reaction was conducted in a medium 1.00 *M* in NaOH, KCN, or KCl. The rate of hydride formation was found to follow third-order kinetics, first order with respect to molecular hydrogen and second order with respect to cobalt concentration. Equations (32) and (33) have been proposed. This reaction sequence is in accord with the kinetic and

$$2\,[Co(CN)_5]^{3-} \underset{k_2}{\overset{k_1}{\rightleftharpoons}} \underset{\mathbf{21}}{[Co_2(CN)_{10}]^{6-}} \tag{32}$$

$$[Co_2(CN)_{10}]^{6-} + H_2 \underset{k_4}{\overset{k_3}{\rightleftharpoons}} 2\,[Co(CN)_5H]^{3-} \tag{33}$$

equilibrium studies of Burnett *et al.* [142]. However, it is not possible to distinguish between this reaction sequence and the one proposed by King and Winfield [136], as described above, since the intermediate species in the two alternative mechanisms, $[Co_2(CN)_{10}]^{6-}$ and $[H_2Co(CN)_5]^{3-}$, are considered to be in very low, steady-state concentrations.

Thus the combination of molecular hydrogen and pentacyanocobaltate(II) involves homolytic fission of the hydrogen molecule and the reversible formation of a relatively stable hydrido complex. It may be noted that analogous reactions, in which metal ions [e.g., silver(I) and copper(I)] induce homolytic fission of H_2, also follow third-order kinetics—first order with respect to molecular hydrogen and second order with respect to the metal ion.

Simandi and Nagy [143,144] confirmed the results of deVries [141] by measuring the rate of absorption of molecular hydrogen by pentacyanocobaltate(II) solution under nonequilibrium conditions. A simplified form of the rate equation [Eq. (34)] involving only the forward reaction was used to calculate the rate constant k,

$$\text{Rate} = k[\text{Co}]^2[\text{H}_2] \tag{34}$$

where [Co] is pentacyanocobaltate(II). The rate constant for the dissociation of the hydridopentacyanocobaltate(II) complex was indirectly obtained from the equilibrium constant of the reaction. The simplified rate expression is applicable only in the initial states of the reaction (up to 40 minutes) where the back reaction is not appreciable. The equilibria between pentacyanocobaltate(II) and molecular hydrogen are given by Eqs. (35)–(37). The mechanism proposed

$$[\text{Co(CN)}_5]^{3-} + \text{H}_2 \underset{k_{-1}}{\overset{k_1}{\rightleftharpoons}} [\text{H}_2\text{Co(CN)}_5]^{3-} \tag{35}$$

$$[\text{H}_2\text{Co(CN)}_5]^{3-} + [\text{Co(CN)}_5]^{3-} \underset{k_{-2}}{\overset{k_2}{\rightleftharpoons}} \underset{\mathbf{22}}{[(\text{CN})_5\text{Co}\cdots\text{H}\cdots\text{H}\cdots\text{Co(CN)}_5]^{6-}} \tag{36}$$

$$[(\text{CN})_5\text{Co}\cdots\text{H}\cdots\text{H}\cdots\text{Co(CN)}_5]^{6-} \underset{k_{-3}}{\overset{k_3}{\rightleftharpoons}} 2\,[\text{HCo(CN)}_5]^{3-} \tag{37}$$

is very similar to that of King and Winfield [136] with the only difference that a binuclear hydrogen complex is proposed [143,144] as an intermediate. In this reaction with molecular hydrogen, pentacyanocobaltate(II) was described by the term "hydrogen carrier" [143].

In reviewing the three mechanisms proposed above for the formation of the hydridopentacyanocobalt(III) intermediate, serious objections may be raised concerning that of King and Winfield [Eqs. (30) and (31)], and that of Simanid and Nagy [Eqs. (35)–(37)] on the grounds that chemical bonding of the intermediates **19** and **22** is virtually impossible. Except for a very unstable transition state and hydrogen bonding to electronegative atoms, bis-covalent hydrogen is not considered possible. Thus the intermediate **19** and **22** do not serve the desired purpose of eliminating a termolecular step in the process.

On the other hand, the binuclear intermediate **21** proposed by deVries seems quite feasible and has analogies in the dicobalt and dimanganese carbonyls. Reaction of **21** with hydrogen could occur through a transition state **23** to give the final product **20**.

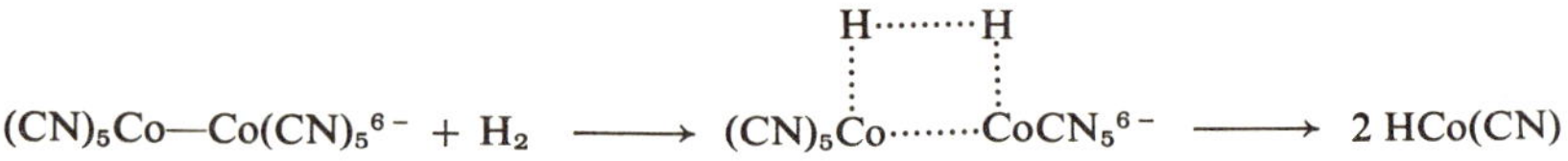

21 **23** **20**

(38)

Halpern and Pribanic [145] have measured the rate constant k_1 for the formation of hydridopentacyanocobaltate(III) from a solution of pentacyanocobaltate(II) at hydrogen pressures ranging from 6.8 to 27.2 atm. The third-order rate constant corresponding to the forward reaction of the equilibrium [Eq. (39)] at 25°C and $\mu = 0.1\ M$ has been reported [145] as 40 M^{-2} sec^{-1}. The enthalpy and entropy corresponding to k_1 are 0.7 ± 0.5 kcal/mole and -55 ± 5 eu, respectively. The equilibrium constant corresponding to reaction (39) in the hydrogen pressure range of 6.8–27.2 atm is $(1.0 \pm 0.1) \times 10^5\ M^{-1}$ at $\pm 0.2°$ and $\mu = 0.1\ M$. The study of equilibrium (39) at high

$$2\ Co(CN)_5^{3-} + H_2 \underset{k_{-1}}{\overset{k_1}{\rightleftharpoons}} 2\ Co(CN)_5H^{3-} \tag{39}$$

pressures of hydrogen is especially advantageous as it slows down considerably the competing decomposition, reaction (40).

$$2\ Co(CN)_5^{3-} + H_2O \longrightarrow Co(CN)_5H^{3-} + Co(CN)_5OH^{3-} \tag{40}$$

The splitting of hydrogen by pentacyanocobaltate(III) has been considered as a concerted process with a transition state illustrated by formula **23**.

A. Catalysis of Hydrogenation Reactions by Pentacyanocobaltate(II)

Catalytic hydrogenation of various substrates by pentacyanocobaltate(II) in solution has been reported by a number of investigators [123,127–129,131, 132]. Kwiatek *et al.* [123,128] reported that the catalyst is selective not only with respect to the functional groups of the substrate, but also to other structural features. A requirement [128] for the reduction of olefins is the presence of conjugated double bonds. Besides conjugated olefins, α,β-unsaturated carboxylic acids and carbonyl compounds are reduced by pentacyanocobaltate(II) [128]. A representative list of compounds reduced by $HCo(CN)_5^{3-}$ is given in Table VI.

The isomer distribution in the case of butadiene strongly depends on the molar ratio of $[CN^-]/[Co]$, particularly in the range 5.5–6.0. At a $[CN^-]/[Co]$ ratio below 5.5, *trans*-2-butene is the main (86%) reaction product. With a $[CN^-]/[Co]$ ratio of 6 or more, however, 1-butene predominates. The dependence of isomer distribution on $[CN^-]/[Co]$ ratio was explained by Kwiatek [127] on the basis of the formation of different σ and π intermediates in the reaction of $[HCo(CN)_5]^{3-}$ with butadiene. At a low $[CN^-]/[Co]$ ratio (less than 5.5) a π-allyl complex, **25** (see Scheme I) is formed which gives *trans*-2-butene as the main reaction product. The formation of the π-allyl complex **25** requires that the conjugated diene should be able to assume a "cissoid" conformation. This conformation of the substrate is sterically required to facilitate the formation of a π-allylic intermediate with $Co(CN)_5^{3-}$ prior to hydrogenation.

At a high $[CN^-]/[Co]$ ratio, however, a σ-Co(III) complex, **24**, is formed either directly from the diene and $Co(CN)_5H^{3-}$ or by the rearrangement of

TABLE VI

CONJUGATED AND α,β-UNSATURATED COMPOUNDS REDUCED BY HYDRIDOPENTACYANOCOBALTATE(III)

Substrate	Product	References
$CH_2{=}CH{-}CH{=}CH_2$	$CH_3{-}CH_2{-}CH{=}CH_2$	128
	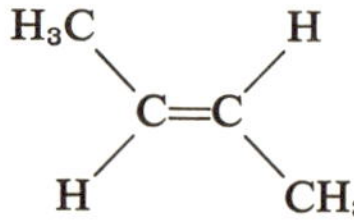	
$CH_2{=}CH{-}C(CH_3){=}CH_2$	$CH_3{-}CH{-}C(CH_3){=}CH_2$	128,129, 146
$C_6H_5CH{=}CHCOOH$	$C_6H_5CH_2{-}CH_2COOH$	
$CH_3{-}CH{=}CH{-}CH{=}CHCOOH$	$CH_3{-}CH{=}CH{-}CH_2{-}CH_2COOH$	129
$C_6H_5CH{=}CH_2$	$C_6H_5CH_2{-}CH_3$	128
$C_6H_5C(CH_3){=}CH_2$	$C_6H_5CH(CH_3){-}CH_3$	128
$C_6H_5C(COO^-){=}CH_2$	$C_6H_5CH(COO^-){-}CH_3$	128
$C_6H_5CH{=}CHCH_2OH$	$C_6H_5CH_2CH_2CH_2OH$	128
$CH_2{=}C(CH_3)COO^-$	$CH_3{-}CH{-}COO^-$ | CH_3	128
$CH_3CH{=}CH{-}CHO$	$CH_3CH_2CH_2{-}CHO$	128
$CH_3CH{=}C(CH_3){-}CHO$	$CH_3CH_2{-}CH(CH_3){-}CHO$	128
$CH_2{=}C{-}COO^-$ | CH_2COO^-	$CH_3CH{-}COO^-$ | CH_2COO^-	128
		128
		128
$C_6H_5CH_2{-}C{-}COOH$ ‖ NOH	$C_6H_5CH_2{-}CH{-}COOH$ | NH_2	129
$CH_3{-}C{-}COOH$ ‖ NOH	$CH_3{-}CH{-}COOH$ | NH_2	129

the π-allyl complex **25** leading to the formation of 1-butene. In the mechanism shown in Scheme I, suggested by Kwiatek, the dependence of the nature of the product on the number of cyano groups in the mixed intermediate complex may be visualized as related to differences in steric requirements of π-allyl and σ-bonded butenyl groups. At low $[CN^-]/[Co]$ ratios the hindrance for the addition of a negative π-allyl group which can possibly occupy two coordination positions on cobalt(III) is much lower than at high $[CN^-]/[Co]$ ratios where the butenyl group possibly has to displace a coordinated CN^-. In the latter case σ coordination of the butenyl group can take place at only

$$2\,Co(CN)_5^{3-} + H_2 \rightleftharpoons 2\,Co(CN)_5H^{3-} \tag{41}$$

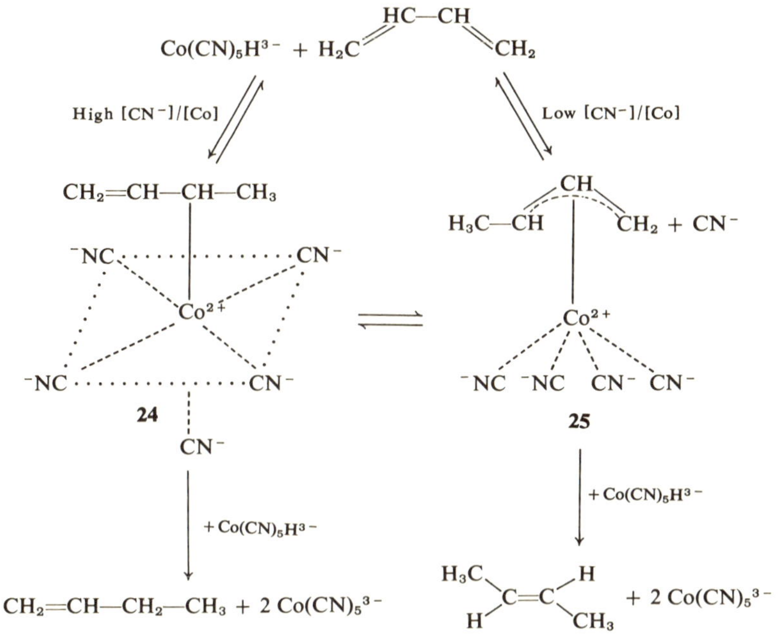

(42)

$$Co(CN)_5C_4H_7^{3-} + H_2O \longrightarrow C_4H_8 + Co(CN)_5OH^{3-} \tag{43}$$

$$Co(CN)_5H^{3-} + Co(CN)_5OH^{3-} \longrightarrow 2\,Co(CN)_5^{3-} + H_2O \tag{44}$$

Scheme I. Mechanism of hydrogenation of butadiene with pentacyanocobaltate(II).

one coordination position of cobalt(III) with the formation of 1-butene as the hydrogenation product. Presumably, formation of *trans*- rather than *cis*- 2-butene is due to the configuration of **24**, in which the CH_3 group becomes *trans* to the hydrogen atom inserted by the second mole of $HCo(CN)_5^{3-}$.

In the mechanism shown, hydridopentacyanocobaltate(III) reacts reversibly with hydrocarbon to give a mixed hydrocarbon–cyano complex, the formation of which involves 1 mole of butadiene per mole of hydride. The reversibility of this reaction has been established by Kwiatek [127] using deuterium oxide as a solvent. The product, 1-butene, was 66% dideuterated, 15% was a combination of mono- and trideuterated, and the remainder (19%) was tetra-, penta-, and undeuterated. Since uncoordinated butene itself does not take part in the exchange reaction with D_2O, the observed exchange may be explained by assuming that equilibrium takes place between the coordinated butenyl groups in **24** and **25** and the deuterated form of the catalyst by the dissociative equilibria.

The presence of butenyl groups in $[Co(CN)_5C_4H_7]^{3-}$ (**24**) was established by Kwaitek [127] by studies of the solid compound dissolved in D_2O. Pentacyanobutenylcobaltate(III) may react further with either hydridopentacyanocobaltate(III) or water to give butene. In the case of hydrolysis of butenylpentacyanocobaltate(III), hydroxopentacyanocobaltate(III) is first formed and reacts further with hydridopentacyanocobaltate(III) to regenerate the catalyst. The final step is thus the reverse of the "aging" reaction for $[Co(CN)_5]^{3-}$.

In the case of allylic compounds in general, including halides, acetates, alcohols, ester, and amines, cleavage of the substrate takes place by reaction with $[Co(CN)_5]^{3-}$ to yield the allylpentacyanocobaltate(III) complex [127], as indicated in Scheme II. The rearrangement of the σ-bonded complex to a π-allyl complex has been substantiated by measurement of the PMR spectra of mixtures of these complexes in D_2O Kwiatek [127] tentatively assigned structure **28** to π-allyltetracyanocobaltate(III), $C_3H_5Co(CN)_4^{2-}$. The assignment is based on an analogy between the σ–π allyl conversion in allylpentacyanocobaltate(III) and the corresponding σ–π allyl conversion in cobalt and manganese carbonyls [147].

$$2\,Co(CN)_5^{3-} + H_2 \rightleftharpoons 2\,Co(CN)_5H^{3-} \quad (45)$$

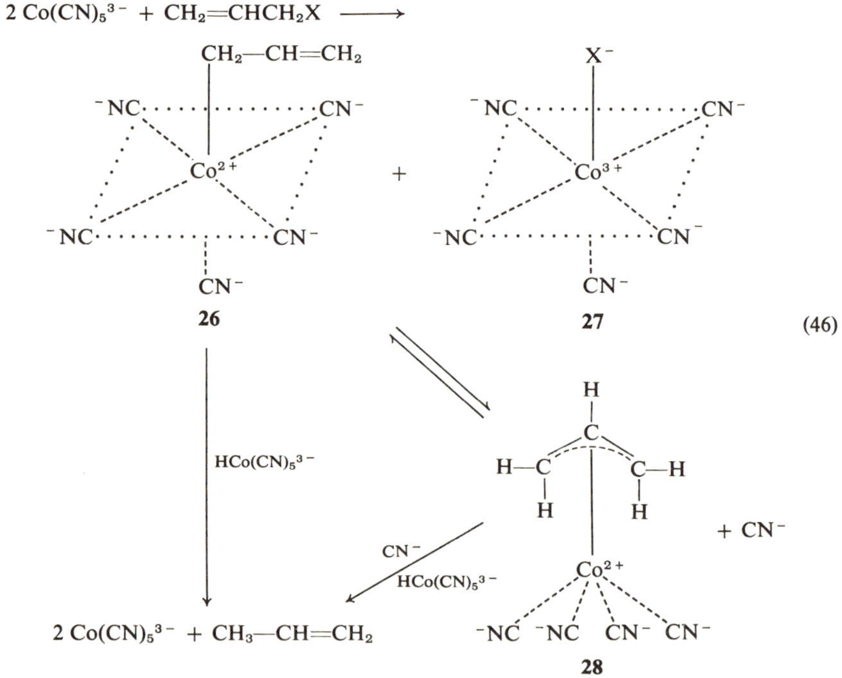

Scheme II. Hydrogenation of allylic compounds with pentacyanocobaltate(II).

The conformations assigned to σ- and π-allyl complexes in Schemes I and II are indirectly supported by structural and constitutional studies of related complexes. The formation of σ-bonded alkyl and aryl pentacyanocobaltate(III) complexes has recently been reported by Halpern and Maher [148]. The complexes are formed by the reaction of pentacyanocobaltate(II) with alkyl or aryl halides and are isolated in the form of their sodium salts. Alkyl cobalt carbonyls, $RCo(CO)_4$, are formed by the interaction of hydridocobalt tetracarbonyl, $HCo(CO)_4$, and alkenes, and have been throughly characterized by Heck and Breslow [149]. Such complexes play an important role in isomerization and hydroformylation reactions of unsaturated hydrocarbons, and will be discussed in detail in Volume II.

Simandi and Nagy [143,144,146] have recently reported the catalytic hydrogenation of cinnamic acid by $[Co(CN)_5]^{3-}$. The reaction was carried out under a constant pressure of molecular hydrogen. The mechanism shown in Eqs. (47)–(50) was proposed

$$2\,[Co(CN)_5]^{3-} + H_2 \underset{}{\overset{K_1}{\rightleftharpoons}} 2\,[Co(CN)_5H]^{3-} \tag{47}$$

$$[Co(CN)_5H]^{3-} + S \xrightarrow{k_1} [Co(CN)_5]^{3-} + SH\cdot \tag{48}$$

$$[Co(CN)_5H]^{3-} + SH\cdot \xrightarrow{k_2} [Co(CN)_5]^{3-} + SH_2 \tag{49}$$

$$[Co(CN)_5]^{3-} + SH\cdot \xrightarrow{k_3} [Co(CN)_5SH]^{3-} \tag{50}$$

where S is cinnamic acid; $SH\cdot$, cinnamate free radical; SH_2, hydrogenated cinnamic acid. In Eq. (47) hydridopentacyanocobaltate(III) is formed from $Co(CN)_5{}^{3-}$ and molecular hydrogen. Cinnamic acid then reacts with $[Co(CN)_5H]^{3-}$ in a rate-determining step to form cinnamate free radical, which reacts with another mole of $Co(CN)_5H^{3-}$ to give the hydrogenated species. Cinnamate free radical can also react with $[Co(CN)_5]^{3-}$ to give hydrocinnamylpentacyanocobaltate(III). The mechanism of Simandi and Nagy [143,144,146] differs from that of Kwiatek *et al.* [123] in that the hydrogenerated substrate is formed through a free-radical intermediate. The proposal of a free-radical substrate intermediate by Simandi and Nagy [144] is based on the observation of Takahashi [150] that solutions of $Co(CN)_5{}^{3-}$ saturated with hydrogen initiate radical polymerization. The reaction of pentacyanocobaltate(II) with the cinnamate free radical to form an organopentacyanocobaltate(II) is the chain-terminating step.

Murakami [129] and Murakami and Kang [131,132] used $Co(CN)_5{}^{3-}$ as a homogeneous hydrogenation catalyst for the preparation of α-amino acids either by reduction of the oximes of α-keto acids or reductive amination of the parent keto acid in the presence of 6% ammonia solution. The reactions were usually carried out at 30°–70°C under a hydrogen pressure of 50 kg/cm^2. The yield of the amino acids obtained varied from 33 to 82%. In all cases, a mixed α-amino acid–cobalt cyanide complex was postulated as an intermediate, which then reacts with hydrogen in a subsequent step. Conjugated acids such as cinnamic acid and sorbic acid were hydrogenated at 70°C under a hydrogen pressure of 50 kg/cm^2. The products, dihydrocinnamic acid and γ,δ-hexenoic acid, were obtained in good yields. In accordance with the observation of Kwiatek *et al.* [123] the presence of conjugation was found to be essential for the hydrogenation of olefins. Chloropentamminecobalt(III) was also found to be a hydrogenation catalyst but was somewhat less effective than pentacyanocobaltate(II).

VIII. Hydrogenation of π Complexes

A. Catalysis by Metal Ions

Metal ions that activate both π bonds and molecular hydrogen are usually effective as catalysts for the hydrogenation of olefins to saturated products. Activation of the π bond of an olefin or of an aromatic compound by π-complex formation seems to be essential in both homogeneous and heterogeneous hydrogenation reactions. For aromatic systems, the concept of participation of metal π complexes in hydrogenation seems to be supported by the activation energies of hydrogenation of benzene, toluene, ethylbenzene, *p*-xylene, and mesitylene on a nickel–magnesium oxide catalyst [151]. The decrease in the activation energies in the metal-catalyzed hydrogenation of substituted benzenes parallels their relative donor tendencies toward metal ions.

For homogeneous systems, the requirement for coordination of both π bonds and molecular hydrogen is satisfied by the platinum group metal ions platinum(II) [152,153], ruthenium(II) [154–158], palladium(II) [159], and rhodium(I) [160,161]. Metal ions of copper(II), silver(I), and mercury(II) are ineffective in catalytic hydrogenation because they are preferentially reduced to the metal by hydrogen in the presence of the olefin.

1. *Ruthenium(II)-Catalyzed Reactions*

Halpern *et al.* [154–158] used ruthenium(II) complexes as catalysts in the homogeneous hydrogenation of conjugated unsaturated acids, such as maleic, fumaric, and acrylic acids, in 3 *M* hydrochloric acid at 70° to 90°C. Catalytically active ruthenium(II) chloride was generated in the reaction mixture by the reduction of a ruthenium(IV) compound, $(NH_4)_2RuCl_6$, with titanium trichloride. Catalytic hydrogenation did not take place with olefins containing isolated double bonds, such as ethylene and propylene. Participation of the carboxylic group in the activation of the double bond and in the overall reaction mechanism is not well understood, but it is tempting to speculate that it functions as a coordinating group in the bonding of the olefinic acid to the ruthenium(II) ion. The rate law [156] may be expressed as shown in Eq. (51).

$$-\frac{d[\text{olefin}]}{dt} = k[H_2][Ru^{2+}] \tag{51}$$

The following mechanism [156] is in accord with the observed dependence of the reaction kinetics on hydrogen and metal ion concentrations:

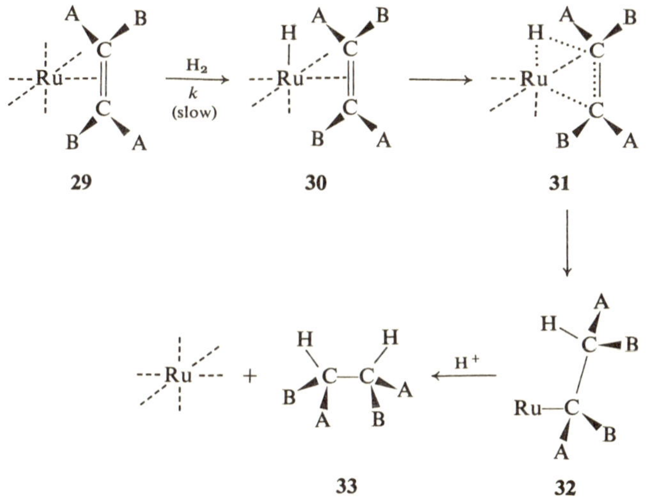

A ruthenium(II)–olefin complex **29** is depicted in the first step. In the case of maleic acid, the formation of this 1:1 olefin–meta complex was demonstrated spectrophotometrically, and a stability constant of 5×10^3 was reported [154]. The metal–olefin complex then splits hydrogen heterolytically in a slow step with the formation of a ruthenium–olefin–hydrido complex, **30**, which rearranges to the ruthenium–alkyl complex, **32**, through an intermediate, **31**. Hydride addition postulated in steps **30** → **31** → **32** is well supported by studies of Green and Nagy [162]. Complex **32** is cleaved by a proton to form the saturated product **33** and free ruthenium(II), which forms the olefin complex **29** again in a fast equilibrium step. The reaction is stereospecific and the addition of hydrogen to the double bond is always *cis*. This is confirmed by the infrared spectrum of the *dl* isomer (*dl*-2,3-dideuterosuccinic acid) formed in the hydrogenation of fumaric acid in D_2O. Reduction of the olefin in water with deuterium gave no deuterated product, whereas reduction with hydrogen or deuterium in D_2O yielded the *dl*-2,3-dideuterosuccinic acid. It was therefore concluded by Halpern and James [156] that the hydrogen that adds to the olefin is derived predominantly from the solvent. Since the hydrogen atoms originally bonded to the olefin do not exchange with the solvent, and since D_2–H_2O exchange does not take place during hydrogenation of the olefin, the deuterium in the product must originate by the exchange of hydride in complex **30** with the solvent. This type of exchange has not yet been well established with respect to all its mechanistic details. The second deuterium atom incorporated in step **32** → **33** may come either directly from the solvent or by exchange with a deuteron in the solvation sphere of the metal ion.

Tris(triphenylphosphine)dichlororuthenium(II) and tetrakis(triphenylphosphine)dichlororuthenium(II) catalyze the hydrogenation of alkenes and alkynes at room temperature and 1 atm pressure [163]. 1-Heptene and 1-hexyne are rapidly hydrogenated by a 10^{-3} *M* solution of these ruthenium complexes in a 1:1 benzene–ethanol solution. Reactions of $RuCl_2(PPh_3)_3$ or a concentrated solution of $RuCl_2(PPh_3)_4$ with molecular hydrogen gave a hydrido complex, $RuClH(PPh_3)_3$, which is considered to be an active intermediate in the catalytic hydrogenation reaction. The co-catalyst, ethanol, may possibly be the source of hydride ion for the formation of the $RuClH(PPh_3)_3$ species, as evidenced by the very slow rate of hydrogenation in the absence of ethanol [163].

The ruthenium complexes $RuCl_2(PPh_3)_3$ and $RuCl_2(PPh_3)_4$ dissociate in methanol–benzene solution to form solvated species according to the equilibria

$$RuCl_2(PPh_3)_4 \rightleftharpoons RuCl_2(PPh_3)_3S + PPh_3 \qquad (52)$$

$$RuCl_2(PPh_3)_3 \rightleftharpoons RuCl_2(PPh_3)_2S_2 + PPh_3 \qquad (53)$$

shown in Eqs. (52) and (53) where S represents solvent molecules The coordinated solvent is displaced subsequently on reaction with molecular hydrogen or olefins. Whether a mixed hydrido–olefin–Ru(II) complex is formed as an intermediate in the ruthenium(III)-catalyzed hydrogenation of alkenes and alkynes is uncertain at this stage.

When deuterium is used in place of hydrogen, partially deuterated hydrocarbons are obtained [163]. In the absence of the olefin, fast exchange between the hydroxyl proton of ethanol and deuterium takes place. The mechanism of hydrogenation with these ruthenium(II) catalysts thus seems to be similar to the proposed by Halpern and James [156,164].

Jardine and McQuillin [165] reported the hydrogenation in methanol solution of cyclopentene, cyclohexene, cyclooctene, and diphenylacetylene with tris(triphenylphosphine)chlorohydridoruthenium(II) as the catalyst. Except for diphenylacetylene (which gave *cis*-stilbene), these substrates gave saturated products. The catalyst was formed *in situ* by the treatment of ruthenium trichloride with triphenylphosphine in methanol. The mechanism of Eq. (54) has been advanced [165] for these ruthenium(II)-catalyzed hydrogenations:

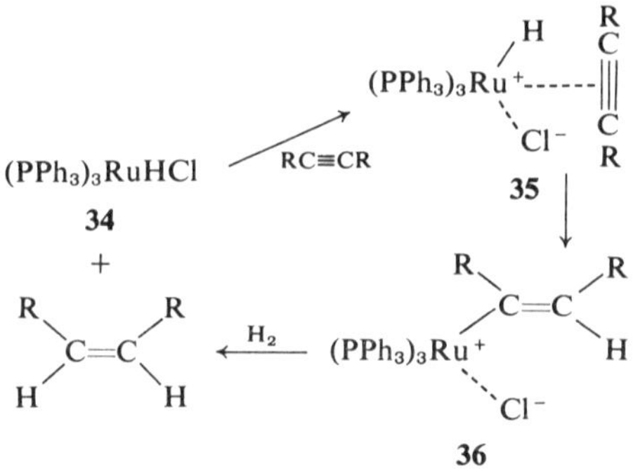

(54)

Formation of a hydrido ruthenium complex, **34**, with the alkylacetylene to give **35** is followed by insertion of the acetylene in the ruthenium-hydrogen bond to give the σ-bonded alkene complex, **36**, which then undergoes hydrogenolysis to give the alkene and regenerate the catalyst.

Taqui Khan and Vanchesan [166] have studied the homogeneous hydrogenation of alkenes catalyzed by ruthenium(II). Complexes of the type L_3RuClX {L = $AsPh_3$, $SbPh_3$, or $\frac{1}{2}$ diphos [1,2-bis(diphenylphosphino-ethane]; X = Cl^- or $SnCl_3^-$} at 35°C and 1 atm H_2. In benzene solution, the initial rates of hydrogenation increased considerably in the presence of a basic co-solvent such as ethanol. With cyclohexene or 1-heptene as substrates, the rate of hydrogenation was first order with respect to alkene and catalyst concentrations. The catalytic activity of the ruthenium(II) complexes mentioned above was very much dependent on the π-acceptor character of the neutral ligand L, decreasing in the order $AsPh_3 > SbPh_3 >$ diphos. The same order of reactivity was observed [115] for $L_3RuCl(SnCl_3^-)$ complexes. The catalytic activity of the complexes is much enhanced when X is a coordinated $SnCl_3^-$ group as expected from the higher π-acidic character of $SnCl_3^-$ as compared to chloride.

The active intermediate in the catalytic hydrogenation of alkenes was suggested to be a monohydrido species of ruthenium(II) of the type $L_3HRuL'X$ (L′ = olefin; X = $SnCl_3^-$ or Cl^-, formula **38**) formed *in situ* from L_3RuXCl, olefin, and molecular hydrogen. The rate-determining step of the reaction [Eq. (55)] appears to be the formation of this intermediate by the displacement of the coordinated chloride by a hydride ion. Reaction of $(AsPh_3)_3RuCl_2$ with hydrogen in benzene containing ethanol or triethylamine gave a hydrido complex of the composition $(AsPh_3)_3RuClH$ which has been isolated and characterized.

$$L_3RuClX + \text{olefin} \rightleftharpoons L_3RuClX(\text{olefin})$$

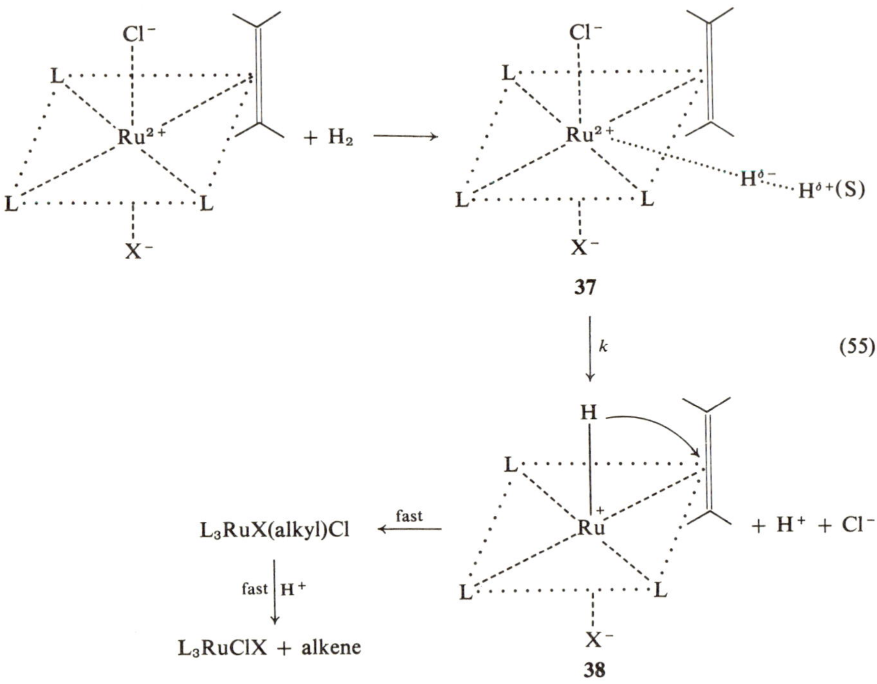

(55)

The enthalpy and entropy of activation for the hydrogenation of cyclohexene catalyzed by $L_3RuCl(SnCl_3)$ ($L = PPh_3$, $AsPh_3$) have been found to be 2.1 kcal/mole and −53 eu ($L = PPh_3$) and 1.3 kcal/mole and −56 eu ($L = AsPh_3$), respectively. The high entropy of activation of the reaction suggests a polar transition state such as **37**, resulting from the combination of neutral molecules $L_3RuCl(SnCl_3)$, molecular hydrogen, and the alkene. The polarization of the hydrogen molecule in **37** is favored by the presence of a basic co-solvent in the reaction medium.

2. *Rhodium(I)- and Iridium(I)-Catalyzed Hydrogenation*

Tris(triphenylphosphinechlororhodium(I) has been reported [29,160,161, 167–169] as a very good catalyst for the rapid hydrogenation of alkenes and alkynes at 25°C in ethanol–benzene solution. Solutions of $(PPh_3)_3RhCl$ in benzene or deuterochloroform take up molecular hydrogen, and the NMR spectra show three high-field lines attributed to the presence of *cis*-dihydrorhodium(III) species in solution [161,162]. Hydrogen is completely displaced as hydrogen gas from the dihydrido complex by passing gaseous nitrogen through the solution. Young [161] postulated that $(PPh_3)_3RhCl$ dissociates in the solvent to give a solvated species $(PPh_3)_2RhCl(S)$ which can in turn reversibly add hydrogen and ethylene, as shown in Eqs. (56)–(58):

$$RhCl(PPh_3)_3 + S \underset{}{\overset{K_1}{\rightleftharpoons}} RhCl(PPh_3)_2(S) + PPh_3 \tag{56}$$

$$RhCl(PPh_3)_2(S) + H_2 \underset{}{\overset{K_2}{\rightleftharpoons}} RhCl(PPh_3)_2(S)(H)_2 \tag{57}$$

$$RhCl(PPh_3)_2(S) + C_2H_4 \underset{}{\overset{K_3}{\rightleftharpoons}} RhCl(PPh_3)_2(C_2H_4)(S) \tag{58}$$

The second and third equilibria are shifted to the left as the basicity of the solvent increases, resulting in a reduction in the rate of hydrogenation.

There has been some controversy regarding the extent of dissociation of $RhCl(PPh_3)_3$ in solution. Arai and Halpern [170] have reported the equilibrium constant K_1 ($=7 \times 10^{-5}$ M) corresponding to reaction (56) for a dilute ($<10^{-3}$ M) solution of $RhCl(PPh_3)_3$ in benzene. The degree of dissociation is thus quite low and becomes negligible at higher concentrations of the catalyst. A ^{31}P spectrum of an 0.14 M solution of $RhCl(PPh_3)_3$ in methylene chloride did not show the presence of any dissociated phosphine [171]. The ^1H and ^{31}P NMR spectra of the *cis*-dihydride obtained in equilibrium (57) is in accord with a six-coordinated rhodium(III) complex $RhCl(H)_2(PPh_3)_3$ with a meridional arrangement of the three phosphine groups. On passing nitrogen through a solution of this complex no free triphenylphosphine was detected in solution. Thus it seems that the six-coordinate *cis*-dihydrido complex in equilibrium (57) does not contribute significantly to the reaction pathway, if it exists at all. Consequently the most probable equilibrium for formation of the dihydro species is represented by Eq. (59).

$$RhCl(PPh)_3 + H_2 \rightleftharpoons RhCl(H)_2(PPh_3)_3 \tag{59}$$

Benzene solutions of tris(triphenylphosphine)chlororhodium(I) react with ethylene to form a bright yellow crystalline complex, $RhCl(PPh_3)_2(C_2H_4)$. ^{19}F NMR spectra of the corresponding fluoro complex indicate that the fluorine is *trans* to the olefin and *cis* to two mutually *trans* triphenylphosphine groups. The equilibrium constant K_3 for the reaction [Eq. (58)] of $RhCl(PPh_3)_2(S)$ with ethylene at 25°C (at 1 atm of C_2H_4) is 100 M^{-1}. The value of K_3 decreases with increasing molecular weight of the olefin and with an increase in the *trans* effect of the coordinated halide, in the order chloride < bromide < iodide. For propylene, K_2 is less than 0.05 M^{-1}. The decrease in the stability of rhodium(I) complexes of ethylene homologs may be partly steric and partly inductive (through weakening of the π-acid character of the olefin). Migration of the double bond in terminal olefins and *cis*–trans isomerization in internal olefins has not been observed with the rhodium catalyst over a period of 24 hours [29]. The complex $RhCl(PPh_3)_2$ has been found to be an efficient catalyst for the hydrogenation of nonconjugated alkenes and alkynes at ambient temperatures and at hydrogen pressures of 1 atm or below. Ethylene and acetylene are not hydrogenated by the catalyst, probably because they form very stable rhodium(I) complexes. Terminal and *cis*-olefins are hydrogenated more rapidly than are the internal and *trans*-olefins.

Kinetic studies of the hydrogenation of alkenes and alkynes were conducted by the addition of the solid catalyst $RhCl(PPh_3)_3$ to a 1:1 benzene–ethanol solution saturated with hydrogen and followed by the addition of the unsaturated compound. This order of addition of catalyst to the solution ensures better dissolution of the catalyst without dimerization to $[RhCl(PPh_3)_2]_2$. The solution is maintained saturated with respect to hydrogen at all times. The addition of a polar solvent such as ethanol or a ketone to benezene sufficiently accelerates the rate of hydrogenation, although no correlation could be found between the dielectric constant of the polar solvent (co-catalyst) and the reaction rate.

Catalysis of the hydrogenation of the alkenes by rhodium(I) may take place by two paths, A and B, indicated in the following reaction scheme. The first step may be equilibration of the catalyst with molecular hydrogen (pathway A) or the alkene (pathway B), respectively. The hydride **40** or the rhodium (I)-olefin complex **41** then reacts with olefin or hydrogen, respectively, to form an intermediate, **42**, in which both hydrogen and olefin are coordinated to the metal ion, followed by rearrangement to the products.

It is noted that the dihydro intermediate **40** is formulated on the basis that Eq. (57) is correct. If on the other hand Eq. (59) is more appropriate, as suggested above, formula **40** would be modified by replacing the solvent ligand S with a triphenylphosphine group.

The rate constant k' for the disappearance of the olefin is given by Eq. (60).

$$\text{Rate} = \left[\frac{k'K_1 + k''K_2}{1 + K_1[H_2] + K_2\,[\text{olefin}]}\right][H_2][\text{olefin}][\text{Rh}] \tag{60}$$

$$k' = 1.1 \times 10^{16} \exp(-22{,}900/RT)\ M^{-1}\ \text{sec}^{-1}$$

(at a constant hydrogen concentration of 1.0 atm)

Rhodium(I)-catalyzed hydrogenation of alkenes thus involves (1) activation of molecular hydrogen, (2) activation of the olefin, and (3) hydrogen transfer to the olefin. Activation of molecular hydrogen requires homolytic cleavage of the hydrogen molecule via oxidative addition to form the dihydride. If hydrogen activation is to be reversible, the heat of formation of the two metal–hydrogen bonds in the dihydride should be comparable to the bond dissociation energy of the hydrogen molecule plus the energy resulting from expansion of the coordination shell of the rhodium complex. For complexes of

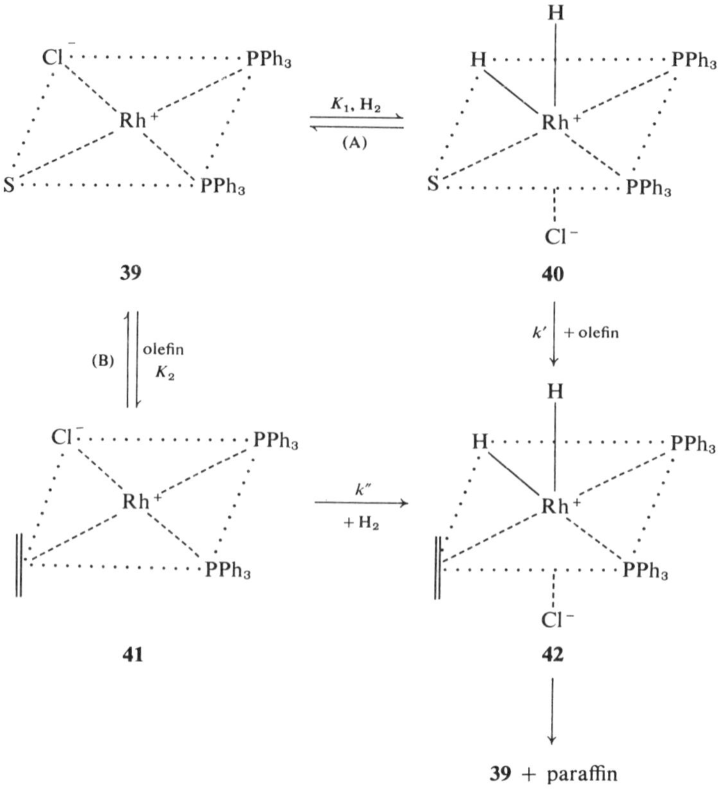

the type $[RhCl(PPh_3)_2L]$, where L is carbon monoxide or ethylene (strong π-acids), the reduction of the electron density around the metal ion by back donation raises the oxidation potential of the Rh(I)/Rh(III) couple; hence activation of molecular hydrogen is inhibited. The ability of $IrX(PPh_3)_2CO$ to activate molecular hydrogen [14] is due to a lower oxidation potential of the Ir(I)/Ir(III) couple as compared to the Rh(I)/Rh(III) couple. The intermediate steps indicated above for the homolytic cleavage of molecular hydrogen and formation of the metal dihydride are purely speculative at the present stage.

The olefin activation step requires coordination of an olefin either to a rhodium(I) complex or to its corresponding dihydride. Although there is no conclusive evidence in favor of either of the pathways, the latter (pathway A) is preferred [29] on the basis of low activity of rhodium(I)-olefin complexes in the activation of molecular hydrogen. For both rhodium(I) complex and its dihydride, a vacant or relatively labile coordination site in the metal complex is necessary for effective catalysis. Substituted olefins which cannot readily approach the labile coordination site on the metal ion are not readily reduced.

The importance of the steric factors in the complex formation between substituted olefins and the rhodium hydrogenation catalysts **40** or **41** is indicated by increasing negative values of the entropy of activation of hydrogenation with increasing substitution on the substrate [168]. Thus for cyclohexene and 1-methylcyclohexene, $\Delta S\ddagger$ values are 1.3 and −26.0 eu, respectively, the corresponding $\Delta H\ddagger$ values being 18.6 and 12.7 kcal/mole, respectively. A similar effect of the substituent has been noted with 1-hexene and 2-methyl-1-pentene, the $\Delta S\ddagger$ and $\Delta H\ddagger$ values being, respectively, 1.1 eu and 18.6 kcal/mole for 1-hexene and −18.5 eu and 12.7 kcal/mole for 2-methyl-1-pentene. Steric factors are also indicated by the entropies of activation of *cis* and *trans* isomers of internal olefins. Reduction of *cis*-4-methyl-2-penetene has an entropy of activation about 18 eu more positive than that of *trans*-4-methyl-2-pentene. The more favorable enthalpy usually observed for substituted olefins suggests stronger coordination between metal ion and the substituted olefins, for reasons that are not yet entirely clear.

The step involving hydrogen addition to the olefin requires the coordination of both molecular hydrogen and olefin to the metal ion. A mixed hydrogen–metal–olefin complex, **42**, is suggested in the mechanism, although such a complex has not been isolated or characterized [29]. Since molecular hydrogen adds stereospecifically to rhodium(I) to form a *cis*-dihydride (confirmed by infrared and NMR), it may be concluded that the vacant for site coordination with the olefin is *cis* to the two metal hydride bonds as shown in **42**. Location of an olefin *cis* to the two hydrogen atoms in this mixed-ligand complex facilitates transfer of both hydrogens to the olefinic bond to form the saturated product. The transfer of two hydrogen atoms to the olefin has been postulated to take place [Eq. (61)] through a pentagonal transition state, **43**, involving the participation of both hydrogen atoms, the metal ion, and the olefin.

The distorted hexagonal transition state **43** has been suggested [29] because there is no evidence of a stepwise transfer of hydrogen atoms to the coordinated olefin through a metal–alkyl complex. In the presence of pure deuterium the rate of the reaction k_D was almost equal to that observed with hydrogen ($k_H/k_D = 0.9$). It was therefore inferred that the slow step in the hydrogenation process is possibly the formation of the mixed-ligand hydridoolefin complex of the type indicated by **43**. This conclusion is partly supported by the observation of almost equal rates of reaction for endocyclic olefins, cyclopentene, cyclohexene, and cycloheptene, indicating that the strain energies are not released during the rate-determining step (i.e., during the complex formation step).

When $(PPh_3)_3RhCl$ is used as a catalyst, 1-hexyne is reduced in ethanol–benzene solution to hexane under a hydrogen pressure of less than 1 atm at 20°C. If the hydrogenation is interrupted by using a quantity of hydrogen less

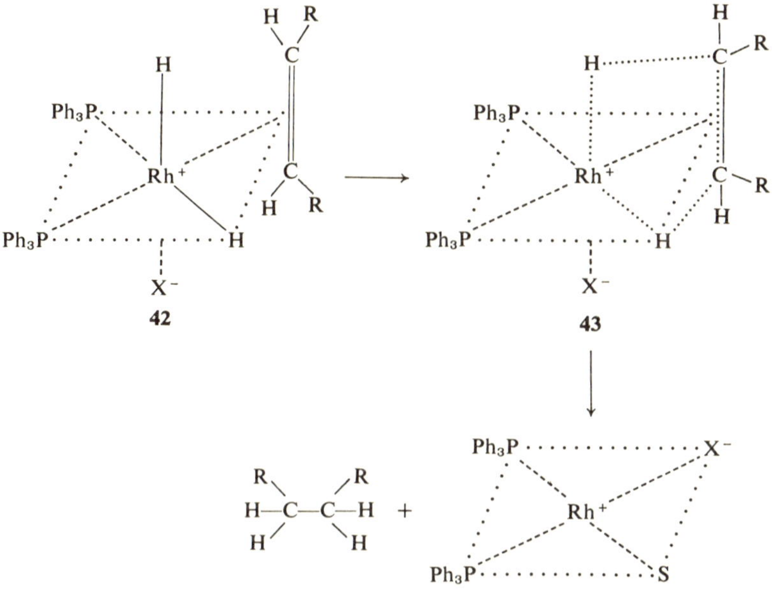

42 43 (61)

39

than that needed for complete reduction of the substrate, both 1-hexene and *n*-hexane can be detected in the solution by gas–liquid chromatography [160]. The reduction therefore takes place most probably in two steps, as indicated by Eq. (62). It has therefore been proposed by Jardine *et al.* [160] that the

$$RC{\equiv}CR' \longrightarrow RCH{=}CHR' \longrightarrow RCH_2{-}CH_2R' \qquad (62)$$

overall rate of reduction of an alkyne is much slower than that of the corresponding alkene. The alkynes studied [110] include phenylacetylene, diphenylacetylene, 3-methyl-1-pentyne-3-ol, 3-methyl-1-bytyne-3-ol, 2-pentyne-1,4-diol, and 1-ethynylcyclohexane-1-ol.

Mague and Wilkinson [172] have studied the catalytic properties of tris-(triphenylarsine)chlororhodium(I) and tris(triphenylstibine)chlororhodium-(I) in the catalytic hydrogenation of alkenes. The catalytic activities of these rhodium(I) complexes and of the related phosphine complex decrease in the order $(PPh_3)_3RhCl > (AsPh_3)_3RhCl > (SbPh_3)_3RhCl$. The stibine complex was also the least stable of the three complexes. The *cis*-dihydridorhodium(III) complexes of triphenylarsine and triphenylstibine are nevertheless very stable with respect to the dissociation of hydride groups as compared to the triphenylphosphine complex, as indicated by the fact that bound hydrogen cannot be displaced by sweeping with nitrogen. This greater stability of $(AsPh_3)_2RhH_2Cl$ and $(SbPh_3)_2RhH_2Cl$ is considered the reason for their lower catalytic activity in the hydrogenation of alkenes.

The effect of the variation of alkyl and aryl groups of the tertiary phosphine bound to these rhodium(I) catalysts on the rate of hydrogenation of a single olefin substrate, cyclohexene in benzene solution, has been reported by Montalatici *et al.* [169]. The effect of the phosphine ligand on the catalytic activity of the species $RhX(PR_3)_2(S)$ was considered to be due mainly to the tendency of the PR_3 groups to dissociate in the presence of the solvent and to provide sites for the coordination of hydrogen and olefin.

p-Methoxy as a substituent on the phenyl rings of the triphenylphosphine ligand increases the rate of hydrogenation by 50% over that of triphenylphosphine by facilitating factors 1 and 2 (p. 48). The presence of electron-withdrawing groups such as fluoride ion on the phenyl rings of triphenylphosphine causes a decrease in the rate of hydrogenation-reducing effects 1 and 2. For the alkyl phosphines, the rates of hydrogenation are much lower than for the aryl phosphines. The alkyl phosphines are much more basic (good σ donors and poor π acceptors) than the aryl phosphines, so that the dissociation of the former in solvents is much lower than that of the latter; hence vacant sites on the metal ion are not as readily available for coordination with hydrogen or an olefin. The trialkylphosphine–dihydridorhodium(III) complexes, once formed, are nevertheless very stable, and no hydrogen is lost on passing nitrogen through their solutions. The high stability of the dihydrido complexes may also be partly responsible for their lower activity in transferring hydrogen atoms to olefins.

Unsaturated aldehydes have been reduced to saturated aldehydes with catalysis by tris(triphenylphosphine)chlororhodium(I) [173]. Thus the specific reduction of the olefinic bond in propenal, but-2-enal, and *trans*-2-methylpent-2-enal results in the formation of the corresponding saturated aldehydes. A complication in the use of $(PPh_3)_3RhCl$ as the hydrogenation catalyst for unsaturated aldehydes is the side decarbonylation reaction of the aldehyde, which accelerates with increasing concentration of the substrate. The carbonyl complex $RhCl(CO)(PPh_3)_2$ formed during the course of this side reaction is an ineffective catalyst, and the catalytic activity gradually decreases as hydrogenation proceeds. To suppress decarbonylation of the substrate, the ratio of catalyst to substrate is increased to as high as 2:1, thus greatly reducing its efficiency and corresponding interest in this reaction for catalysis.

Specific deuteration of the double bonds in cyclohexene, oleic acid, and linoleic acid without introduction of additional label in the saturated product has been carried out by the use of $(PPh_3)_3RhCl$ as catalyst [174,175]. Cyclohexene and methyl oleate were employed to form 1,2-dideuterocyclohexene and the 9,10-dideuterated methyl stearate, respectively. With ergosterol, the 5,6-dideuterated compound was obtained, the double bonds at positions 7 and 22 being unaffected.

O'Connor *et al.* [176,177] have reported the catalytic activity of hydridocarbonyltris(triphenylphosphine)rhodium(I) in the hydrogenation of α-olefins. The catalyst forms the solvated species $H(CO)(PPh_3)_2Rh(S)$ in mixed solutions of hydrocarbons and protic solvents, thus providing a labile site for coordination with hydrogen or the olefin in a manner similar to that described for the catalyst $RhX(PPh_3)_3$ [Eqs. (57) and (58)]. The reactivity of $HCO(PPh_3)_3Rh(I)$ is, however, much less than that of **42** in the hydrogenation of α-olefins. The complex loses its catalytic activity in benzene solution due to the formation of a dimer, dicarbonyltetrakis(triphenylphosphine)dirhodium, $Rh_2(CO)_2(PPh_3)_4$. The active catalyst $RhH(CO)(PPh_3)_3$ can be regenerated by boiling the dimer with an excess of triphenylphosphine in benzene solution under hydrogen.

James *et al.* [178] have reported catalysis by tris(diethyl sulfide)trichlororhodium(III), $RhCl_3(SEt_2)_3$, in *N,N*-dimethylacetamide for the hydrogenation of ethylene and substituted ethylenes. It may be recalled (p. 48) that $(PPh_3)_3RhCl$ is ineffective for the hydrogenation of ethylene because of the formation of a very stable olefin complex. The replacement of the tertiary phosphines by alkyl sulfides in the coordination sphere of rhodium thus alters to a considerable extent its catalytic properties. With maleic acid as a substrate, in tenfold excess, the rate of hydrogenation is first order with respect to the rhodium(III) complex and molecular hydrogen, and independent of the substrate concentration. In the absence of maleic acid, rhodium(III) is rapidly reduced to the metallic state. The initial reaction seems to involve the reduction of the rhodium(III) complex to a rhodium(I) by molecular hydrogen [179]. The rhodium(I) complex then forms an olefin complex and reacts with molecular hydrogen in a subsequent step to yield succinic acid, regenerating the rhodium(I) catalyst, as shown in Eqs. (63)–(65).

$$\mathrm{Rh(III)} + \mathrm{H_2} \xrightarrow{k_1} \mathrm{Rh(I)} + 2\,\mathrm{H^+} \tag{63}$$

$$\text{Maleic acid} + \mathrm{Rh(I)} \rightleftharpoons \text{complex} \tag{64}$$

$$\text{Complex} + \mathrm{H_2} \xrightarrow{k_2} \mathrm{Rh(I)} + \text{succinic acid} \tag{65}$$

where Rh(III) = $[RhCl_3(SEt_2)_3]$ and Rh(I) = $[RhCl(SEt_2)_3]$. The constants k_1 and k_2 have been reported [180] as 1.85 and 0.36 $M^{-1}\ \mathrm{sec}^{-1}$, respectively, at 80°C. The activation parameters corresponding to k_1 and k_2 are, respectively, $\Delta H_1^\ddagger = 12.9 \pm 0.1$ kcal/mole and $\Delta S_1^\ddagger = -21 \pm 1$ eu, and $\Delta H_2^\ddagger = 21.5 \pm 1$ kcal/mole and $\Delta S_2^\ddagger = -1.0 \pm 3$ eu. Because of the poor π-acceptor capacity of the sulfide ligands, rhodium(I) is not stabilized to the same extent as with triphenylphosphine as a ligand in $(PPh_3)_3ClRh(I)$, further coordination to a π-acceptor olefin is therefore essential to stabilize rhodium(I) in solution. Excess diethyl sulfide inhibits the reaction, possibly by displacing the olefin from the coordination sphere of the rhodium(I) complex,

$$RhCl(SEt_2)_2(S) + SEt_2 \rightleftharpoons RhCl(SEt_2)_3 \quad (66)$$

where S = solvent. The initial reduction of rhodium(III) species to rhodium(I) by molecular hydrogen provides a facile route for the preparation of other catalytic rhodium(I) complexes *in situ*.

An interesting study of the use of $(PPh_3)_3RhCl$ as a catalyst in the presence of sulfur ligands has been conducted by Birch and Walker [181]. Trace amounts of thiophenol added to $(PPh_3)_3RhCl$ had no effect on the rate of hydrogenation of 1-octene. The rate of hydrogenation, however, was reduced when thiophenol was added to the catalyst in the ratio of 5:2. The inhibition of the catalytic effect of the rhodium(I) complex may be due to the equilibrium between the normal catalytic species, $(PPh_3)_3RhCl$, and a thiol complex of rhodium(I), only the former species contributing to the catalytic reaction. In contrast to thiophenol, alkyl and aryl sulfides do not affect the activity of the catalyst. Further study is required to elucidate the nature of the intermediate complexes involved.

Square-planar d^8 complexes of iridium(I), e.g., *trans*-$[IrX(CO)(PPh_3)_2]$, X = halogen, catalyze the hydrogenation of ethylene and acetylene at 40°–60°C at subatmospheric pressure of the reacting gases ($H_2 + C_2H_4$), ($H_2 + C_2H_6$) [14]. The rates of hydrogenation of alkenes and alkynes are very slow with the iridium(I) catalyst. The yield of ethane from ethylene at 60°C was only 40% in 18 hours and acetylene gave 10 and 5%, respectively, of ethylene and ethane under the same conditions. The catalytic experiments were conducted by passing the reactants (hydrogen + ethylene or acetylene) at subatmospheric pressure through an air-free solution of the complex in benzene or toluene and stirring the solution at 20°–80°C for 7–26 hours. At the conclusion of the experiment, the sample of gas was analyzed chromatographically.

The iridium(I) complexes *trans*-$[IrXCO(PPh_3)_2]$ react reversibly in toluene solution with molecular hydrogen to form the *cis*-dihydride adducts $[(H)_2IrXCO(PPh_3)_2]$, the configurations of which were confirmed by their infrared and NMR spectra. The iodo complex *trans*-$[IrICO(PPh_3)_2]$ absorbs ethylene at 20°C and 700 mm pressure to form a colorless adduct that has not been isolated and was assumed to be $(C_2H_4)Ir(CO)(PPh_3)_2$. The reactive intermediates in the hydrogenation of ethylene and acetylene in the presence of iridium(I) complexes as catalysts, though uncertain, may well be hydrido-olefin mixed-ligand complexes formed by oxidative addition to the catalyst.

A five-coordinate monohydrido d^8 complex of iridium(I), $IrH(CO)(PPh_3)_3$, catalyzes the hydrogenation of ethylene and acetylene at much faster rates than does the corresponding square-planar complex, $Ir(CO)X(PPh_3)_2$ [182].

The formation of a seven-coordinate complex by oxidative addition of hydrogen to the iridium(I) complex was suggested [Eq. (67)].

$$IrH(CO)(PPh_3)_3 + H_2 \rightleftharpoons Ir(H)_3(CO)(PPh_3)_3 \tag{67}$$

A six-coordinate iridium(I) complex with the olefin as a ligand was also proposed [Eq. (68)].

$$IrH(CO)(PPh_3)_3 + C_2H_4 \rightleftharpoons Ir(C_2H_4)H(CO)(PPh_3)_3 \tag{68}$$

The mechanism of the addition of the hydride ion to the olefin in the monohydrido complex should be different from that of the dihydrido complex, $IrX(CO)(PPh_3)_2(H)_2$, as indicated by the observed difference in the rates of hydrogenation in systems containing these complexes. The corresponding pentacoordinate rhodium(I) complex, $RhH(CO)(PPh_3)_3$, does not form the hydrogen adduct, but catalyzes the hydrogenation of ethylene and acetylene [182].

Hydridodichlorotris(dimethyl sulfoxide)iridium(I) (**44**) catalyzes the hydrogenation of alkynes to alkenes. A stable alkenyl intermediate (**45**) has been obtained by the reaction of diphenylacetylene with $HIrCl_2(Me_2SO)_3$ in isopropanol at 73°C [183]. The structure of the intermediate **45** has been

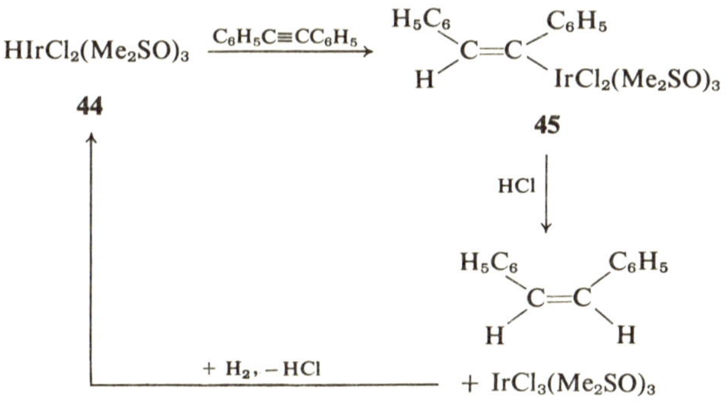

(69)

confirmed by infrared and NMR measurements. On boiling with methanol and concentrated hydrochloric acid, **45** is quantitatively converted into *cis*-stilbene with regeneration of the catalyst, indicating *cis*-addition of hydrogen to the alkyne. The alkenyl complex **45** is the first intermediate isolated thus far in a hydrogenation reaction with a platinum group metal complex as catalyst.

Tris(triphenylphosphine)chloroiridium(I), $IrCl(PPh_3)_3$, unlike its rhodium(I) analog, $RhCl(PPh_3)_3$, is inactive as a hydrogenation catalyst [63]. However, an active bis(triphenylphosphine)chloroiridium(I) species was obtained [184] *in situ* by the addition of two equivalents of triphenylphosphine to a benzene solution of $[IrCl(C_8H_{14})_2]_2$, where C_8H_{14} is cyclooctene. This iridium complex has been found [184] to be ten times as reactive as the rhodium(I) complex, $RhCl(PPh_3)_3$. Unlike $RhCl(PPh_3)_3$, the iridium(I) complex $IrCl(PPh_3)_2$ causes extensive isomerization to terminal olefins to a mixture of internal olefins.

3. *Palladium(II)- and Platinum(II)-Catalyzed Reactions*

Bis(ethylene)platinum(II) dichloride was first reported as a catalyst for the hydrogenation of simple monoolefins by Flynn and Hulbust [153]. The reduction of ethylene to ethane takes place homogeneously in the presence of platinum(II) chloride and excess ethylene at −40°C, in acetone solution, with an activation energy of about 2 kcal/mole. The reaction sequence of Eqs. (70)–(72) was suggested [154].

$$2\,PtCl_2(C_2H_4)_2 + 2\,H_2 \longrightarrow (PtCl_2C_2H_4)_2 + 2\,C_2H_6 \qquad (70)$$

$$(PtCl_2C_2H_4)_2 + 2\,C_2H_4 \longrightarrow 2\,PtCl_2(C_2H_4)_2 \qquad (71)$$

Overall reaction

$$C_2H_4 + H_2 \xrightarrow{Pt(II)} C_2H_6 \qquad (72)$$

Cramer *et al.* [152] investigated the hydrogenation of ethylene and acetylene with a Pt(II)–$SnCl_3$ complex obtained by dissolving stannous chloride ($SnCl_2 \cdot 2\,H_2O$, 10 mmole) and chloroplatinic acid (H_2PtCl_6, 1 mmole) in 120 ml of methanol. Platinum(IV) is reduced to platinum(II) by the excess of stannous chloride and a platinum(II)–stannous chloride complex is formed. The structure of this complex, $[Pt(SnCl_3)_5]^{3-}$, [see **47**, Eq. (74)] has been elucidated recently by X-ray diffraction studies and is shown to be a trigonal bipyramid with platinum in the center coordinated to five $SnCl_3^-$ groups through platinum–tin bonds. Pentakis(trichlortin)platinate(II) is the first example of a five-coordinate platinum(II) complex with monodentate ligands. It was shown by Lindsey *et al.* [185] that the trichlorotin group is a strong *trans*-activating ligand, with a *trans* effect second only to that of the cyanide ion. It is a weak σ donor and a strong π acceptor. The strong *trans* activation by $SnCl_3^-$ is considered to play an important role in the catalytic activity of $[Pt(SnCl_3)_5]^{3-}$, probably by facilitating ligand-exchange reactions. A similar complex of rhodium(III), tetrakis(trichlorotin)-$\mu\mu'$-dichlorodirhodium(III) $[Rh_2(SnCl_3)_4Cl_2]$ (**46**) has been prepared by Davies *et al.* [186] by the interaction of rhodium trichloride and stannous chloride in hydrochloric acid.

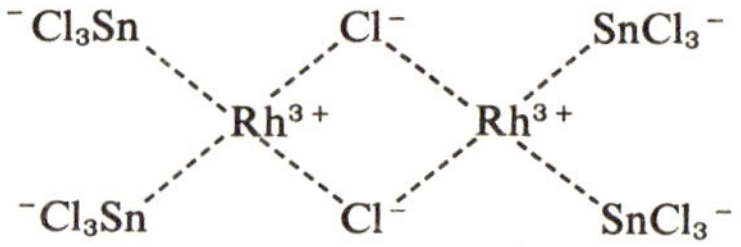

46

Diamagnetic tetraalkyl ammonium salts of the ions $[Ru(SnCl_3)_2Cl_2]^{2-}$, $[Ir(SnCl_3)_2Cl_3]^{2-}$, and $[Pt(SnCl_3)_2Cl_2]^{2-}$ have been obtained by a similar interaction of the corresponding higher valence metal chlorides with stannous chloride [187,188]. The catalytic activities of these complexes of trichlorotin, with the exception of that of platinum(II), have not yet been investigated.

When a 1 : 1 mixture of ethylene and hydrogen at a pressure of 1 atm is admitted to a solution of $[Pt(SnCl_3)_5]^{3-}$ (**47**), rapid and quantitative hydrogenation of ethylene takes place. Similarly, a 1 : 1 mixture of acetylene and hydrogen produced ethane and ethylene in about a 3 : 1 molar ratio. There is evidence [152] that the trichlorotin ligand promotes the coordination of ethylene to platinum(II). Thus a solution of potassium chloroplatinate containing 5 mole% stannous chloride in hydrochloric acid is quantitatively and rapidly converted to Zeise's salt, $KPtCl_3C_2H_4 \cdot H_2O$. In the absence of stannous chloride, Zeise's salt could not be obtained by this reaction even after a considerable period of time. Rapid exchange of free to coordinated olefin is also verified by the nmr spectra of a solution of this complex in CH_3OD in the presence of excess ethylene.

The following mechanism is suggested for the $[Pt(SnCl_3)_5]^{3-}$ catalyzed hydrogenation of olefins.

Formation of a hydrido complex of platinum(II) (**48**) in the first step seems probable because complexes with platinum(II)–hydrogen bonds such as $HPt(SnCl_3)(PPh_3)_2$ have been isolated and characterized [185]. The Pt—H bond in $HPt(SnCl_3)_4$ would be expected to be very labile because of the π-acid

$$H_2PtCl_6 + 6\,SnCl_2 + 3\,Cl^- \longrightarrow Pt(SnCl_3)_5{}^{3-} + H_2SnCl_6 \qquad (73)$$

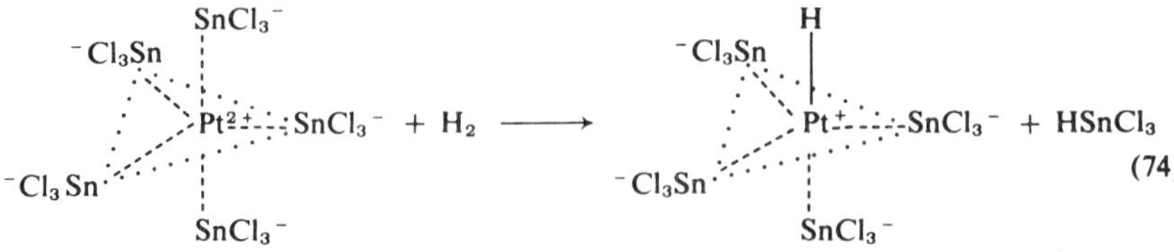

47 **48** (74)

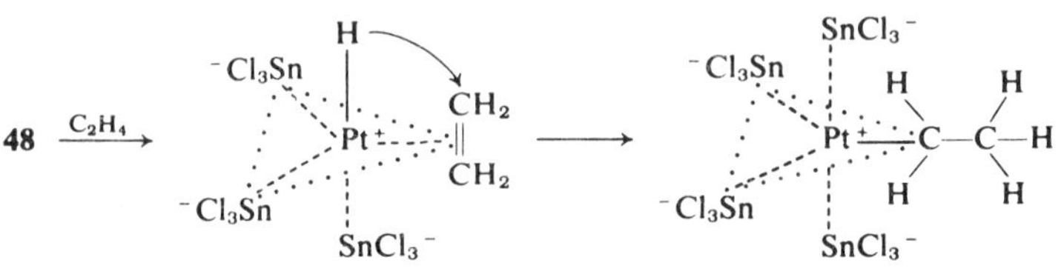

49 **50** (75)

$$\underset{\mathbf{50}}{C_2H_5 \cdot Pt(SnCl_3)_4{}^{3-}} + HSnCl_3 \longrightarrow C_2H_6 + [Pt(SnCl_3)_5]^{3-} \qquad (76)$$

character of four $SnCl_3^-$ groups in the complex. In the subsequent step the formation of an ethylene–platinum(II) complex (**49**) by the displacement of a coordinated $SnCl_3^-$ group is suggested. An internal hydride transfer in the mixed-ligand hydridoolefin complex gives rise to a σ-bonded alkyl complex (**50**) which on further reaction with $HSnCl_3$ forms the alkane and regenerates the catalyst.

An equally feasible mechanism for the hydrogenation of the olefin may be visualized as the initial formation of an olefin complex in the first step, followed by heterolytic splitting of hydrogen by this complex to form the mixed-ligand hydridoolefin complex **49** with the subsequent formation of the alkyl complex **50** in a manner analogous to Halpern and James' mechanism [164]. These two reaction sequences, the formation of a hydrido or an olefin complex in the first step, followed by the formation of a mixed-ligand hydridoolefin complex of platinum(II), are, however, kinetically equivalent processes and cannot be distinguished by measurement of reaction rate.

The presence of a coordinated $SnCl_3^-$ group in platinum(II) complexes seems to prevent the reduction of platinum(II) to metallic platinum by molecular hydrogen, a process that usually takes place readily at the temperature of this reaction (25°C). It may be recalled that Flynn and Hulbust [153] avoided this reduction of the platinum(II)–olefin complex by working in the temperature range of $-40°$ to $-10°C$ in the presence of excess olefin. The hydrogenation of cottonseed oil by a Pt(II)–$SnCl_2$ mixture reported by Bailar and Itatani [189] may be catalyzed in solution by the same complex, $[Pt(SnCl_3)_3]^{3-}$.

Carboxylic acids varying from acetic acid to stearic acid, and their esters, have been found to be good solvents for the hydrogenation of alkenes and alkynes with the platinum(II)–tin(II) complex ($H_2PtCl_6 \cdot 6\,H_2O + SnCl_2 \cdot 2\,H_2O$) as the catalyst [190]. With acetic acid as solvent, the hydrogenation of 1-hexene was found to proceed smoothly at an optimum temperature of 80°C, with a platinum/tin ratio of 5:1. Above 80°C, metallic platinum separates as a heterogeneous phase. The addition of chloride or bromide ions accelerates the rate of the reaction. The accelerating effect of halide ions on the rate of platinum(II)–tin(II)-catalyzed hydrogenation has also been reported by Van Bekkum *et al.* [191]. The nature of participation of halide ions in the reaction has not yet been explained in mechanistic terms.

The source of the hydrogen in the Pt(II)–Sn(II)-catalyzed hydrogenation reaction has been determined [192] by passing a mixture of deuterium and ethylene through the reaction mixture with alcohol as the solvent. Monodeuteroethane is the main product in the initial phases of the hydrogenation reaction. Thus in the hydrogenation of ethylene with this catalyst, one of the hydrogen atoms seems to come from the molecular hydrogen and the other seems to be derived from the hydroxyl group of the solvent, by a mechanism similar to that suggested by Halpern and James (p. 44) for ruthenium(II) catalysis [164].

Catalysis of the hydrogenation of alkenes by complexes of the type $(R_3Q)_2MX_2$ (R = alkyl or aryl; Q = P, As, or Sb; M = Pt or Pd; X = halogen or pseudohalogen) have been studied by Bailar and Itatani [193]. The catalytic effects of the platinum or palladium complexes are significantly enhanced by the addition of Group IV halides, MX_2 or MX_4 [M = tin(II), tin(IV), germanium(IV), lead(II), and silicon(IV)]. These metal halides influence the platinum catalyst mainly through their π-accepting ability (π-acidity). Thus for $PtCl_2(PPh_3)_2$, the catalytic efficiency increases in the order $SiHCl_3 < PbCl_2$, $SiCl_4 < GeCl_4 < SnCl_4 < SnCl_2$. Since $SnCl_3^-$ is known to be bound to platinum(II) in the complex $HPt(SnCl_3)(PPh_3)_2$, it is possible that silicon, germanium, and lead are similarly bonded to platinum(II), though direct evidence is lacking for complexes of this type.

The catalytic activities of platinum(II) complexes are very dependent on the nature of the neutral ligand L coordinated to the platinum atom. For hydrogenation in the presence of $PtCl_2L_2$ and $SnCl_2$, the catalytic activities decrease with L substituents in the order, $(PhO)_3P > Ph_3As > Ph_3P > Ph_3Sb > (C_4H_9)_3P$, in the approximate order (excepting for the stibine) of increasing π-acceptor activity. Since it is usually considered that a strong π-acceptor ligand labilizes the metal–hydrogen bond in the hydrido complex species formed as an intermediate, thus enhancing its catalytic ability, it seems that the above correlation is not well understood. This interpretation is verified by the increasing stretching frequencies of the platinum–hydrogen bonds in the above series of complexes with either a decrease in the π-acidity or increase in σ-basicity of the neutral ligands, L [193].

The catalytic activities of palladium(II) complexes decrease in the order, $(PPh_3)_2PdCl_2 + SnCl_2 > (PPh_3)_2PdCl_2 + GeCl_2 > (PPh_3)_2Pd(CN)_2 > (AsPh_3)_2Pd(CN)_2 > (PPh_3)_2PdCl_2 \gg (AsPh_3)_2PdCl_2$ [194]. The combination of K_2PdCl_4 and $SnCl_2$ was completely ineffective, as was the thiocyanato complex $(PPh_3)_2Pd(SCN)_2$. A remarkable feature of the palladium(II) catalysts is the activity of the cyano complexes $(PPh_3)_2Pd(CN)_2$ and $(AsPh_3)_2Pd(CN)_2$ in the presence and absence of $SnCl_3^-$ ligands. Corresponding cyano complexes of platinum(II) are completely inactive in the presence or absence of $SnCl_3^-$. The factors responsible for the reactivity of the cyano complexes of palladium(II) are not very clear and may be due to the ability of the coordination shell of palladium(II) to expand to accommodate hydride ion without displacement of the cyano ligand, resulting in the formation of a five-coordinate complex, $L_2Pd(CN)_2H$. Expansion of the coordination shell in this manner has been found to be common for palladium(II) complexes but not for platinum(II) complexes.

Hydrogenation of alkenes with the platinum(II)–tin(II) or palladium(II) catalysts described above takes place both in the presence and in the absence of molecular hydrogen. When hydrogen is absent, the alcoholic solvent is oxidized to the corresponding aldehyde. Aldehydes have been detected in the hydrogenation reaction with palladium(II) and platinum(II) complexes as catalysts by Itatani and Bailar [193,194]. This indicates that a hydrido–metal complex intermediate **53** may be formed directly from molecular hydrogen or by hydride abstraction from a coordinated solvent molecule through

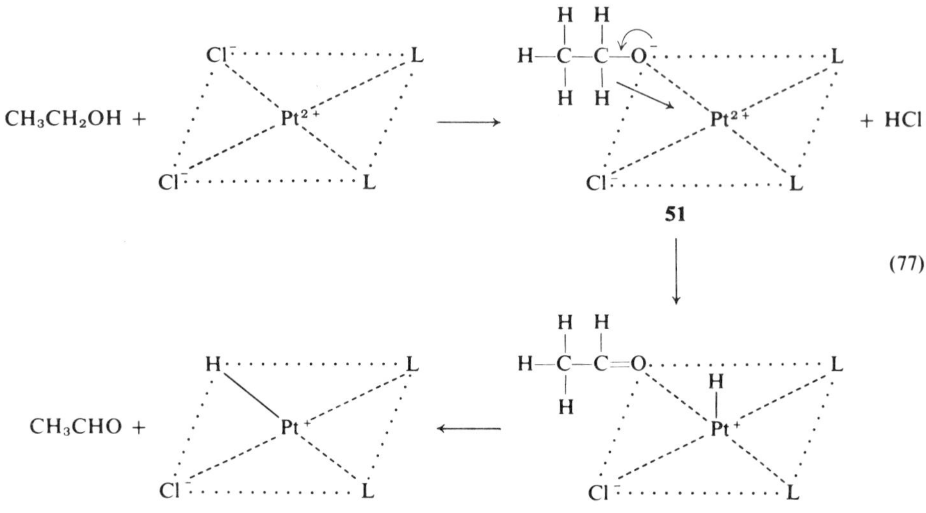

51 **52** **53** (77)

51 and **52**. Though *cis*- and *trans*-hydridochlorobis(triphenylphosphine) platinum(II) complexes have been prepared by hydride displacement from an alcoholic solvent [189], the corresponding hydrido–palladium(II) complexes could not be prepared and as yet are unknown [194].

The participation of Group IV metal halides in the hydrogenation catalysis occurs through the formation of mixed-ligand complexes with palladium(II)–M and platinum–M bonds [M = tin(II), germanium(IV), lead(II), and silicon-(IV)]. The introduction of another π-acid into the complex (in addition to the neutral tertiary phosphines or arsines) further increases the lability of the metal–hydride bond in the intermediate hydrido complexes. Excess of stannous chloride or a Group IV halide favors the reaction by shifting the equilibrium [Eq. (78)] to the right:

$$MHCl(PPh_3)_2 + SnCl_2 \rightleftharpoons MHSnCl_3(PPh_3)_2 \qquad (78)$$

The formation of a mixed-ligand complex of platinum, containing both H^- and $SnCl_3^-$ in the coordination sphere of the metal ion, is indicated by the isolation of hydrido species from the hydrogenation reaction media. Thus, complexes with the formula $[H(SnCl_3)Pt(PPh_3)_2]$ having *cis* and *trans* phosphine groups [193] have been prepared and characterized [193] from hydrogenation reaction media consisting of *cis*- and *trans*-[$PtHCl(PPh_3)_2$] and stannous chloride [189].

Tayim and Bailar [196] have conducted hydrogenation studies with catalysts such as platinum(II) and palladium(II) diaryl sulfide and selenide complexes of the type $MCl_2(QPh_2)_2$ (Q = S or Se). The reactions were conducted at 50°–100°C in a 1:1 methanol–benzene solution under a hydrogen pressure of 5000 lb/inch^2 and with a 10:1 molar ratio of stannous chloride to platinum or palladium. The catalysts were ineffective in the absence of stannous chloride. For platinum(II) complexes the catalytic activities of the complexes decrease in the order, ($AsPh_3$) > (PPh_3) > ($SePh_2$) > (SPh_2), in the same order as decreasing π-acid character of the neutral ligand. The catalytic activities of platinum(II) catalysts decrease with increase in the σ-donor capacity of the solvents in the order, methylene chloride, ethylene dichloride > acetone > tetrahydrofuran (THF) ≈ methanol > pyridine. The yield of monoenes from a polyene was accordingly found to range from 100% in methylene chloride to zero in pyridine.

The catalytic properties of the platinum and palladium catalysts [193–195] described above may be summarized as follows:

1. Isomerization proceeds much more rapidly than hydrogenation as indicated by the fact that conjugated and unconjugated dienes are hydrogenated at the same rate.
2. The migration of the double bond is stepwise, as is found in the isomerization of 1,5-cyclooctadiene by $Pd(PPh_3)_2(CN)_2$.
3. Selective hydrogenation of the polyene to a monoene takes place.
4. The double bond of a monoene is not hydrogenated; α-olefins are isomerized to *trans* internal olefins, and *cis* internal olefins are isomerized to the *trans* form.
5. Hydrogen for the metal hydride formation may be derived either from molecular hydrogen or from the solvent.

The mechanism outlined in Eqs. (79) and (80) has been suggested [196] for the hydrogenation of a polyene by platinum(II) and palladium(II) of the type described above.

The hydride derived by the addition of molecular hydrogen to the platinum(II) or palladium(II) complex with trichlorostannite ligands forms a π-complex, **54**, with the polyene. If the starting polyene is a nonconjugated diene, it is isomerized very rapidly to a conjugated diene. Formation of the metal hydride may also take place by hydride abstraction from the solvent as discussed above. Metal hydride formation is not rate-determining since the rate of reaction is identical with $MCl(SnCl_3)L_2$ and $M(SnCl_3)HL_2$ [M = platinum(II) or palladium(II)]. The π complex then rearranges to the σ complex **55** by metal hydride addition. Such reversible addition reactions of

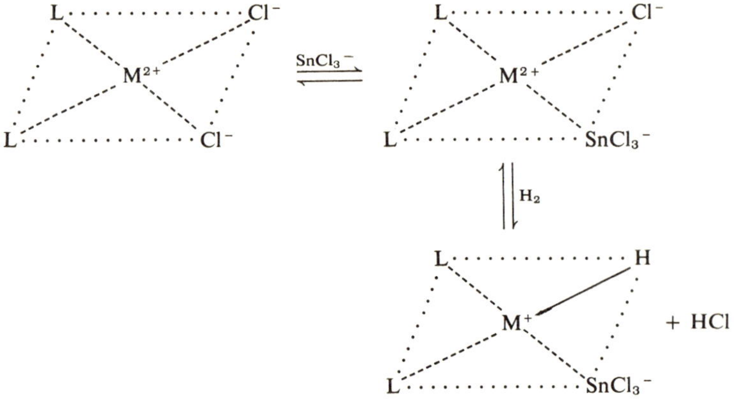

(79)

MH

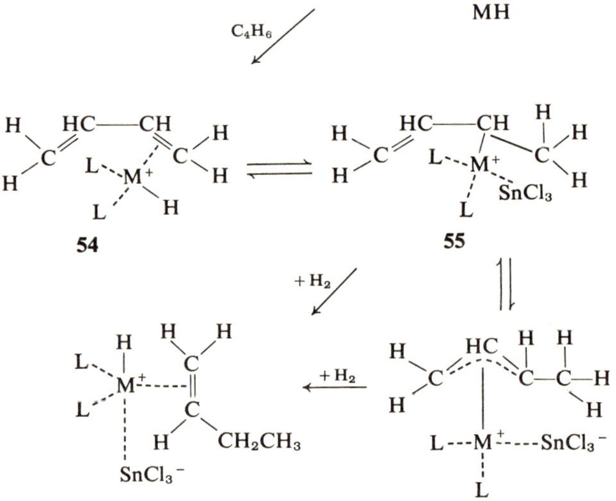

54 55

(80)

57 56

[L = Ph_3P, Ph_3As; M = platinum(II), palladium(II)]

olefins to platinum hydride have been well established by Chatt *et al.* [197]. The position of equilibrium in the reversible reaction between olefins and platinum hydrides depends on the nature of the olefin and the group bound to platinum(II) *trans* to the hydride [198]. The platinum(II)–alkyl complex **55** can either rearrange to form a π-allyl complex (**56**) or add molecular hydrogen to form a mixed-ligand metal–olefin–hydride complex (**57**). Such mixed-ligand complexes have been isolated from the reaction medium and characterized, as in the case of the 4-octene complex [189], **58**. Displacement of the coordinated alkene in **57** by a diene results in the formation of the diene complex **54** and free monoene. Exchange of the coordinated monoene with diene–metal complex takes place at every stage of the reaction. Thus, isomerized or partly hydrogenated olefins may be obtained at each stage of the reaction. Once the monoene is formed, it may undergo isomerization to an equilibrium mixture of terminal and internal olefins.

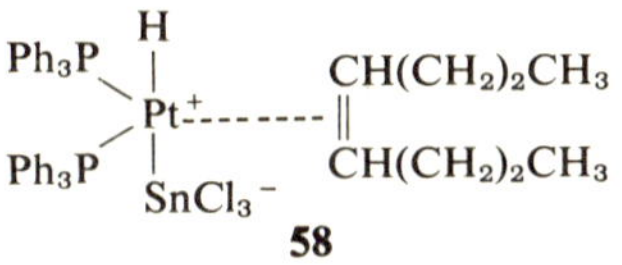

58

Catalytic hydrogenation in good yields of norbornadiene to norbornane and 2,5-dimethyl-3-hexyne-2,5-diol to 2,5-dimethylhexane-2,5-diol by hydridochlorobis(triphenylphosphine)platinum(II) in the presence of stannous chloride has been reported by Jardine and McQuillin [165]. The catalyst was totally ineffective in the absence of stannous chloride, indicating that the active catalyst in the reaction is probably $(PPh_3)_2PtHSnCl_3$.

Homogeneous hydrogenation of ethyl crotonate at 30°C and at 1 atm of hydrogen is catalyzed [159] by palladium(II) chloride. Although a precipitate of metallic palladium appeared during the course of the reaction, the major pathway for the reaction was, however, the catalysis by a soluble palladium-(II) complex with a negligible contribution by the heterogeneous catalytic reaction involving metallic palladium. Acetates of copper(II), nickel(II), zinc(II), silver(I), cadmium(II), sodium(I), calcium(II), magnesium(II), and chromium(III) have been used as promoters for this reaction. Nyholm [199] has postulated the reaction mechanism shown in Eq. (81). The Pd(II)–olefin

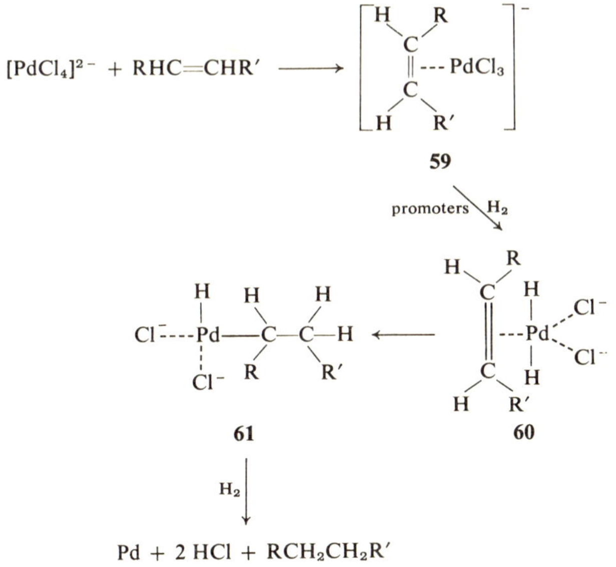

(81)

complex **59** is first formed, followed by the formation of a palladium dihydride complex, through the homogeneous splitting of hydrogen assisted by the promoters. The dihydride **60** then transfers one of the hydrogen atoms to the coordinated olefin with the formation of the palladium–alkyl complex, **61**. Reduction of this complex by molecular hydrogen leads to the formation of metallic palladium and the alkane. Since palladium(II) is not regenerated in the reaction, it acts as a reactant rather than as a catalyst. The rate of the reaction accordingly decreases as palladium(II) precipitates out of the solution in the metallic form.

4. *Chromium(II)-Catalyzed Reactions*

An interesting example of homogeneous hydrogenation in which the metal ion has the dual role of catalyst and reductant is provided by the reduction of alkynes in the presence of chromium(II) sulfate. Chromium(II) sulfate in aqueous solution or aqueous dimethylformamide reduces alkynes to *trans* olefins, a reaction that can be cleanly terminated at the olefin stage. The kinetics and mechanism of this stereospecific reduction have been elucidated by Castro and Stephens [200]. The reduction of acetylene to ethylene with heterogeneous catalysts [201] yields only the *cis*-olefins. The *trans* addition of hydrogen in the reduction by chromium(II) requires a specific orientation of the metal ion with respect to the acetylenic bond, and this is only possible through the participation of a π-complex intermediate. Various substituted alkynes were reduced by chromium(II), and the relative order of reactivity decreases in the following series:

$$HC{\equiv}CCH_2OH \simeq HOCH_2C{\equiv}CCH_2OH \simeq HOOCC{\equiv}CCOOH >$$
$$H_3CC{\equiv}CCH_2OH > HC{\equiv}CC(CH_3)_3 \gg H_3CC{\equiv}CCH_2CH_3 \sim C_6H_5C{\equiv}CH \simeq$$
$$(CH_3)_2COHC{\equiv}CC(CH_3)_3 \simeq C_6H_5C{\equiv}CC_6H_5$$

Third-order kinetics is followed in the reduction of alkynes by chromium(II) and this has been verified with excess as well as with equal concentration of substrate and chromium(II). The rate law is expressed by the equation

$$-\frac{d[\text{alkyne}]}{dt} = k[Cr^{2+}]^2[\text{alkyne}]$$

Castro and Stephens [200] proposed the formation of a 1:1 complex of chromium(II) and acetylene in the first step [Eq. (82)], followed by rate-determining attack of chromous ion on the chromous acetylene complex **62** [Eq. (83)]. The sequence of reactivities of various alkynes given above illustrates the importance of accessible coordination sites on the alkyne so that it may function as a ligand for the chromous ion. The formation of the 1:1

complex **62** is observed by the red color imparted to the solution on mixing di(*o*-carboxyphenyl)acetylene (*o*-carboxytolane) and chromium(II). A similar complex is not formed by *p*-carboxytolane. The rate-determining transfer of hydrogen to the triple bond for these chromium(II)-catalyzed reactions is

$$Cr^{2+} + RC{\equiv}CR' \rightleftharpoons \underset{\mathbf{62}}{\underset{Cr^{2+}}{\overset{RC\equiv CR'}{\vdots}}} \qquad (82)$$

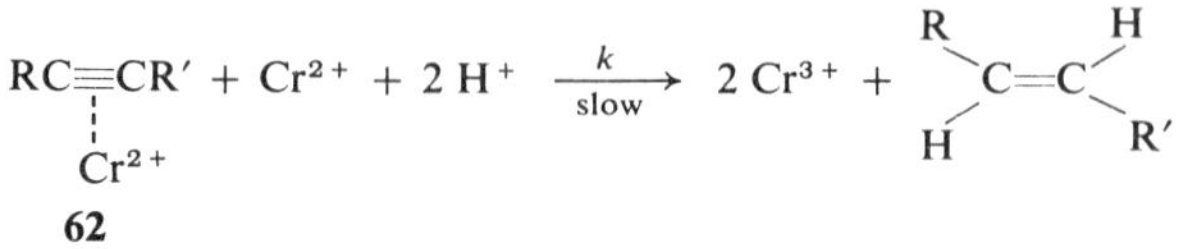

(83)

visualized as taking place through the formation of a transition-state complex (**65**) involving transfer of two protons from coordinated water and an electron from each of two chromium(II) atoms to the acetylene.

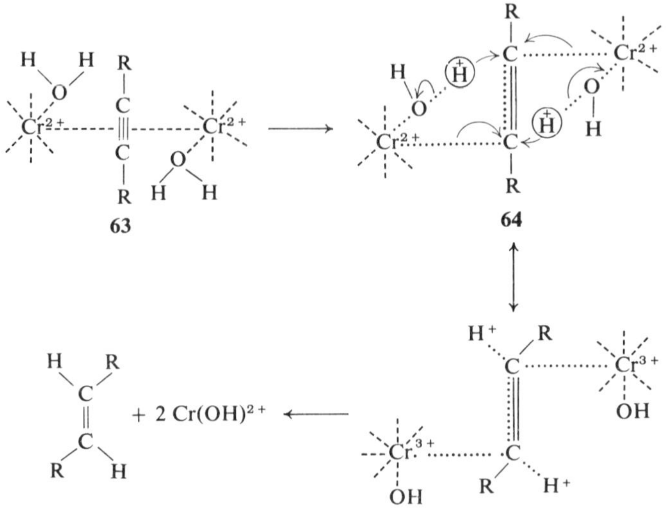

(84)

The second chromous ion must approach the π complex from a position opposite to that of the metal ion already present in the complex to form intermediate **63**. In the intermediates **64** and **65** transfer of two electrons, one from each chromium(II) ion, to suitable antibonding orbitals of the alkyne and concomitant back-sided proton transfer from the solvation spheres of chromium(II) ions results in the formation of the alkene and chromium(III). The reduction of the alkyne to the alkene is thus stereospecific, with the formation of only the *trans*-alkene. It was proposed [200] that the transfer of hydrogen ions to the triple bond in the transition complex takes place from the solvation sphere of the metal ion. This hypothesis is similar to that suggested by Halpern and James [164] for the ruthenium(II) catalyzed reduction of fumaric acid by molecular hydrogen. The highly charged activated complexes **64** and **65** are in accord with the relatively high negative entropy of activation, -60 eu, observed for this reaction.

The chromium(II)–EDTA chelate has also been used as a reductant [200] in the reaction. The rate of the reduction of alkyne in this case is independent of the alkyne concentration and depends only on third power of the total chromium concentration. The rate dependence is given by Eq. (85).

$$-\frac{d[\text{alkyne}]}{dt} = k[\text{Cr(II)}]^3_{\text{total}} \qquad (85)$$

This rate law holds through 85% completion of the reaction, In this case, the formation of a mixed complex seems improbable because of the saturation of the coordination positions of chromium(II) by the multidentate ligand EDTA, so that a somewhat different mechanism from that depicted for the chromium(II) redox reaction has been suggested. Although details of such a mechanism are lacking [200], it may very well involve outer orbital transfer of electrons from chromium(II) to the alkyne. The fact that stereospecificity of the chromium(II) reaction is lost in the reduction of the alkyne by the chromium(II)–EDTA complex is a further indication of the difference of mechanism of the chromium(II)– and the chromium(II)–EDTA-catalyzed reaction.

Isolated double bonds are not reduced by chromium(II) in acidic solution, but activated double bonds where the olefin forms part of an enedione system are readily reduced in acidic dimethylformamide solution to give the corresponding saturated compounds [202]. Thus maleic acid, fumaric acid, cinnamic acid, and acrylonitrile have all been reduced to the corresponding saturated products. Castro *et al.* [202] have proposed the formation of a 1:1 chromium(II)–alkene complex in the first step, followed by rate-determining attack of chromium(II) ion on the chromium(II)–alkene complex in a manner analogous to the reduction of alkynes by chromium(II) [200], as indicated above by Eq. (84).

5. *Catalysis by Chromium(II), Iron(II), Nickel(II), Manganese(II), and Titanium(III)*

Tulupov [203] has described catalysis by chromium(III), iron(II), and nickel(II) in the homogeneous hydrogenation of cyclohexene and by manganese(II) in the corresponding hydrogenation reaction of cyclopentene. He has classified metal ions that catalyze homogeneous hydrogenation of olefins into two groups depending on whether they react either with molecular hydrogen or with the olefin in the rate-determining step. Ions of the first group—copper(I), copper(II), silver(I), mercury(II), and mercury(I)—preferentially activate molecular hydrogen by forming an intermediate hydride which may either react with the substrate or become reduced to the free metal. Metal ions of the second group, which include chromium(II), iron(II), nickel(II), manganese(II), platinum(II), and rhodium(I), react with the olefin to form a complex which is then reduced by hydrogen. According to Tulupov [203], the main difference between these two reaction types is in their kinetic dependencies on reactant concentrations and their activation parameters. Reactions of the first group seem to be characterized by higher activation energy and negative activation entropy, while the second group seems to have lower activation energy and positive activation entropy. How general this observation will prove to be is not clear at the present time, since more data are needed on activation parameters of these catalytic reactions.

Itatani and Bailar [195] described the hydrogenation of methyl linoleate with bis(triphenylphosphine)dihalonickel(II) as the catalyst in tetrahydrofuran or benzene solvent at 90°C under 39 atm of hydrogen. The catalytic efficiency of the catalyst decreases with decreasing electronegativity of the halide ion, $Cl^- > Br^- > I^-$, which is coordinated to nickel(II). Bis(triphenylphosphine)dithiocyanatonickel(II) was found to be less effective than the halo complexes in the catalytic hydrogenation of methyl linoleate. The efficiency of these nickel(II) catalysts is not enhanced by the addition of trichlorostannite ($SnCl_3^-$), in contrast to the behavior of the palladium(II) and platinum(II) catalysts described above [193]. In contrast to the platinum(II) or palladium(II) triphenylphosphine complexes, hydrogenation of olefins with nickel(II) complexes is more rapid than isomerization of the olefins. No hydrido reaction intermediates have been isolated in these nickel(II)-catalyzed hydrogenation reactions, and further work is needed to establish the reaction mechanisms involved.

The homogeneous hydrogenation of alkynes with bis(cyclopentadienyl)dicarbonyltitanium(II) as the catalyst has been studied by Sonogashira and Hagihara [204]. The reaction takes place at 50°–60°C under an initial hydrogen pressure of 50 atm in a hydrocarbon solvent. The relative activating effects of substituents on the hydrogenation of the alkynes, RC≡CH, are

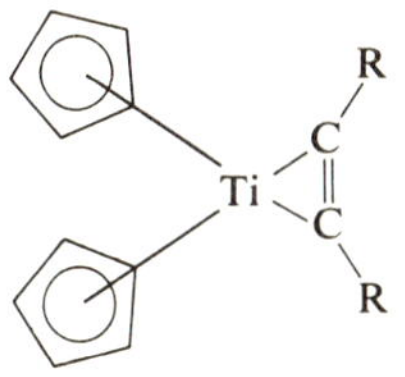

66

C_6H_5 > *t*-butyl > *n*-alkyl for R. Thus phenylacetylene was completely reduced to ethylbenzene whereas *t*-butylacetylene gave a mixture of *t*-butylethylene and *t*-butylethane. 1-Pentyne and 1-hexyne were reduced only to the alkene state, 1-pentene and 1-hexene. Conjugated olefins were not found to be reduced with this catalyst. Although a complete mechanism for the hydrogenation of alkynes with titanium(II) catalyst has not been proposed, intermediates similar to **66** have been postulated.

In the absence of molecular hydrogen, diphenylacetylene and phenylacetylene react with the titanium catalyst in a nitrogen atmosphere to form either a tetraphenylcyclobutadiene derivative (**67**) or its isomer (**68**). Phenylacetylene undergoes similar reactions. Compounds **67** and **68** did not undergo hydrogenation when treated with hydrogen under the conditions that resulted in hydrogenation of alkynes in the presence of the bis(cyclopentadienyl)dicarbonyltitanium catalyst.

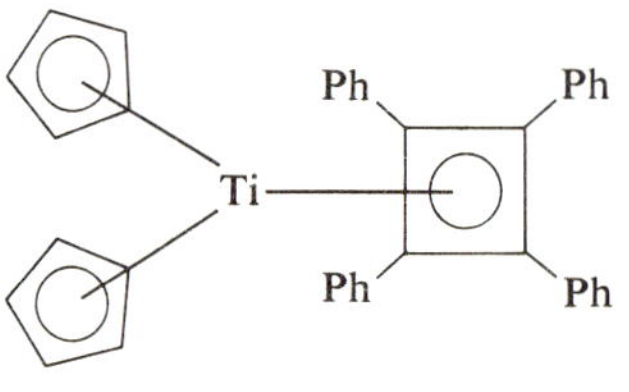

67

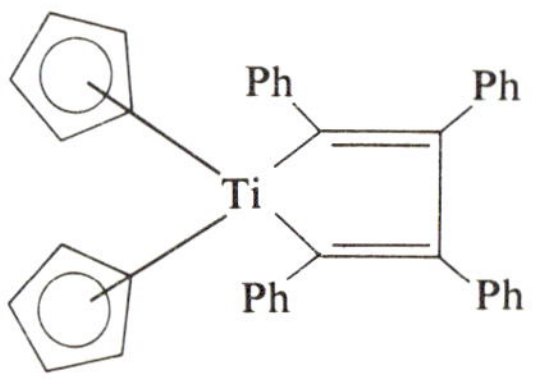

68

B. Catalysis by Metal Carbonyls

The metal carbonyls considered in this section are restricted to those that form π complexes with olefins as indicated either by isolation of the π complexes or by kinetic data indicating their presence as intermediates. Catalysis of olefin hydrogenation by metal carbonyl hydrides such as hydridocobalt tetracarbonyl, $HCo(CO)_4$, that may or may not involve prior formation of π complexes, is described in Vol. II, Chapter 3.

The best known example of hydrogenation catalysis through the activation of π bonds by a metal carbonyl is the iron pentacarbonyl–catalyzed hydrogenation of polyunsaturated fats, soybean oil, or cottonseed oil. In addition to hydrogenation of the double bonds, these reactions are characterized by *cis–trans* isomerization and migration of π bonds to form conjugated dienes from unconjugated trienes [205,206]. It may be seen in Vol. II, Chapter 2 that metal-catalyzed migration of π bonds and isomerization of nonconjugated dienes involve the formation of π complexes as intermediates. Observation of isomerization reactions in the hydrogenation of olefins with iron pentacarbonyl, $[Fe(CO)]_5$, as the catalyst is in accord with the concept of π-complex formation prior to hydrogenation. In some cases such intermediates have been actually isolated [201,207–209]. A π-diene–iron tricarbonyl complex has been isolated [207,208] by heating castor oil and iron pentacarbonyl at 180°C in a stainless-steel bomb under nitrogen until evolution of carbon monoxide ceases. Analyses by UV, IR, and NMR spectroscopy show that the compound contains iron coordinated to three carbon monoxides and to a methyl ester of conjugated octadecadienoic acids formed by isomerization of methyl linoleate. These π complexes readily decompose in solution, even under a nitrogen atmosphere, to give brown or black precipitates. Similar mixed diene–tricarbonyliron complexes have been isolated by Frankel *et al.* [205] on heating methyl linoleate and iron pentacarbonyl at 180°C under nitrogen or hydrogen pressure. These mixed-ligand complexes have been assigned structure **69**. Pure diene–iron tricarbonyl complexes have been

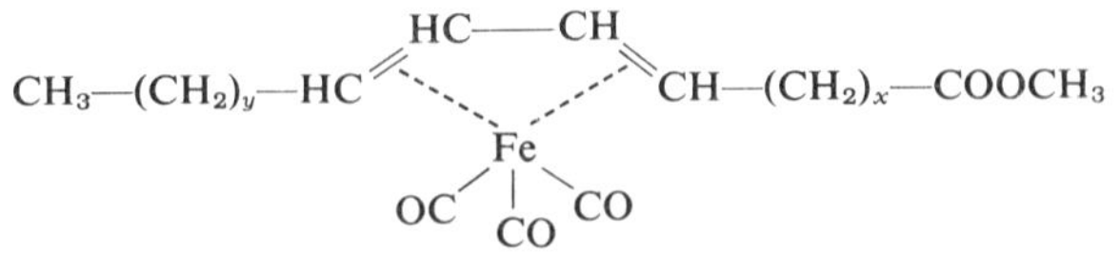

69

(x, y = 4, 8; 5, 7; 6, 6; 7, 5; 8, 4; 9, 3; and 10, 2, respectively)

separated from the mixture indicated by **69** by countercurrent distribution and alumina chromatography. The individual dienes were characterized by decomposing the π-diene–iron tricarbonyl complex with potassium hydroxide or ferric chloride and comparing the liberated diene with an authentic sample.

Hashimoto and Shiina [210,211] have reported the hydrogenation of soybean oil with iron pentacarbonyl as a catalyst at 180°C under a hydrogen pressure of 280 lb/inch^2. Hydrogenation under these conditions was very sluggish and the products are largely unsaturated. However, when the mixture of catalyst and the substrate was heated at 180°–200°C under nitrogen at high pressure for 1 hour and the nitrogen then replaced by hydrogen, subsequent hydrogenation at 180°C and 280 lb/inch^2 of hydrogen proceeded smoothly to the extent that 60% of the products were saturated hydrocarbons. These observations lead to the conclusion that a π-diene–iron tricarbonyl complex is formed *in situ* by the reaction of iron pentacarbonyl with the polyene and acts as a catalyst for the subsequent hydrogenation reaction of the polyenes in soybean oil. When the π-diene–iron tricarbonyl complex was used as the substrate the hydrogenation [205,210,211] reaction proceeded smoothly at 180°C under moderate hydrogen pressure to give good yields of monoene and methyl stearate. These results provide further support for the π-diene–iron tricarbonyl as the active catalyst for the hydrogenation of polyenes in soybean oil.

The hydrogenation of the olefin is not observed below a hydrogen pressure of 50 lb/inch^2. This failure of olefins to hydrogenate at low hydrogen pressures indicates that the reaction of hydrogen with the π complex is incomplete. At hydrogen pressures of 280 lb/inch^2 and above, however, inhibition of the reaction by carbon monoxide is largely suppressed and the reaction probably proceeds through the formation of a diene–iron tricarbonyl π complex **71** with conversion to a hydrido-π-allyl complex **73**. It is also possible that the diene reacts to give the dihydrido tricarbonyl species **72** but the formation of **73** by these alternative routes would be kinetically equivalent. The intermediate **73** undergoes a hydrogen shift and forms the monoene complex **74**, which either dissociates to **75** or undergoes further hydrogenation to the saturated product **76**. The iron tricarbonyl moiety may then react with hydrogen and diene to form complexes **71** and **72**.

The yield of hydrogenated product increased in the presence of cyclohexane as solvent due to lowering of the viscosity of the medium, thus facilitating better diffusion of hydrogen gas through the reaction mixture. Formation of the proposed hydrido intermediate **73** is supported by the work of Wender and Sternberg [212], who suggested a hydride of cyclopentadienyliron carbonyl, $(C_5H_5)Fe(CO)_2H$, as an intermediate in the hydrogenation of cyclopentadiene with iron carbonyls. An intramolecular hydrogen transfer similar to that suggested in the conversion of **73** to **74** was proposed by Brown *et al.* [213] for the reduction of olefinic compounds by chromium hexacarbonyl and by tricarbonylarenechromium.

Cobalt octacarbonyl, $Co_2(CO)_8$, has been used by Frankel *et al.* [206] as a catalyst in the homogeneous hydrogenation of unsaturated esters such as methyl linoleate. The cobalt carbonyl catalyst is more active at lower temperatures than is iron carbonyl, catalytic hydrogenation having been carried out successfully at 75°–80°C under a hydrogen pressure of 250–3000 lb/inch^2. No evidence of a stable π-diene–cobalt carbonyl complex was obtained in this case. Although the products of hydrogenation of methyl linoleate were found to be similar to those obtained with iron carbonyls, there was less double-bond migration, more *trans* isomerization, and absence of fully saturated products when cobalt carbonyl was employed. The reaction also requires a much higher hydrogen pressure than was the case for iron carbonyls. It is, however, possible that a π-diene–cobalt carbonyl complex is formed as a kinetic intermediate in the reaction, but is too unstable under the conditions used for separation of the product. The mechanism of catalysis by dicobalt octacarbonyl may thus be visualized as similar to that suggested for iron pentacarbonyl, $Fe(CO)_5$, although evidence for the various mechanistic steps and intermediates is not yet available.

A very interesting example of catalysis by dicobalt octacarbonyl is provided by the intermolecular hydrogenation of fulvenes reported by Altman and Wilkinson [214]. α,α-Dialkylfulvenes interact with dicobalt octacarbonyl to

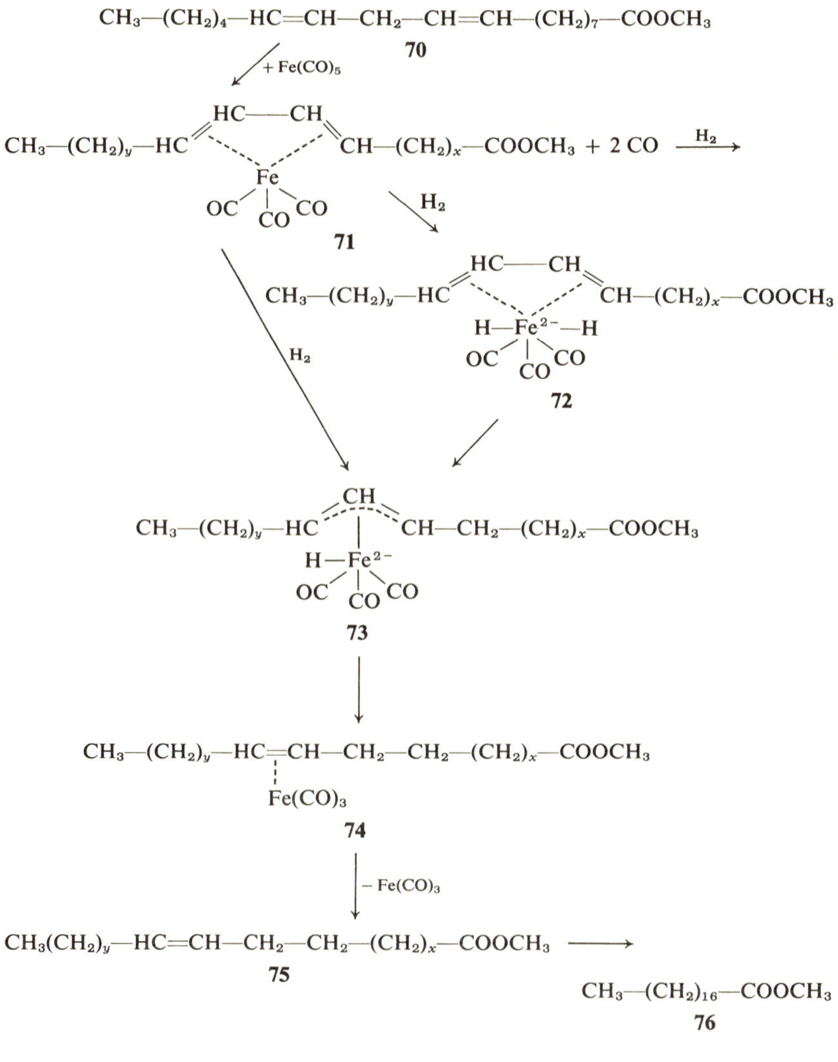

give a mixture of ring-substituted π-cyclopentadienyldicarbonylcobalt(I) compounds, **77** and **78** in equimolar proportion [Eq. (86)]. α,α-Diarylfulvenes with no α-hydrogen atoms react in a different manner to form a binuclear olefinic derivative, indicated by **79**.

The formation of a 1:1 mixture of saturated and unsaturated products **77** and **78** is unaffected by changing the solvent from light petroleum to ethylene-glycol dimethyl ether. This observation rules out the possibility of solvent participation as source of hydrogen. The formation of adduct **79** with aryl-fulvenes possessing no aliphatic or side-chain hydrogen atoms suggest that

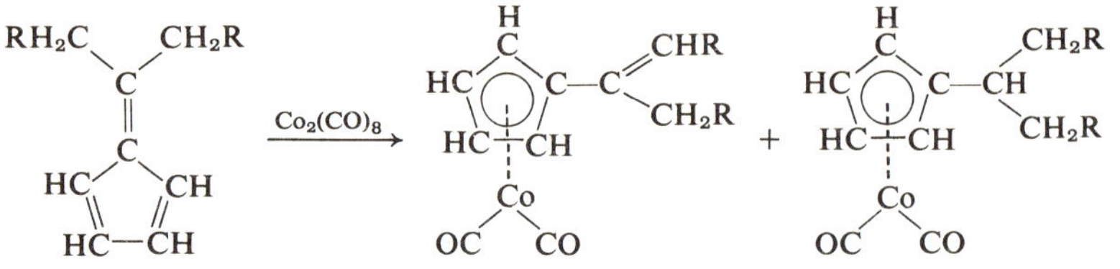

77 78

(86)

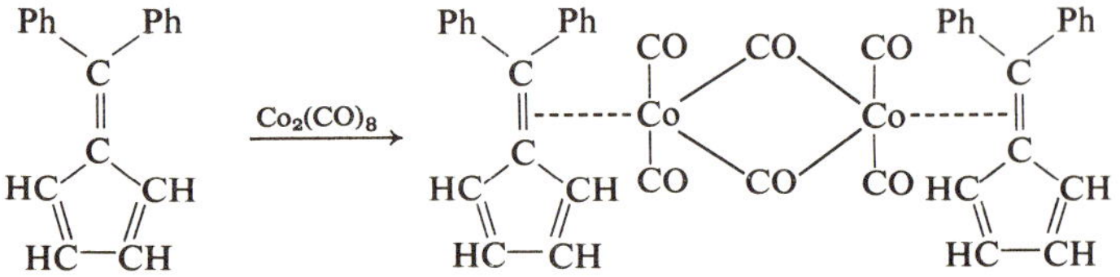

79

(87)

disproportionation of the fulvenes with simultaneous formation of equimolar quantities **77** and **78** takes place through the formation of intermediate complexes that undergo intermolecular hydrogen atom transfer. Carbonyls of molybdenum and tungsten undergo interaction with fulvenes to give binuclear π-cyclopentadienyl complexes [210] in the manner analogous to that indicated by Eq. (87) for dicobalt octacarbonyl.

References

1. J. Halpern, *Quart. Rev.* **10**, 463 (1956).
2. J. Halpern, *Advan. Catal. Relat. Subj.* **9**, 302–312 (1957).
3. J. Halpern, *Advan. Catal. Relat. Subj.* **11**, 301–370 (1959).
4. J. Halpern, *Collect. Pap. Symp. Coord. Chem.*, p. 351 (1964).
5. H. D. Kaesz and R. B. Saillant, *Chem. Rev.* **72**, 231 (1972).
6. G. Wilkinson, in "Advances in the Chemistry of Coordination Compounds (S. Kirschner, ed.), p. 50. Macmillan, New York, 1961.
7. W. P. Griffith and G. Wilkinson, *J. Chem. Soc., London* p. 2757 (1959).
8. P. G. Owston, J. M. Partridge, and J. M. Rowe, *Acta Crystallogr.* **13**, 246 (1960).
9. S. J. Laplaca, J. A. Ibers, and W. C. Hamilton, *J. Amer. Chem. Soc.* **86**, 2288 (1964).
10. P. L. Orioli and L. Vaska, *Proc. Chem. Soc., London* p. 333 (1962).
11. S. J. Laplaca and J. A. Ibers, *J. Amer. Chem. Soc.* **85**, 3501 (1963).
12. L. Knox and A. P. Ginsberg, *Inorg. Chem.* **3**, 555 (1964).
13. M. L. H. Green and C. J. Jones, *Advan. Inorg. Chem. Radiochem.* **7**, 115 (1965).
14. L. Vaska, *Inorg. Nucl. Chem. Lett.* **1**, 89 (1965).
15. L. Vaska and D. L. Catone, *J. Amer. Chem. Soc.* **88**, 5324 (1966).
16. J. Chatt, *Proc. Chem. Soc., London* p. 318 (1962).
17. J. Chatt and B. L. Shaw, *J. Chem. Soc., London* p. 705 (1959).
18. J. Chatt and B. L. Shaw, *J. Chem. Soc., London* p. 1718 (1960); *Chem. Ind. (London)* p. 931 (1960).
19. L. Orgel, "Transition Metal Chemistry, Ligand Field Theory." Methuen, London, 1960.
20. R. D. Gillard and G. Wilkinson, *J. Chem. Soc., London* p. 3594 (1963).
21. H. B. Gray and C. J. Ballhausen, *J. Amer. Chem. Soc.* **85**, 260 (1963).

22. L. Vaska and R. E. Rhodes, *J. Amer. Chem. Soc.* **87**, 4970 (1965).
23. M. Calvin, *Trans. Faraday Soc.* **34**, 1181 (1938).
24. M. Calvin, *J. Amer. Chem. Soc.* **61**, 2230 (1939).
25. M. Calvin and M. K. Wilmarth, *J. Amer. Chem. Soc.* **78**, 1301 (1956).
26. J. F. Harrod, S. Ciccone, and J. Halpern, *Can. J. Chem.* **39**, 1372 (1961).
27. J. F. Harrod and J. Halpern, *Can. J. Chem.* **37**, 1933 (1959).
28. J. Halpern, J. F. Harrod, and P. E. Potter, *Can. J. Chem.* **37**, 1446 (1959).
29. J. A. Osborn, F. H. Jardine, J. F. Young, and G. Wilkinson, *J. Chem. Soc., A* p. 1711 (1966).
30. J. Chatt and S. A. Butler, *Chem. Commun.* 501 (1967).
31. R. F. Heck, *J. Org. Chem.* **28**, 604 (1963).
32. J. Chatt, N. P. Johnson, and B. L. Shaw, *J. Chem. Soc., A* p. 604 (1967).
33. J. P. Collman and C. T. Sears, Jr., *Inorg. Chem.* **7**, 27 (1968).
34. R. G. Pearson and W. R. Muir, *J. Amer. Chem. Soc.* **92**, 5519 (1970).
35. J. A. Labinger, R. J. Braus, D. Dolphin, and J. A. Osborn, *Chem. Commun.* p. 612 (1970).
36. A. J. Deeming and B. L. Shaw, *J. Chem. Soc., A* p. 1562 (1969).
37. B. L. Shaw and R. E. Stainbank, *J. Chem. Soc.* (*Dalton*) p. 223 (1972).
38. M. Kubota and D. M. Blake, *J. Amer. Chem. Soc.* **93**, 1368 (1971).
39. C. A. Reed and W. R. Roper, *J. Chem. Soc., A* p. 3054 (1970).
40. M. A. Bennett and R. Charles, *J. Amer. Chem. Soc.* **94**, 666 (1972).
41. D. Foster, *Inorg. Chem.* **11**, 473 (1972).
42. A. J. Hart-Davis and W. A. G. Graham, *Inorg. Chem.* **9**, 2658 (1970).
43. A. J. Hart-Davis and W. A. G. Graham, *Inorg. Chem.* **10**, 1653 (1971).
44. D. N. Lawson, J. A. Osborn, and G. Wilkinson, *J. Chem. Soc., A* p. 1733 (1966).
45. R. F. Heck, *J. Amer. Chem. Soc.* **86**, 2796 (1964).
46. R. V. Parish and R. G. Simms, *J. Chem. Soc.* (*Dalton*) p. 809 (1972).
47. J. P. Collman, N. W. Hoffman, and D. E. Morris, *J. Amer. Chem. Soc.* **91**, 5659 (1969).
48. J. T. Mague and J. P. Mitchener, *Chem. Commun.* p. 911 (1968).
49. J. R. Shapeley, R. R. Schrock, and J. A. Osborn, *J. Amer. Chem. Soc.* **91**, 2816 (1969).
50. F. L'Eplattenier, *Inorg. Chem.* **8**, 965 (1969).
51. A. J. Cheney and B. L. Shaw, *J. Chem. Soc., A* p. 3545 (1971).
52. A. J. Cheney and B. L. Shaw, *J. Chem. Soc., A* p. 3549 (1971).
53. K. Kudo, M. Hidai, and Y. Uchida, *J. Organometal. Chem.* **33**, 393 (1971).
54. B. Hessett, J. H. Morris, and P. G. Perkins, *Inorg. Nucl. Chem. Lett.* **7**, 1149 (1971).
55. J. L. Burmeister and L. E. Edwards, *J. Chem. Soc., A* p. 1663 (1971).
56. E. O. Greaves, C. J. T. Lock, and P. M. Maitlis, *Can. J. Chem.* **46**, 3879 (1968).
57. K. S. Wheelock, J. H. Nelson, and H. B. Jonassen, *Inorg. Chim. Acta* **4**, 399 (1970).
58. F. L'Eplattenier and M. C. Pelichet, *Helv. Chim. Acta* **53**, 1091 (1970).
59. J. F. Harrod and C. A. Smith, *J. Amer. Chem. Soc.* **92**, 2699 (1970).
60. J. A. Osborn, *Amer. Chem. Soc. Symp. Activation Small Mol.*, (1972).
61. G. Hatta, H. Kondo, and A. Miyake, *J. Amer. Chem. Soc.* **90**, 2278 (1968).
62. J. Chatt and J. M. Davidson, *J. Chem. Soc., London* p. 843 (1965).
63. M. A. Bennett and D. L. Milner, *J. Amer. Chem. Soc.* **91**, 6983 (1969).
64. E. W. Ainscough and S. D. Robinson, *Chem. Commun.* p. 863 (1970).
65. E. W. Ainscough and S. D. Robinson, *Chem. Commun.* p. 130 (1971).
66. A. J. Cheney, B. E. Mann, B. L. Shaw, and R. M. Slade, *J. Chem. Soc., D* p. 1176 (1970).

67. G. E. Hartwell, R. V. Lawrence, and M. J. Smas, *Chem. Commun.* p. 912 (1970).
68. J. Halpern, C. Czapski, J. Hortner, and G. Stern, *Nature (London)* **196**, 629 (1960).
69. J. Halpern, E. R. MacGregor, and E. J. Peters, *J. Phys. Chem.* **60**, 1455 (1956).
70. J. Halpern, *J. Phys. Chem.* **63**, 398 (1959).
71. G. J. Korinek and J. Halpern, *Can. J. Chem.* **34**, 1372 (1956); *J. Phys. Chem.* **60**, 285 (1956).
72. A. J. Chalk and J. Halpern, *J. Amer. Chem. Soc.* **81**, 5846 and 5842 (1959).
73. J. Halpern, U.S. Patent 2,879,134 (1959).
74. W. K. Wilmarth, J. C. Dayton, and J. H. Fluoronoy, *J. Amer. Chem. Soc.* **75**, 4549 (1953).
75. E. J. De Witt, R. L. Ramp, and L. E. Trapasso, *J. Amer. Chem. Soc.* **83**, 4672 (1961).
76. R. Kostner, *Angew. Chem.* **68**, 383 (1956).
77. R. Kostner, B. Gunter, and B. Paul, *Justus Liebigs Ann. Chem.* **644**, 1 (1961).
78. C. Walling and L. Bollyky, *J. Amer. Chem. Soc.* **83**, 2968 (1961).
79. C. Walling and L. Bollyky, *J. Amer. Chem. Soc.* **86**, 3750 (1964).
80. A. H. Webster and J. Halpern, *J. Phys. Chem.* **60**, 280 (1956).
81. F. Nagy and L. Simandi, *Magy. Kem. Foly.* **69**, 433 (1963); *Acta Chim. Acad. Sci. Hung.* **38**, 213 and 373 (1963).
82. S. W. Weller and G. A. Mills, *Advan. Catal. Relat. Subj.* **8**, 163–205 (1956).
83. A. H. Webster and J. Halpern, *Trans. Faraday Soc.* **53**, 51 (1957).
84. A. H. Webster and J. Halpern, *J. Phys. Chem.* **61**, 1239 and 1245 (1957).
85. J. Halpern and J. B. Milne, *Actes Congr. Int. Catal., 2nd, 1960* Vol. 1, p. 445 (1961).
86. M. Beck, *Rec. Chem. Progr.* **27**, 37 (1966).
87. M. Beck and I. Gimesi, *Magy. Kem. Foly.* **69**, 552 (1963).
88. R. E. Connick and A. D. Paul, *J. Phys. Chem.* **65**, 1216 (1961).
89. M. Beck, I. Gimesi, and J. Farkas, *Nature (London)* **73**, 197 (1963).
90. J. Halpern, *142nd Meet., Amer. Chem. Soc., Atlantic City, N.J.*, p. 23N (1962).
91. J. Halpern and J. P. Maher, *J. Amer. Chem. Soc.* **86**, 2311 (1964).
92. R. D. Gillard, J. A. Osborn, P. B. Stockwell, and G. Wilkinson, *Proc. Chem. Soc., London* p. 284 (1964).
93. W. P. Griffith and G. Wilkinson, *J. Chem. Soc., London* p. 1629 (1959).
94. F. Basolo, J. Chatt, H. S. Gray, R. G. Pearson, and B. L. Shaw, *J. Chem. Soc., London* p. 2207 (1961).
95. L. Vaska, *J. Amer. Chem. Soc.* **83**, 756 (1961).
96. L. Vaska, *Proc. Int. Conf. Coord. Chem., 7th*, p. 266 (1962).
97. L. Vaska, *Proc. Int. Conf. Coord. Chem., 8th, 1964* p. 99 (1964).
98. L. Vaska and J. W. Diluzio, *J. Amer. Chem. Soc.* **83**, 1262 (1961).
99. L. Vaska and J. W. Diluzio, *J. Amer. Chem. Soc.* **84**, 679 (1962).
100. S. S. Bath and L. Vaska, *J. Amer. Chem. Soc.* **85**, 3500 (1963).
101. A. Sacco, M. Rossi, and C. F. Nobile, *Chem. Commun.* p. 589 (1966).
102. A. Sacco and M. Rossi, *Chem. Commun.* p. 316 (1967).
103. L. M. Haines, *Inorg. Chem.* **10**, 1693 (1971).
104. R. C. Taylor, J. F. Young, and G. Wilkinson, *Inorg. Chem.* **5**, 20 (1966).
105. M. J. Mays, R. N. F. Simpson, and F. P. Stefanini, *J. Chem. Soc., A* p. 3000 (1970).
106. A. J. Deeming and B. L. Shaw, *J. Chem. Soc., A* p. 3356 (1970).
107. J. Y. Chen and J. Halpern, *J. Amer. Chem. Soc.* **93**, 4439 (1971).
108. J. F. Biellmann and M. J. Jung, *J. Amer. Chem. Soc.* **90**, 1673 (1968).
109. L. Malatesta, G. Caglio, and M. Angoletta, *J. Chem. Soc., London* p. 6974 (1965).
110. F. L'Eplattenier and F. Calderazzo, *Inorg. Chem.* **7**, 1290 (1968).
111. F. L'Eplattenier and F. Calderazzo, *Inorg. Chem.* **6**, 2092 (1967).

112. J. C. Collman and W. R. Roper, *Advan. Organometal. Chem.* **7**, 54 (1968).
113. P. B. Chock and J. Halpern, *J. Amer. Chem. Soc.* **88**, 3511 (1966).
114. L. Vaska and J. W. Diluzio, *J. Amer. Chem. Soc.* **83**, 2786 (1961).
115. L. Vaska, *Accounts Chem. Res.* **1**, 335 (1968).
116. P. W. Schneider, P. F. Phelan, and J. Halpern, *J. Amer. Chem. Soc.* **91**, 71 (1969).
117. L. G. Marzilli, P. A. Marzilli, and J. Halpern, *J. Amer. Chem. Soc.* **91**, 5752 (1970).
118. L. G. Marzilli, P. A. Marzilli, and J. Halpern, *J. Amer. Chem. Soc.* **93**, 1374 (1971).
119. J. Halpern and P. F. Phelan, *J. Amer. Chem. Soc.* **94**, 1881 (1972).
120. G. W. Parshall, *Accounts Chem. Res.* **3**, 139 (1970).
121. W. Beck, K. Schorph, and F. Kern, *Angew. Chem., Int. Ed. Engl.* **10**, 66 (1971).
122. R. J. Hodges, D. E. Webster, and P. B. Wells, *J. Chem. Soc., A* p. 3230 (1971).
123. J. Kwiatek, I. L. Mador, and J. K. Seylor, *Advan. Chem. Ser.* **37**, 201 (1963).
124. J. Kwiatek, *Catal. Rev.* **1**, 37 (1967).
125. M. E. Winfield, *Aust. J. Sci. Res., Ser. A* 4, 385 (1951).
126. M. E. Winfield, *Rev. Pure Appl. Chem.* **5**, 217 (1955).
127. J. Kwiatek, *Proc. Int. Conf. Coord. Chem., 8th 1964* p. 308 (1964).
128. J. Kwiatek, I. L. Mador, and J. K. Seylor, *J. Amer. Chem. Soc.* **84**, 304 (1962).
129. M. Murakami, *Proc. Int. Conf. Coord. Chem., 7th, 1962* p. 268 (1962).
130. J. Bayston, N. K. King, and M. E. Winfield, *Advan. Catal. Relat. Subj.* **9**, 312–318 (1957).
131. M. Murakami and J. W. Kang, *Bull. Chem. Soc. Jap.* **35**, 1243 (1962).
132. M. Murakami and J. W. Kang, *Bull. Chem. Soc. Jap.* **36**, 763 (1963).
133. M. Iguchi, *J. Chem. Soc. Jap. Pure Chem. Sect.* **63**, 634 (1943).
134. G. A. Mills, S. Wheeler, and A. Wheeler, *J. Phys. Chem.* **63**, 403 (1959).
135. N. K. King and M. E. Winfield, *J. Amer. Chem. Soc.* **80**, 2060 (1958).
136. N. K. King and M. E. Winfield, *J. Amer. Chem. Soc.* **83**, 3366 (1961).
137. R. G. S. Banks and J. M. Pratt, *J. Chem. Soc., A* p. 854 (1968).
138. W. P. Griffith and J. R. Lane, *J. Chem. Soc. (Dalton)* p. 158 (1972).
139. F. D. Tsay, H. B. Gray, and J. Danon, *J. Chem. Phys.* **54**, 3760 (1971).
140. B. deVries, *Proc. Kon. Ned. Akad. Wetensch. Ser. B* **63**, 443 (1960).
141. B. deVries, *J. Catal.* **1**, 489 (1962).
142. M. G. Burnett, P. J. Conolly, and C. Kemball, *J. Chem. Soc., A* p. 800 (1967).
143. L. Simandi and F. Nagy, *Magy. Kem. Foly.* **71**, 6 (1965).
144. L. Simandi and F. Nagy, *Acta Chim. Acad. Sci. Hung.* **46**, 101 and 137 (1965).
145. J. Halpern and M. Pribanic, *Inorg. Chem.* **9**, 2616 (1970).
146. L. Simandi and F. Nagy, *Collect. Pap. Symp. Coord. Chem.*, p. 329 (1964).
147. W. R. McClellan, H. N. Holhn, H. N. Cripps, E. L. Muetterties, and B. W. Howk, *J. Amer. Chem. Soc.* **83**, 1601 (1961).
148. J. Halpern and J. P. Maher, *J. Amer. Chem. Soc.* **86**, 2311 (1964).
149. R. F. Heck and D. S. Breslow, *J. Amer. Chem. Soc.* **83**, 4023 (1961).
150. M. Takahashi, *Bull. Chem. Soc. Jap.* **36**, 622 (1963).
151. J. Volter, *J. Catal.* **3**, 297 (1964).
152. R. D. Cramer, R. V. Lindsey, Jr., E. L. Jenner, and U. G. Stolberg, *J. Amer. Chem. Soc.* **85**, 1691 (1963).
153. J. H. Flynn and H. M. Hulbust, *J. Amer. Chem. Soc.* **76**, 3393 (1954).
154. J. Halpern, J. F. Harrod, and B. R. James, *J. Amer. Chem. Soc.* **83**, 753 (1961).
155. J. Halpern, B. R. James, and A. L. W. Kemp, *J. Amer. Chem. Soc.* **88**, 5150 (1966).
156. J. Halpern and B. R. James, *Abstr. 142nd Meet., Amer. Chem. Soc., Atlantic City, N.J.* p. 23N (1962).
157. J. Halpern, *Abstr. 141st Meet., Amer. Chem. Soc., Washington, D.C.* p. 10Q (1962).

158. J. Halpern, *Proc. Int. Conf. Catal., 3rd*, p. 146 (1965).
159. E. B. Maxted and S. M. Ismail, *J. Chem. Soc., London* p. 1750 (1964).
160. F. H. Jardine, J. A. Osborn, G. Wilkinson, and J. F. Young, *Chem. Ind.* (*London*) p. 560 (1965).
161. J. F. Young, J. A. Osborn, F. H. Jardine, and G. Wilkinson, *Chem. Commun.* p. 131 (1965).
162. M. L. H. Green and P. L. K. Nagy, *J. Amer. Chem. Soc.* **84**, 1310 (1962).
163. D. Evans, J. A. Osborn, F. H. Jardine, and G. Wilkinson, *Nature* (*London*) **208**, 1203 (1965).
164. J. Halpern and B. R. James, *Can. J. Chem.* **44**, 495 (1966).
165. I. Jardine and F. J. McQuillin, *Tetrahedron Lett.* 4871 (1966).
166. M. M. Taqui Khan and S. Vaucheson, unpublished data.
167. J. A. Osborn, G. Wilkinson, and J. F. Young, *Chem. Commun.* p. 17 (1965).
168. F. H. Jardine, J. A. Osborn, and G. Wilkinson, *J. Chem. Soc., A* p. 1574 (1967).
169. S. Montalatici, A. Van der Ent, J. A. Osborn, and G. Wilkinson, *J. Chem. Soc.*, A p. 1054 (1968).
170. H. Arai and J. Halpern, *Chem. Commun.* p. 1571 (1971).
171. P. Meakin, J. P. Jesson, and C. A. Tolman, *J. Amer. Chem. Soc.* **94**, 3240 (1972).
172. J. T. Mague and G. Wilkinson, *J. Chem. Soc., A* p. 1736 (1966).
173. F. H. Jardine and G. Wilkinson, *J. Chem. Soc., C* p. 270 (1967).
174. A. J. Birch and K. A. M. Walker, *Tetrahedron Lett.* p. 4939 (1966).
175. A. J. Birch and K. A. M. Walker, *J. Chem. Soc., C* p. 1894 (1966).
176. C. O'Connor, G. Yagupsky, D. Evans, and G. Wilkinson, *Chem. Commun.* p. 420 (1968).
177. M. Yagupsky, C. K. Brown, G. Yagupsky, and G. Wilkinson, *J. Chem. Soc., A.* p. 937 (1970).
178. B. R. James, F. T. T. Ng, and G. L. Rampel, *Inorg. Nucl. Chem. Lett.* **4**, 197 (1968).
179. B. R. James and G. L. Rampel, *Can. J. Chem.* **44**, 233 (1966).
180. B. G. James and F. T. T. Ng, *J. Chem. Soc.* (*Dalton*) p. 355 (1972).
181. A. J. Birch and K. A. M. Walker, *Tetrahedron Lett.* p. 1935 (1967).
182. L. Vaska, *Inorg. Nucl. Chem. Lett.* **1**, 89 (1965).
183. J. Trocha-Grimshaw and H. B. Henbest, *Chem. Commun.* 757 (1968).
184. H. Van Gaal, H. G. A. N. Cuppers, and A. Van der Ent, *Chem. Commun.* p. 1694 (1971).
185. R. D. Cramer, R. F. Lindsey, C. T. Prewitt, and U. G. Stolberg, *J. Amer. Chem. Soc.* **87**, 658 (1965).
186. A. G. Davies, G. Wilkinson, and J. F. Young, *J. Amer. Chem. Soc.* **85**, 1692 (1963).
187. J. F. Young, R. D. Gillard, and G. Wilkinson, *J. Chem. Soc., London* p. 5176 (1964).
188. R. C. Taylor, J. F. Young, and G. Wilkinson, *Inorg. Chem.* **5**, 20 (1966).
189. J. C. Bailar, Jr. and H. Itatani, *Inorg. Chem.* **4**, 1618 (1965).
190. L. P. Vant'Hof and B. G. Linsen, *J. Catal.* **7**, 295 (1967).
191. H. Van Bekkum, J. Van Gogh, and G. Van Minnenpathuis, *J. Catal.* **7**, 292 (1967).
192. A. P. Khrusch, A. A. Tokina, and A. E. Shilov, *Kinet. Catal* (*USSR*) **7**, 793 (1966).
193. J. C. Bailar, Jr. and H. Itatani, *J. Amer. Chem. Soc.* **89**, 1592 (1967).
194. H. Itatani and J. C. Bailar, Jr., *J. Oil Chem. Soc.* **44**, 147 (1967).
195. H. Itatani and J. C. Bailar, Jr., *J. Amer. Chem. Soc.* **99**, 1600 (1967).
196. H. A. Tayim and J. C. Bailar, Jr., *J. Amer. Chem. Soc.* **89**, 4330 (1967).
197. J. Chatt, R. S. Coffey, A. Gough, and D. T. Thompson, *J. Chem. Soc., A* p. 190 (1968).
198. J. Chatt and B. L. Shaw, *J. Chem. Soc., London* p. 5075 (1962).

199. R. S. Nyholm, *Proc. Int. Conf. Catal., 3rd*, p. 25 (1965).
200. C. E. Castro and R. D. Stephens, *J. Amer. Chem. Soc.* **86**, 4358 (1964).
201. B. S. Rabinovitch and F. S. Lodney, *J. Amer. Chem. Soc.* **75**, 2652 (1953).
202. C. E. Castro, R. D. Stephens, and S. Mose, *J. Amer. Chem. Soc.* **88**, 4964 (1966).
203. V. A. Tulupov, *Collect. Pap. Symp. Coord. Chem., 1964* p. 577 (1964).
204. K. Sonogashira and N. Hagihara, *Bull. Chem. Soc. Jap.* **39**, 1178 (1966).
205. E. N. Frankel, E. A. Emken, H. M. Peters, V. L. Davison, and R. O. Butterfield, *J. Org. Chem.* **29**, 3292 (1964).
206. E. N. Frankel, H. M. Peters, E. P. Jones, and H. J. Dutton, *J. Amer. Oil Chem.* **41**, 186 (1964).
207. I. Ogata and A. Misono, *Bull. Chem. Soc. Jap.* **37**, 439 (1964).
208. I. Ogata and A. Misono, *Bull. Chem. Soc. Jap.* **37**, 900 (1964).
209. I. Ogata and A. Misono, *Nippon Kagaku Zasshi* **85**, 748 (1964).
210. T. Hashimoto and H. Shiina, *Yukagu* **8**, 259 (1959).
211. T. Hashimoto and H. Shiina, *Tokyo Kogyo Skikensho Hokoku*, **57**, 284 (1962).
212. I. Wender and H. W. Sternberg, *Proc. Int. Conf. Coord. Chem.* p. 53 (1959).
213. D. A. Brown, J. P. Hargaden, C. M. McMullen, N. Gogan, and M. Sloan, *J. Chem. Soc., London* p. 4914 (1963).
214. J. Altman and G. Wilkinson, *J. Chem. Soc., London* p. 5654 (1964).

2

Activation of Molecular Oxygen

I. Introduction

This chapter reviews the various types of metal ion-catalyzed reactions of molecular oxygen and attempts to provide conceptual interpretations of these reactions. The mechanisms of metal ion and metal chelate catalysis of chemical oxidation reactions are examined from the point of view of applying these mechanisms toward a better understanding of the corresponding reactions in biological systems.

II. Reaction Types

As indicated in Table I, oxygen reactions may be classified in two general ways: those in which oxygen itself combines with the substrate, called insertion reactions, and those in which the oxygen does not combine with the substrate, but merely serves as an oxidizing agent and becomes reduced to hydrogen peroxide or water, depending on the nature of the substrate and the reaction conditions. The latter are called non-insertion reactions. Because of the importance of peroxides and superoxides as intermediates in the reduction of oxygen to water, pertinent examples of metal-catalyzed reactions of peroxides and superoxides are included in this chapter.

Non-insertion reactions are of special interest because of the importance of catalysis by metal ions and complexes both in biological and in strictly chemical systems. Many biological reactions of molecular oxygen, such as those catalyzed by xanthine oxidase, uric acid oxidase, and other oxidases

TABLE I

METAL-CATALYZED REACTIONS OF MOLECULAR OXYGEN

NON-INSERTION			INSERTION		
			Insertion with O—O bond cleavage		
Reduction of O_2 to H_2O_2	Reduction of O_2 to H_2O	Non-Insertion Reactions of H_2O_2	Partial Insertion	Total Insertion	Insertion without O—O Bond Cleavage
1. Oxidation by oxygen carriers 2. Metal ion-catalyzed oxidation of ascorbic acid, catechols, etc. 3. Oxidation of metal ions by O_2 4. Amine oxidase models and enzymes 5. Enzyme reactions (uricase, amine oxidase, oxalic acid oxidase, xanthine oxidase, etc.)	1. Oxidases (tyrosinase, laccase, polyphenol oxidases, etc.) 2. Free-radical reactions 3. Enzymatic coupling reactions	1. Model catalase systems 2. Catalase enzymes	1. Reactions of H_2O_2 a. Fenton's reagent b. Model peroxidase systems c. Enzyme systems 2. Reactions of O_2 a. Udenfriend's system b. Monooxygenases c. Free radical reactions	1. Dioxygenases 2. Metal ion-catalyzed free-radical oxidation	1. Metal-catalyzed formation of organoperoxides 2. Addition to organometallic compounds

seem to involve reduction of oxygen to the peroxide state. Such reactions are usually coupled with other enzyme systems, such as peroxidases or catalase, that complete the reduction to water. There are many examples of similar reactions in model systems, such as the redox reactions of coordination compounds that act as oxygen carriers, and metal ion-catalyzed oxidation of such two-electron donors as ascorbic acid and pyrocatechol.

Complete reduction of oxygen to water is accompanied by oxidation of ascorbic acid in the presence of certain copper(II) and iron(III) chelates as catalysts. Examples of this type of oxidation in biological systems are the reactions catalyzed by cytochrome oxidase systems and others that convert the oxygen to water by a series of complex reaction steps.

Many non-insertion reactions involve the use of hydrogen peroxide as the oxidant. Thus hydrogen peroxide oxidation of ascorbic acid is catalyzed by iron(III) and copper(II) chelates of ligands such as EDTA. The cupric ion also catalyzes the oxidation of catechol to the semiquinone by hydrogen peroxide.

Many metal ions and complexes catalyze the disproportionation of hydrogen peroxide to oxygen and water. This reaction is promoted by the catalase enzymes in biological systems and is very important since it serves to prevent the accumulation of hydrogen peroxide as the result of oxidase action. Catalase-controlled conversion of hydrogen peroxide to oxygen and water is often the second step in the oxidation of a substrate by reduction of oxygen to water, with the oxidase promoting only the first step by which oxygen is reduced to peroxide. Because of the special nature of catalase action, it is given separate treatment, as indicated in the third column of Table I.

Partial insertion reactions are many and varied, and usually involve the initiation and propagation of free-radical reactions by metal ions. The best known example of partial insertion of molecular oxygen is Udenfriend's system, which uses a suitable two-electron reductant such as a catechol and a chelate compound of iron(II) for the conversion of alkenes, alkanes, and aromatic compounds to epoxides, alcohols, and phenols. Most other partial insertion reactions are metal ion-catalyzed free-radical reactions in which hydrogen peroxide is the source of oxygen. There are many examples of such reactions, such as Fenton's reagent, model peroxidase systems, and reactions of olefins with pertungstic acid.

Since it is logical to expect nature to avoid free-radical reactions that occur in a random manner, it is not surprising that such partial insertion reactions have yet to be demonstrated for biological systems. On the other hand, there are many examples of biological partial insertion reactions (e.g., those of monooxygenases) involving stoichiometric insertion of one oxygen with reduction of the other to water.

Total insertion reactions, in which both oxygens combine with the substrate, are frequently of a relatively specific nature, rather than of general interest in the area of metal ion-catalyzed oxidations. An interesting example of a total insertion reaction in biological systems is the pyrocatechase, Fe(II)-catalyzed oxidation of pyrocatechol to muconic acid. Even here the reaction may be complex, consisting of a non-insertion-type oxidation of pyrocatechol to quinone, followed by insertion of enzyme-bound peroxide to the substrate and subsequent ring fission. Biological oxidations of catechol derivatives apparently follow similar pathways (e.g., oxygen transferase action).

Insertion reactions of molecular oxygen resulting in the formation of metal salts of organic peroxides and organic hydroperoxides are often catalyzed by metal ions. Many such reactions also occur in the absence of metal ions, such as, for example, the 1,4-addition of activated (singlet) oxygen to a diene to give a bridged peroxy organic compound. Addition reactions of this type seem to be important in the biosynthesis of natural products.

III. Energy Requirements of Oxygen Reduction

Before considering reactions of oxygen in detail, it is of interest to consider the nature of the oxidation states that are possible between oxygen and water. Bond distances, dissociation energies, and redox potentials are given in Table II.

The data in the second column of Table II give the O—O bond distance from oxidation state zero (molecular oxygen) to -1 (H_2O_2). The O—O distance increases as electrons are successively added to the oxygen molecule. The bond dissociation energies in column 4 correlate well with the bond lengths. For the ground state, $^3\Sigma$, of oxygen the dissociation energy according to Pauling [1] and Pitzer [2] is 116.84 kcal/mole. For the singlet state of oxygen, $^1\Delta$, there are no unpaired electrons, and the bonding may be considered to be analogous to that in ethylene. Since the energy needed to convert triplet to singlet oxygen is 22.4 kcal [3], the dissociation energy of the singlet oxygen must be 116.8 − 22.4 or 94.4 kcal/mole [4]. With the addition of one electron to molecular oxygen its dissociation energy is reduced to 88.8 kcal/mole in the superoxide anion [5]. For the addition of two electrons, in the peroxide anion O_2^{2-}, the dissociation energy of the O—O bond is further reduced to 51.4 kcal/mole [5]. For the dissociation of H_2O_2 to OH in aqueous solution, Yatsimirski [5] has reported a value of 46 kcal/mole, which is about the same as the dissociation of H_2O_2 to H^+(aq), O, and OH^-(aq) [6].

The dissociation energies listed in the right-hand column of Table II show that the OH bonds in oxidized forms, OH, H_2O_2, and HO_2, are always weaker than those in water [7]. It is also seen that H_2O_2 will break more

TABLE II
PROPERTIES OF OXYGEN AND ITS REDUCTION PRODUCTS

	Oxidation state	Bond distance (Å)	Dissociation energies: O—O Bond dissociation reactions	ΔH (kcal/mole)	H—O Bond dissociation reactions	ΔH (kcal/mole)
O_2	0	1.20[a,b]	$O_2 \rightarrow 2\,O$	116.84[a,d]	—	—
O_2^*	0	—	$O_2^* \rightarrow 2\,O$	94.4[c,d]	—	—
			$HO_2 \rightarrow HO + O$	64.9[e]	$HO_2 \rightarrow H + O_2$	47.2 ± 4[g]
O_2^-	$-\frac{1}{2}$	1.27[a,b]	$O_2^- \rightarrow O + O^-$	88.8[e]	—	—
O_2^{2-}	−1	1.35[a,b]	$O_2^{2-} \rightarrow O^- + O^-$	51.4[e]	—	—
			$H_2O_2 \rightarrow HO + OH$	52.3[e]	$H_2O_2 \rightarrow H + HO_2$	89.5 ± 4[g]
			$H_2O_2(aq) \rightarrow H^+(aq) + O(aq) + OH^-(aq)$	46[f]	—	—
			$H_2O_2(aq) \rightarrow 2\,OH(aq)$	46[f]	—	—
O^-	−1	—	—	—	$HO \rightarrow H + O$	102.4 ± 0.5[g]
O^{2-}	−2	—	—	—	$H_2O \rightarrow H + OH$	118.8 ± 0.5[g]

[a] Reference [1].
[b] Reference [2].
[c] Reference [3].
[d] Reference [4].
[e] Reference [5].
[f] Reference [6].
[g] Reference [7].

easily at the O—O bond than at the O—H bond, but that HO_2 will dissociate a hydrogen atom more easily than an oxygen atom. This trend is what one would expect on the basis of the increasing multiplicity of the bonding in the series H_2O_2, HO_2, O_2.

IV. Non-Insertion Reactions Involving the Reduction of Oxygen to Hydrogen Peroxide

Reduction of oxygen to hydrogen peroxide may take place in one or two steps via two- or one-electron reductions. It will be seen that the nature of the reductant, and sometimes the reaction conditions, may determine which of the two mechanistic reaction types apply in each case. In the reactions catalyzed by metal ions, a metal–substrate–oxygen complex is frequently formed, followed by one- or two-electron transfer through the metal ion to oxygen.

Although oxygen carriers by definition are compounds which reversibly combine with oxygen and thus do not effect oxygen reduction, it seems logical to discuss the formation and properties of oxygen carriers at this point. There is now considerable evidence indicating that the chemical state of the oxygen in the oxygen carrier is generally that of a peroxide anion, or in some cases as a superoxide moiety. Although biological oxygen carriers are involved only in oxygen transport and do not effect chemical transformations, it is now clear that certain synthetic oxygen carriers under favorable conditions may oxidize substrates with the release of hydrogen peroxide and may even act as catalysts for the oxidation of the substrates by molecular oxygen.

A. Reactions of Oxygen Carriers

1. *Natural Oxygen Carriers*

a. Composition, Structure, and Properties. Natural oxygen complexes play a significant role in the absorption, transport, and storage of molecular oxygen for respiration in biological systems. These oxygen carriers, including hemoglobin, myoglobin, hemerythrin, and hemocyanin, are generally oxygen complexes in which oxygen is coordinated to a metal ion firmly bound in a multidentate ligand. The oxygen-free metal complex has been named a "prosthetic group." Hemoglobin and myoglobin consist of an iron(II)-protoporphyrin [iron(II)-heme or ferro-heme, see Fig. 1] prosthetic group associated with a protein part called globin [8]. All vertebrates have hemoglobin in their blood cells and myoglobin in muscular tissue. Hemoglobin is also found in the root nodules of leguminous plants. Hemerythrin and hemocyanin [9], which are nonheme oxygen-carrying proteins containing iron and

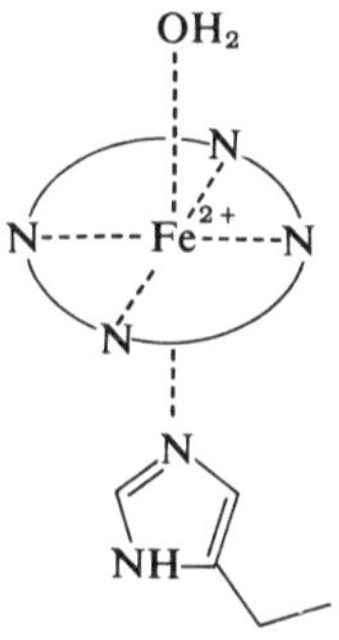

1

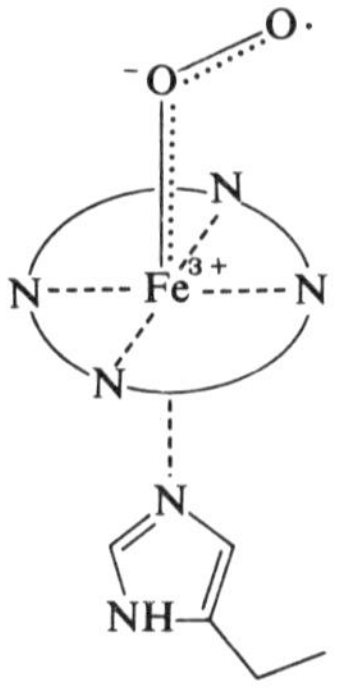

2

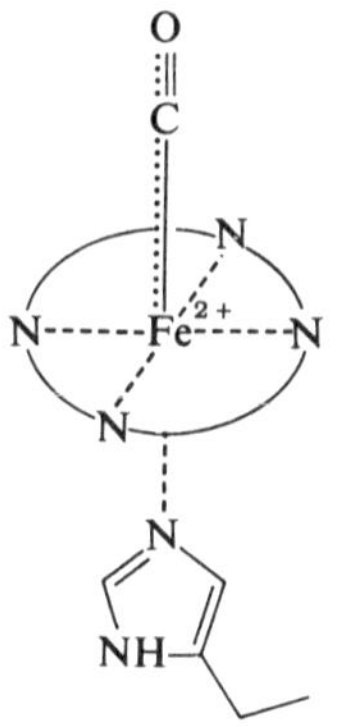

3

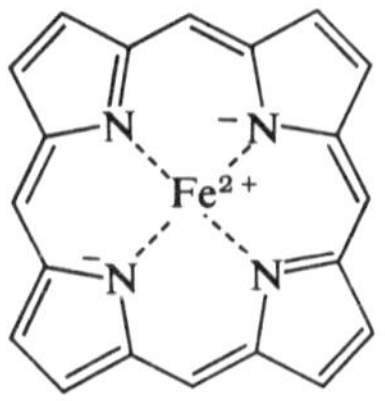

4

Fig. 1. Typical porphyrin complexes in hemoglobin. **1**, Prosthetic group (ferroheme) of hemoglobin (schematic). Ligand is binegative, complex is paramagnetic; **2**, oxygen complex of ferroheme (in oxyhemoglobin) complex is diamagnetic; **3**, carbon monoxide complex of ferroheme (in carbonmonoxyhemoglobin); complex is diamagnetic, **4**, one resonance form of Fe(II)–protoporphyrin ring. (Substituents omitted.)

copper, respectively, in their corresponding prosthetic groups, are found in various types of terrestial and marine invertebrates [9]. A survey of the composition and properties of some natural oxygen carriers is given in Table III.

Myoglobin and hemoglobin (**1**) have one and four heme groups, respectively, per molecule. The X-ray structure of myoglobin shows the globin part to be in the form of α helices with axial repeats of about 5.4 Å. Each segment of the right-handed α helices contains between 7 and 20 amino acid residues [10]. Hemoglobin is an ellipsoid of dimensions 64 × 55 × 50 Å, and the four iron hemes are present as symmetrically related pairs lying at the corners of an irregular tetrahedron with distances of 33.4 and 36.0 Å between the pairs.

TABLE III
PROPERTIES OF NATURAL OXYGEN CARRIERS

	Hemoglobin[a]	Myoglobin[a]	Hemerythrin[b]	Hemocyanin[b]
1 Molecular weight	66,000–68,000	16,900–17,600	Average 66,000	300,000–10 million
2 Metal percent	Fe, 0.33–0.34	Fe, 0.345	Fe, 0.81–0.87	Cu, 0.24–0.26, 0.17–0.18
3 Nature of the prosthetic group	Heme	Heme	Metal-bound polypeptide (nonheme)	Metal-bound polypeptide (nonheme)
4 Metal–oxygen combining ratio	1:1	1:1	2:1	2:1
5 Absorption bands of the oxygenated product (nm)	412–415 540–542 576–578	Absorbs at slightly longer wavelengths than oxyhemoglobin	335 365 (shoulder) 500	278 345 550–560
6 Heat of oxygenation (kcal/mole)	−9 to −14	−8 to −14	−12 to −20	−9 to −16

[a] Reference [8].
[b] Reference [9].

The polypeptide chain appears as four separate units that are identical in the symmetrical pairs. The ferrous ions in myoglobin and hemoglobin are firmly coordinated to the four planar pyrrole nitrogens of the porphyrin ring, and in addition each ferrous ion has a fifth coordination to a nitrogen atom of the imidazole group of a histidine residue [11] of the polypeptide globin chain. The sixth coordination position of the iron atom is coordinated to the oxygen atom or a water molecule. The iron–oxygen distance is close to 2.1 Å and that between iron and the bound imidazole nitrogen atom is 1.9 Å. Hemoglobin and myoglobin are paramagnetic, with the magnetic moment corresponding to four unpaired electrons per heme [10].

The iron content of hemerythrin varies within the range 0.81–0.87% depending on the source of the pigment, which differs from phylum to phylum and even genus to genus [12]. About 0.6 mole fraction of the iron atoms in hemerythrin are in the iron(II) state bound to cysteine side chains of the polypeptide. One oxygen molecule is bound per two iron atoms. The remaining 0.4 mole of iron atoms remains in the trivalent state and appears to be either nonessential or acts in some unknown way. Hemerythrin is paramagnetic.

On the basis of their copper content hemocyanins may be divided into two groups. Those obtained from mollusks contain 0.24–0.26% copper, and those obtained from arthropods contain 0.17–0.18% copper (Table III). The sulfur content of hemocyanin is fairly low (0.7–1.2%). The site of copper binding in the polypeptide chain is still controversial, though the bulk of evidence available to date favors binding to the histidine residues. The copper in deoxygenated hemocyanin is monovalent. Despite the attempts of several investigators, the valence state of copper in oxyhemocyanin [13] is still obscure. Since two copper atoms are involved in binding one oxygen molecule, it is believed that a copper atom in each valence state is involved in the binding of each oxygen molecule. Thus the oxygen molecule could take up one electron to form a superoxide ion, $\dot{O}_2^-$, coordinated to a copper(II) atom.

The nature of the metal–oxygen bond (**2**) in hemoglobin has recently been elucidated by Griffith and Weiss [10,14]. Weiss [10] has suggested an oxidation state of 3+ for iron in oxyhemoglobin, based on (1) magnetic properties (diamagnetic), (2) absorption spectra, and (3) acid dissociation constant of the oxygenated complex. According to this view, the ferric ion is stabilized by complexing with oxygen in the superoxide anionic form. This model is used

$$Fe^{2+} + O_2 \rightleftharpoons Fe^{3+} + \cdot O_2^- \qquad (1)$$

as the basis for explaining the diamagnetism of oxyhemoglobin. In the strong field of the ligand, the ferric ion has one unpaired electron which couples with the unpaired electron of $\dot{O}_2^-$ to give the observed diamagnetism. The

$$\text{globin}\ Fe^{2+} + O_2 \rightleftharpoons \text{globin}\ Fe^{3+}\text{---}\cdot O_2^- \qquad (2)$$

formation of carbomonoxyhemoglobin has been shown by Weiss [10] to take place by the same mechanism as does the formation of oxyhemoglobin.

b. Reactivities of Hemoglobin and Myoglobin. Ferroheme (the protein-free prosthetic group, **1**) and hemoglobin show wide differences in their reactivities toward molecular oxygen, the oxygen complex of the former being irreversibly oxidized to the iron(III) form at a rate 10^8 faster than that of the latter [15]. Apparently the globin interacts with the oxygen-binding site so as to stabilize oxyhemoglobin, in which the metal–oxygen coordination may be considered to involve the group $Fe^{3+}\text{---}\dot{O}_2^-$. According to Wang [8] heme residues in hemoglobin are surrounded by hydrophobic groups of the globin molecule, reducing the effective dielectric constant of the oxygen coordination site and hence retarding oxidation of oxyhemoglobin. This hypothesis is supported by the fact that a benzene or ether solution of ferrohemochrome containing an excess of pyridine is resistant toward oxidation to the ferric form.

The four heme residues (Hb) in hemoglobin combine reversibly with molecular oxygen to form successively four complexes with increasing degrees of oxygenation: $Hb_4(O_2)$, $Hb_4(O_2)_2$, $Hb_4(O_2)_3$, and $Hb_4(O_2)_4$ [8]. There is interaction between the heme groups which results in decreasing the binding energy as the four groups are successively oxygenated. The exact nature of this interaction has not been satisfactorily explained.

Hemoglobin also combines with carbon monoxide and nitric oxide to form carbomonoxyhemoglobin (**3**) and nitrosohemoglobin, respectively. The equilibrium constant for the combination of nitric oxide with hemoglobin is 15,200 times greater than that of carbon monoxide. The interaction of the four heme residues with carbon monoxide or nitric oxide decreases successively as in the case of the reaction with oxygen, because of heme–heme interaction. The equilibria of hemoglobin with oxygen and with carbon monoxide are strongly pH-dependent, a characteristic which is usually referred to as the Bohr effect [8].

As in hemoglobin, the heme residue in myoglobin may be considered as a ligand reversibly bound to globin [11]. Since myoglobin contains only one heme per molecule, its reaction with a ligand (O_2, CO, NO, CN^-, RNC, RNO) may be written as

$$\mathrm{Mb + L \underset{k'}{\overset{k}{\rightleftharpoons}} MbL} \qquad (3)$$

For the oxygenation reaction, the values of k and k' at 20°C are, respectively, $1.4 \times 10^7\ M^{-1}\ \mathrm{sec}^{-1}$ and $11\ \mathrm{sec}^{-1}$. Though there is a large difference in the amino acid composition of the myoglobins derived from different animal sources, the oxygen affinity is very nearly the same for all of them. Myoglobin combines with oxygen at a rate faster than hemoglobin, although oxygenated myoglobin is much less stable than oxyhemoglobin. The oxygen affinity of myoglobin, unlike that of hemoglobin, is independent of pH, salt concentration, and concentration of the pigment. For further details on the structure and properties of natural oxygen carriers, the reader is referred to recent reviews [10–12,16–20].

c. Oxidation of Substrates by Natural Oxygen Complexes. The oxidation of ascorbic acid by molecular oxygen in the presence of oxyhemoglobin was reported by Lemberg *et al.* [21]. The oxidation of ascorbic acid to dehydroascorbic acid takes place through the intermediate formation of oxyhemoglobin. Hemoglobin itself is also oxidized during the reaction with the formation of methemoglobin. Methemoglobin then interacts irreversibly with globin to form a product named choleglobin. Thus hemoglobin is not regenerated in the reaction. The mechanism proposed is as follows:

$$\mathrm{Hb + O_2 \rightleftharpoons HbO_2} \qquad (4)$$

$$\mathrm{HbO_2 + H_2A \longrightarrow HbH_2O_2 + A} \qquad (5)$$

$$\mathrm{HbH_2O_2 + Hb \longrightarrow 2\,HbOH} \qquad (6)$$

$$\mathrm{HbOH + globin \longrightarrow Hb^*(HbOH \cdot globin)} \qquad (7)$$

where HB indicates hemoglobin; HbO_2, oxyhemoglobin; HbOH, methemoglobin; and Hb*, choleglobin.

In the reaction of oxyhemoglobin with ascorbic acid, two hydrogen atoms are apparently transferred from the latter and a product, HbH_2O_2, is believed to be formed, as indicated by Eq. (5). Direct evidence for hydrogen atom transfer in this reaction, however, is not yet available. The stoichiometry of the reaction with respect to oxygen and ascorbic acid indicates that ten molecules of ascorbic acid are oxidized per mole of hemoglobin. The complete reaction mechanism therefore must be quite complex.

2. *Synthetic Dioxygen Carriers*

Synthetic dioxygen carriers have been of considerable interest not only as models of natural dioxygen carriers but also as a possible means for separating pure oxygen from the air, and for enrichment of ^{18}O [22]. Synthetic dioxygen carriers have been prepared from copper(II), cobalt(II), iron(II), nickel(0), platinum(0), palladium(0), iridium(I), rhenium(I), ruthenium(II), osmium(0), and manganese(II), combined with suitable complexing agents. All metal ions whose complexes combine with molecular oxygen are those that exist in more than one oxidation state. Apparently, the oxidation potential of the metal ion is suitably modified by the chelating agents so that combination with molecular oxygen is effected without appreciable irreversible oxidation of the metal ion. Metal complexes and chelates of cobalt(II) are the best known of the synthetic dioxygen carriers and many examples have been studied, such as Co(II)–bis(salicylaldehyde imine), Co (II)–glycylglycine, Co(II)–histidine, and a series of Co(II)–polyamine complexes. The oxygenated forms of most synthetic dioxygen carriers seem to be polynuclear complexes in which two atoms of the metal ion are coordinated to a molecule of oxygen.

In recent years a number of mononuclear dioxygen complexes have been synthesized and studied. Information now available on these compounds is discussed in the following sections.

a. Binuclear Dioxygen Complexes. A compound that has recently been thoroughly investigated is the decaammine-μ-peroxodicobalt(III) cation, which is formed directly from cobalt(II), ammonia, and oxygen and has an ionic charge of 4+. This complex may be oxidized to a complex with a charge of 5+. The species with 4+ charge is brown and diamagnetic, while that with 5+ charge is green and paramagnetic.

Vleck [23] first suggested perpendicular bonding of dioxygen to the *d* orbitals of both the cobalt ions of the complex, similar to the bonding found in olefin–metal complexes. X-Ray analysis of the nitrate salt of the oxidized (paramagnetic) form of the decaammine–cobalt oxygen carrier by Brosett and Vannerberg [24,25] seemed to support the theoretical conclusion of Vleck. Recently, however, Schaefer and Marsh [26] determined the X-ray structure

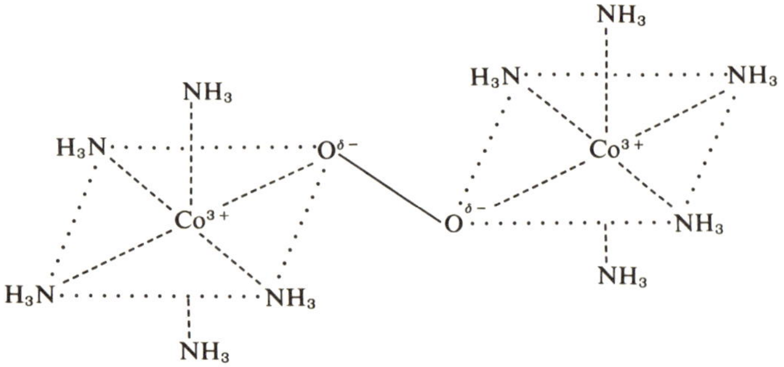

5

Fig. 2. Suggested bonding in $[(NH_3)_5Co—O_2—Co(NH_3)_5]^{5+}$; d_{O-O} = 1.31 Å. From Schaefer and Marsh [26].

of a crystalline sample of the paramagnetic form, $[(NH_3)_5Co—O_2—Co(NH_3)_5]$-$(SO_4)_2(HSO_4)\cdot 3\ H_2O$ (**5**, Fig. 2), and concluded that the coordination of two cobalt ions in the paramagnetic cation is nearly exactly octahedral, with an average deviation of the ligand–cobalt–ligand angle from 90° of less than 2°. The axis of the bridging dioxygen group was found not to be perpendicular to the Co—Co axis and not bonded to the cobalt ions by $d\pi$ bonds. As indicated in formula **5** each oxygen atom of the bridging dioxygen, designated as a peroxo group, is σ-bonded to each of the cobalt atoms, resulting in a staggered arrangement of the Co—O—O—Co bonds. The X-ray analysis also indicates that the horizontal planes of the coordination spheres are nearly coplanar as illustrated by formula **5**. The O—O distance in the paramagnetic complex **5** was found to be 1.31 Å, considerably shorter than that reported by Brosett and Vannerberg (1.45 Å), for the paramagnetic $[(NH_3)_5CO—O_2—Co(NH_3)_5](NO_3)_5$ (**6**) and slightly longer than the O—O distance in the superoxide anion (1.28 Å). On the basis of the O—O bond distance in $[(NH_3)_5Co—O_2—Co(NH_3)_5]^{5+}$, it is concluded [26] that the dioxygen in this complex is a superoxide anion, $\dot{O}_2^-$. Electron spin resonance (ESR) spectra [27] of the decaammine-μ-superoxodicobalt(III) cation (**5**) indicate that the unpaired electron interacts equally with both cobalt nuclei and resides primarily on the oxygen molecule. On the basis of this structure, one would expect the bridging oxygen atoms in the diamagnetic complex of 4+ charge to correspond closely to a peroxo group, O_2^{2-}.

Although examples of the formation of dioxygen carriers in solution have been known for some time [28] only during the past few years have a sufficient number of complexes been investigated to indicate the generality of the oxygenation–coordination reaction with synthetic chelate compounds. Fallab [29,30] and Miller and Wilkins [31,32] have described the composition and kinetics of formation of some oxygenated complexes of cobalt(II)–polyamine

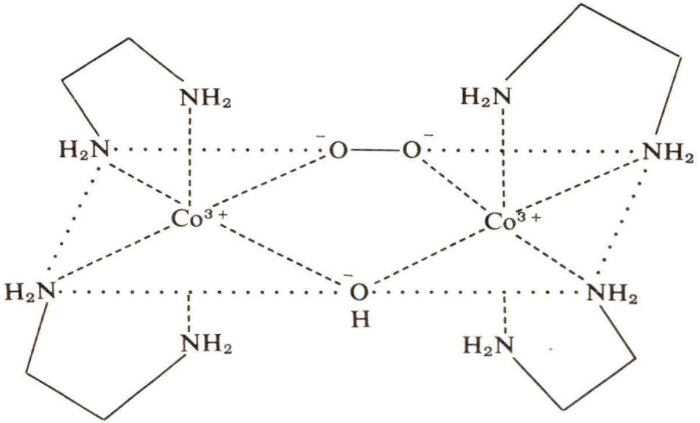

6

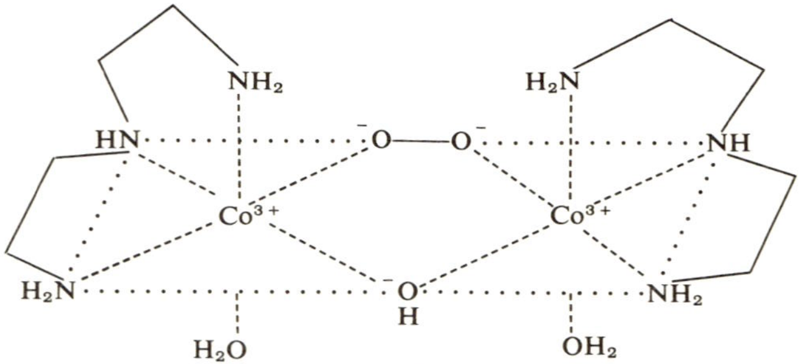

7

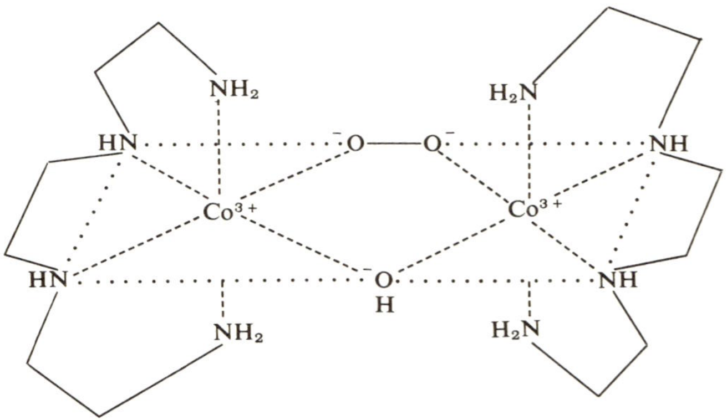

8

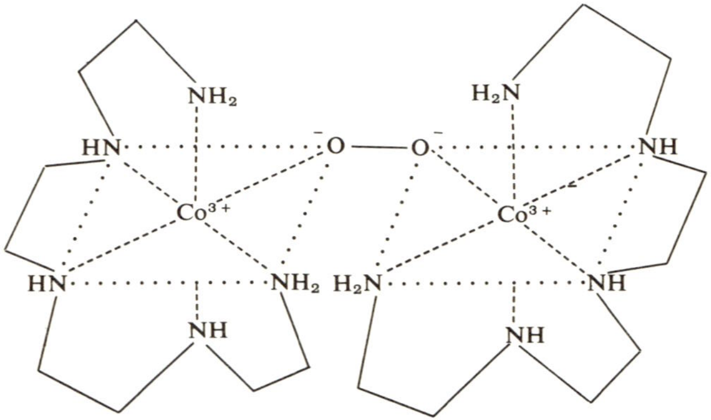

9

Fig. 3. Some binuclear complexes formed from cobalt(II) and polyamines. **6**, Tetrakis(ethylenediamine)-μ-hydroxo-μ-peroxo-dicobalt(III) ion, $Co_2L_4O_2OH^{3+}$ (ethylenediamine = L); **7**, Bis(diethylenetriamine)-μ-hydroxo-μ-peroxo-dicobalt(III) ion, $Co_2L_2O_2OH^{3+}$ (diethylenetriamine = L); **8**, Bis(triethylenetetramine)-μ-hydroxo-μ-peroxo-dicobalt(III) ion, $Co_2L_2O_2OH^{3+}$ (triethylenetetramine = L); **9**, Bis(tetraethylenepentamine)-μ-peroxo-dicobalt(III) ion, $Co_2L_2O_2^{4+}$ (tetraethylenepentamine = L).

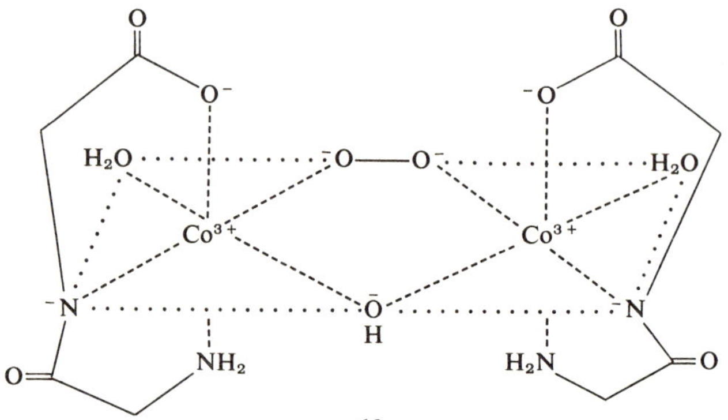

10

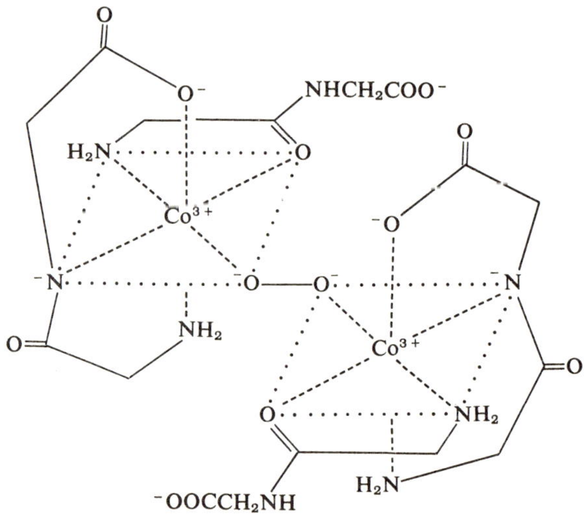

11

Fig. 4. Suggested ligand arrangements in binuclear oxygen complexes of cobalt(II). **10**, Bis(glycylglycinato)-μ-hydroxo-μ-peroxodicobalt(III) ion, $Co_2(H_{-1}L)_2O_2(OH)^-$ (glycylglycine = HL); **11**, Tetrakis(glycylglycinato)-μ-peroxodicobalt(III) ion, $Co_2\cdot(H_{-1}L)_2L_2O_2^{2-}$; **12**, Tetrakis(histidinato)-μ-peroxodicobalt(II), $Co_2O_2L_4$; **13**, Bis-(diglycylethylenediaminetetraacetato)-μ-peroxodicobalt(III) ion, $Co_2O_2L_2^{8-}$.

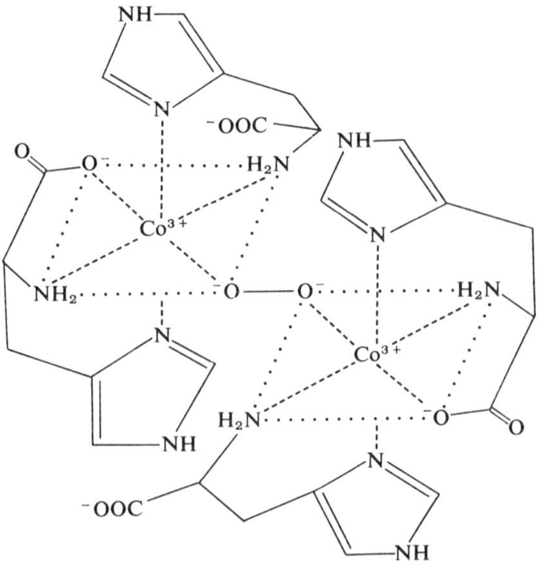

12

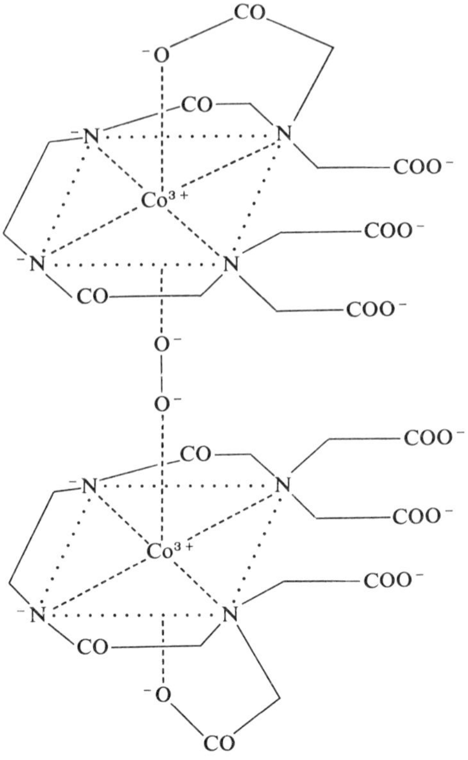

13

chelates at high pH. An excellent review by Sykes and Weil [33] covers the recent literature on μ-peroxo and μ-superoxo dicobalt complexes.

The equilibria between oxygen and cobalt(II) complexes, described by Nakon and Martell [34–36,90], have revealed the acid–base stoichiometry for the dioxygen complexes in solution, and resulted in the determination of the relatively high stability constants for their formation from gaseous oxygen, the metal ion, and the ligand. The stoichiometries of oxygenation are indicated in Figs. 3 and 4 for complexes **6–13**. Potentiometric data suggested the existence of the hydroxo bridging group in the binuclear dioxygen complexes of ethylenediamine (**6**), diethylenetriamine (**7**), triethylenetetramine (**8**), and the 1:1 glycylglycine–cobalt chelate (**10**). For glycylglycine and histidine, a 2:1 or higher molar ratio of ligand to metal ion gave binuclear dioxygen complexes **11** and **12**, respectively, without hydroxo bridges, presumably because all of the available coordination positions of the cobalt atoms are occupied by the ligand donor groups and the peroxo bridge. A binuclear μ-peroxo cobalt(III) complex, **13**, $LCoO_2CoL$ (L = *N,N,N‴,N‴*-diglycylethylenediaminetetraacetic acid) without the hydroxo bridges has been recently reported [36]. Saturation of the coordination positions on cobalt with ethylenediamine in a 1:3 cobalt–ethylenediamine complex gave [37] a μ-peroxodicobalt(III) complex, $[(en)_3CoO_2Co(en)_3](ClO_4)_4$, without hydroxo bridges. This interpretation was supported by the fact that the 1:1 tetraethylenepentamine–cobalt complex **9** combines with oxygen to give a binuclear dioxygen-bridged complex without a hydroxo bridge, whereas hydroxo bridges are formed in all cobalt–dioxygen complexes containing lower polyamines. In the latter the binuclear dioxygen complexes would have cobalt(III) atoms with one or more coordination sites occupied by water molecules, which would be very reactive toward hydroxo group coordination since such groups are formed by dissociation of a proton from a coordinated water molecule.

Powell and Nancollas [38] have measured the enthalpy and entropy changes for reaction (8)

$$2\,CoL_2{}^{n+}(aq) + O_2(aq) + H_2O \rightleftharpoons [(CoL_2)_2O_2(OH)]^{(2n-1)+} + H_3O^+ \qquad (8)$$

for L = histidinate and ethylenediamine, $\Delta H = -30.1 \pm 1.3$ and -20.4 ± 0.6 kcal/mole, and $\Delta S = -70 \pm 5$ and -49 ± 4 eu, respectively.

The cobalt atom in all of the oxygen complexes shown (formulas **6–13**) may be considered to be in a +3 oxidation state, with the coordination oxygen reduced to a binegative peroxide anion. Evidence pointing to this type of electronic structure include diamagnetism of the complex, the unusually high stability of the dioxygen complex, and the fact that coordination with dioxygen greatly increases the interaction of the metal ion with the ligand, as indicated by potentiometric titration curves and the calculated stability constants.

b. Binuclear Superoxide Complexes. μ-Superoxo complexes of cobalt(III) are comparatively less stable in solution than the μ-peroxo complexes. The structure and reactions of a number of μ-superoxo complexes of cobalt(III) have been described by Sykes and Weil [33]. Woods and Weil [39] have reported a μ-superoxobis[bis(L-histidinato)cobalt(III)] chloride, which was obtained by the oxidation of the corresponding μ-peroxo complex with nitric acid. The complex has been characterized in solution by ESR, NMR, and electronic spectra. Reduction of the superoxide complex with hydroquinone in ethanol resulted in quantitative release of the bridging superoxide moiety as molecular oxygen.

μ-Peroxo and μ-superoxo complexes of biological interest are the bis(dimethylglyoximato)cobalt(II) complex [cobaloxime(II)] and its various derivatives obtained by modification of the dimethylglyoxime ring system. Cobaloxime(II) reacts [40] with molecular oxygen in the presence of a base such as pyridine to form intensely colored μ-peroxocobaloximes, Py—$Co(D_2H_2)O_2Co(D_2H_2)$—Py (D = dianion of dimethylglyoxime), which on subsequent oxidation gave μ-superoxocobaloxime, **14**, [Py—$Co(D_2H_2)$—O_2—$Co(D_2H_2)$—$Py]^+$. An ESR study of **14** revealed that the unpaired electron is symmetrically delocalized over the Co—O—O—Co system. A study of cobaloxime(II) and its analogs is of special interest from a biochemical viewpoint since these compounds resemble vitamin B_{12} (cobalamine) in their oxygen-carrying properties. The μ-superoxocobaloxime **14** decomposes with pyridine into the pyridine derivative of "cobaloxime(III) anhydrobase," $Co(D_2H)\cdot 2\,Py$ (**16**), and the mononuclear peroxo complex radical Py—$Co(D_2H_2)$—$\dot{O}_2$ (**15**) in accordance with reaction (9). The ESR spectrum of **15**

$$[\text{Py—Co}(D_2H_2)\text{—}O_2\text{—Co}(D_2H_2)\text{—Py}]^+ \underset{\mathbf{14}}{} + 2\,\text{Py} \longrightarrow$$

$$\underset{\mathbf{15}}{\text{Py—Co}(D_2H_2)O_2\cdot} + \underset{\mathbf{16}}{\text{Co}(D_2H)\cdot 2\,\text{Py}} + \text{PyH}^+ \qquad (9)$$

resembles that of peroxocobalamine, indicating a similar electronic environment of the cobalt–oxygen system in both compounds. The unpaired electron density in **15** resides [40] in a molecular orbital which has 10–20% metal character, thus suggesting that the odd electron in **15** remains primarily on oxygen. The structure of **15** is indicated on page 98.

It has been recently shown by Calligaris *et al.* [41,42] that the presence of σ donors such as pyridine in the coordination sphere of cobalt(II) complexes such as *N,N'*-ethylenebis(salicylideneiminato)cobalt(II), or Co(salen), improves considerably its reversible oxygen-binding properties. Basic σ donors

TABLE IV

PROPERTIES OF MONONUCLEAR DIOXYGEN COMPLEXES

	Complex[a]	Color	Stability	Solvent	ν_{M-O_2} Group stretching frequency (cm^{-1})	References
1	$IrX(CO)(O_2)(PPh_3)_2$ $X = Cl^-, Br^-, I^-$	Brown	Photosensitive stable	Benzene	858–862	43–45
2	$[Ir(diphos)_2(O_2)]PF_6$	Brown	Photosensitive stable	Benzene	845	46,47
3	$IrCl(CO)(AsPh_3)_2(O_2)$	Orange	Photosensitive stable	Benzene	860	48
4	$IrI(CO)(AsPh_3)_2(O_2)$	Pink	Photosensitive stable	Benzene	850	48
5	$IrCl(CO)(PBu^tMe_2)(O_2)$	Brown	Photosensitive stable	Benzene	824	49
6	$IrBr(CO)(PBu^tMe_2)(O_2)$	Brown	Photosensitive stable	Benzene	848	49
7	$IrCl(O_2)(PPh_3)_3$	Salmon	Photosensitive stable	Benzene	852	50–52
8	$IrCl(CO)(PPh_2Et)_2(O_2)$	Brown	Photosensitive stable	Benzene	856	53
9	$[Ir(CO)(O_2)(PPh_2Me)_3]^+BF_4^-$	Pale brown	Photosensitive stable	Acetonitrile	840	54
10	$[Ir(CO)(O_2)(PPh_2Et)_3]^+BF_4^-$	Pale brown	Photosensitive stable	Acetonitrile	845	54
11	$[Ir(O_2)(2{=}phos)_2]^+BF_4^-$	Brown	Photosensitive stable	Chlorobenzene	840	55
12	$[Rh(diphos)_2(O_2)]PF_6$	Brown	Photosensitive stable	Methanol	880	50
13	$[Rh(2{=}phos)_2(O_2)]BF_4$	Brown	Photosensitive stable	Chlorobenzene	880	55
14	$[Rh(dmpe)_2(O_2)]Cl$	Brown	Photosensitive stable	Benzene	845	56
15	$RhCl(O_2)(PPh_3)_2$	Brown	Photosensitive stable	Benzene	900	57
16	$[Rh(PMe_2Ph)_4(O_2)]PF_6$	Brown	Photosensitive stable	Dichloromethane	841, 870	58
17	$[Rh(AsMe_2Ph)_4(O_2)]PF_6$	Brown	Photosensitive stable	Dichloromethane	862, 867	58
18	$RhCl(O_2)(PPh_3)_2(Bu^tNC)$	Brown	Very stable	Benzene	892	59
19	$RhBr(O_2)(PPh_3)_2(Bu^tNC)$	Brown	Very stable	Benzene	893	59

20	$Rh(I)(O_2)(PPh_3)_2(Bu^tNC)$	Brown	Very stable	Benzene	893	59
21	$RhCl(O_2)(AsPh_3)_2(Bu^tNC)$	Brown	Very stable	Benzene	884	59
22	$[RhCl(O_2)(PPh_3)_2]_2$	Brownish green	Very stable	Dichloromethane	845	60
23	$[RhL(O_2)]Cl$	Light brown	Very stable	Benzene	862	61
24	$[Co(2{=}phos)_2(O_2)]BF_4$	—	Stable photosensitive	Chloroform	909	62
25	$RuCl_2(O_2)(AsPh_3)_3$	Dark brown	Very stable	Benzene	880	63
26	$RuCl(SnCl_3)(O_2)(AsPh_3)_3$	Yellowish brown	Very stable	Dimethylformamide	805	64
27	$RuHCl(O_2)(AsMePh_2)_3$	Brown	Very stable	Dimethylformamide	855	64
28	$RuCl_2(PMePh_2)_2(AsPh_3)(O_2)$	Brown	Very stable	Benzene	835	64
29	$RuCl_2(PPr_2{}^nPh)_2(AsPh_3)(O_2)$	Brown	Very stable	Benzene	845	64
30	$RuCl_2(PPh_3)_2(AsPh_3)(O_2)$	Brown	Very stable	Benzene	850	64
31	$Ru(NO)(O_2)(NCS)(PPh_3)_2$	Brown	Very stable	Benzene	860	65
32	$Ru(O_2)(CO)_2(PPh_3)_2$	Brown	Very stable	Benzene	849	66
33	$Os(O_2)(CO)_2(PPh_3)_2$	Brown	Very stable	Benzene	820	66
34	$Pt(PPh_3)_2O_2$	Yellow	Stable	Benzene	818	67–69
35	$Pd(PPh_3)_2O_2$	Yellow	Stable	Benzene	875	67–69
36	$Ni(PPh_3)_2O_2$	Green	Stable below −35°C	Benzene	875	68
37	$Pd(Bu^tNC)_2(O_2)$	Colorless	Stable in the solid state, decomposes in solution	Tetrahydrofuran or ether	893	70
38	$Ni(Bu^tNC)_2(O_2)$	Pale green	Stable in the solid state, decomposes in solution	Tetrahydrofuran or ether	898	70
39	$Ni(cyclo\text{-}C_6H_{11}NC)_2(O_2)$	Pale green	Stable in the solid state, decomposes in solution	Tetrahydrofuran or ether	904	70

[a] L = $P(CH_2CH_2CH_2PPh_2)_3$; 2=phos = $Ph_2PCH{=}CHPPh_2$; diphos = $Ph_2PCH_2CH_2PPh_2$; dmpe = $(CH_3)_2PCH_2CH_2P(CH_3)_2$.

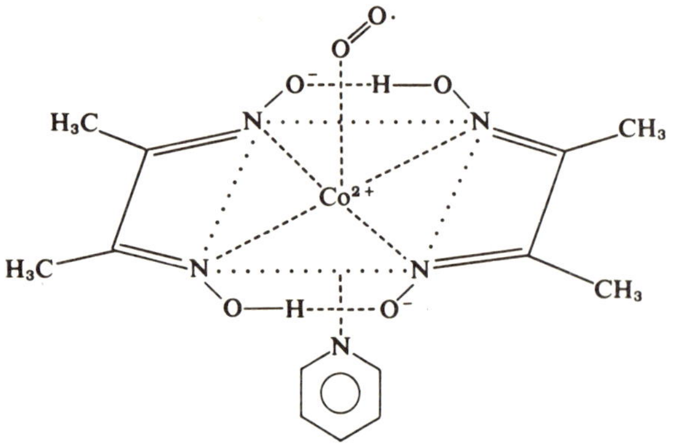

15

coordinated to the metal atom affect its charge density in such a way that reversible oxygenation is possible. Thus cobaloxime(II) [40] and Co(salen) [41,42] bind oxygen only in the presence of pyridine.

c. Mononuclear Metal–Dioxygen Complexes. Table IV presents data on 1:1 metal–dioxygen complexes that have been described in the literature, including complexes of iridium(I) [43–55], rhodium(I) [56–61], cobalt(I) [62], ruthenium(II) [63,64], ruthenium(0) [65,66], osmium(0) [66], platinum(0) [67–69], palladium(0) [67–70], and nickel(0) [68–70]. Only those complexes that have been well characterized are listed. In all cases, the dioxygen complexes have been prepared by passing molecular oxygen through a solution of the parent compound in an appropriate organic solvent. Coordination of the metal with aralkyl arsines or phosphines results in the formation of more stable oxygen complexes with respect to the dissociation of the metal–oxygen bond than do aryl phosphines or arsines. Tertiary phosphines or arsines give more stable dioxygen complexes than mixed ligand carbonyl arsine or phosphine complexes. Thus it seems that low π acidity generally favors the formation of stable metal–dioxygen complexes.

The structures of oxygen complexes determined by X-ray diffraction are illustrated in Figs. 5–10 for $IrCl(CO)(O_2)(PPh_3)_2$ [44], $[M(diphos)_2(O_2)]BPh_4$ [47] [M = iridium(I), rhodium(I); diphos = $Ph_2PCH_2CH_2PPh_2$], $IrCl(CO)(O_2)(PPh_2Et)_2$ [53], $[RhCl(O_2)(PPh_3)_2]_2$ [60], $[Co(Ph_2PCH{=}CHPPh_2)_2(O_2)]ClO_4$ [62], and $Pt(PPh_3)_2(O_2)$ [71].

Vaska reported the synthesis of the first mononuclear iridium(I)–dioxygen complex, having the formula $IrCl(CO)(PPh_3)_2(O_2)$, **17**, by the reaction of bis(triphenylphosphine)chlorocarbonyliridium(I) with oxygen in benzene solution. The brown diamagnetic complex has a dipole moment of 5–9 D in benzene. The structure of this iridium–dioxygen complex, determined by Laplaca and Ibers [44] by X-ray analysis, indicates a coplanar arrangement

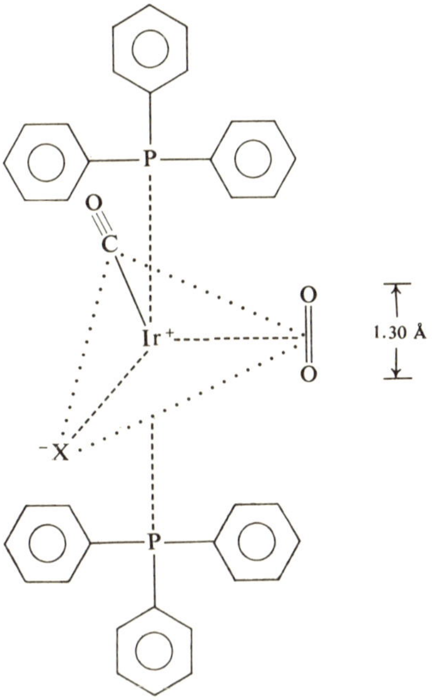

17 X = Cl⁻

18 X = I⁻

Fig. 5. Arrangement of ligands in $Ir(PPh_3)_2O_2XCO$. After Laplaca and Ibers [44].

of chloride, carbon monoxide, iridium, and dioxygen, with the triphenylphosphine donors above and below the plane, as indicated by **17**, Fig. 5. This structure may be described as a trigonal bipyramid if the oxygen is considered as a single ligand, or as a distorted octahedron if the two oxygen atoms are considered to be separately bonded to iridium. The two oxygen atoms are equidistant from iridium, with an Ir—O distance of 2.06 ± 0.02 Å, and an O—O bond distance of 1.30 Å, which is characteristic of a superoxide radical, $\dot{O}_2^-$. The observed diamagnetism of the complex may be due to spin coupling of the unpaired electron in iridium(II) with that of $\dot{O}_2^-$.

An alternate arrangement of bonds would be σ bonding of iridium(I) by donation of the π electrons of the dioxygen to the metal ion, with back donation of electrons from the appropriate *d* orbitals of the metal to the antibonding π orbitals of the oxygen molecule. The removal of electrons from the bonding π orbitals would lengthen the O═O bond of molecular oxygen to the value observed (1.30 Å) [44]. These concepts of dioxygen binding are similar in that they end up with the same result, although they have quite different starting points.

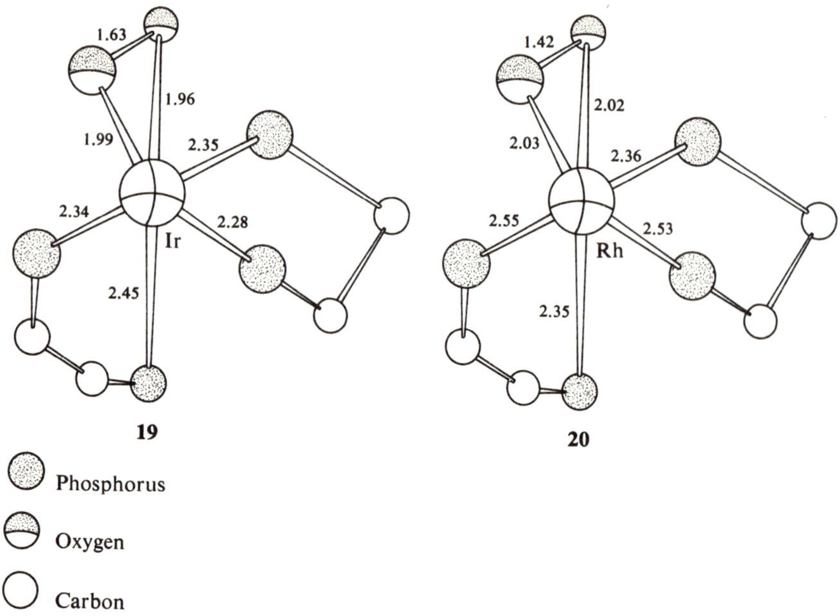

Fig. 6. Structures of **19**, $[Ir(O_2)(diphos)_2]^+$, and **20**, $[Rh(O_2)(diphos)_2]^+$: Diphos = $Ph_2PCH_2CH_2PPh_2$. After McGinnety *et al.* [47].

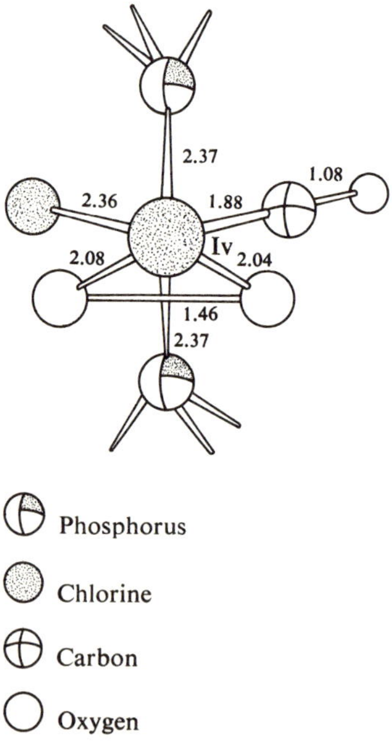

Fig. 7. Structure of $IrCOCl(O_2)(PPh_2Et)_2$. After Weininger *et al.* [53].

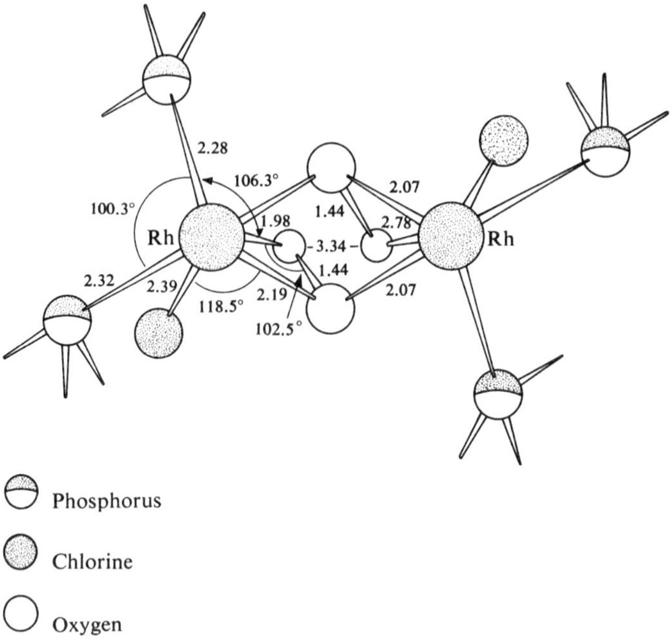

Fig. 8. Structure of $[Rh(PPh_3)_2Cl(O_2)]_2$. After Bennett and Donaldson [60].

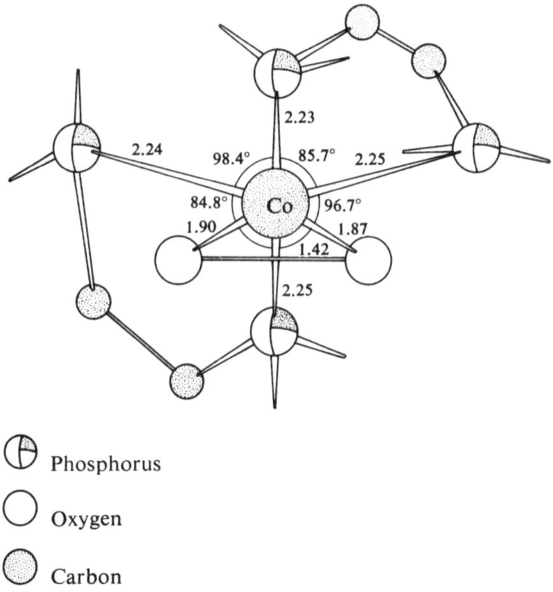

Fig. 9. Structure of $[Co(O_2)(Ph_2PCH{=}CHPPh_2)_2]^+$. After Terry *et al.* [62].

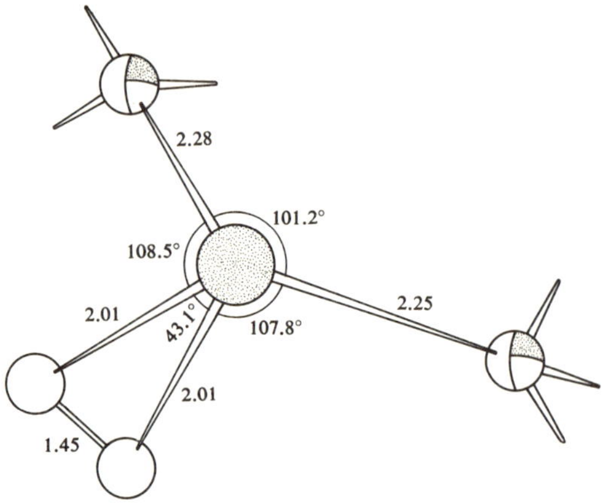

Fig. 10. Structure of $Pt(O_2)(PPh_3)_2$. After Kashiwazi *et al.* [71].

Molecular oxygen is displaced from complex **17** on boiling the benzene solution. The corresponding iodide dioxygen complex **18** requires 150°C for the displacement of molecular oxygen in dimethylformamide solution [55].

Complexes of the composition $[M(diphos)_2(O_2)]PF_6$ [where M = rhodium(I) or iridium(I)] have been obtained [46,47] as brown photosensitive crystals by passing molecular oxygen through a benzene solution of $[M(diphos)_2]PF_6$. The X-ray structure of the complexes is illustrated in Fig. 6. In the iridium(I) complex, **19**, for which addition of dioxygen is irreversible, the strong metal–oxygen bond is accompanied by a weaker O—O bond, the O—O distance being 1.625 Å, considerably longer than a normal O—O single bond (1.49 Å). In the case of the analogous rhodium complex, **20**, however, the dioxygen is bound reversibly. The weaker metal–oxygen bond in this complex results in a stronger O—O bond having an O—O distance of 1.418 Å. The X-ray structure of the complex $[Rh(diphos)_2]ClO_4$ has been reported by Hall *et al.* [72]. A characteristic feature of the diphos chelate rings has been shown to be staggered conformation about the CH_2—CH_2 bond, with the barrier to torsion about M—P and C—C bonds lower than for conventional CH_2—CH_2 bonds. The flexibility of the chelate rings in $[M(diphos)]^+$ complexes thus seems to facilitate [72] the formation of dioxygen complexes. The replacement of the —CH_2—CH_2— group in diphos by an unsaturated bridge (CH═CH) does not appear to alter the flexibility of the chelate rings in *cis*-Ph_2PCH═$CHPPh_2$ (diphos-2). Thus stable dioxygen complexes of the general formula $[M(diphos\text{-}2)_2(O_2)]BF_4$ [M = iridium(I), rhodium(I), and cobalt(I)] have been reported by Vaska [43] and Terry *et al.* [62]. The ligand bis[1,2-(dimethylphosphino)ethane] (dmpe) also appears to behave in a manner similar to diphos with respect to the flexibility of the —CH_2—CH_2—

group. Thus $[Rh(dmpe)_2(O_2)]Cl$ is a stable complex in which molecular oxygen is bound in a reversible manner [56]. Quadridentate phosphines with long methylene chains, such as $P(CH_2CH_2CH_2PPh_2)_3$, seem to form a flexible chelate ring system with rhodium(I), resulting in the formation of a very stable dioxygen complex of the composition $\{Rh[P(CH_2CH_2CH_2PPh_2)_3](O_2)\}Cl$ [64].

Other steric factors also appear to be very important in the formation of metal–dioxygen complexes. In some cases overcrowding of the ligands around the metal ion reduces the reactivity of the complexes with regard to the formation of dioxygen complexes. Clark *et al.* [54] have shown by X-ray structural evidence that steric factors play a dominant role in the reactivity of the complex $[Ir(PPh_2Me)_4]BF_4$ towards combination with dioxygen.

Taqui Khan *et al.* [63] prepared a Ru(II)–dioxygen complex having the composition $RuCl_2(AsPh_3)_3(O_2)$, **21**, by the interaction of a benzene solution of dichlorobis(triphenylarsine)ruthenium(II) with molecular oxygen at room temperature and 1 atm. The infrared spectrum of the complex exhibits a band at 880 cm^{-1}, assigned to O—O stretch of the triangular ruthenium–dioxygen group. The NMR spectrum of the complex has a doublet centered at δ 7.4 ppm, which indicates the presence of two mutually *trans*- and one *cis*-triphenylarsine ligands. The dioxygen complex is paramagnetic with $\mu_{eff} = 2.9$ Bohr magnetons, corresponding to two unpaired electrons. From the ESR studies of this complex it seems probable that the unpaired electrons are mostly centered on a molecular orbital which is predominantly $\pi^*(O_2)$. Complex **21** thus seems to be the first example of a well-characterized paramagnetic 1:1 dioxygen complex of a d^6 system.

Oxygen complex **21** reacts with sulfur dioxide to give a sulfato complex, which was confirmed by its infrared spectrum in benzene solution. A solution of **21** rapidly takes up carbon monoxide to give a dicarbonyl species (IR, 1950 and 2100 cm^{-1}) with the displacement of molecular oxygen. The dicarbonyl complex is quite stable and fails to react with molecular oxygen. In benzene solution the dioxygen complex oxidizes triphenylphosphine to triphenylphosphine oxide [73]. On passing molecular hydrogen into a solution of the dioxygen complex, the O—O band at 880 cm^{-1} disappears, the solution becomes wine red, and a band at 1960 cm^{-1} due to hydride stretch appears, resulting from the formation of $RuClH(AsPh_3)_3$. The hydrido complex formed *in situ* is a good hydrogenation catalyst [73]. The dioxygen complex **21** reacts [64] with nitric oxide in benzene solution to form the nitrosyl dioxygen complex $RuCl_2(AsPh_3)_2(O_2)(NO)$, **22**. Complex **22** shows intense bands at 1870 and 880 cm^{-1} assigned to nitrogen–oxygen (of NO^+) and ruthenium–oxygen (of Ru—O_2) stretching frequencies, respectively. There was no oxidation of nitric oxide to nitrogen dioxide in the formation of this complex.

Oxygenation of $RuCl(SnCl_3)(AsPh_3)_3$ in benzene solution gave [64] a 1:1

dioxygen complex of the composition, $RuCl(SnCl_3)(AsPh_3)_3(O_2)$, **23**, with a ruthenium–oxygen stretching frequency of 805 cm^{-1}. The presence of a strong π acid $SnCl_3^-$ lowers the $Ru—O_2$ frequency in **23** by about 75 cm^{-1}, as compared to complex **21**. The dioxygen seems to be irreversibly bound in **23**, since it is not displaced by boiling in benzene solution.

A novel dioxygen complex with coordinated hydride and dioxygen, $RuHCl(AsMePh_2)_3(O_2)$, **24**, has been obtained [64] on reaction of the hydrido complex, $RuHCl(AsMePh_2)_3$, **25**, in chloroform solution with molecular oxygen. The infrared spectrum of **24** shows a band of medium intensity at 1955 cm^{-1}, assigned to the ruthenium–hydrogen stretching frequency, and a high intensity band at 855 cm^{-1} assigned to the ruthenium–dioxygen group. It may be recalled that the dioxygen complex **21** reacts with molecular hydrogen to form the hydrido complex $RuHCl(AsPh_3)_3$ and hydrogen peroxide. (The dioxygen complex **21** can be reformed by reaction of the hydride with excess molecular oxygen.) Thus the coordinated dioxygen in **24** seems to be more stable to substitution (and reduction) than the dioxygen in complex **21**.

A benzene solution of $RuCl_2(PPh_3)_3$, **26**, rapidly absorbs dioxygen at room temperature and pressure to form a green complex **27**, which rapidly turns brown with the simultaneous oxidation of triphenylphosphine to triphenylphosphine oxide [64]. Cenini *et al.* [74] have measured the rate of oxygen uptake by a benzene solution of complex **26** but were unable to determine the nature of the species formed in solution. The green (**27**, **28**) and brown (**29**) complexes obtained by the oxidation of complex **26** have been identified [75] and the mechanism of Eq. (10) proposed [75] for the oxidation of triphenylphosphine to phosphine oxide. The green complexes **27** and **28** obtained by the initial absorption of molecular oxygen in **26** gave an infrared peak at 805 cm^{-1}, which may be assigned to the ruthenium–dioxygen stretching frequency in **27** or the ruthenium–oxygen bond in **28**. The green complexes turn brown with the loss of the absorption at 805 cm^{-1} and appearance of a new absorption at 970 cm^{-1}, which was assigned to the dimer species **29**. The formation of the phosphine oxide in reaction (10) was confirmed by the appearance of a new infrared absorption at 1190 cm^{-1}, assigned to the phosphorus–oxygen stretch of the phosphine oxide. The marked decrease of oxygen uptake [74] after reaction with about 0.8 mole of oxygen per mole of complex is in accord with the formation of the unreactive dimer **29**.

The reactivity of mixed-ligand tertiary phosphine–tertiary arsine complexes with respect to the oxidation of the phosphine group is much lower than that of **26**. A number of stable mixed-ligand phosphine–arsine dioxygen complexes of ruthenium(II) have been described in Table IV (entries 27–30).

Ruthenium(0) and osmium(0) dioxygen complexes have been reported by Graham *et al.* [65] and Cavil *et al.* [66]. These complexes show the usual

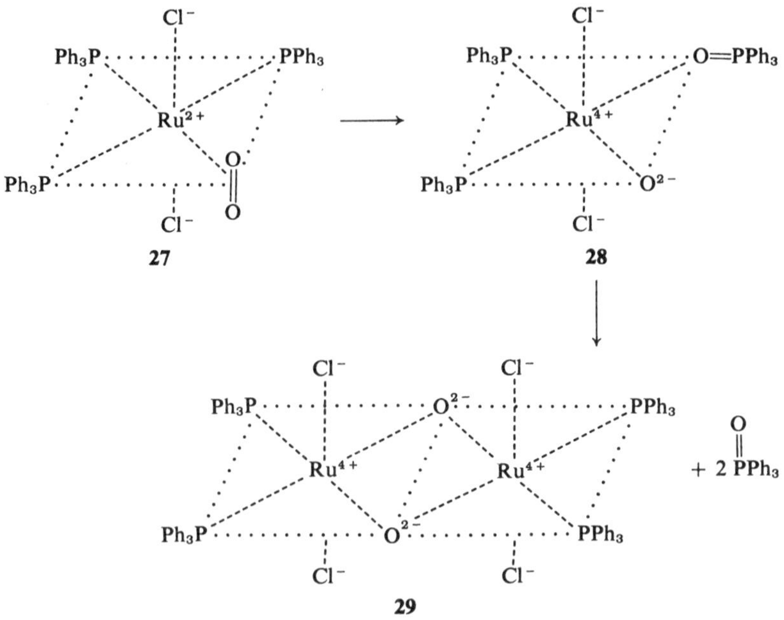

(10)

reactivity pattern towards oxidative additions, which are generally characteristic of d^8 complexes.

Molecular oxygen reacts rapidly with a benzene solution of tetrakis-(triphenylphosphine)platinum(0) to yield the complex $(PPh_3)_2PtO_2$, **30**, and two equivalents of triphenylphosphine oxide [67–69] [reaction (11)]. The

$$(PPh_3)_4Pt + 2\,O_2 \longrightarrow \underset{\textbf{30}}{(PPh_3)_2PtO_2} + 2\,Ph_3PO \qquad (11)$$

complex $(PPh_3)_2Pt(O_2)$ reacts with iodine in chloroform solution to give a quantitative yield of $(PPh_3)_2PtI_2$ and molecular oxygen (identified by mass spectral measurements). Complex **30** has an infrared band at 818 cm^{-1} with a shoulder at 824 cm^{-1} assigned to coordinated O—O stretch. Heating the dioxygen–platinum(0) complex **30** in vacuum does not liberate oxygen, indicating an irreversible combination of molecular oxygen with platinum(0). In the reaction of $(PPh_3)_4Pt(0)$ with molecular oxygen or alkenes, two molecules of triphenylphosphine are lost with the formation of a dioxygen or alkene complex. It has therefore been proposed [69] that $(PPh_3)_2Pt$ acts as a coordinately unsaturated species toward reaction with donors having multiple bonds, such as molecular oxygen, carbon monoxide, alkenes, and alkynes.

Wilke *et al.* [68] prepared both the palladium(0) and the platinum(0) dioxygen complexes by the reaction of $(Ph_3P)_4Pd(0)$ and $(Ph_3P)_4Pt(0)$ in benzene solution with molecular oxygen at 20°C. When their benzene solutions are boiled, the complexes decompose to form triphenylphosphine oxide (equation 12). The palladium(0)–O_2 complex shows an infrared absorption band characteristic of O—O stretch at 875 cm^{-1}, while the corresponding band of the platinum(0)–oxygen complex absorbs at 830 cm^{-1} [68].

$$(Ph_3P)_2MO_2 \longrightarrow M + 2\ Ph_3PO \tag{12}$$

The nickel(0) complex $(Ph_3P)_2Ni(O_2)$ was prepared by the reaction of oxygen with tetrakis(triphenylphosphine)nickel(0) at −78°C in ether and in toluene solutions [68]. The complex is stable in solution only below −35°C. Analogous nickel(II) complexes of molecular oxygen, $[(C_6H_{11})_3P]_2NiO_2$ and $[(C_5H_{10}N)_3P]_2NiO_2$, stable up to −5° and −30°C, respectively, were prepared by the action of oxygen on toluene solutions of ethylenebis(tricyclohexylphosphine)nickel(0) and ethylenebis(tripiperidinophosphine)nickel(0), respectively. The O—O vibrational stretching frequency of these nickel(II) complexes could not be determined because of overlap with the IR bands of the phosphine ligands.

The nickel(0), platinum(0), and palladium(0) complexes of triphenylphosphine and dioxygen described above may also be catalysts for oxidation reactions. These complexes decompose at room temperature to yield the oxidized ligand and zero-valent metal. When triphenylphosphine is the ligand, triphenylphosphine oxide is obtained. The zero-valent metal ion may be trapped immediately with additional phosphine ligand, and a new cycle of oxygenation starts. In this manner one nickel atom can oxidize 50 molecules of triphenylphosphine, and one palladium atom, 500 molecules of triphenyl phosphine [68].

Otsuka *et al.* [70] have reported dioxygen complexes of the composition $M(O_2)(RNC)_2$ [R = *t*-butyl and cyclohexyl for M = nickel(II), R = *t*-butyl for M = palladium(II)]. The complexes are diamagnetic, insoluble in nonpolar solvents, dimeric in the solid state, and are catalysts for the oxidation of alkyl isocyanides to alkyl isocyanates and of triphenylphosphine to triphenylphosphine oxide.

d. Bonding in 1:1 *Metal–Dioxygen Complexes.* Two types of bonding schemes may be visualized [76] for molecular oxygen complexes: a monodentate model, **31**, and the "isosceles" or triangular model, **32**. The difference

31 **32**

between the two models is that of a localized π bond in **31** (with some d–π^*

backbonding) and extensive delocalization of the π electrons of oxygen into the metal, resulting in the formation of a structure that approaches **32**, containing two σ bonds between the metal and the oxygen atoms.

The force constant of the O—O bond in a complex approaching the isosceles configuration in **32** would be lowered to a greater extent from that of ground-state oxygen (11.5 mdyne/Å) than would that of a complex more closely approximating the monodentate model. Nakamura *et al.* [76] have prepared and studied spectrally the dioxygen complexes $Pt(O_2)(PPh_3)_2$, $M(O_2)(t\text{-}BuNC)_2$ [M = nickel(0), palladium(0)], and $RhX(O_2)L_2(t\text{-}BuNC)$ ($X = Cl^-$, Br^-; L = PPh_3, $AsPh_3$) with $^{16}O_2$, ^{16}O—^{18}O, and $^{18}O_2$. From the splitting pattern of the high-intensity infrared band in the 800–900 cm^{-1} range, the vibration was assigned to an O—O stretch of coordinated dioxygen. The O—O stretching force constants were calculated on the basis of an isosceles model and for the complexes studied were in the range 3.0–3.5 mdyne/Å.

A molecular orbital description of bonding in dioxygen complexes has recently been proposed by McGinnety *et al.* [77] on the basis of the isosceles model (local C_{2v} symmetry of the metal–dioxygen grouping).

Molecular oxygen seems to be a poor σ donor and a good π acceptor, the strength of the metal–dioxygen bond depending on the extent of the back donation of electrons from the filled molecular orbitals of the metal ion to the antibonding orbitals of the ligand. Any factor that increases the available electron density on the metal ion for back donation increases the strength of the metal–dioxygen bond and weakens the oxygen–oxygen bond. It has been found by McGinnety *et al.* [77] that a decrease in the electronegativity of the σ-donor group X in the iridium–oxygen complexes $IrX(CO)(PPh_3)_2(O_2)$ increases the strength of the iridium–dioxygen bond, as indicated by an increase in the O—O distance and a decrease in the reversibility of the bonding of dioxygen to iridium. Thus for $X = Cl^-$ and I^-, **17** and **18**, the O—O distance in the iridium complexes is 1.30 and 1.51 Å, respectively, indicating that the iodide ion in complex **18** imparts greater electron density to the metal ion for back donation.

Vaska *et al.* [55] have reported that the activation energy for the addition of molecular oxygen (reaction 13) to isostructural complexes of the type $[M(diphos\text{-}2)_2]BF_4$ [M = cobalt(I), rhodium(I), iridium(I), diphos-2 = $Ph_2PCH{=}CHPPh_2$] follow the order cobalt(I) ≪ iridium(I) < rhodium(I).

$$[M(diphos\text{-}2)_2]BF_4 + O_2 \longrightarrow [M(diphos\text{-}2)_2(O_2)]BF_4 \qquad (13)$$

For reaction (13) the activation energies for cobalt(I), iridium(I), and rhodium(I) have been reported as 11.0, 17.8, and 18.8 kcal/mole. The activation of small molecules usually follows the order third group > second group > first group. The basis for the anomalous position of cobalt in the above activation sequence is not understood and quantitative kinetic measurements on a series of isostructural complexes of these metal ions is needed.

e. Reactions of 1:1 *Dioxygen Complexes.* Reactions of 1:1 dioxygen–metal complexes may be classified into four groups [78]:

1. Intramolecular atom-transfer redox reactions with conversion of ligand to an oxygenated product and formation of a deoxygenated metal complex
2. Formation of an oxygenated ligand and the oxidation of the metal
3. $M—O_2$ bond cleavage and displacement of coordinated dioxygen by another ligand
4. Oxidative addition to the metal complex and formation of substrate peroxide

Reactions of the first type have been observed with zero-valent complexes of platinum, palladium, and nickel [68,70]. Reaction of MO_2L_2 (M = platinum, palladium, and nickel; L = PPh_3, Bu^tNC, cyclohexyl-NC) with excess L proceeds at ambient temperature and pressure with the formation of a zero-valent metal complex ML_n and the oxidized ligand LO in accordance with reaction (14). The reaction involves incipient formation of the oxygenated ligand LO in the complex, followed by displacement of LO by additional

$$L_2MO_2 + nL \longrightarrow ML_n + 2\,LO \tag{14}$$

free ligand, L. Examples of (14) for L = PPh_3 and L = CO are given in Eqs. (15)–(18).

Halpern and Pickard [79] have advanced the mechanism shown in Eqs. (15)–(17) for the oxidation of triphenylphosphine to triphenylphosphine oxide by molecular oxygen with tris(triphenylphosphine)platinum(0) as catalyst.

$$(PPh_3)_3Pt + O_2 \xrightarrow{k_1} (PPh_3)_2PtO_2 + PPh_3 \tag{15}$$

$$(PPh_3)_2PtO_2 + PPh_3 \xrightarrow{k_2} (PPh_3)_3PtO_2 \tag{16}$$

$$(PPh_3)_3PtO_2 + 2\,PPh_3 \xrightarrow{\text{fast}} (PPh_3)_3Pt + 2\,Ph_3PO \tag{17}$$

The constants k_1 and k_2 have been reported as 2.6 ± 0.1 and 0.15 ± 0.1 M^{-1} sec^{-1} in benzene at 25°C.

Carbon monoxide reacts [78] with $Ni(Bu^tNC)_2(O_2)$ at 20°C in chlorobenzene to give $Ni(CO)_2(Bu^tNC)_2$ and carbon dioxide, in accordance with reaction (18).

$$Ni(Bu^tNC)_2O_2 + 4\,CO \longrightarrow Ni(CO)_2(Bu^tNC)_2 + 2\,CO_2 \tag{18}$$

Reactions of the second type have been observed with substrates that have high affinity towards molecular oxygen such as dinitrogen tetroxide, carbon dioxide, and sulfur dioxide. These reactions are illustrated in Eqs. (19)–(21) without the metal moiety of the metal–oxygen complex represented. These reactions occur with oxygen complexes of iridium(I) [48,50,51,80], rhodium(I) [48,80], ruthenium(0) [51], platinum(0) [48], palladium(0) [48,50,78], and nickel(0) [78].

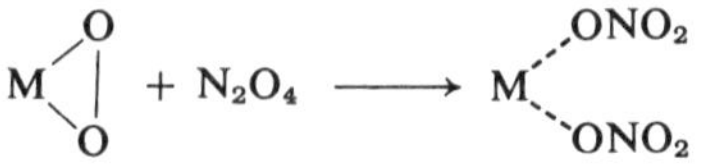
(19)

(20)

(21)

Horn *et al.* [50] have established by ^{18}O studies that the reaction with sulfur dioxide occurs at the metal–oxygen and not the oxygen–oxygen bond, as shown in reaction (22).

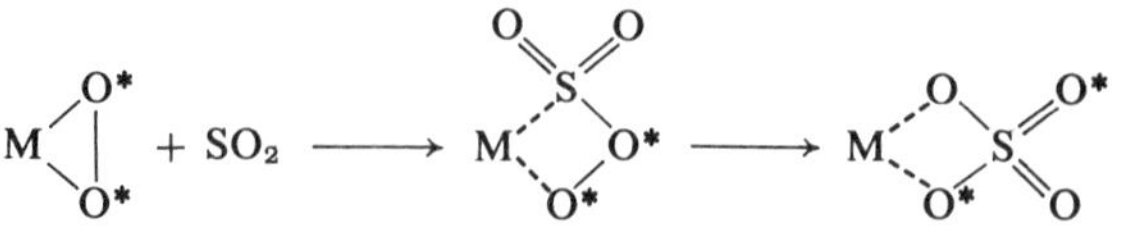
(22)

The third group of reactions has been observed [78] with those electrophiles that react with the metal complex by an oxidative substitution process with the resulting displacement of molecular oxygen (reaction 23). In this case, it is seen that the bonding of dioxygen to the metal, and of the molecule that displaces it, are very similar.

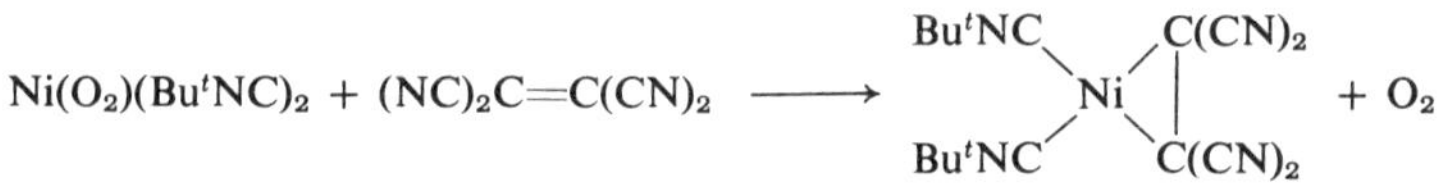
(23)

In the fourth type of reaction, oxidative addition of the anion of a reactant such as an acyl halide to the complex is accompanied by insertion of molecular oxygen into the acceptor center of the reactant with the formation of a peroxide [78]. The reaction may proceed through an intermediate, as shown in Eq. (24), in which peroxide formation and oxidative addition are concerted processes.

A hydroperoxide intermediate has been postulated [81,82] in the oxidation of cyclooctene to cyclooctenone catalyzed by a chlorocyclooctene–rhodium complex, **33**. Oxygenation of **33** in *N,N*-dimethylacetamide solution results in the formation of a dioxygen–alkene complex **34**. The mechanism proposed [81] for the oxidation of the coordinated cyclooctene is illustrated by reaction sequence (25) and formulas **33** through **38**.

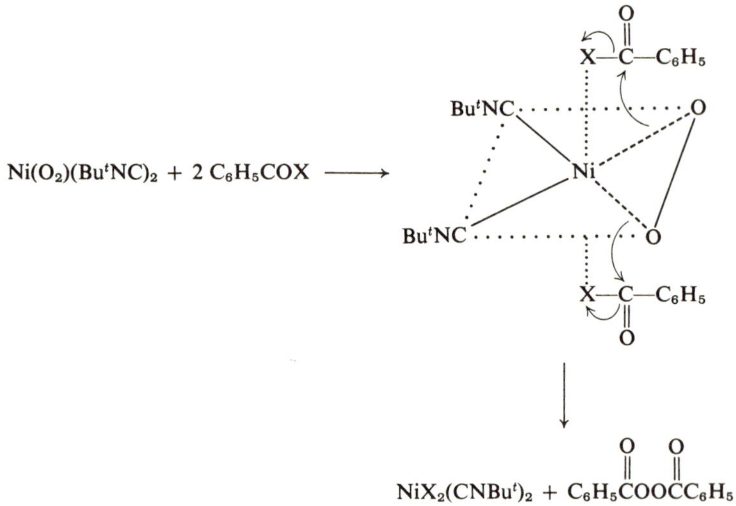

(24)

The rhodium(I)–dioxygen complex **34** abstracts a hydride ion from the coordinated cyclooctene, or alternatively the coordinated peroxo group abstracts a proton from the octene, to give the hydroperoxide complex **36**. Complex **36** then rearranges to the cyclooctene hydroperoxide complex **37** that decomposes to the product complex **38**. Displacement of cyclooctenone in **38** by two molecules of cyclooctene gives back the complex **34** and the product.

f. Reactions of Binuclear Dioxygen Carriers. Two reasonable pathways for oxidation of substrates by oxygen carriers are the following: (1) Oxidation of the substrate may take place by direct reaction with the coordinated dioxygen of the dioxygen carrier. (2) The metal ion itself may act as an electron acceptor, by virtue of its tendency to be oxidized by the coordinated dioxygen. The possibility of the metal ion serving as the reactive center would be enhanced by the presence of one or more free coordination sites about the metal ion.

One would expect that coordinated dioxygen would be less reactive than free oxygen in oxidation reactions. On the other hand, as indicated above, coordinated dioxygen may vary between two extremes, molecular oxygen and the peroxo dianion, with superoxide as an intermediate state, so that enhanced reactivity is not unreasonable. In the following paragraphs a few examples of the reactions of dioxygen carriers are discussed.

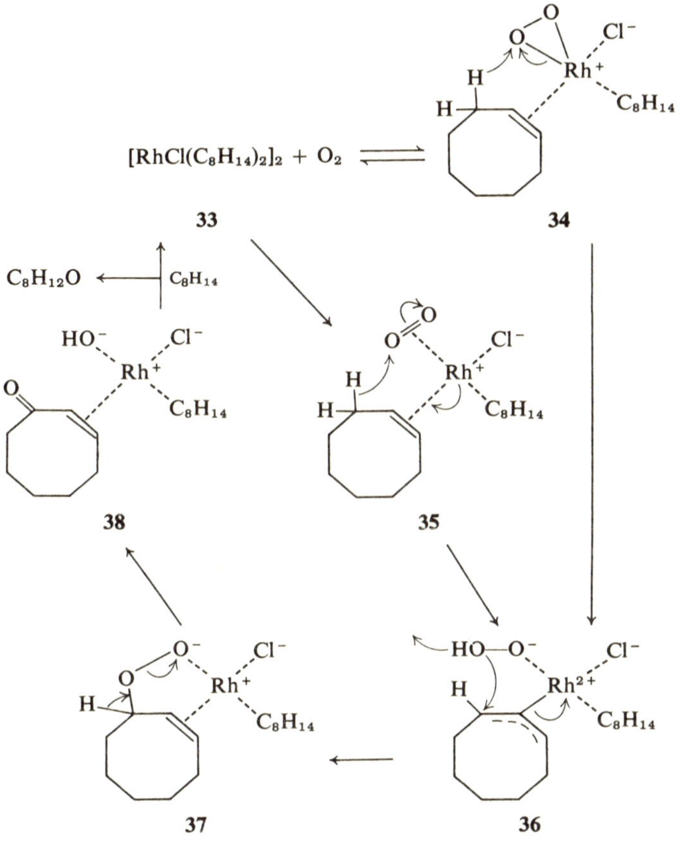

(25)

Oxidation of iron(II) with the complex $[(NH_3)_5Co—O_2—Co(NH_3)_5]^{5+}$ as oxidant has been described by Sykes [83–85]. The rate-determining step in the reaction is the formation of the normal diamagnetic dioxygen carrier from the paramagnetic species:

$$[(NH_3)_5Co—O_2—Co(NH_3)_5]^{5+} + Fe^{2+} \xrightarrow{\text{slow}} Fe^{3+} + [(NH_3)_5Co—O_2—Co(NH_3)_5]^{4+} \quad (26)$$

$$[(NH_3)_5Co—O_2—Co(NH_3)_5]^{4+} \xrightarrow{\text{fast}} 2\,[Co(NH_3)_5]^{2+} + O_2 \quad (27)$$

The Mn(II)–phthalocyanine complex in pyridine solution binds oxygen to form a peroxy product with 2 moles of phthalocyanine–Mn(II) per mole of oxygen [86]. It has been shown recently by Przywarska-Boniecka [87] that oxygen binding by Mn(II)–phthalocyanine also takes place in other bases such as aniline, imidazole, and quinoline. In pyridine, the Mn(III)–peroxy-phthalocyanine gradually decomposes to a blue product containing manganese(IV). The ligands are phthalocyanine, a pyridine molecule on one side of the planar aromatic ring, and an oxide anion on the other. Under constant oxygen concentration, the rate of decomposition of the peroxy compound to manganese(IV) follows first-order kinetics with respect to manganese(II) concentration [87]. The activation energy of the reaction is 20 kcal/mole and the activation entropy is −5.94 eu. The suggested mechanism [87] is shown in Eqs. (28)–(31).

$$(Mn) + Py \rightleftharpoons (Mn)Py \quad (28)$$

$$(Mn)Py + O_2 \rightleftharpoons O_2(Mn)Py \quad (29)$$

$$Py(Mn)O_2 + (Mn)Py \rightleftharpoons Py(Mn){-}O_2{-}(Mn)Py \quad (30)$$

$$Py(Mn){-}O_2{-}(Mn)Py + Py(Mn)O_2 \xrightarrow{k} 2\,Py(Mn)O + (Mn)Py + O_2 \quad (31)$$

[(Mn) = Mn(II)–phthalocyanine complex; Py = pyridine]

According to Elvidge and Lever [86] the O–Mn(IV)–phthalocyanine complex formed in step (31) loses pyridine and molecular oxygen on boiling and goes back to the original Mn(II)–phthalocyanine complex. Crandall [16], however, observed no oxygen evolution in the decomposition reaction of Phth–Mn(IV)O. According to Crandall [16], pyridine is oxidized in the reduction of the manganese(IV) complex to Mn(II)–phthalocyanine. Similar oxidation reactions of manganese(II) to manganese(IV) through manganese(III) also occur in the analogous Mn(II) porphyrins and may play an important role in photosynthesis in the evolution of oxygen [87a].

An Fe(II)–phthalocyaninetetrasulfonic acid dioxygen carrier, **39** (Fig. 11), has been recently described by Vonderschmitt *et al.* [88] and by Weber and Busch [88a]. At a neutral pH (6.5) both mononuclear and binuclear Fe(II)–phthalocyanine tetrasulfonic acid–dioxygen complexes are formed in solution, in accordance with the equilibria given in Eqs. (32) and (33). At 20°C,

$$Fe(II){-}PTS + O_2 \underset{}{\overset{K_1}{\rightleftharpoons}} Fe(II){-}PTS \cdot O_2 \quad (32)$$

$$Fe(II){-}PTS \cdot O_2 + Fe(II){-}PTS \overset{K_2}{\rightleftharpoons} PTS \cdot Fe(II) \cdot O_2 \cdot Fe(II) \cdot PTS \quad (33)$$

(PTS = phthalocyaninetetrasulfonic acid)

the values of K_1 and K_2 are reported to be 2×10^4 and 4×10^7 mole^{-1} liter, respectively. A tentative structure for the binuclear complex [88] is shown in formula **39**. In highly alkaline or in strongly acidic solutions, the reaction between Fe(II)–phthalocyaninetetrasulfonic acid and dioxygen becomes irreversible. Reversible binding of oxygen in the binuclear complex was demonstrated [88] by the total expulsion of oxygen when nitrogen is passed

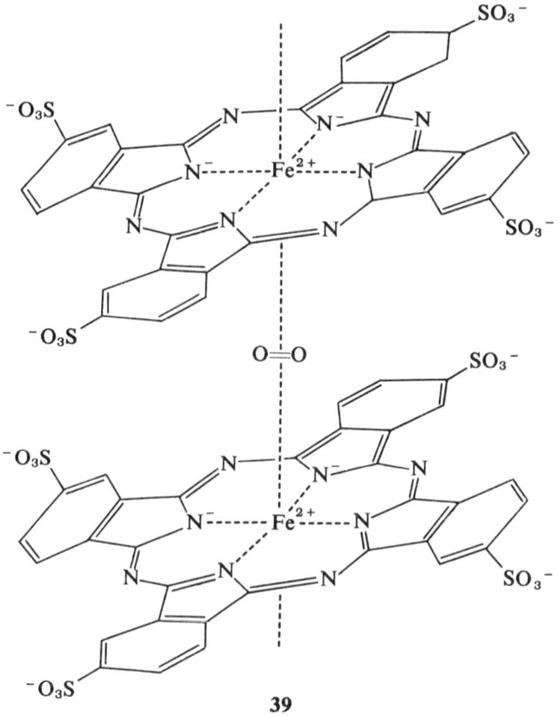

39

Fig. 11. Proposed structure for Fe(II)-phthalocyaninetetrasulfonic acid oxygen carrier in aqueous solution [44].

through solutions of the dioxygen complex. Deoxygenation was indicated by disappearance of the characteristic band (632 nm) of the binuclear oxygenated species, with the simultaneous appearance of a band at 670 nm characteristic of the oxygen-free Fe(II)–phthalocyaninetetrasulfonic acid.

Beck and Gorog [89] have reported the concerted oxidation of the Co(II)–glycylglycine (GG) complex and ascorbic acid by molecular oxygen. The rate of formation of the oxygen carrier $[GG—Co—O_2—Co—GG]^{2+}$ increases in the presence of ascorbic acid. At the end of the reaction Co(II)–glycylglycine is irreversibly oxidized to the cobalt(III) complex. According to the authors, molecular oxygen is initially activated by ascorbic acid, resulting in more complete formation of the Co(II)–glycylglycine–oxygen complex. The oxygen carrier thus oxidizes ascorbic acid more effectively than does molecular dioxygen.

Further study of this reaction by Nakon [90] revealed a number of significant new facts and suggests an alternative mechanism. Only the glycylglycine–cobalt–dioxygen systems with molar ratios of glycylglycine to cobalt greater than 1:1 accelerate the rate of ascorbic acid oxidation. The 2:2:1 dioxygen complex **10** (Fig. 4), which has a hydroxo bridge as well as a bridging peroxo group, is inactive as a catalyst. When catalytic activities of the polyamine oxygen complexes **6–9** (Fig. 3) were examined, it was found that only the 2:2:1 diethylenetriamine–cobalt dioxygen complex, which has two free (aquo) positions about the metal ions, is active. In this case, it is seen that coordination sites are available for combination with the bidentate substrate.

In the case of the 4:2:1 glycylglycine–Co(II)–dioxygen complex, **11** (Fig. 4, Fig. 12), only one ligand per metal ion is tightly bound, the other being weakly bound by only the amino group and the adjacent amide carbonyl oxygen donor, which could be displaced readily by the substrate. It should be noted that structural assignments for the cobalt–glycylglycine–oxygen complexes **10** and **11** are tentative and should be considered definite only from the point of view of stoichiometry. These systems are characterized by side reactions leading to irreversible cobalt (III) complexes.

These observations [90] led to the postulation of direct transfer of electrons from the ligand to the metal ion. This reaction may take place in one two-electron transfer process, or in two successive one-electron steps. A probable mechanism for the two-electron transfer process, involving no intermediate free-radical species, is illustrated in Fig. 12. In this process donation of an electron to the remote cobalt(III) not coordinated to the substrate may occur to produce a superoxide-bridged intermediate, which may then accept an electron from the substrate through the cobalt(III) ion to which it is coordinated by going through a cobalt(II) intermediate. A second electron transfer from the substrate would then form a thermodynamically unstable chelate which would spontaneously and rapidly dissociate to the cobalt(II) complex **42**, ligand, oxidized substrate, and peroxide.

A similar mechanism can be written involving a one-electron transfer to convert one of the cobalt(III) ions to cobalt(II), followed by dissociation into components including the ascorbate semiquinone type intermediate, followed by further oxidation and reduction in solution to stable products. Similar mechanisms may also be written for the oxidation of substrate by the 2:2:1 diethylenetriamine–cobalt(II)–oxygen complex (formula **7**).

B. Metal Ion- and Metal Chelate-Catalyzed Oxidation of Ascorbic Acid

Metal ion and metal chelate catalysis of the oxidation of ascorbic acid, which have been known for a long time, provide interesting examples of non-insertion reactions in which the oxygen is reduced to peroxide or water, and the metal ion serves as an electron-transfer agent between the substrate and the oxidant. This reaction has recently been given detailed study by the

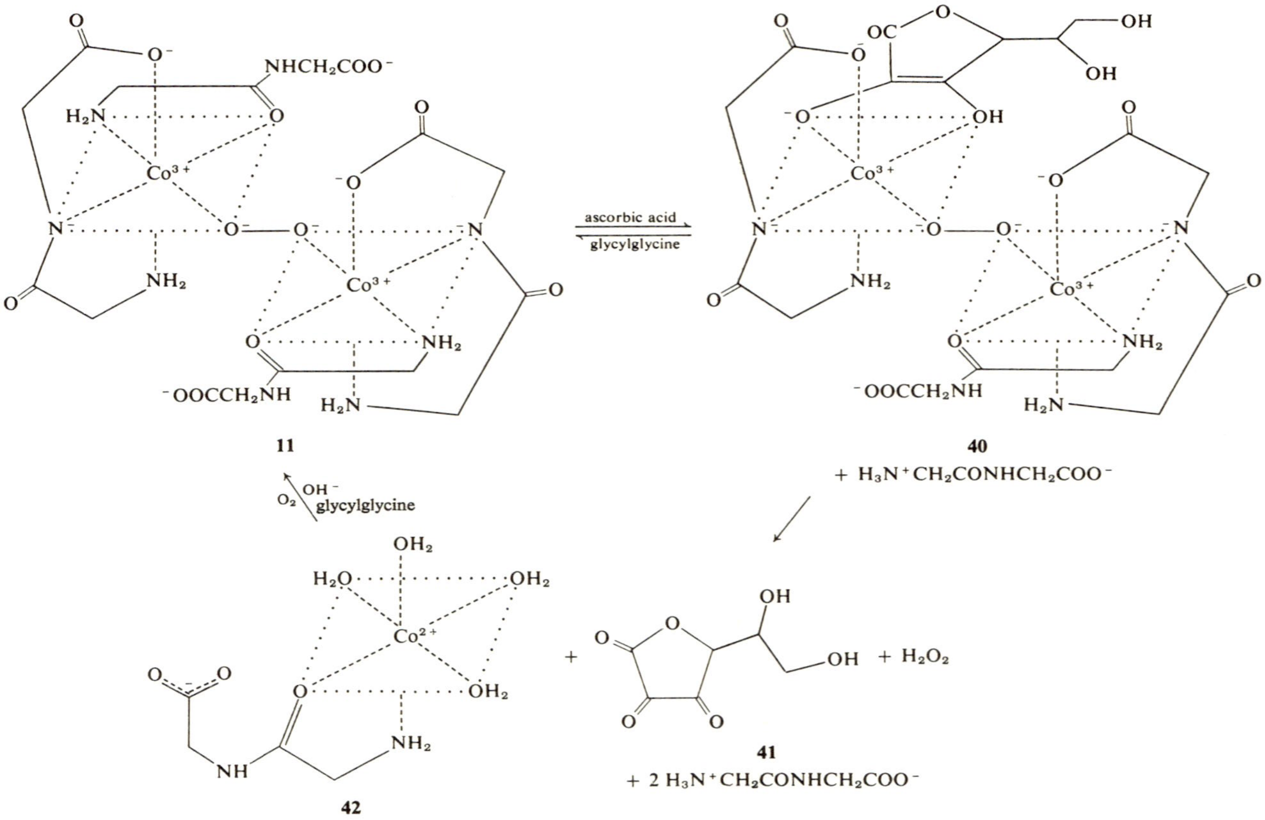

Fig. 12. Oxidation of ascorbic acid by O_2 in the presence of Co^{2+} and glycylglycine.

authors of this book and will be described in some detail. The copper(II) ion-catalyzed oxidation of ascorbic acid by molecular oxygen was reported by Weissberger *et al.* [91] and by Nord [92,93]. The Fe(III)–EDTA-catalyzed oxidation of ascorbic acid by hydrogen peroxide, a different reaction, was described by Grinstead [94].

1. *Copper(II) and Iron(III) Ion Catalysis*

In the case of the copper(II) and iron(III) ion-catalyzed oxidation of ascorbic acid, the rate of the reaction has been found [95] to be first order with respect to the ascorbate anion, metal ion catalyst, oxygen concentration, and the reciprocal of the hydrogen ion concentration. The kinetic observations for copper(II) and iron(III) ion-catalyzed oxidation of ascorbic acid support the mechanism illustrated in Fig. 13. The intermediate dioxygen complex can exist in several resonance forms, two of which are shown. The free-radical species can be considered as simply one of the resonance forms since no atoms have actually moved. Transfer of a proton to the coordinated dioxygen can assist the two-electron transfer process, which results directly in the formation of reaction products.

Alternatively, if dissociation to free-radical intermediates of the type suggested in formula **45** were to occur, the mechanisms could be expressed by the reaction sequence shown in Eqs. (34)–(40).

$$M^{n+} + HA^- \rightleftharpoons MHA^{(n-1)+} \quad (34)$$

$$MHA^{(n-1)+} + O_2 \rightleftharpoons MHA(O_2)^{(n-1)+} \quad (35)$$

$$MHA(O_2)^{(n-1)+} \xrightarrow{\text{slow}} M^{n+}(HA\cdot)(\dot{O}_2^-) \quad (36)$$

$$M^{n+}(HA\cdot)(\dot{O}_2^-) \xrightarrow{\text{fast}} M^{n+} + HA\cdot + \dot{O}_2^- \quad (37)$$

$$HA\cdot + O_2 \xrightarrow{\text{fast}} HO_2\cdot + A \quad (38)$$

$$HA\cdot + M^{n+} \xrightarrow{\text{fast}} M^{(n-1)+} + A + H^+ \quad (39)$$

$$M^{(n-1)+} + \dot{O}_2^- \xrightarrow{\text{fast}} M^{n+} + O_2^{2-} \quad (40)$$

Both mechanisms are identical up to the rate-determining step and therefore would have the same kinetics. Differentiation between the two processes would require a sensitive technique, such as ESR, to detect free-radical intermediates, or the use of suitable reagents that would trap or detect these intermediates.

Both mechanisms suggested above involve electron transfer from π orbitals of σ-bonded ascorbate dioxygen, through appropriate d orbitals of the ion, to the vacant antibonding orbitals of oxygen, as indicated by Fig. 14. In order that a metal ion function in this manner as a catalyst for ascorbic acid oxidation, it would seem helpful if there were a reasonably stable lower valence state to facilitate electron transfer through the metal ion. In this connection

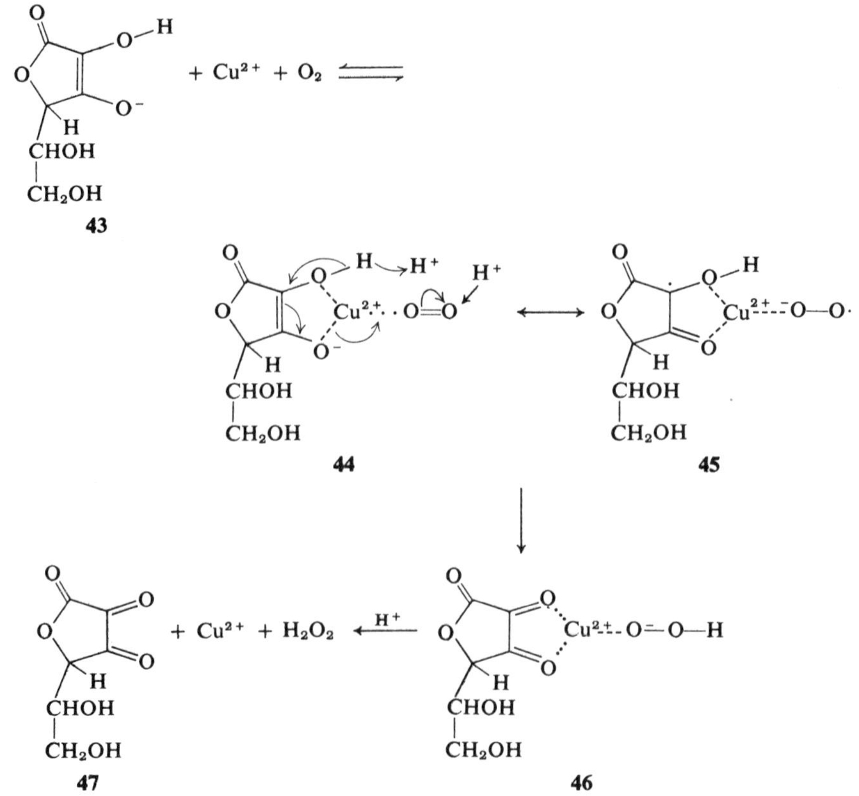

Fig. 13. Copper(II) ion-catalyzed oxidation of ascorbic acid.

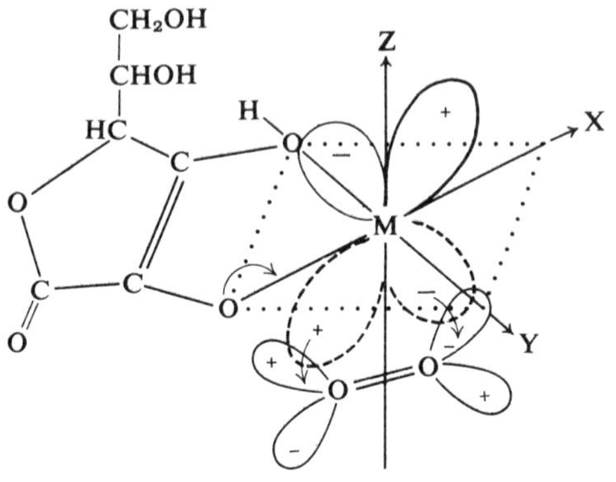

Fig. 14. Schematic representation of electron-transfer step in Cu(II)- and Fe(III)-catalyzed oxidation of ascorbic acid by molecular oxygen.

it is interesting that metal ions such as zinc(II), manganese(II), cobalt(II), and nickel(II), for which there are no stable lower oxidation states, have been found to be inactive as catalysts. The kinetic evidence presently available cannot differentiate between the two reaction sequences following the initial rate-determining electron transfer [Eqs. (37)–(40) or (41)–(43)].

$$MHA(O_2)^{(n-1)+} \xrightarrow{k} MHA\cdot(O_2\cdot)^{(n-1)+\cdot} \quad (41)$$

$$MHA\cdot(O_2\cdot)^{(n-1)+} \xrightarrow{\text{fast}} MA(HO_2^-)^{(n-1)+} \quad (42)$$

$$MA(HO_2^-)^{(n-1)+} \xrightarrow{\text{fast}} M^{n+} + HO_2^- + A \quad (43)$$

2. *Vanadyl Ion Catalysis*

Vanadyl ion catalysis [96] differs from copper(II) and iron(III) catalysis in that while the reaction is first order with respect to vanadyl ion and oxygen concentrations, it is directly proportional, rather than inversely proportional, to the hydrogen ion concentration. This indicates a difference in the composition of the activated complex by two hydrogen ions. The mechanism shown in Eqs. (44)–(48) is in accord with the kinetic observations.

$$VO^{2+} + HA^- \rightleftharpoons VOHA^+ \quad (44)$$

$$VOHA^+ + O_2 \rightleftharpoons VOHA(O_2)^+ \quad (45)$$

$$VOHA(O_2)^+ + 2\,H^+ \rightleftharpoons (VOH^{3+})(HA^-)(HO_2^+)^{3+} \quad (46)$$

$$(VOH^{3+})(HA^-)(HO_2^+)^{3+} \xrightarrow{\text{slow}} VOH^{3+}(HA\cdot)(HO_2\cdot) \quad (47)$$

$$VOH^{3+}(HA\cdot)(HO_2\cdot) \xrightarrow{\text{fast}} VO^{2+} + A + H_2O_2 + H^+ \quad (48)$$

A vanadyl–ascorbate complex is probably formed in the first preequilibrium step, followed by formation of a vanadyl–ascorbate–dioxygen complex. Apparently before the rate-determining electron transfer can occur two protons must be added to this complex. It is, of course, not possible to determine the acceptor sites for these protons, but one could be the oxo oxygen of the vanadyl ion. The other proton may combine with the coordinated oxygen molecule to assist the concerted electron transfer from ascorbate oxygen to molecular oxygen through the vanadyl ion. A possible structure of the complex involved in the rate-determining step is schematically shown in Fig. 15.

3. *Oxidation in the Presence of Copper(II) and Iron(III) Chelates*

In the case of cupric and ferric chelate-catalyzed oxidation of ascorbic acid [97], the rate of oxidation has been found to depend on the first power of ascorbate and metal chelate concentrations and the reciprocal of the hydrogen ion concentration. The rate of oxidation, unlike the metal ion-catalyzed oxidation reactions, has been found to be independent of oxygen concentra-

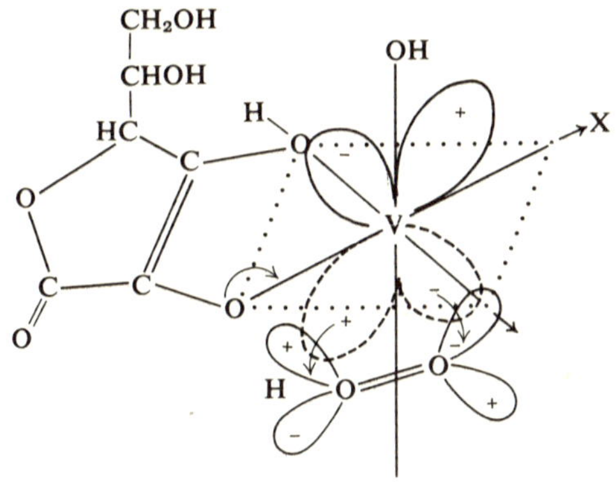

Fig. 15. Schematic representation of electron-transfer step in VO(IV) ion-catalyzed oxidation of ascorbic acid by molecular oxygen.

tion. The catalytic activity of the metal chelate catalyst in both copper(II) and iron(III) chelates of DTPA, EDTA, CDTA, HEDTA, HIMDA, and IMDA (see Appendix II) is found to increase as the stability of the metal chelate species decreases.

The independence of the rate of oxidation of ascorbic acid from oxygen concentration in the case of metal chelate-catalyzed oxidation suggests that the rate-determining step is a one-electron reduction of chelate species by ascorbic acid followed by a fast reoxidation of the lower-valent metal chelate by molecular oxygen. The mechanism shown in Eqs. (49)–(54) is proposed for metal chelate-catalyzed oxidation of ascorbic acid. It is suggested that the

$$ML^{n+} + HA^- \rightleftharpoons ML(HA)^{(n-1)+} \tag{49}$$

$$ML(HA)^{(n-1)+} \xrightarrow{\text{slow}} ML^{(n-1)+} + HA\cdot \tag{50}$$

$$HA\cdot + O_2 \xrightarrow{\text{fast}} A + HO_2\cdot \tag{51}$$

$$ML^{n+} + HA\cdot \xrightarrow{\text{fast}} ML^{(n-1)+} + A + H^+ \tag{52}$$

$$ML^{(n-1)+} + O_2 \xrightarrow{\text{fast}} ML^{n+} + \dot{O}_2^- \tag{53}$$

$$ML^{(n-1)+} + H_2O_2 \xrightarrow{\text{fast}} ML^{n+} + OH^- + OH\cdot \tag{54}$$

$$ML^{(n-1)+} + OH\cdot \xrightarrow{\text{fast}} ML^{n+} + OH^- \tag{54a}$$

mixed-ligand complex is either very unstable and is present in very low concentration, or dissociates immediately after the first electron transfer, which renders the substrate a poorer donor and the metal ion a poorer coordinator. The proposed mechanism described above is further supported by a similar investigation of the rate of oxidation of ascorbic acid by copper(II) and iron(III) chelates in the absence of oxygen. Close correspondence of the initial rates to the rates of catalysis in the presence of oxygen indicated a similarity in mechanism and the fact that the role of oxygen is simply that of reoxidizing the metal chelate in non-rate-determining steps.

No hydrogen peroxide was detected in the metal chelate-catalyzed oxidation of ascorbic acid, in contrast to the corresponding reaction involving metal ion catalysis. In the case of metal chelate-catalyzed oxidation, it is probable that reduction of hydrogen peroxide is completed by further reaction with the lower-valence form of the metal chelate, converting it to the original higher-valence form.

The enzyme ascorbic acid oxidase seems to catalyze the oxidation of ascorbic acid by a radical mechanism consisting of two successive one-electron steps [98]. The ascorbic acid is pictured as interacting with the oxidized form of the copper enzyme through mixed-ligand formation. A one-electron transfer results in the formation of a free radical, which may readily dissociate because of its lower donor activity, as well as the lower acceptor activity of the metal ion. The presence of a free-radical intermediate has been demonstrated by ESR spectroscopy [99]. The oxidation of the free radical may occur very rapidly at another enzymatic site, while the reduced metallo-enzyme sites are rapidly regenerated by reaction with molecular oxygen.

C. Oxidation of Catechols

1. *Nonenzymatic Systems*

The nonenzymatic oxidation [13] of 3,5-di-*t*-butylpyrocatechol in mildly alkaline solution by molecular oxygen is catalyzed by manganese(II), cobalt(II), iron(II), nickel(II), and zinc(II). The products of oxidation are the corresponding quinone and 2,4-di-*t*-butyl-4,5-dihydroxy-α-hydromuconic acid γ-lactone. The latter compound has been considered to be formed by ring cleavage and further oxidation of the quinone. The catalytic activities of the metal ions decrease in the order manganese(II) > cobalt(II) > iron(II) > copper(II) > zinc(II) > nickel(II). The free-radical mechanism illustrated in Fig. 16 was proposed by Grinstead [13] for the oxidation of di-*t*-butylpyrocatechol. The reaction is visualized as occurring through the formation of a metal–substrate complex, **51**, followed by successive one-electron transfers to molecular oxygen from the metal ion. Intramolecular electron transfers in the metal complex result in the formation of the quinone and regeneration of the lower valence state of the metal ion. In view of the implications of this mechanism that redox reactions of the metal ion are important, it is perhaps worthwhile to study in detail the mechanisms of reaction with the various catalytically active metal ions listed. It is noted that the two least active metals cannot participate in electron transfer.

No suggestion was made for the nature of the intermediate steps in going from the *o*-quinone, **49**, to the 4,5-dihydroxy-α-hydromuconic acid γ-lactone,

Overall Reaction

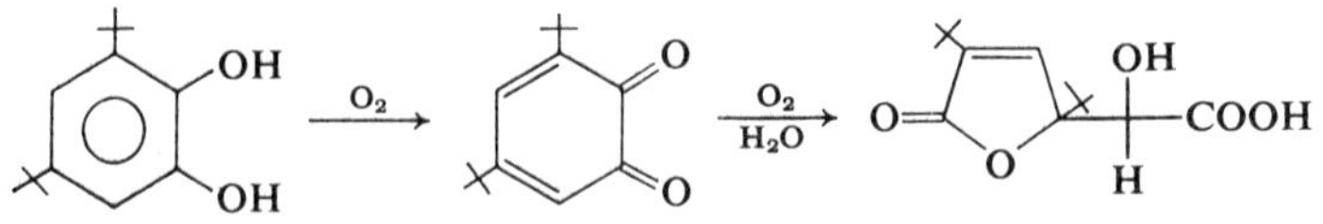

48 **49** **50**

Mn(II) > Co(II) > Fe(II) > Cu(II) > Zn(II) > Ni(II)

Mechanism of First Step

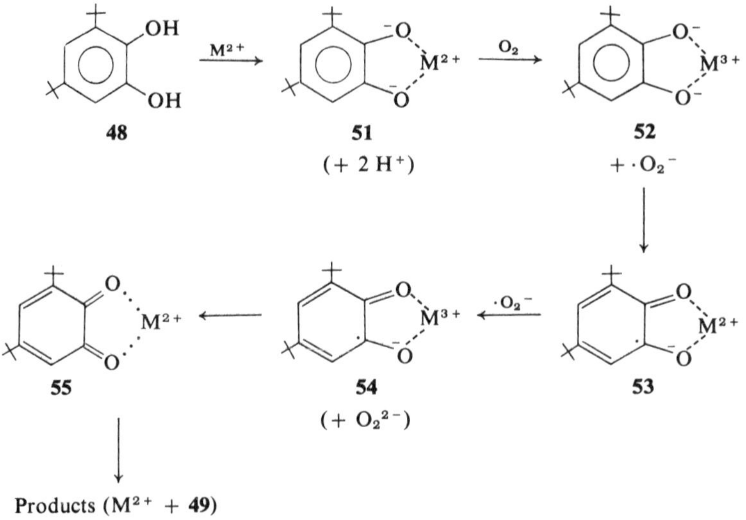

Fig. 16. Metal ion-catalyzed oxidation of di-*t*-butylpyrocatechol.

50 (see Section IX,B). However, it is apparent that the oxidation of a catechol to an *o*-quinone is a typical non-insertion reaction, while insertion may take place in a subsequent step by reaction with the peroxide that is formed in the initial part of the reaction.

A further study of metal chelate-catalyzed oxidation of catechols recently completed in the author's laboratory [100] employed bidentate ligands as carrier ligands. In these studies no appreciable concentration of free-radical intermediates could be detected, and a variable amount of hydrogen peroxide was formed, depending on the metal ion employed as catalyst and its degree of catalase activity. The bidentate ligands employed, tetrabromocatechol and 4-nitrocatechol, having the formula H_2L, were found to form stable complexes, ML, with divalent metal ions, and these complexes did not dissociate during the course of the reaction under investigation. These observations, together with the dependence of the initial rates on oxygen concentration, on the concentration of ML, and on the concentration of the monoprotonated form of the substrate HA^- (where H_2A is catechol or 3,5-di-*t*-butylcatechol) led to the suggestion of the ionic mechanism illustrated in Fig. 17. A two-electron transfer process is visualized as occurring in the metal–ligand–substrate oxygen complex, **57**. The shift of negative charge from the ligand to the coordinated oxygen is balanced by dissociation of a proton from the oxygen donors of the ligand and combination of a proton from the solvent with the coordinated oxygen molecule to yield intermediate **58**, which quickly dissociates to reaction products and catalyst because of the poor donor qualities of the quinone.

2. *Enzymatic Systems*

The oxidation of catechols to quinones (catecholase activity) is catalyzed by polyphenol oxidases and tyrosinases [101–107]. The metal ion at the active site, copper(I), does not change its valence during the reaction so that an ionic mechanism similar to those illustrated in Figs. 13 and 17 probably applies. Although the oxygen is reduced all the way to water, the mechanism most likely involves a two-electron (and proton) transfer to give an intermediate

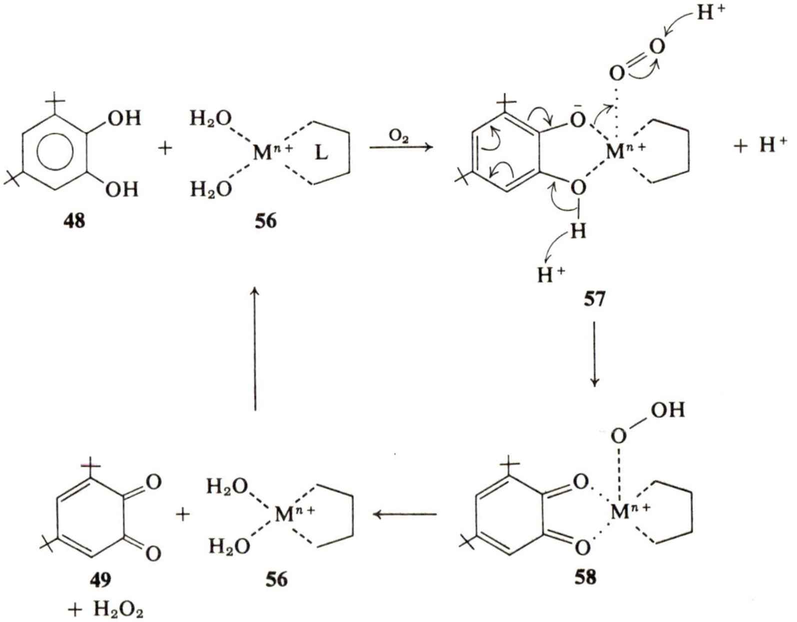

Mn(II) > Co(II) > Fe(II) > Cu(II) > Ni(II), Zn(II)

Fig. 17. Metal chelate-catalyzed oxidation of di-*t*-butylcatechol.

peroxo complex. The latter could then oxidize another mole of complex in the manner indicated in Section V,A and Fig. 19 of this chapter.

D. Oxidation of Metal Ions

Non-insertion reactions of molecular oxygen also include oxidation of lower valence metal ion species to higher valence forms in one-electron or two-electron oxidations. Some of these reactions are listed in Table V [94,95,108–119]. A recent review by Fallab [29] describes metal ion-catalyzed reactions of molecular oxygen in considerable detail. In cases of the oxidation of iron(II) [108], vanadium(III) [109], and uranium(IV) the reaction is postulated to take place by electron transfer to molecular oxygen in one-electron transfer steps. The oxidation of vanadium(II) [110] and chromium(II) to vanadium(III) and chromium(III) is considered to take place in two steps, a rapid two-electron transfer to molecular oxygen to form the tetravalent metal species, followed by a slow reaction between the tetravalent and divalent forms to yield the trivalent species. It is noted that some of the oxidation reactions in Table V show the formation of superoxide, $HO_2\cdot$. In such cases, because of the greater reactivity of hydrogen superoxide, rapid further oxidation of the metal ion results in its removal in non-rate-determining steps.

The reactions of the type listed in Table V are also of interest because of their possible participation in catalytic systems. In any of the examples shown, or with many other metal ions that exist in two or more oxidation states, facile oxidation of a substrate by one (or more) of the higher valence forms can serve as the basis of a catalytic system consisting of substrate, metal ion, and oxygen. Oxidation of the substrate by the metal ion is then followed by rapid reoxidation of the metal ion by molecular oxygen. Thus the reaction may proceed rapidly even when only a trace of metal ion is present. This type of mechanism is favored when the substrate itself is a ligand and has considerable tendency to combine with the metal ion in dilute solution. Similar oxidation reactions also occur with metal complexes, as well as with simple aquo ions, in which case the metal complex is the catalytic species.

An example of this type of catalysis has been given above (Section IV,B) in the description of the metal chelate-catalyzed oxidation of ascorbic acid.

E. Oxidation of α-Amino Acids

1. *Model Systems*

A model reaction for the oxidation of α-amino acids to α-keto acids has recently been reported by Hamilton and Revesz [120], by Hill and Mann [121], and has been studied in the author's laboratory [121a]. Many amino acids are oxidized to α-keto acids and ammonia by molecular oxygen in the

TABLE V

Oxidation of Metal Ions by Molecular Oxygen

	Metal ion	Medium	Reaction	Rate law	References
1	Cr^{2+}(aq)	$HClO_4$	$Cr^{2+} + O_2 \rightarrow Cr^{4+} + O_2^{2-}$ $Cr^{4+} + Cr^{2+} \rightarrow 2\ Cr^{3+}$		111
2	Cu^{+}(aq)	HCl	$Cu^{+} + O_2 \rightleftharpoons CuO_2^{+}$ $CuO_2^{+} + H^{+} \rightarrow Cu^{2+} + HO_2\cdot$		94,95
3	Cu^{+}(aq)	CH_3COOH	$O_2 + 2\ LiCuCl_2 \rightarrow LiCl_2CuOOCuCl_2Li\ (=A)$	$-\frac{d[Cu(I)]}{dt} = k[Cu(I)][O_2]$	112
		LiCl	$A + CH_3COOH \rightarrow$ $2\ CH_3COOLi + 2\ CuCl_2 + H_2O_2$		
4	Fe^{2+}(aq)	Cl^-	$H^{+} + Fe^{2+} + O_2 \rightarrow Fe^{3+} + HO_2\cdot$	$-\frac{d[Fe^{2+}]}{dt} = k[Fe^{2+}][O_2]$	113
		PO_4^{3-}	$2\ Fe^{2+} + O_2 + 4\ H^{+} \rightarrow 2\ Fe^{4+} + 2\ H_2O_2$		114
5	Fe^{2+}(aq)	SO_4^{2-}	$2\ Fe^{2+} + 2\ H^{+} + O_2 \rightarrow 2\ Fe^{3+} + H_2O_2$	$-\frac{d[Fe^{2+}]}{dt} = k[Fe^{2+}]^2[O_2]$	115,116
		ClO_4^-			108
6	Fe^{2+}(aq)	H_2SO_4	$4\ Fe^{2+} + O_2 + 4\ H^{+} \rightarrow 4\ Fe^{3+} + 2\ H_2O_2$	$-\frac{d[Fe^{2+}]}{dt} = k_1[Fe^{2+}][O_2] + k_2[Fe^{2+}]^2[O_2]$	117
7	Pu^{3+}(aq)	SO_4^{2-}	$2\ Pu^{3+} + 2\ H^{+} + O_2 \rightarrow 2\ Pu^{4+} + H_2O_2$	$-\frac{d[Pu^{3+}]}{dt} = k[Pu^{3+}][O_2]$	118
8	U^{4+}(aq)	$HClO_4$	$2\ U^{4+} + O_2 + 2\ H_2O \rightarrow 2\ UO_2^{2+} + 4\ H^{+}$	$-\frac{d[U^{4+}]}{dt} = k\frac{[U^{4+}][O_2]}{[H^{+}]}$	119
9	V^{3+}(aq)	$HClO_4$	$2\ VOH^{2+} + O_2 \rightarrow 2\ VO^{2+} + H_2O_2$	$-\frac{d[V^{3+}]}{dt} = k\frac{[V^{3+}][O_2]}{[H^{+}]}$	109
10	V^{2+}(aq)	$HClO_4$	$2\ V^{2+} + O_2 \rightarrow 2\ VO^{2+}$ $VO^{2+} + V^{2+} \rightarrow VOV^{4+} \rightarrow 2\ V^{3+} + O^{2-}$		110

presence of pyridoxal and transition metal ions at room temperature and somewhat elevated pH (pH ~9). The mechanism of this reaction seems to be related to those of other pyridoxal-catalyzed reactions [122,123] in that a Schiff base chelate is formed as an intermediate. However, the reaction seems to occur without change of valence of the metal ion and without the formation of free-radical intermediates. Moreover, pyridoxamine is not produced in the reaction, so that the keto acid is not formed by the familiar pyridoxal-metal ion-catalyzed transamination reaction.

In view of this information, the ionic mechanism outlined in Fig. 18 is now proposed. According to this mechanism, an electron shift from the substrate in the Schiff base form to loosely coordinated oxygen occurs through the metal ion, producing a quinonoid (oxidized) structure in the organic ligand while reducing the dioxygen ligand to hydroperoxide. The mixed-ligand complex **60** probably has only transient existence. Metal-catalyzed hydrolysis of the Schiff base and subsequent hydrolytic removal of ammonia regenerates pyridoxal and the original catalyst.

It is seen that this internal oxidation–reduction reaction involves a two-electron shift and does not require that the metal be in the higher valence form to any considerable extent. After the redox shift the oxidized and reduced moieties (keto acid and peroxide) dissociate, leaving the metal-pyridoxal catalyst at its original oxidation level.

The requirement that the metal ion exist in a higher valence state is implied by the intermediate dioxygen complex **60**. Before the reaction occurs (i.e., before the movement of protons) this mixed-ligand complex is stabilized by several resonance forms, only one of which is shown. Most of these resonance forms involve one and two oxidation levels of the metal ion above the 2+ state illustrated.

2. *Enzymatic Systems*

The mechanisms of reaction of amine and amino acid enzymes that require copper(II) and pyridoxal phosphate may be similar to the ionic mechanism suggested in Fig. 18 for model reactions. It is interesting that these enzymes are the only pyridoxal phosphate-containing enzymes that require a metal ion for activity and also the only ones having oxygen as a reactant. From ESR measurements [124–126] it is known that free-radical intermediates are not formed in detectable amounts, thus lending greater plausibility to the proposed ionic type of mechanism.

F. Enzyme-Catalyzed Two-Electron Oxidations by Molecular Oxygen

We have seen in Section IV, B, C, and E that the oxidation of various substrates by molecular oxygen occurs through biological and nonbiological

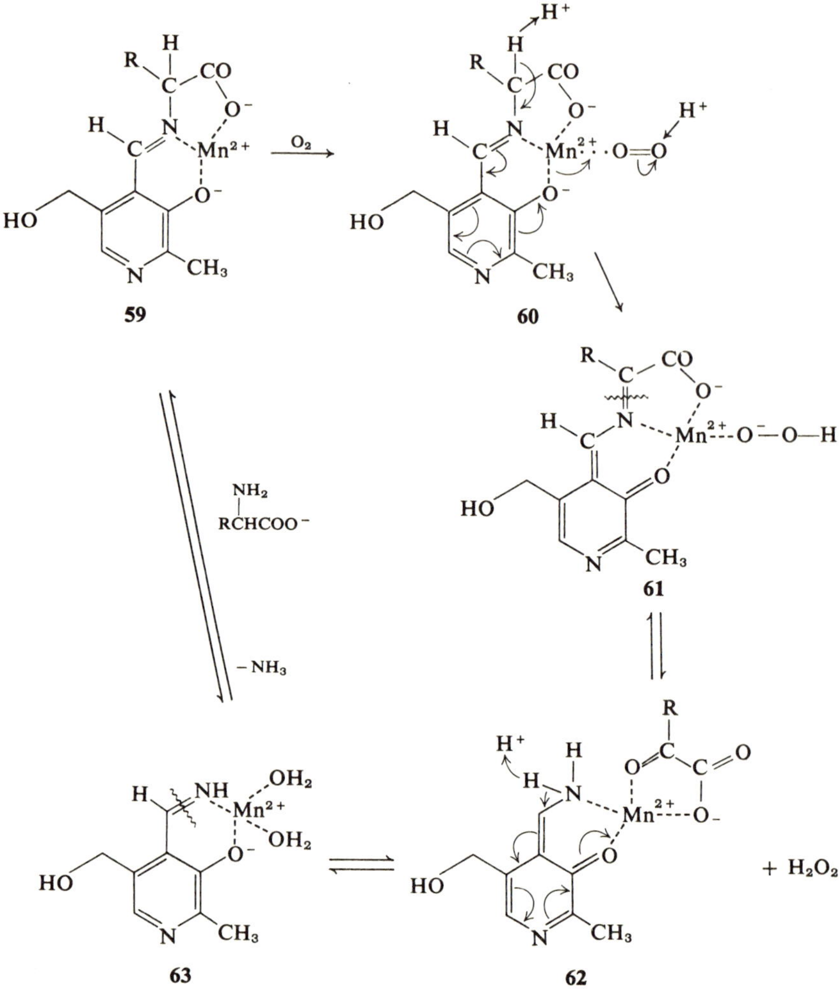

Overall Reaction

$$\underset{|}{\overset{NH_2}{}}RCHCOO^- + O_2 + H_2O \longrightarrow RCOCOO^- + NH_3 + H_2O_2$$

Fig. 18. Amine and amino acid oxidase model.

pathways which frequently have striking similarities, but sometimes differ in some fundamental way. There are many more enzymes that catalyze two-electron non-insertion oxidation reactions in which the oxidant, molecular oxygen, is reduced to hydrogen peroxide. Many of these reactions are believed to involve ionic mechanisms similar to those described above for the metal ion-catalyzed oxidation of ascorbic acid, catechols, and amino acids.

Some of these enzymes, together with experimental observations indicating possible reaction mechanisms, are listed in Table VI [87,98,99,124–132].

TABLE VI
EXAMPLES OF NON-INSERTION TWO-ELECTRON REACTIONS OF MOLECULAR OXYGEN CATALYZED BY OXIDASE ENZYMES

Enzyme	Redox reactions	Mechanistic information	References
Ascorbic acid oxidase	Ascorbic acid → dehydro-ascorbic acid	Semiquinone-type intermediates	98
	$O_2 \rightarrow H_2O_2$	Copper reversibly oxidized and reduced	99
Amino acid oxidases	Amino acid → keto acid + NH_3	Cu(II) and pyridoxal phosphate required	124 125
	$O_2 \rightarrow H_2O_2$	Cu(II) does not change valence	126
Uricase	Uric acid → intermediates → allantoin	One active site per molecule containing Cu(II)	127
	$O_2 \rightarrow H_2O_2$	Cu(II) does not undergo redox reactions	
		No free-radical intermediates	
Galactose oxidase	Galactose → corresponding aldehyde	Cu(II) at active site does not change valence	128
	$O_2 \rightarrow H_2O_2$	No free radicals detected	
Inositol oxygenase	Myoinositol → dialdehyde → D-glucuronic acid	One atom of iron per mole	87 129
	$O_2 \rightarrow H_2O_2$ (H_2O_2 adds to aldehyde)		130
Oxalic acid oxidase	Oxalic acid → CO_2 $O_2 \rightarrow H_2O_2$	Metal ion possible but not detected	131
Impure enzyme preparation	Side chain of cholesterol cleaved to two compounds	Metal ion not known	132
		No free-radical intermediates	
	$O_2 \rightarrow H_2O_2$ (NADPH required) (H_2O_2 reacts with aldehyde)		

V. Non-Insertion Reactions Involving Complete Reduction of Oxygen to Water

With the exception of the reactions given above, metal-catalyzed (or uncatalyzed) oxidation of organic compounds with molecular oxygen produces intermediates, generally of a free-radical nature, that are usually more reactive as reducing agents than the original compound or substrate. Under such conditions the intermediate peroxide formed by reduction of oxygen is usually reduced further to water. At the present time there do not seem to be any examples of the ionic type mechanism (in nonenzyme systems) by which oxygen is converted to water in two successive two-electron steps without free-radical intermediates.

A. Enzyme-Catalyzed Reactions

In enzymatic systems oxidation by molecular oxygen seems to occur both with and without free-radical intermediates. The hydroperoxide intermediate is usually postulated in most of the mechanisms proposed and may (or may not) remain bound to the original site in undergoing the second step in its conversion to water.

There are also examples of enzymatic systems which catalyze one-step redox reactions involving transfer of two electrons and the formation of peroxide. The latter is then converted to water and oxygen by catalase, or by an adjacent site having catalase activity, with the result that water is the only reduction product formed.

1. *Enzymatic Oxidation of Diphenols to Quinones*

The oxidation of *o*-diphenols to quinones, with the corresponding complete conversion of oxygen to water, occurs in biological systems through reaction mechanisms that seem to be both of the ionic and free-radical type. These reactions usually occur through an ionic-type mechanism, but free radicals (semiquinones) may or may not be formed depending on the enzymes and substrates involved. Thus the catecholase action of tyrosinase seems not to involve the formation of semiquinone intermediates and the metal ion [Cu(I)] does not seem to change valence [133]. A probable reaction mechanism, illustrated in Fig. 19, involves two successive two-electron transfers by shifts of electron pairs through the conjugated system that includes the metal ion. The intermediate hydroperoxide (**67**) is pictured as remaining bound to the same metal center. On the other hand, the possibility of the second half of the reaction taking place by coordination of the hydroperoxide groups at an adjacent active site of the enzyme is not precluded. Thus the catecholase activity of tyrosinase can be explained by a series of electronic shifts through the metal ion from the catechol derivative to the oxygen, with redistribution of protons through the solvent to neutralize the negative charges that are transferred. There is no valence change in the metal species shown, it being remembered that the combination of copper(II) with a superoxide anion **65** is electronically equivalent to the coordination of copper(I) with molecular oxygen.

The enzyme laccase, which also catalyzes the dehydrogenation of *o*-diphenols to quinones through the reduction of oxygen to water, is known [134,135] to involve semiquinone radicals as intermediates, and the metal ion

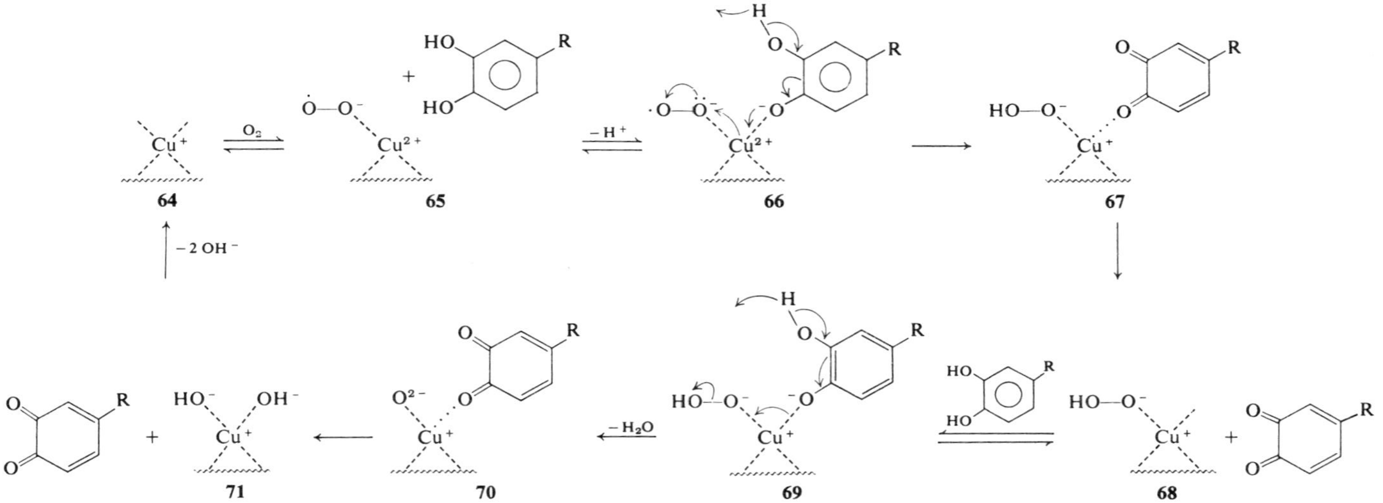

Fig. 19. Enzymatic dehydrogenation of *o*-diphenols to quinones.

is found to be reversibly oxidized and reduced. The mechanism of laccase action may be considered somewhat similar to that given above for catecholase, with the insertion of intermediate steps involving the formation of semiquinone structures between **66** and **67** and between **70** and **71**. In order that these free-radical intermediates and metal redox reactions be detectable, it is necessary that they have appreciable concentrations and considerable lifetimes, and be at least partially dissociated from the metal site. Such considerations indicate that the semiquinone may be formed at a different metal center from that to which the oxygen is bound and that the relatively stable semiquinone intermediate may migrate from one site to the other in order to complete the second phase of the redox process.

Apparently with substrates that can form relatively stable free-radical intermediates such as ascorbic acid and catechols, both the radical and ionic mechanisms may occur. The question of whether free radicals may or may not be formed depends on the nature of the coordination sphere of the metal ion at the active sites of the enzyme, where the semiquinone may be generated and released or further oxidized.

B. Nonenzymatic Free-Radical Reactions

There are many examples of metal catalysis of free-radical reactions of organic substances in which the metal ion is the electron acceptor and oxygen serves to regenerate the oxidized form of the metal ion. In these reactions the metal ion functions as the free-radical initiator and thus serves as the electron carrier in the transfer of electrons from the substrate to the oxygen.

1. *Oxidation of Benzoic Acid to Phenol*

A free-radical reaction of commercial interest is the copper(II)-catalyzed oxidation of benzoic acid to phenol by molecular oxygen [136]. The stoichiometry of the overall reaction is indicated by Eq. (55). Cu(II)–benzoate is

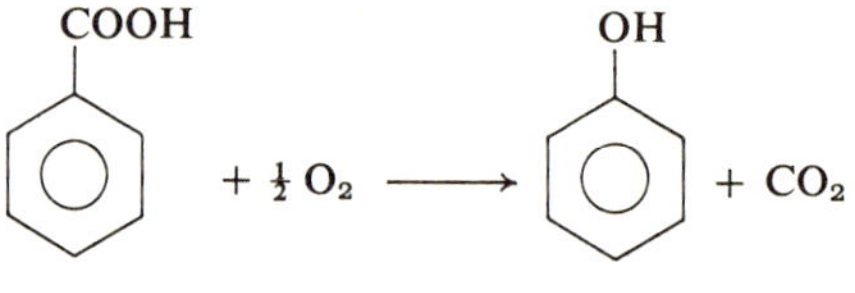

(55)

melted in benzoic acid and a mixture of air and steam is bubbled into the melt at a temperature of 200°–240°C. (Benzoic acid is itself a suitable solvent for the reaction because its high boiling point permits the reaction to be conducted at atmospheric pressure.) The sequence given in Eqs. (56)–(58) has been proposed for the reaction.

The significant step in this reaction is the Cu(II)-catalyzed reaction of benzoate to give salicyl benzoate [Eq. (56)]. As indicated, the reaction probably goes through two successive electron transfers to copper(II) to form copper(I). In the first step the benzoic radical $C_6H_5\dot{C}OO$ is formed. This species then attacks an adjacent benzoate ion at the *ortho* position. The metal ion is believed to play an important part in the direction of this reaction so as to give primarily *ortho* substitution in the formation of the intermediate.

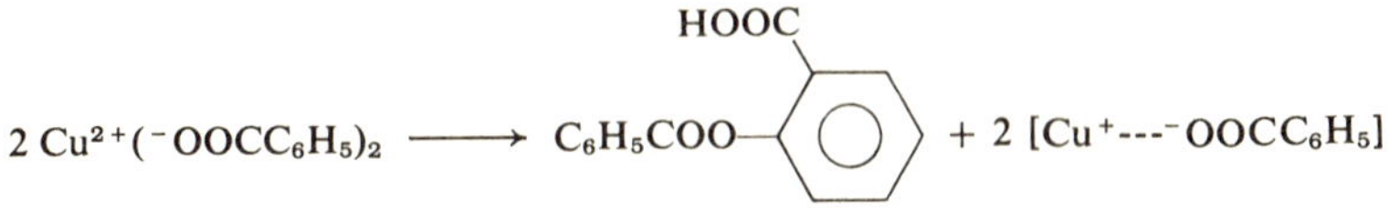

(56)

$$2\,CuOOCC_6H_5 + 2\,C_6H_5COOH + \tfrac{1}{2}\,O_2 \longrightarrow 2\,Cu(OOCC_6H_5)_2 + H_2O \quad (57)$$

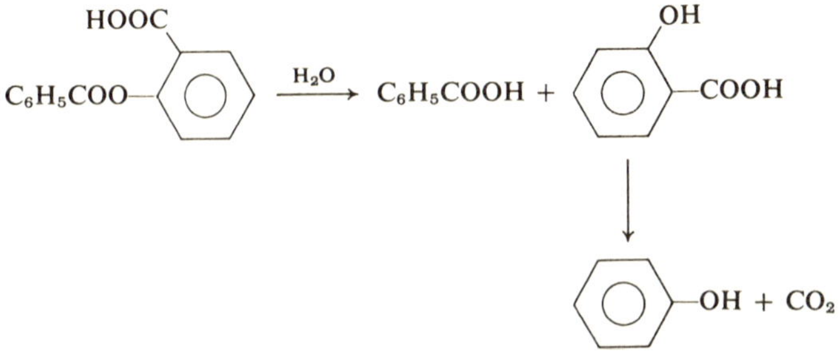

(58)

However, the *para* substituent would also be converted to the same final product. Completion of the reaction is accomplished by a second electron transfer to copper(II). The copper(II) is regenerated by direct oxidation [Eq. (57)] and the phenol is formed by hydrolysis of the salicyl benzoate and subsequent decarboxylation [reaction (58)].

2. *Oxidation of Methanol*

Copper(II) phenanthroline in basic solution catalyzes the oxidation of methanol to formaldehyde in the presence of molecular oxygen [137] by a free-radical mechanism [Eqs. (59)–(62)]. The reaction shows an induction period of a few minutes unless initiated by the addition of a small amount of hydrogen peroxide or cuprous salt. In the first step hydrogen peroxide probably reduces copper(II) phenanthroline to the copper(I) complex, the presence of which seems to be necessary for the continuation of oxidation. In the following steps, the free-radical ion $\dot{O}_2^-$ abstracts a hydrogen atom from copper(II) phenanthroline-bound methanol with subsequent transfer of an electron

$$HO_2^- + \text{Cu(II) complex} \longrightarrow HO_2\cdot + \text{Cu(I) complex} \quad (59)$$

$$HO_2\cdot + OH^- \longrightarrow H_2O + \dot{O}_2^- \quad (60)$$

$$\dot{O}_2^- + CH_3OH\text{Cu(II) complex} \longrightarrow HO_2^- + H^+ + HCHO + \text{Cu(I) complex} \quad (61)$$

$$\text{Cu(I) complex} + O_2 \longrightarrow \text{Cu(II) complex} + \dot{O}_2^- \quad (62)$$

to the copper(II), resulting in the formation of a copper(I) complex and formaldehyde. The copper(I) complex is reoxidized by molecular oxygen to the copper(II) form and reactive $\dot{O}_2^-$.

3. *Carbon-Chain Degradation*

Coordination by metal ions has an electron-withdrawing effect on functional groups, so that they become somewhat protected from direct attack by oxidizing (electron-seeking) agents. If one or more of the functional groups in a coordinated ligand are themselves not coordinated to the metal ion, they may be selectively oxidized (e.g., with molecular oxygen), since these groups are much more vulnerable to attack than the donor groups in the same molecule that are more tightly bound to the metal ion. Although there are few examples of this type of selectivity (and only one will be given here), the principle is a general one, and many more examples can be expected when this type of selective inhibition of oxidation is studied further.

a. Oxidation of N-Hydroxyethylethylenediamine. Huggins and Drinkard [138] have described an interesting reaction in which Co(II)-*N*-hydroxyethylethylenediamine is oxidized by molecular oxygen in the presence of activated carbon to Co(III)–ethylenediamine, considerable free ethylenediamine, and its oxidation products, HCHO, HCOOH, and NH_3. During the course of the reaction, a steady-state concentration of Co(II)–Co(III) is established in which the ratio of Co(III)/Co(II) is 1.2. On the basis of kinetic observations, it was proposed that the cobalt(III) ion acts as a catalyst for the cleavage of *N*-hydroxyethylethylenediamine by bringing together oxygen and the ligand in a mixed-ligand complex. This reaction is apparently an example of selective stabilization of one moiety (the ethylenediamine part) of the ligand against oxidation through strong metal coordination, while another less strongly coordinated part of the ligand is oxidized.

In the absence of activated carbon the cobalt(II) ion was completely oxidized to cobalt(III) by the air stream in 4 hours. In the presence of the activated carbon, however, the cobalt(II) concentration reached a steady-state value of approximately one-fourth of the initial concentration of cobalt(II). It has been suggested by the authors that the oxidation of cobalt(II) to cobalt(III) occurs in conjunction with the oxidative cleavage of the C—C bond of the hydroxyethyl group and the formation of ethylenediamine. The mechanism illustrated in Fig. 20 was proposed.

4. *Oxidation of Mercapto Groups*

The oxidation of mercaptides with molecular oxygen generally is catalyzed by metal ions, and the reactions therefore probably occur in the coordination

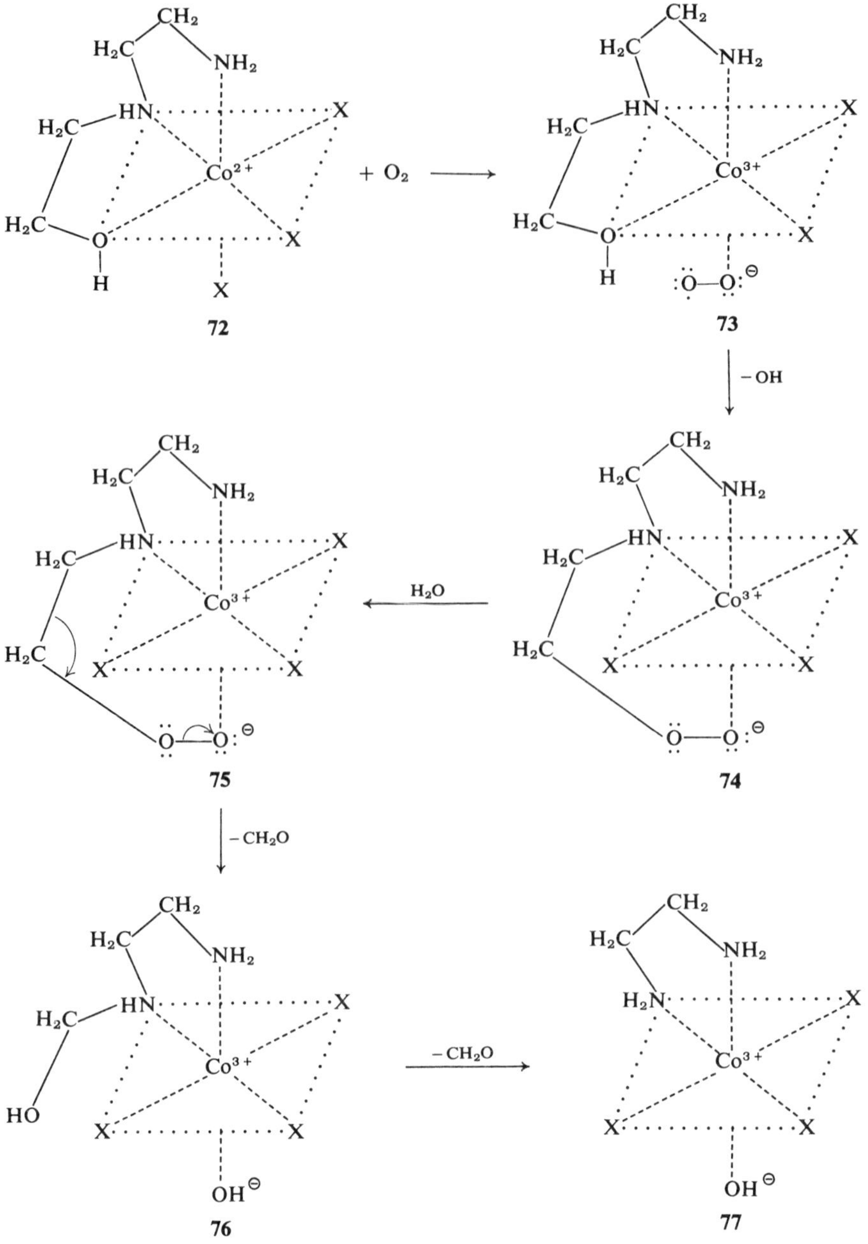

Fig. 20. Metal-controlled oxidation of the hydroxyethyl group of *N*-hydroxyethyl-ethylenediamine.

spheres of the metal ions. Interaction with the metal ions results in one-electron transfer steps to produce mercaptide radicals that then combine in pairs to form disulfides. A disulfide generally has lower coordinating affinity than do the original mercaptides toward the metal ion, so that the complex dissociates and the metal ion is released and reoxidized by molecular oxygen in the body of the solution. Although the reaction sequences shown in this section do not indicate metal complexes, it is considered highly probable that the electron transfer from the negative mercapto group to the metal ion occurs in the appropriate metal–mercaptide complex.

a. Formation of Cystine from Cysteine. Taylor *et al.* [139] have studied the Fe(III)-catalyzed oxidation of cysteine to cystine. The reaction is zero order with respect to molecular oxygen and cysteine but two-thirds order with ferric ion. There is a considerable pH effect, with a rate maximum at pH 8.1. A tris(cysteine)–Fe(III) complex is proposed as the species that undergoes dissociation into a monocysteine–Fe(I) complex and two cysteine radicals that combine to form cystine [see Eqs. (63)–(66)]. In a fast reaction, the iron(I) complex species is reoxidized to tris(cysteinato)–Fe(III). The role of oxygen in the reaction is not clear. In an alternate scheme [Eqs. (67)–(72)], oxygen is involved in forming a complex either with bis(cysteine)–Fe(II) or tris(cysteine)–Fe(III), displacing in both cases cysteine radicals that combine to form cystine. The nature of the cysteine residues that combine to form cystine has not been demonstrated. They may be free-radical intermediates or the electron-paired species, L, L^{2-} (H_2L = cysteine). There are other

Proposed Mechanism

$$Fe^{3+} + 3\,HL \rightleftharpoons Fe(L)_3 + 3\,H^+ \tag{63}$$

$$Fe(L)_3 \rightleftharpoons Fe(L) + 2\,L\cdot \tag{64}$$

$$2\,L\cdot \longrightarrow L_2(\text{cystine}) \tag{65}$$

$$Fe(L) + 2\,HL + \tfrac{1}{2}\,O_2 \longrightarrow Fe(L)_3 + H_2O \tag{66}$$

Alternate Scheme

$$Fe^{3+} + 3\,HL \rightleftharpoons Fe(L)_3 + 3\,H^+ \tag{67}$$

$$Fe(L)_3 \rightleftharpoons Fe(L)_2 + L\cdot \tag{68}$$

$$Fe(L)_2 + O_2 \rightleftharpoons FeL\cdot(O_2) + L\cdot \tag{69}$$

$$Fe(L)_3 + O_2 \rightleftharpoons FeL\cdot(O_2) + 2\,L\cdot \tag{70}$$

$$2\,L\cdot \longrightarrow L_2(\text{cystine}) \tag{71}$$

$$2\,HL + FeL(O_2) \longrightarrow Fe(L)_3 + H_2O + \tfrac{1}{2}\,O_2 \tag{72}$$

possible reaction mechanisms, including electron transfer to oxygen in the metal complex to produce a superoxide ion or, if two electrons are transferred, a peroxide dianion directly. The latter mechanism would be aided by attack of the electron-deficient mercaptide linkage by a coordinated or uncoordinated negative mercaptide anion.

Similar mechanistic possibilities also apply to the cobalt(II)–cobalt(III)-catalyzed oxidation of cysteine to cystine, reported by McCormick and Gorin [140].

b. Oxidation of Mercaptoacetic Acid. Leussing and Tischer [141] have studied the oxidation of ferrous mercaptoacetate in solution by oxygen and H_2O_2. It was concluded that a free-radical reaction is involved. At relatively high concentration of iron(II) and mercaptoacetate the results are in accord with the mechanism of Eqs. (73)–(77).

$$\mathrm{Fe(II)} + O_2 + H_2O \xrightarrow{k_1} \mathrm{Fe(III)} + HO_2\cdot + OH^- \tag{73}$$

$$HO_2\cdot + RSH^- \longrightarrow H_2O_2 + R\dot{S}^- \tag{74}$$

$$\mathrm{Fe(II)} + H_2O_2 \longrightarrow \mathrm{Fe(III)} + OH\cdot + OH^- \tag{75}$$

$$OH\cdot + RSH^- \longrightarrow H_2O + R\dot{S}^- \tag{76}$$

$$2\,R\dot{S}^- \longrightarrow RSSR^{2-} \tag{77}$$

5. *Oxidative Coupling of Organic Compounds*

a. Nonenzymatic Coupling Reactions. Metal ions may act as one-electron oxidants to generate organo free radicals that then undergo coupling reactions. The oxygen serves to reoxidize the metal ion, so that the latter serves as a catalyst for the overall reaction of the organic compound with oxygen.

An example of this type of reaction is the conversion of 2,6-di-*t*-butylphenol to 3,3′,5,5′-tetra(*t*-butyl)diphenoquinone by oxygen with copper(I) as catalyst, described by Ochiai [142]. Amines such as pyridine, ethylenediamine, diethylenetriamine, or triethylenetetramine were employed as coordinating ligands for the copper(I)/copper(II) system. The mechanism given in Eqs. (78)–(80) was suggested for this reaction.

$$2\,CuL_n^{\,+} + \tfrac{1}{2}O_2 + 2\,(m-n)L + 2\,H^+ \longrightarrow 2\,CuL_m^{\,2+} + H_2O \tag{78}$$

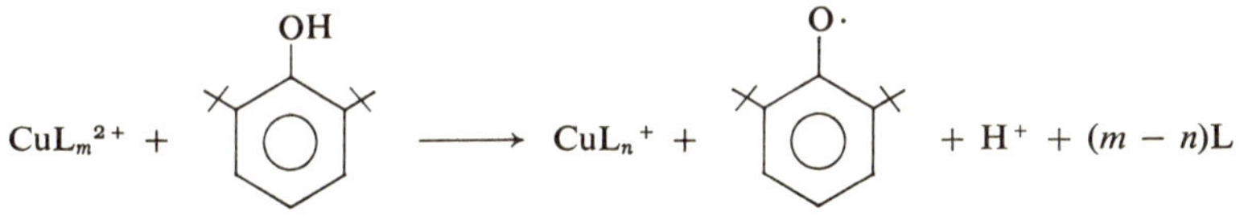

(79)

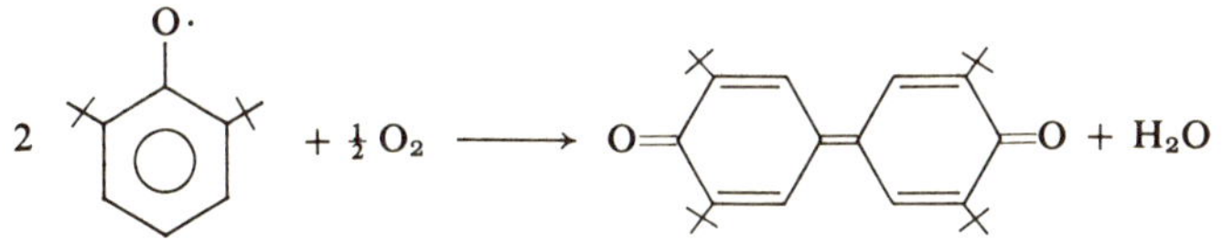

(80)

b. Enzyme-Catalyzed Coupling Reactions. It is now known that oxidative coupling reactions are the basis for the biosynthesis of many natural products [143]. The mechanisms for such reactions are probably similar to the free-radical process described above for the nonenzymatic system, although it has been suggested [144] that an ionic mechanism may be appropriate for biological coupling reactions. An observation that favors the free-radical process is that both enzymatic and nonenzymatic reactions give very low yields of coupled reaction products. Thus far no enzymes have been isolated that are specific for the individual coupling reactions, but it is expected that such enzymes will eventually be isolated and identified.

VI. Non-Insertion Reactions of Hydrogen Peroxide

Two-electron non-insertion reactions in which hydrogen peroxide is an oxidant are of interest because hydrogen peroxide, or a complex peroxide species, is frequently an intermediate in the four-electron oxidations in which the oxidant is molecular oxygen. As pointed out above (Section V, page 128), there are two general pathways for completion of the reaction with conversion of hydrogen peroxide to water. A common route would be disproportionation of the hydrogen peroxide to oxygen and water (catalase action), with the oxygen thus formed recycled to the reaction. Appropriate substrates can also be dehydrogenated by catalase enzymes. Examples of these reactions are given in this section.

A second route for the second two-electron oxidation of a four-electron sequence is retention of the peroxide at the reactive site with subsequent conversion to water through reaction with the substrate. An example of this type of mechanism has been illustrated in Fig. 19, in which the substrate, catechol, is a two-electron reductant, with 2 moles of product, *o*-quinone, formed for each mole of oxygen reduced. The initial two-electron reduction of oxygen produces a hydroperoxide anion at the reactive site. The latter then is the oxidant for a second mole of substrate.

Peroxidase enzymes, which are very similar to, and have some reactions in common with, the catalases, generally involve one-electron transfers and free-radical intermediates. Although these catalysts are also discussed in this section, the intermediate (oxidized) forms of these enzymes also catalyze oxygen insertion and are therefore considered in Section VII.

A. Catalase Models

a. Copper(II)-Polyamine Complexes. In a recent review of model catalase and peroxidase reactions, Siegel [145] has discussed the catalase

$$CuL^{n+} + H_2O_2 \longrightarrow CuL \cdot H_2O_2^{n+} \longrightarrow CuL^{n+} + \tfrac{1}{2}\,O_2 + H_2O$$

Polyamine–Cu(II) Chelates

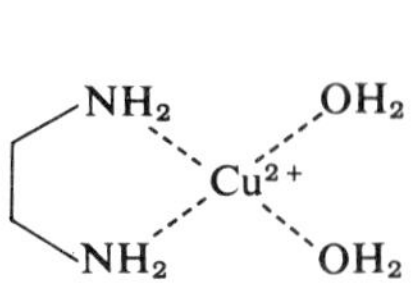

78

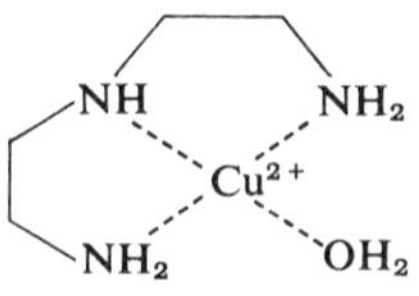

79

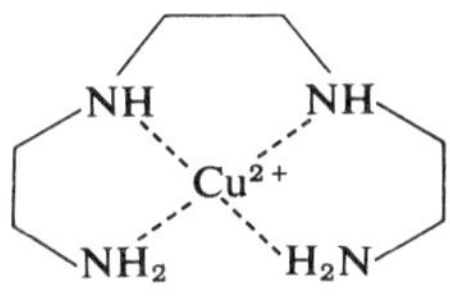

80

$$enCu^{2+} \gg dienCu^{2+} > trienCu^{2+} \sim 0$$

Diglycylalkylenediamine–Cu(II) Chelates

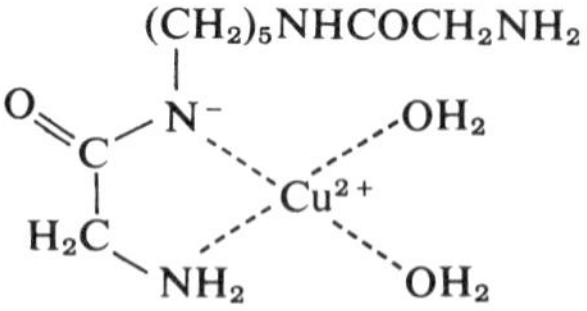

81

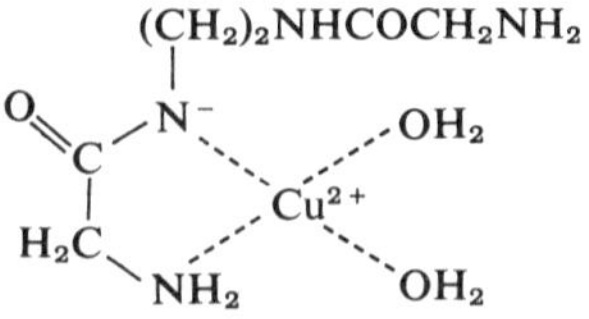

82

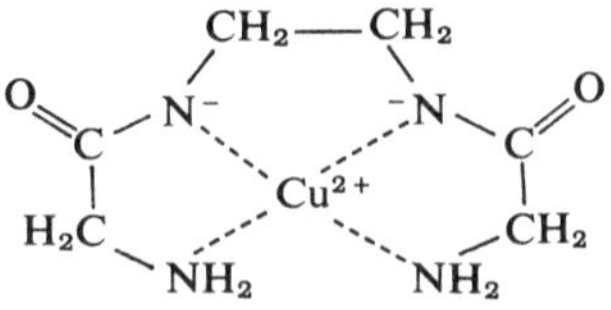

83

81 > **82** ≫ **83** ~ 0

Fig. 21. Cu(II) chelates as models of catalase activity.

activities of a number of copper(II) chelates, some of which are listed in Fig. 21. From the data given, it is seen that for the polyamine–Cu(II) chelates two free coordination sites such as in formulas **78** and **81** are required for full catalytic activity. The fully coordinated copper(II) complexes, e.g., **80** and **83,** have been found to be inactive.

Similarly, of the diglycyl derivatives of diamines, in which the metal ion is complexed with displacement of protons from the amide nitrogens, the most active species are those in which the ligand occupies two coordination sites on the metal ion, and in which the probability of forming more completely coordinated species is relatively low. The highly stable complex **83,** in which the ligand is tetradentate, is inactive as a catalyst.

The reaction mechanism thus seems to involve the combination of the metal ion with two hydroperoxide anions, followed by an intramolecular two-electron shift and a corresponding transfer of protons through the solvent, in a manner similar to that indicated in Fig. 22 for a metal ion of coordination number 6.

b. Manganese(II)– and Iron(III)–Trien Systems. The concept of the requirement of two free coordination positions about the metal ion is

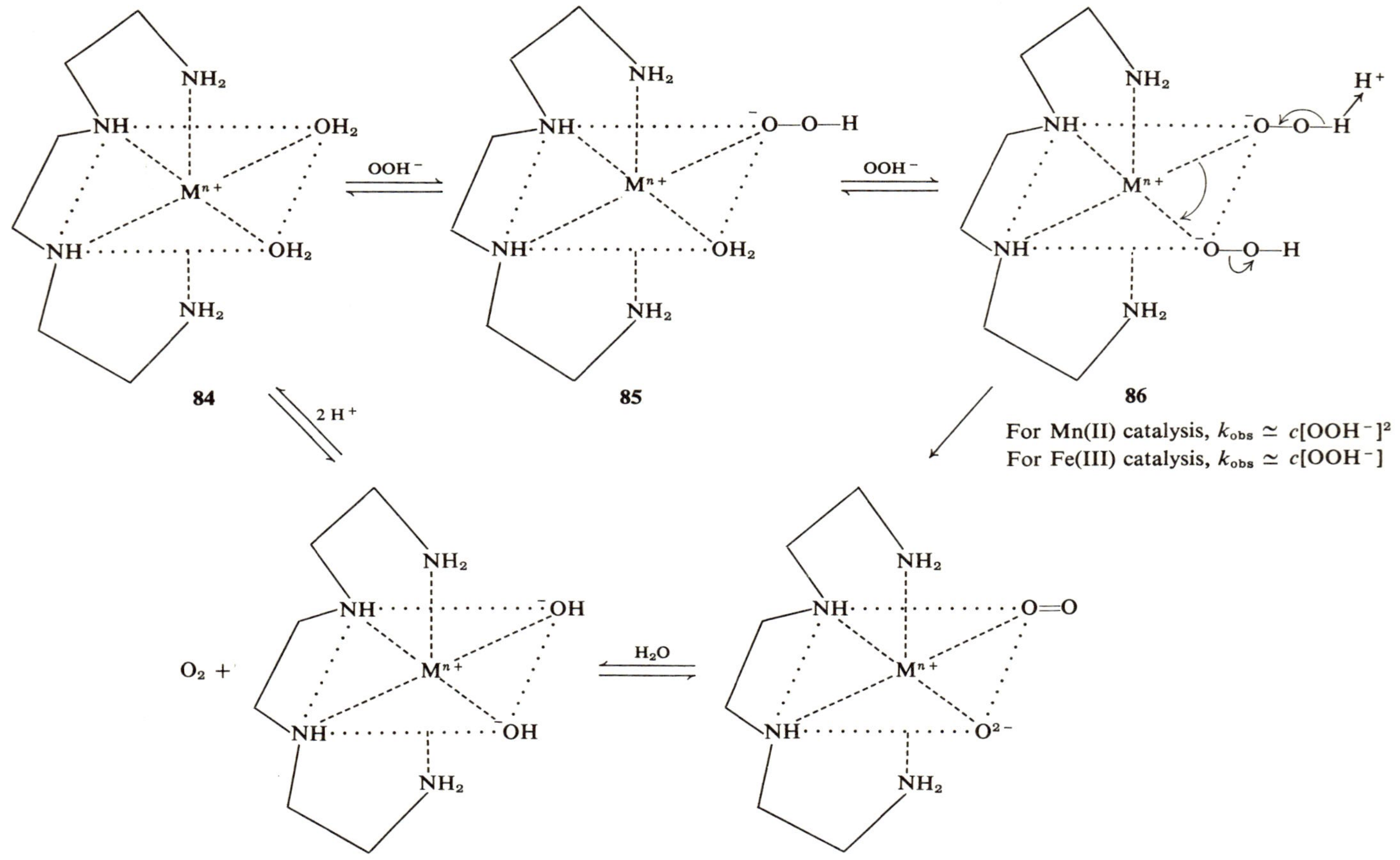

Fig. 22. Octahedral triethylenetetramine chelates as models of catalase action.

strengthened by the observation that for manganese(II) and iron(III), which have coordination numbers of 6, the tetradentate diethylenetriamine chelate is again an active catalyst, while it was found to be inactive with copper(II) as the central metal ion. The catalase activities of these metal chelate systems were first described by Wang [146,147], who proposed that the reaction takes place through metal coordination such that one peroxide anion is bidentate and occupies two *cis* positions of the metal ion. In view of the fact that the reaction is second order in peroxide concentration when manganese(II) is the catalytic metal ion, Hamilton [144] suggested the mechanism described in Fig. 22 in which two hydroperoxide groups are simultaneously coordinated to the two available sites on the metal ion (**86**). In this complex, the electron transfer could take place smoothly through the metal ion to give oxygen and water, the protons being transferred through the aqueous solvent. The first-order dependence of the Fe(III)–trien reaction rate on peroxide concentration was considered due to the fact that the higher coordinating ability of the iron(III) ion resulted in one of the peroxide anions being bound to the ferric ion at the beginning of the reaction, leaving only one more to be added prior to the rate-determining step. In view of all the facts available this reaction mechanism seems to be reasonable. A similar mechanism may apply to the active copper(II) catalase model catalysts described in Fig. 21.

B. Catalase Enzymes

The catalases [148], which have iron(III)–porphyrin prosthetic groups at the active sites, are catalysts for the disproportation of hydrogen peroxide to water and oxygen, a process known as the catalatic reaction. Catalases are very closely related to the peroxidases [149,150] and the two types of enzymes have certain mechanistic steps in common. An oxidized form of the enzyme, **93**, arises when the free enzyme reacts with hydrogen peroxide or an organic hydroperoxide. Compound **93**, which has an oxidation state two units higher than the original active center of the enzyme, then reacts directly with hydrogen peroxide to give oxygen. As suggested by George [151], **93** probably does not contain coordinated peroxide, but consists of an oxidized form of the iron–porphyrin system as indicated by the reaction sequence shown in Eqs. (81)–(85)

$$\mathrm{E{-}Fe^{3+} + H{-}O{-}OH \rightleftharpoons E{-}Fe^{3+}{\cdots}O^{-}{-}O{-}H + H^{+}} \qquad (81)$$

$$\underset{}{\mathrm{E{-}Fe^{3+}{\cdots}O^{-}{-}O{-}H}} \longrightarrow \underset{\textbf{89a}}{\mathrm{E{-}Fe^{3+}\cdot O^{0}}} \longleftrightarrow \underset{\textbf{89b}}{\mathrm{OxE{-}Fe^{3+}{\cdots}O^{2-}}} + \mathrm{OH^{-}} \qquad (82)$$

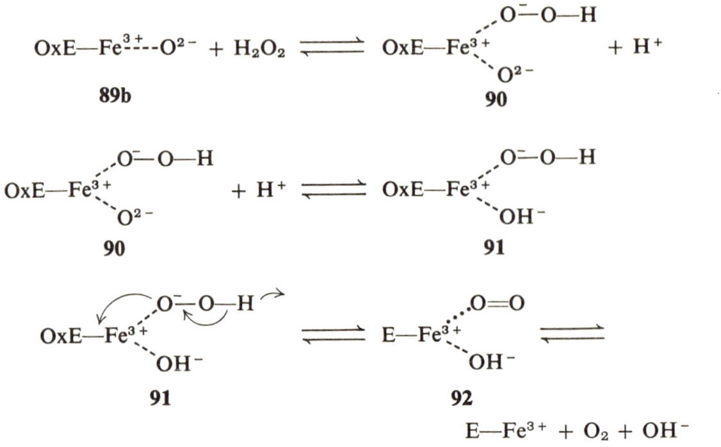

(83)

(84)

(85)

where OxE represents a two-electron oxidized form of the enzymatic active site resulting from the removal of two electrons from the porphyrin ring coordinated to the iron(III) ion. Thus the intermediate of catalase is presented by the two extreme formulas **89a** and **89b**, which are resonance forms of a single complex. Recognizing the ability of the metal ion to also exist in the 2+ state, and the resonance-stabilized free-radical intermediates of the conjugated porphyrin ligand, it is seen that formula **89** is stabilized by considerable resonance energy. Similar arguments may be used for predicting stabilities of other intermediate complexes, **90**, **91**, and **92**, formed with a second hydroperoxide anion.

The reversible oxidation and reduction of the porphyrin ring system leading to the resonance-stabilized oxidation intermediate of catalase, is represented by formulas **93a** and **93b**. It is seen that the removal of two electrons from the porphyrin ring produces a quinonoid structure analogous to the formation of a quinone from a hydroquinone or catechol, or the formation of dehydroascorbic acid from ascorbic acid. In the structures shown, one cannot discriminate between the oxidized and reduced forms of the metal ion, of the coordinated oxygen, or of the porphyrin ring. The only meaningful interpretation is one that considers the oxidation state of the complex as a whole. This complex can also exist in a number of additional free-radical-type resonance forms, as pointed out above, involving different charges on the metal ion-coordinated oxygen, and the porphyrin ring. The extent of the contributions of the individual resonance forms to the overall structure of **93** (i.e., the electron and charge distribution in the complex) will depend on the ligand X in the remote axial position (which is not known for the catalases), as well as on the placement of hydrophobic and polar groups about the active site.

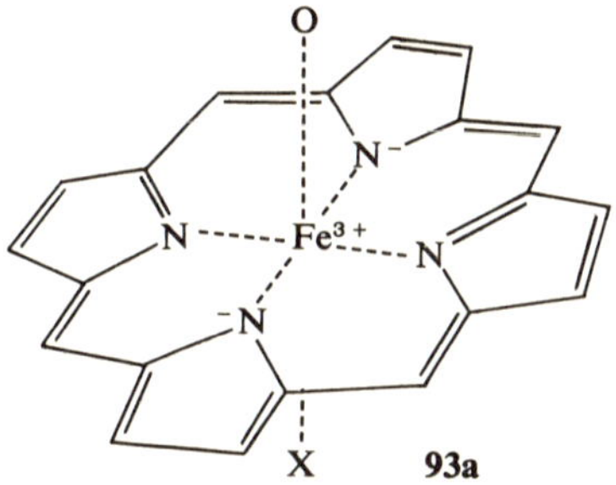

93a

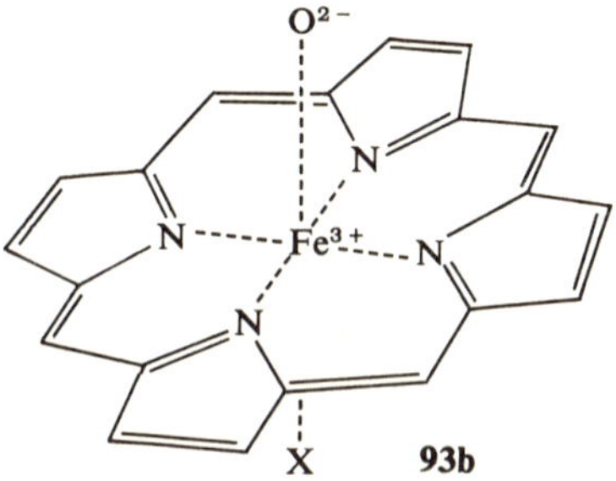

93b

Oxidized intermediate in catalase and peroxidase enzymes

a. Oxidation of Alcohols. As noted above, the catalase enzymes oxidize alcohols as well as promote the disproportionation of hydrogen peroxide. A possible mechanism for such reactions that are analogous to the mechanism proposed above for hydrogen peroxide is shown in Eqs. (86)–(88).

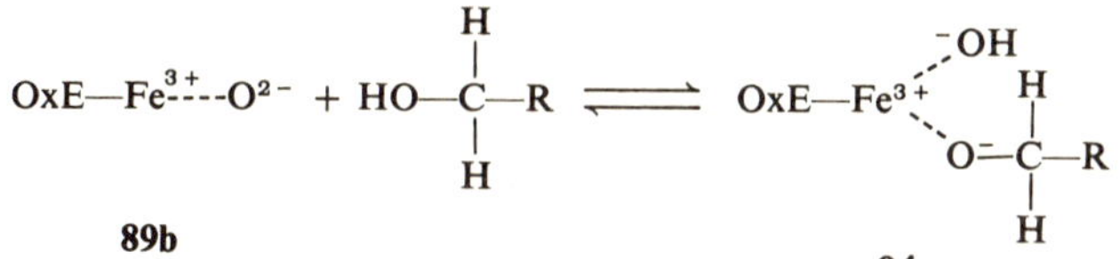

(86)

89b **94**

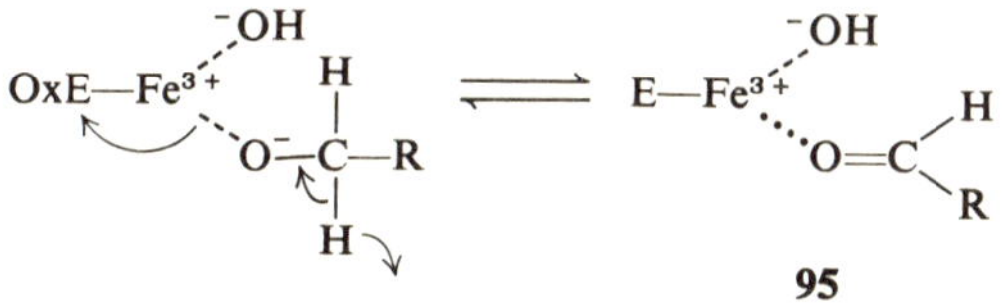

(87)

95

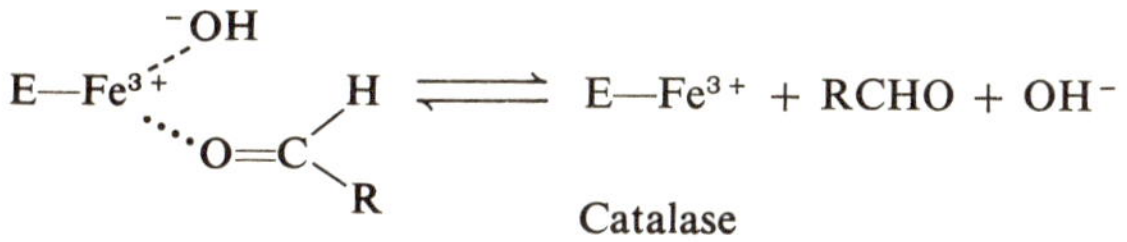

(88)

Overall Reaction

$$\mathbf{89b} + RCH_2OH \longrightarrow \text{Catalase} + RCHO + H_2O$$

It is seen that intermediate compounds **94** and **95** are analogous to intermediates **91** and **92** in the reaction scheme [Eqs. (81)–(85)] proposed for disproportionation of hydrogen peroxide.

Many of the reactions presented above, Eqs. (81)–(85) and (86)–(88), for the mechanism of catalase action may very well occur in a concerted fashion, so that two or more successive reactions may be combined. Each individual chemical transformation is represented as a separate reaction, however, to enable the reader to more readily perceive the changes taking place. While there seems to be no proof at the present time for these mechanisms, they seem to represent the most attractive possibilities in view of the evidence available.

Additional evidence favoring the reaction mechanisms described above for catalase enzymes is their similarity to a model catalase system described recently by Hamilton [144]. It was discovered that a complex of iron(III), hydrogen peroxide, and catechol would oxidize (dehydrogenate) alcohols, just as has been illustrated above (**86–88**) for **93**. The active intermediate suggested for this reaction is similar to **93**, in which the catechol, metal ion, and a single oxygen donor have an oxidation state two units higher than the original ligand and metal ion, but equivalent to the combination of ligand, metal ion, and peroxide. Since this intermediate is an effective catalyst in oxygen insertion reactions, a more complete discussion of its properties is reserved for Section VII.

C. Peroxidase Enzymes

Peroxidases, which are very closely related to the catalases, promote the oxidation by hydrogen peroxide of many organic compounds such as aromatic amines and phenols. The relationships between catalase and peroxidase action, and the nature of the intermediates, have been extensively studied by Chance [152]. Compound **93** (sometimes called Compound I) described above, is common to both types of enzymes. In peroxidase, **93** usually reacts with the substrate (aromatic phenol or amine) to give a second intermediate (Compound II) and an aromatic free radical. Compound II, which has an oxidation state one unit higher than the enzyme, then reacts with substrate to produce the original peroxidase enzyme and a second free radical. Thus Compound II has one oxidizing equivalent per mole, as compared with **93** which has two oxidizing equivalents per mole. This difference in oxidation state has been demonstrated by direct titration as well as by the nature of the reactions which these intermediates undergo. There is no evidence for free-radical intermediates in catalase action, which involves the formation of only **93** as the intermediate oxidant.

That Compounds I and II are not merely peroxide complexes of the iron(II) and iron(III) forms of the enzymes has been demonstrated by preparing these intermediates with a wide variety of oxidizing agents. Thus Compound II must have a structure similar to that for **93** with the exception that an additional electron is present. Such an intermediate would have considerable resonance stabilization.

Since peroxidase enzymes are also active as catalysts for the introduction of oxygen into organic substrates, further discussion will be reserved for Section VII,A,4, which deals with oxygen insertion reactions of peroxides.

VII. Partial Insertion Reactions

A. Partial Insertion Reactions with Peroxide as an Oxidant

Reactions in which the O—O bond is broken and the oxygen enters the substrate to form hydroxyl groups are generally termed hydroxylation reactions. One or both of the fragments may then enter the substrate, depending on the reaction mechanism. Since the steps leading to such hydroxyl insertion reactions may or may not go through hydrogen peroxide, it is of interest to also discuss hydroxylation reactions in which hydrogen peroxide is itself a reactant, as is done below.

1. *Fenton's Reagent*

In order to understand the nature of the hydroxylating species in the metal-catalyzed reactions of molecular oxygen and hydrogen peroxide, it is important to consider the reactions of Fenton's reagent. The use of Fenton's reagent for hydroxylation has been widely explored by a number of investigators [153–159]. The presence of hydroxyl (OH) radicals as the active species in the reactions of Fenton's reagent has been substantiated by a parallelism in the nature of the products, their yields, and isomer distribution in the corresponding reactions involving ionizing radiation [159–165]. Strong evidence for the presence of the OH radical as the active hydroxylating species of Fenton's reagent comes from the work of Baxendale *et al.* [166], in which the polymerization of monomers by Fenton's reagent indicates the presence of free radicals.

a. The Nature of Fenton's Reagent. The oxidation of iron(II) by H_2O_2 was first studied by Haber and Weiss [167]. Their mechanism is based on the concept of one-electron transfer as shown in Eqs. (89)–(92).

$$Fe^{2+} + H_2O_2 \xrightarrow{k_1} Fe^{3+} + OH\cdot + OH^- \tag{89}$$

$$Fe^{2+} + OH\cdot \xrightarrow{k_2} Fe^{3+} + OH^- \tag{90}$$

$$OH\cdot + H_2O_2 \xrightarrow{k_3} HO_2\cdot + H_2O \tag{91}$$

$$HO_2\cdot + H_2O_2 \xrightarrow{k_4} O_2 + H_2O + OH\cdot \tag{92}$$

The relative extent of the oxidation of iron(II) is measured by a factor $\bar{n}$ [168] defined by

$$\bar{n} = \frac{\text{number of moles of } H_2O_2 \text{ consumed}}{\text{number of moles of ferrous salt oxidized}} \tag{93}$$

The value of $\bar{n}$ depends on the concentration of the reactants and the speed of mixing. In the case where the speed of mixing was rapid compared to the rate of the reaction, reproducible values of $\bar{n}$ were obtained at well-defined initial concentrations of the reactants. With different mixing devices, Haber and Weiss [167] got a value of 0.5 for the ratio $\bar{n}$, under conditions where a steady state had been reached. In such a case

$$\bar{n} = \frac{\Delta(H_2O_2)}{\Delta(Fe^{2+})} = 0.5 \tag{94}$$

At a value of $\bar{n}$ of 0.5, Eq. (89) and (90) are the only important reactions, and 2 moles of ferrous ion are oxidized per mole of hydrogen peroxide consumed. At a value of $\bar{n}$ greater than 0.5, the chain reactions (91) and (92) are superimposed on (89) and (90) and the value of $\bar{n}$ is a complicated function of k_4, k_3, k_2, and k_1 [168].

Studies of the oxidation of iron(II) by H_2O_2 have been reported by Evans *et al.* [169–171], Baxendale [168], Kolthoff and Medalia [156], and also Barb *et al.* [154,155]. Barb *et al.* [155] studied the oxidation of iron(II) by H_2O_2 in considerable detail by varying the concentration ratio H_2O_2/Fe(II) over a wide range in the presence and absence of iron(III) and copper(II) ions. The mechanism proposed is similar to that of Haber and Weiss [167] and includes Eqs. (89), (90), and (91) plus two additional reactions, (95) and (96).

$$HO_2\cdot + Fe^{2+} \xrightarrow{k_5} HO_2^- + Fe^{3+} \tag{95}$$

$$HO_2\cdot + Fe^{3+} \xrightarrow{k_6} O_2 + H^+ + Fe^{2+} \tag{96}$$

At low initial concentrations of iron(II) and hydrogen peroxide ($\approx 10^{-4}$ *M*), the reaction follows second-order kinetics and 2 moles of ferrous ion are oxidized per mole of hydrogen peroxide consumed. Step (89) becomes rate-determining and (90) is a fast subsequent step. The rate of ferrous ion oxidation is then expressed by Eq. (97).

$$-\frac{d[Fe^{2+}]}{dt} = k[Fe^{2+}][H_2O_2] \tag{97}$$

$$k = 2k_1$$

The rate law of Eq. (97) has been confirmed by Kolthoff and Medalia [156] by the study of the oxidation of iron(II) by H_2O_2 in the absence of an organic substrate. As the ratio of $H_2O_2/Fe(II)$ is increased, deviation from second-order kinetics is observed [155]. The course of the reaction is then determined by competition for the OH· radical in steps (90) and (91). At a sufficiently high $H_2O_2/Fe(II)$ ratio, step (90) may be neglected and the competition for $HO_2\cdot$ radical in steps (95) and (96) predominates. The ratio of the rate constants k_6/k_5 may then be determined from a steady-state calculation taking into consideration reactions (89), (91), (95), and (96). The ratio k_6/k_5 has been found to be inversely proportional to hydrogen ion concentration, since both reactions (95) and (96) involve the pH-dependent equilibrium (98).

$$HO_2\cdot \rightleftharpoons H^+ + \dot{O}_2^- \tag{98}$$

In the presence of added ferric ion as catalyst, step (98) becomes insignificant since most of the hydroperoxide free radical, $HO_2\cdot$ will then react with iron(III) and the oxygen evolution reaction becomes important. When cupric ion is added to the system, the rate of disappearance of $HO_2\cdot$ increases because copper(II) reacts more rapidly with $HO_2\cdot$ than does the iron(III) ion. The effect of copper(II) ion is the catalysis of step (96) through reactions (99) and (100).

$$Cu^{2+} + HO_2\cdot \xrightarrow{\text{fast}} Cu^+ + O_2 + H^+ \tag{99}$$

$$Cu^+ + Fe^{3+} \xrightarrow{\text{very fast}} Cu^{2+} + Fe^{2+} \tag{100}$$

In the presence of cupric ion, the ratio of the rates k_3/k_2 is determined from the oxygen evolution data taking into consideration steps (89), (90), (91), (96), and (97) for the steady-state calculation. The value of the ratio k_3/k_2 at 25°C has been reported pH-independent and equal to 0.011. The ratio k_3/k_2 may also be calculated from the quantity $\bar{n}$, and the ratio of the initial concentrations of H_2O_2 and iron(II) by Eq. (101).

$$\bar{n} = 0.5 + \frac{k_3}{k_2}\frac{[H_2O_2]}{[Fe^{2+}]} \tag{101}$$

The calculated value of k_3/k_2 at 25°C is 0.03. Baxendale [168] attributed the discrepancy in the experimental and calculated values of k_3/k_2 to the presence of overlapping reactions other than those represented in steps (90) and (91). One of the possible reactions is the formation of ferryl ion FeO^{2+} from iron(III) and OH· according to Eq. (102). The ferryl ion FeO^{2+} has been

$$Fe^{3+} + OH\cdot \longrightarrow Fe(OH)^{3+} \longrightarrow FeO^{2+} + H^+ \tag{102}$$

originally suggested as an intermediate by Bray and Gorin [172]. It has been pointed out by Kolthoff and Medalia [156] that FeO^{2+} can replace OH· in steps (90) and (91) without altering the overall kinetic picture. The significance of FeO^{2+} in the reactions of H_2O_2 will be discussed elsewhere in this section.

A significant and interesting study of the oxidation of iron(II) by H_2O_2 is the demonstration of the presence of free radicals by Baxendale *et al.* [166]. The reaction was studied in the presence of the monomers acrylonitrile, styrene, and methyl methacrylate. These monomers are rapidly polymerized in the presence of free radicals. The presence of oxygen affects the polymerization reaction because of the reaction of oxygen with the organic radical monomer. There is also competition between iron(II) and the monomer for OH·.

$$OH\cdot + CH_2{=}CHX \longrightarrow HO{-}CH_2{-}\dot{C}HX \longrightarrow \text{polymers} \quad (103)$$

The results obtained on the Fe(II)–H_2O_2 system have been applied to the Fe(III)–H_2O_2 system, and the mechanism shown in the following five equations for the reduction of Fe(III) by H_2O_2 was proposed by Barb *et al.* [154,155]:

$$Fe^{3+} + HO_2^- \xrightarrow{k_7} Fe^{2+} + HO_2\cdot \quad (104)$$

$$Fe^{2+} + H_2O_2 \xrightarrow{k_1} Fe^{3+} + OH\cdot + OH^- \quad (89)$$

$$OH\cdot + H_2O_2 \xrightarrow{k_3} HO_2\cdot + H_2O \quad (91)$$

$$HO_2\cdot + Fe^{2+} \xrightarrow{k_5} Fe^{3+} + HO_2^- \quad (95)$$

$$Fe^{3+} + HO_2\cdot \xrightarrow{k_6} Fe^{2+} + O_2 + H^+ \quad (96)$$

Except for the initial reaction of iron(III) with HO_2^-, the steps in this mechanism are similar to those of the Fe(II)–H_2O_2 system. The enthalpies associated with the various steps in the mechanism have been estimated [15] by taking into consideration the ionization potential of iron(II) in aqueous solution, the electron affinities of the radicals HO· and $HO_2\cdot$, heats of solvation of the corresponding ions OH^- and HO_2^- in water, and the strength of the HO—OH bond. Steps (91), (95), and (96) are exothermic and the reactions proceed very rapidly. Steps (104) and (89) are endothermic to the extent of 28 and 5 kcal/mole, respectively, and are slow steps in the mechanism.

b. Number of Electrons in Electron-Transfer Processes. The question of one-electron or two-electron transfer in oxidation–reduction reactions of hydrogen peroxide has been the subject of much recent interest. The idea of one-electron transfer was originally suggested by Haber and Weiss [167], and later elaborated by Weiss [173]. This concept has been successful in explaining a variety of photochemical and photosensitized reactions [174,175] and also the formation of colored molecular complexes between unsaturated hydrocarbons and quinones or nitro compounds. The driving force for one-electron transfer in these reactions is the entropy gain in the formation of a free-radical complex in the process. If the complex has sufficient stability in solution it may be detected, otherwise rearrangement within the complex may result in the formation of electron-paired species. For a two-electron transfer, very strong interaction is needed between the donor and acceptor since the electrons have to be paired before transfer. This is expected on the basis of the Franck–Condon principle, since the energy difference between the donor and acceptor should be at a minimum. A two-electron transfer usually takes place along with an atom transfer. Thus electron transfer between donor and acceptor by a one-electron or two-electron path depend on the nature of donor and acceptor and the nature of the complex formed between them in the transition state.

Bray and Gorin [172] have proposed a two-electron oxidation of iron(II) to iron(IV) in the reaction of iron(II) and hydrogen peroxide, and proposed the mechanism given in Eqs. (105)–(108). The ferryl ion FeO^{2+} formed either

$$Fe^{2+} + H_2O_2 \longrightarrow FeO^{2+} + H_2O \quad (105)$$

$$FeO^{2+} + H_2O_2 \longrightarrow Fe^{2+} + H_2O + O_2 \quad (106)$$

$$Fe^{3+} + OH\cdot \longrightarrow FeO^{2+} + H^+ \longrightarrow FeOH^{3+} \quad (107)$$

$$Fe(OH)^{3+} + Fe^{2+} \longrightarrow 2\,Fe^{3+} + OH^- \quad (108)$$

in reaction (105) or (107) reacts in subsequent steps either with H_2O_2 or iron(II). The oxidation of iron(II) to iron(IV) by a two-electron reduction of H_2O_2 has also been proposed by Cahill and Taube [176] based on $H_2{}^{18}O_2$ studies. Taking into consideration the ^{18}O enrichment in H_2O_2 and the enrichment of ^{18}O in the products of the reaction, molecular oxygen and H_2O, it was concluded that in a two-electron transfer reaction of $H_2{}^{18}O_2$ the rate is slowed down to the extent of 0.06%, whereas in a one-electron transfer reaction it remains the same as that obtained for H_2O_2. The evidence in favor of a two-electron transfer is thus not overwhelming and the alternate path involving transfer of one electron cannot be ruled out. The presence of iron(IV) in Fe(II)–H_2O_2 systems has recently been disputed by Conocchioli *et al.* [177].

King and Winfield [178] reported the two-electron oxidation of ruthenium(II) to ruthenium(IV) complexes in the catalytic reduction of H_2O_2 by ruthenium(II) in aqueous solution. The complexes of ruthenium(II) studied were $[Ru(acac)(dipy)_2][ClO_4]$, $[Ru(gly)(py)_4][ClO_4]$, $[Ru(dipy)_2(gly)][ClO_4]$, and $[Ru(acac)(py)_4][ClO_4]$.

It has been proposed [178] that for a two-electron oxidation of the metal ion, with no formation of OH· or HOO· free radicals, the metal complex system should have an oxidation potential of +0.8 V for reaction (109).

$$ML^{n+} \longrightarrow ML^{n+2} + 2\,e \quad (109)$$

The value of the oxidation potential of +0.8 V, has been obtained by comparing the standard free energy change ($\Delta G^\circ = -36.9$ kcal/mole) of reaction (109) with the overall standard free energy change for a two-electron reduction of aqueous H_2O_2 ($\Delta G^\circ = -32$ kcal/mole). A two-electron oxidation or reduction of the metal complex system has thus been found to be thermodynamically feasible.

c. *Hydroxylation by Fenton's Reagent.* Kolthoff and Medalia [156] and Barb *et al.* [154,155] studied the reaction of iron(II) with hydrogen peroxide in the presence of an added organic substrate and suggested the following mechanism in Eqs. (89), (90), and (110)–(114) for its oxidation by Fenton's reagent (Fenton's oxidation).

$$Fe^{2+} + H_2O_2 \longrightarrow Fe^{3+} + OH\cdot + OH^- \quad (89)$$

$$Fe^{2+} + OH\cdot \longrightarrow Fe^{3+} + OH^- \quad (90)$$

$$H_2A + OH\cdot \longrightarrow HA\cdot + H_2O \quad (110)$$

$$HA\cdot + Fe^{3+} \longrightarrow HA^+ + Fe^{2+} \quad (111)$$

$$HA^+ + OH^- \longrightarrow HAOH \text{ (primary product)} \quad (112)$$

$$HAOH + OH\cdot \longrightarrow HOA\cdot + H_2O \quad (113)$$

$$HOA\cdot + Fe^{3+} \longrightarrow Fe^{2+} + H^+ + AO \text{ (secondary product)} \quad (114)$$

Cleavage of the HO—OH bond by iron(II) produces $HO\cdot$ and OH^-. The hydroxyl radical then reacts either with iron(II) or the substrate in steps to yield iron(III) or the substrate radical, $HA\cdot$, respectively. The substrate radical then undergoes a series of oxidations in subsequent steps to form the primary product HAOH. Continuation of free-radical reactions results in dehydrogenation to form AO or in the formation of higher hydroxylation products.

Kolthoff and Medalia [179] have proposed the mechanism of Eqs. (89), (90), (110), and (115)–(120) for the oxidation of an organic compound by iron(II) and hydrogen peroxide in the presence of molecular oxygen. The first three reactions are similar to the mechanism suggested above in the absence of oxygen.

$$Fe^{2+} + H_2O_2 \longrightarrow Fe^{3+} + OH\cdot + OH^- \quad (89)$$

$$Fe^{2+} + OH\cdot \longrightarrow Fe^{3+} + OH^- \quad (90)$$

$$H_2A + OH\cdot \longrightarrow HA\cdot + H_2O \quad (110)$$

$$HA\cdot + O_2 \longrightarrow HAO_2\cdot \quad (115)$$

$$HAO_2\cdot + Fe^{2+} + H^+ \longrightarrow HAO_2H + Fe^{3+} \quad (116)$$

$$HAO_2\cdot + H_2A \longrightarrow HAO_2H + HA\cdot \quad (117)$$

$$HAO_2H + Fe^{2+} \longrightarrow HOA\cdot + Fe^{3+} + OH^- \quad (118)$$

$$HOA\cdot + H_2A \longrightarrow HAOH + HA\cdot \quad (119)$$

$$HOA\cdot + H^+ + Fe^{2+} \longrightarrow HAOH + Fe^{3+} \quad (120)$$

The substrate radical formed in reaction (110) can react with molecular oxygen to form the peroxide radical $HAO_2\cdot$, Eq. (115). This in turn can react further with iron(II) or more substrate to form considerable quantities of hydroperoxide HAO_2H by Eqs. (116) and (117). The steps in the presence of molecular oxygen are thus far more numerous than in its absence, and the organic substrate is oxidized to a greater extent compared to the reaction of iron(II) and hydrogen peroxide in the absence of O_2.

2. *Model Peroxidase Systems*

The Fe(II)–EDTA complex has been found to catalyze hydroxylation in the course of the reduction of peroxide to water, with the concomitant oxidation of suitable two-electron donors. The reactions of these systems are considered models for peroxidase action, since the active site of peroxidase is a porphyrin-bound ferrous ion, which may form an intermediate peroxo complex. For the model systems the two-electron donors usually employed are ascorbic acid, isoascorbic acid, dihydroxyfumaric acid, ninhydrin, alloxan, or 2,4,5-triamino-6-hydroxypyridimidine [180]. The strong interaction of EDTA with the ferrous ion is considered necessary to achieve the observed catalytic activity, possibly by increasing the oxidation potential of the Fe(II)/Fe(III) system. The hydroxylation reaction requires equimolar amounts of the two-electron donor and hydrogen peroxide. Hydroxylation of the substrate takes place by a free-radical addition of OH [153,181]. The mechanism proposed by Breslow and Lukens [153] and Grinstead [181] for the hydroxylation of salicylic acid is given in Eqs. (121)–(124).

$$\text{Fe(II)—EDTA} + H_2O_2 \longrightarrow \text{Fe(III)—EDTA} + \cdot OH + OH^- \quad (121)$$

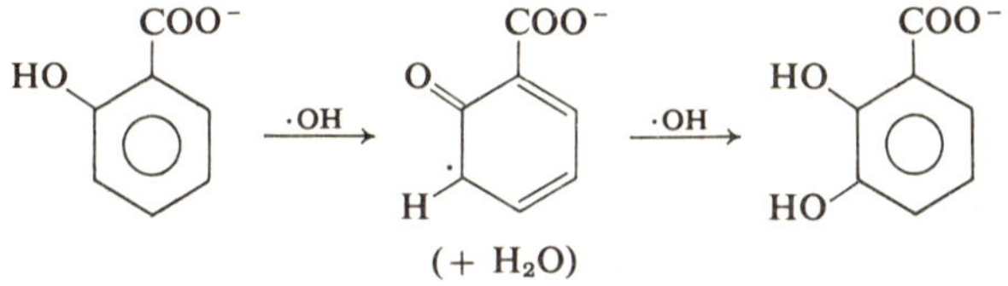

(122)

Regeneration of Iron(II)

$$\text{Fe(III)—EDTA} + OH^- + H_2A \longrightarrow HA\cdot + \text{Fe(II)—EDTA} + H_2O \quad (123)$$

$$\text{Fe(III)—EDTA} + OH^- + HA\cdot \longrightarrow A + \text{Fe(II)—EDTA} + H_2O \quad (124)$$

The first step is the cleavage of the HO—OH bond by the one-electron reduction of hydrogen peroxide. This is followed by reaction of the hydroxyl radical with the substrate to give an organic radical similar to that proposed by Barb *et al.* [155] for hydroxylation by Fenton's reagent. A second attack of the substrate by the hydroxyl radical yields the hydroxylated product. In the last two steps, one-electron reduction of Fe(III)–EDTA to the original catalyst, Fe(II)–EDTA, is accomplished by oxidation of ascorbic acid or some other suitable two-electron donor.

A mechanism of the type indicated above would not provide appropriate models for peroxidase action, in view of the fact that metal ions do not seem to be vitally involved with the rate-determining steps of the reaction. Also, it has been shown [151] that intermediates in enzymatic systems appear to have higher oxidation states of the iron–porphyrin complex of the enzyme. These intermediates are not iron–peroxide complexes because they can be formed by a variety of oxidizing agents in the absence of peroxide.

Hydroxylation of aromatic hydrocarbons and aliphatic alcohols by a model peroxidase system consisting of hydrogen peroxide and catalytic amounts of iron(III) and catechol has been studied by Hamilton *et al.* [149,150,157]. Catechol may be replaced by 1,2-dihydroxy or 1,4-dihydroxy aromatic compounds. Ferric ion was found to be a much better catalyst than copper(II), while chromium(III), cobalt(III), zinc(II), aluminum(III), and magnesium(II) were ineffective as catalysts. The rate of the reaction was found to be first order with respect to ferric ion, hydrogen peroxide, and catechol concentrations, at a low total concentration of catechol ($\sim 2 \times 10^{-4}$ *M*). At higher concentrations of catechol, however, there was a decrease in rate, which may be due to the formation of higher complexes of catechol with iron(III) that are less readily oxidized than the 1:1 Fe(III)–catechol complex. It was found that alcohols and aromatic compounds are oxidized at the same rate, and the rate was independent of the substrate concentration and hydrogen ion concentration in the pH range 3.2–4.5. It was proposed by Hamilton *et al.* [150] that the hydroxylating agent is not the free hydroxyl radical, OH·, but an oxidized metal complex intermediate. The rate data were compatible with the formation of an *o*-quinone–iron oxide complex **98** formed by the elimination of water from a catechol–Fe(III)–hydrogen peroxide, mixed-ligand complex **97**. The mechanism outlined in Fig. 23 was proposed for the hydroxylation of anisole.

The iron(III)–catechol–hydrogen peroxide, mixed-ligand complex **97**, formed in a preequilibrium step, is converted to the *o*-quinone–Fe═O intermediate in the rate-determining step of the reaction, involving the loss of a proton and OH^- ion from the complex accompanied by the transfer of two electrons from the catechol moiety through the metal ion to the oxygen. This process requires fission of the O—O bond in the peroxide ligand, forming a hydroxide ion, and oxidizing the iron–catechol system. The *o*-quinone–iron oxide intermediate can exist in several resonance forms in which the metal ion, ligand, and coordinated oxygen may be considered to exist in a number of oxidation states. The total number of electrons in the complex would be the same in all cases. The iron–oxygen bond in the inter-

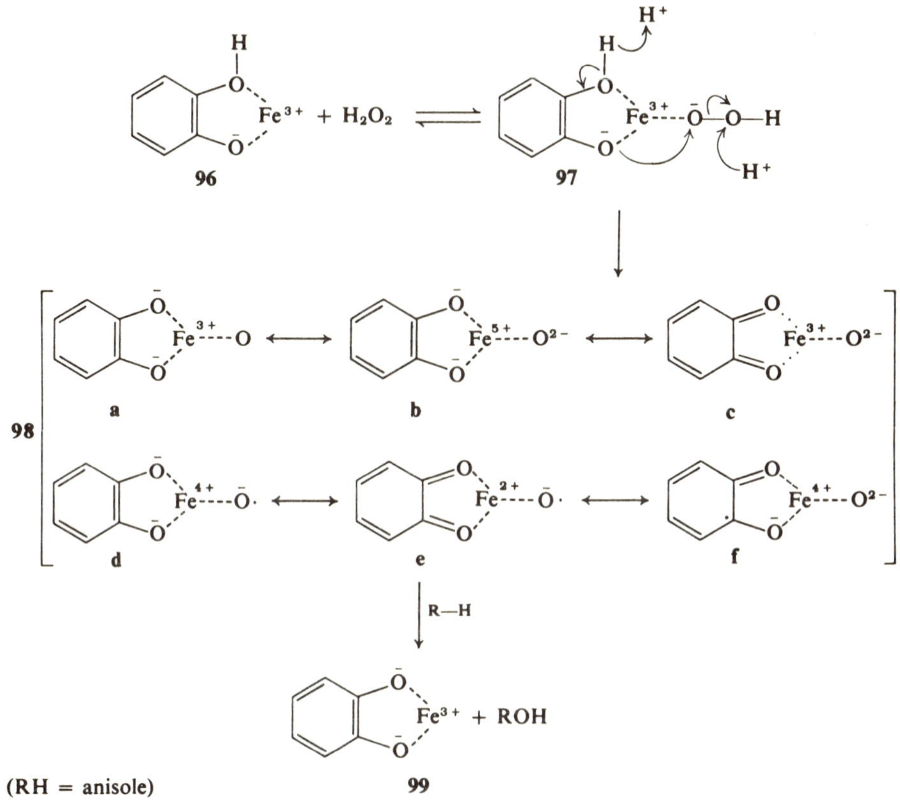

Fig. 23. Hydroxylation of anisole.

mediate would be stabilized by resonance involving these various oxidation states. Only a small number of the possible resonance forms are represented by **98a–f**. The oxidation of anisole is illustrated as taking place through attack by the potentially positive, i.e., electron-seeking, coordinated oxygen of **98** which can feed electrons through the iron atom to the catechol ring. Reaction of **99** with peroxide regenerates the mixed-ligand complex **97** to continue the reaction cycle. Hamilton's mechanism is of special interest in its relationship to the reactions catalyzed by catalase and peroxidase in which resonance forms of the iron–porphyrin ring system can stabilize the Fe═O bond.

3. *Peroxidase Enzymes*

It has been suggested by Hamilton [144] that a structure analogous to the reactive intermediate in Fig. 23 may be assigned to the oxidized intermediate in reaction systems in which peroxidase is acting as the catalyst.

Thus the porphyrin ring coordinated to iron(III) in the active site of catalase, indicated by formula **93a**, may be oxidized to quinoid structure **93b**, two oxidation units higher than the parent compound, through an electron shift without the need to transfer or rearrange any atoms or groups.

Present knowledge of the active sites of peroxidases is incomplete, and the nature of the reactions taking place as well as the most favored electronic configuration around the coordination sphere will depend on the nature of X and on the substituents on the porphyrin ring. It is clear, however, that large conjugated structures such as that in **93** can accept electrons or donate an oxygen atom in their reactions with appropriate substrates. Thus, compound **93b** offers a reasonable resonance form of the oxidized intermediate (Compound II) formed in the reactions catalyzed by peroxidase [151]. It is clear that conjugated systems such as **93** can exist in several electronic states, including a number of radical (unpaired electronic) structures, that offer many possible mechanistic alternatives for the electron-transfer reactions of peroxidase enzymes. Thus similar mechanisms involving one-electron transfers can take place with the catalytic intermediate suggested by formula **93b**, if it were to react with a substrate so as to transfer oxygen in a radical (-1) form rather than as zero-valent oxygen.

Such reactions have excellent analogies in the insertion reactions of the catalase model intermediates suggested by formulas **98a–f**. In the latter some of the resonance forms pictured for the reactive intermediate analogous to compound **93** suggest the possibility of an oxygen in the -1 state. If this oxygen is inserted (e.g., through formula **98e**) the remaining complex could react directly with coordinated water to form another oxidized intermediate, which could then donate a second $\dot{O}^-$ atom. (It is noted that $\dot{O}^-$ is identical to the hydroxyl radical HO·, the intermediate for oxygen insertion by Fenton's reagent.)

For examples of reactions promoted by peroxidase enzymes, the reader is referred to reviews of the subject [182,183].

4. *Oxygen Insertion by Peroxides*

Peroxides may react with olefins to produce addition compounds in which only one of the oxygens of the peroxide enters the substrate. Thus certain metal peroxo acids react with olefins to form an intermediate adduct which then hydrolyzes to a diol. Certain organic peroxides undergo metal complex-catalyzed addition to olefins to form epoxides. Since only one of the oxygen atoms of the peroxo group ends up in the reaction product, these reactions are considered examples of partial insertion reactions.

a. Hydroxylation by Metal Peroxo acids. Pertungstic acid, formed by the reaction of H_2O_2 and WO_3, catalyzes hydroxylation of olefins in aqueous or mixed solvents to form diols [184]. Unlike the hydroxylation by Fenton's reagent that proceeds by a nonspecific free-radical reaction, pertungstic acid-catalyzed hydroxylation is stereospecific, with *trans* addition of OH groups to the olefin. The ionic mechanism of Eqs. (125)–(127) has been proposed, involving the participation of an intermediate hydroperoxide species, $HWO_3O{-}OH$. The participation of free radicals in the catalytic hydroxylation of pertungstic acid has been ruled out on the basis of non-inhibition of the reaction by picric acid, which is known to terminate free-radical chains by combination with the reactive intermediate free radicals.

$$R_2C{=}CR_2 + HWO_3OOH \longrightarrow R_2\underset{\displaystyle OWO_3H}{\underset{|}{C}}{-}\overset{\displaystyle OH}{\overset{|}{C}}R_2 \qquad (125)$$

$$R_2\underset{\displaystyle OWO_3H}{\underset{|}{C}}{-}\overset{\displaystyle OH}{\overset{|}{C}}R_2 + H_2O \longrightarrow R_2\underset{\displaystyle OH}{\underset{|}{C}}{-}\overset{\displaystyle OH}{\overset{|}{C}}R_2 + HWO_3OH \qquad (126)$$

$$HWO_3OH + H_2O_2 \longrightarrow HWO_3OOH + H_2O \qquad (127)$$

Similar additions of H_2O_2 across olefinic double bonds are catalyzed by a variety of metal peroxo acids, such as pervanadic, perchromic, and perosmic acids. The reaction mechanisms are similar to that shown above for pertungstic acid. These reactions are generally stereospecific and produce high yields of the glycol.

b. Epoxidation of Olefins. The metal complex-catalyzed epoxidation of olefins by organic hydroperoxides has recently been reported by Indicator and Brill [185]. Epoxides are formed by the reaction of *t*-butylhydroperoxide with olefins in the presence of the acetylacetonates of chromium(III), vanadium(IV), vanadyl ion, and molybdyl ion in vacuum at 25°C. The yield of the epoxides increases with methyl substitution on the olefin, indicating the nucleophilic nature of the substrate as an important factor in the reaction. The reaction seems to be stereospecific, since the main products are *cis*-epoxides. Catalysis by the metal complex in the reaction has been taken into account by assuming the formation of a metal–acetylacetonate peroxide mixed-ligand complex, and subsequent epoxidation of the olefin by a polar mechanism in a manner similar to that proposed above. Thus the observed stereospecificity and the structural effects in the metal–acetylacetonate-catalyzed epoxidation of the olefins support a polar mechanism for the dissociation of the O—O bond.

B. Partial Insertion of Molecular Oxygen

1. *Model Systems*

The system Fe(II)–EDTA–ascorbic acid–molecular oxygen was first employed by Udenfriend and co-workers [186–188] for the hydroxylation of organic substrates. Udenfriend's system is similar to the model peroxidase systems described above, with oxygen substituting for hydrogen peroxide. Like model peroxidase systems, Udenfriend's system involves a two-electron donor and a suitably complexed transition metal ion catalyst, such as copper(II), cobalt(II), manganese(II), tin(II) [189], and iron(II) [157,158]. The most widely used complexing agent is EDTA, though pyrophosphate, amino acids, F^-, N_3^-, and CN^- have also been employed with varying success. The presence of a complexing agent is not essential, though the reaction seems to be considerably enhanced in its presence. Stoichiometric quantities of oxygen and ascorbic acid (or other electron donor) are required in the reaction. The nature of the hydroxylating species in Udenfriend's system has not been decided conclusively. Grinstead [181] suggested that the hydroperoxide free radical, $HO_2\cdot$, is the reactive species that takes part in hydroxylation by the mechanism of Eqs. (128)–(131).

$$\text{Fe(II)—EDTA} + O_2 + H_2O \longrightarrow \text{Fe(III)—EDTA} + OH^- + HO_2\cdot \quad (128)$$

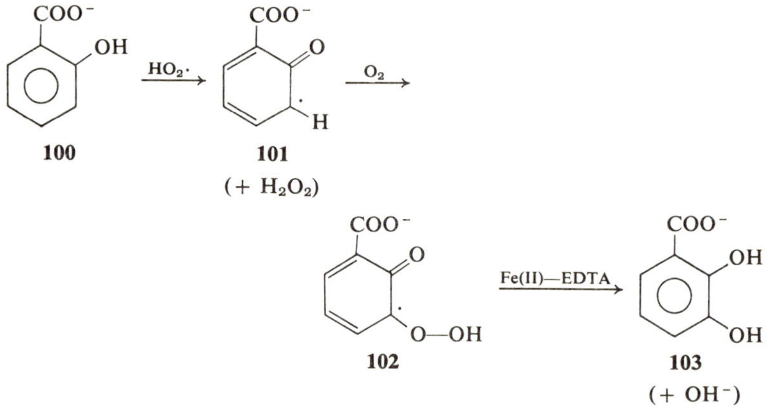

100 **101** (+ H_2O_2) **102** **103** (+ OH^-) (129)

Regeneration of Iron(II)

$$\text{Fe(III)—EDTA} + H_2A + OH^- \longrightarrow \text{Fe(II)—EDTA} + HA\cdot + H_2O \quad (130)$$

$$\text{Fe(III)—EDTA} + HA\cdot + OH^- \longrightarrow \text{Fe(II)—EDTA} + A + H_2O \quad (131)$$

The substrate may be hydroxylated either by addition of molecular oxygen to a substrate radical, followed by further reaction of the intermediate with Fe(II)-EDTA, or alternatively by attack of free $\cdot OH$ radical produced in the last step shown, on the initial substrate radical shown, in a manner analogous to free-radical hydroxylation in the model peroxidases systems described above. The details of this mechanism are far from clear; the rate-determining step is unknown and the reaction steps shown are merely based on the similarity of isomer distribution of the hydroxylation of salicyclic acid to known free-radical reactions, where nearly equal amounts of *ortho*, *meta*, and *para* isomers are obtained.

Subsequent work on isomer distribution of reaction products by Norman and Rodda [190] and by others [191–193] indicates that hydrogen peroxide is not an intermediate in the reaction and that the hydroxylating agent is probably not a free radical. Recently Hamilton and co-workers [158] demonstrated that the Udenfriend system would not only hydroxylate aromatic compounds, but would also oxidize saturated hydrocarbons to alcohols and convert olefins to epoxides. The relationship of these reactions to those of the monooxygenases, described in the next section, was demonstrated by the fact that tetrahydropteridines [194] (which are reducing agents for the oxygenases) and related pyrimidine derivatives [158] may replace ascorbic acid as the reductant in the Udenfriend system.

The ionic reaction mechanism in Fig. 24 was suggested by Hamilton [144] to explain these new observations. The electron-transfer reactions occurring in the intermediate involve concerted oxidation of the ascorbic acid moiety to dehydroascorbic acid, electron transfer through the metal ion to the oxygen atom which remains bound to the metal ion, and transfer of the other oxygen as a neutral electrophilic species (i.e., the oxygen atom) to the substrate. The energy required to break the oxygen–oxygen bond is supplied by the concerted formation of more stable oxygen bonds with other elements.

In addition to holding together the activated complex and providing the pathway for the electron transfer from the reductant to molecular oxygen, the metal ion may also have the function of providing some stability of the oxygen complex prior to reaction with the substrate. Thus the ferrous ion may combine with oxygen in a prior equilibrium to give a complex to which some stability is imparted by the resonance forms **109a** and **109b**.

2. *Monooxygenase or Mixed-Function Oxidase (MFO) Enzymes*

The monooxygenases, described in recent reviews by Mason, Hayaishi, and co-workers [195–197], generally require a transition metal ion and a reducing agent, and catalyze many partial oxygen insertion reactions such as hydroxylation of aromatic and aliphatic hydrocarbons, oxidative decarboxylation, the epoxidation of olefins, and oxidative dehydroxymethylation. The reducing agents required are usually tetrahydropteridines, reduced flavins, or ferridoxins. Reduction may also be mediated by pairs of reduced metal ions at active enzymatic sites [Eq. (136)].

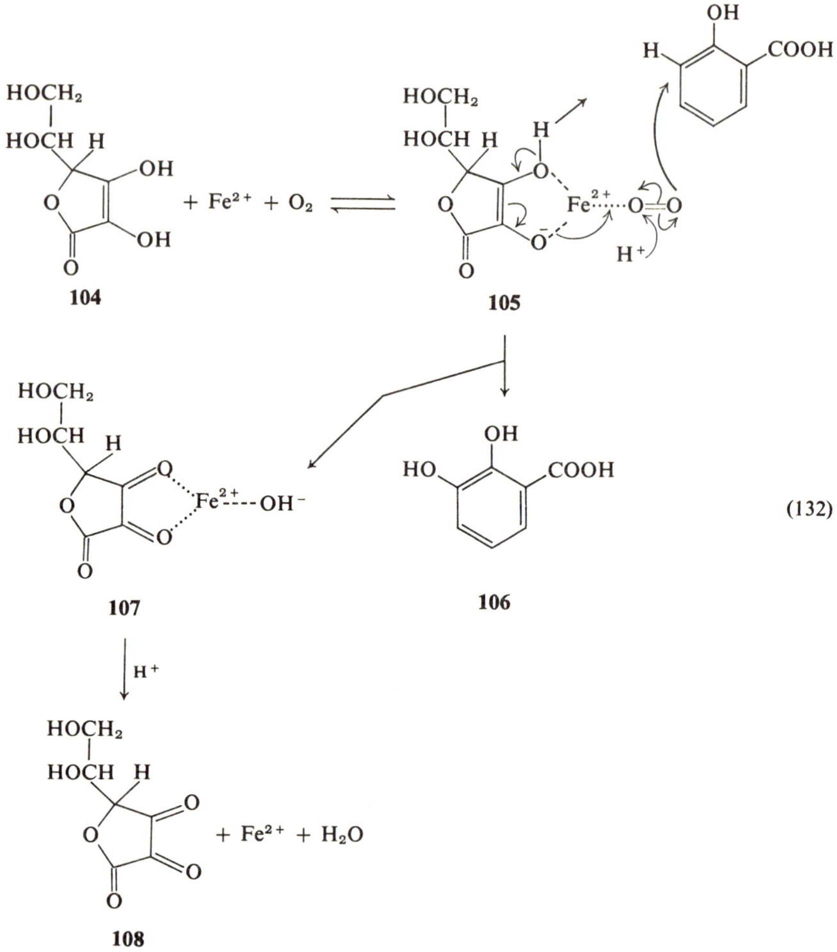

Fig. 24. Mechanism of hydroxylation by a model monooxygenase system.

It is of interest to consider possible reaction mechanisms for the monooxygenase enzymes. It is clear that while molecular oxygen is a four-electron oxidant, only two of the oxidizing equivalents (i.e., one oxygen atom) are used to oxidize the substrate. It has also been found that substrate peroxides are not involved in the mechanisms of reaction of these enzymes. Therefore O—O bond fission must occur prior to combination of one oxygen atom with the substrate, or there must be a concerted reaction for the three processes: O—O fission, insertion of one oxygen atom, and reduction of the other oxygen atom to H_2O. Also, the reduced enzyme is now known

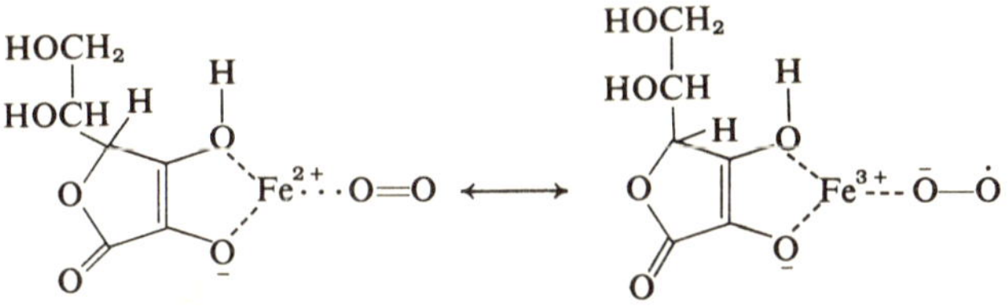

109a **109b**

[195–197] to be the form that reacts with the oxygen and (if the mechanism is concerted) with the substrate.

Possible mechanisms for both the concerted and consecutive processes are provided by the enzyme models described above. The concerted mechanism could be similar to that suggested for the Udenfriend system, whereby the oxygen is bonded simultaneously to the reduced metal ion, iron(II), and to the substrate, at least in the transition state, allowing a concerted electron shift which results in O—O bond fission. Alternatively, it is possible that an oxidized intermediate such as that indicated above for the peroxidase model system could be formed and that this in turn could donate an oxygen atom to the substrate. The two possible mechanisms can therefore be represented by the reaction sequence given in Eqs. (133)–(135)

Concerted Reaction

$$EH_2 + S + O_2 \rightleftharpoons EH_2{\cdot}S{\cdot}O_2 \longrightarrow E{\cdot}SO{\cdot}H_2O \rightleftharpoons E + SO + H_2O \tag{133}$$

Oxidation of Enzyme, Followed by Oxygen Insertion

$$EH_2 + S + O_2 \rightleftharpoons EH_2{\cdot}S{\cdot}O_2 \longrightarrow EO{\cdot}S{\cdot}H_2O \rightleftharpoons EO{\cdot}S + H_2O \tag{134}$$

$$EO{\cdot}S \longrightarrow E{\cdot}SO \rightleftharpoons E + SO \tag{135}$$

where EH_2 = reduced enzyme or enzyme + reduced coenzyme; S = substrate; SO = initial product after oxygen insertion; and $EH_2{\cdot}S{\cdot}O_2$ = ternary enzyme–substrate–oxygen complex (activated complex). Detailed kinetic studies of these enzyme systems should make it possible to distinguish between these two general mechanisms, as well as between the several alternative reaction sequences in the second mechanism. Apparently such experiments have not yet been carried out.

a. Nature of the Reducing Agent in MFO Enzyme Systems. A concerted mechanism similar to that suggested above for Udenfriend's system has been proposed [144,158,180,194,198] for mixed-function oxidases requiring iron(II) and a tetrahydroxypteridine, which functions as a two-electron reductant in the reaction. The suggested reaction sequence in Fig. 25 is analogous to the Udenfriend mechanism.

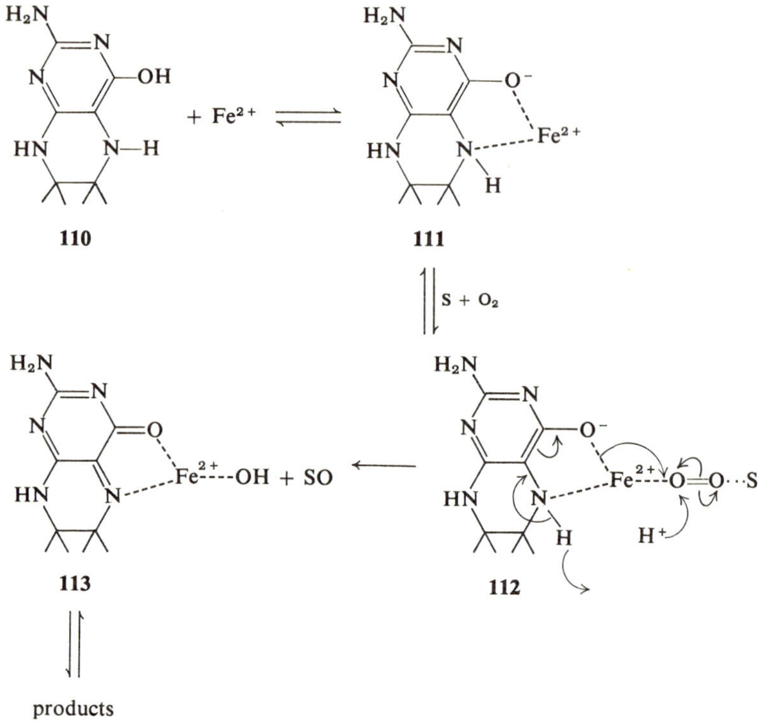

Fig. 25. Mechanism for mixed-function oxidase action.

The metal ion at the active site of the enzyme serves to bring the reactants together, through initial formation of a resonance-stabilized Fe(II)–oxygen complex before or after combination with the reducing agent. The latter is itself a coordinating ligand and may combine initially with the metal ion. Hamilton [144] has suggested that the enzyme also probably assists the proton transfer pictured above through acid–base catalysis mediated by proton donors and acceptors at the active site.

Many other types of reagents could serve the reducing function of the tetrahydropteridine indicated in the above mechanism, the essential requirement being that it be capable of transferring two electrons to one of the oxygen atoms while the other undergoes insertion into the substrate. Thus when the metal ion at the active site is a single electron-transfer agent, such as the copper ions in dopamine-β-hydroxylase, the metal ions are believed to react in electronically linked pairs so that they can transfer two electrons to oxygen, as in Eq. (136). For this copper enzyme it has been suggested that each pair of metal atoms, originally in the $+1$ oxidation state, combines with one oxygen molecule to form a complex analogous to the one described above for the Fe(II)–enzyme (see formula **109**).

(136)

b. Mechanism of Insertion Step. Several possibilities exist for the transfer of an oxygen atom to the substrate in mixed-function oxidases. The transition states suggested above (**112**) may exist in several resonance forms, so that the oxygen moiety being inserted may enter the substrate as a positive, neutral, or even negative species. The particular mechanism that applies to a specific enzyme is expected to be strongly influenced by the type of ligand coordinated to the metal ion. Thus it has been suggested [144] that the epoxidation of olefins, by analogy with known model reactions, involves an electrophilic insertion step. Comparisons with chemical systems for aromatic hydroxylation suggest either an electrophilic or radical addition of oxygen. In either case, the hydrogen shifts detected in these reactions [199], shown in Eq. (137), are consistent with the formation of a benzene opoxide intermediate, **115**.

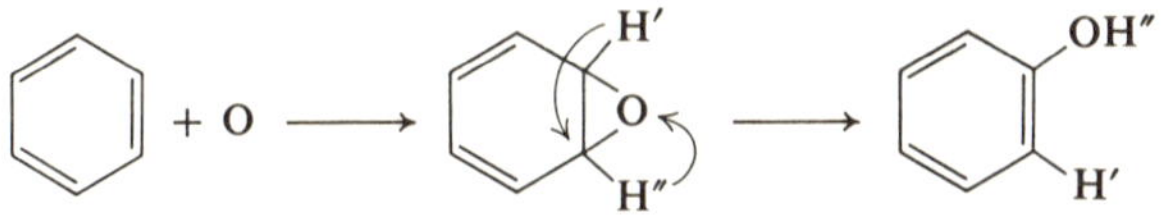

(137)

114 **115** **116**

Another MFO-catalyzed reaction, which can be explained by a mechanism involving a benzene epoxide-type intermediate, is the oxidative dehydroxymethylation of aromatic compounds [200] as indicated by Eq. (138).

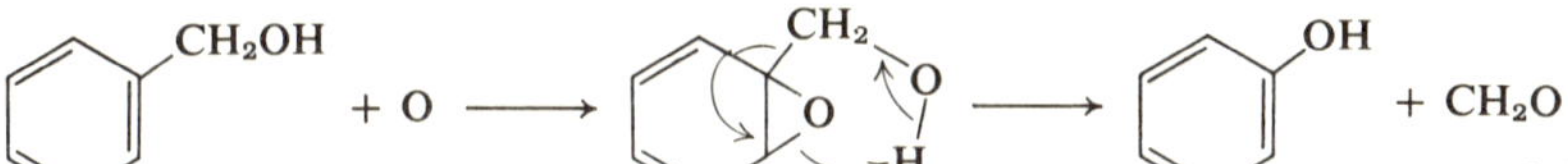

(138)

117 **118** **119**

The resonance energy of the aromatic ring can be recovered if the proton is transferred to the electron pair in the oxygen bridge, thus opening the three-membered ring of the epoxide, **118**.

3. *Partial Insertion Through Free-Radical Addition of Oxygen*

Partial insertion reactions may result from free-radical addition of oxygen to form peroxides, which may then decompose to more stable products. Decomposition of the peroxide is often accomplished by the catalytic action of metal ions. An example of this type of reaction is the oxidation of cumene to acetophenone in the presence of cobalt(III) as a catalyst [201]. The rate of the oxidation of cumene is proportional to the concentrations of catalyst, of substrate, and of molecular oxygen, in accordance with the mechanism

$$C_6H_5\overset{CH_3}{\underset{CH_3}{C}}H + X \longrightarrow C_6H_5\overset{CH_3}{\underset{CH_3}{C}}\cdot + XH \qquad (139)$$

(X is a free-radical initiator)

$$C_6H_5\overset{CH_3}{\underset{CH_3}{C}}\cdot \xrightarrow{O_2} C_6H_5\overset{CH_3}{\underset{CH_3}{C}}\text{—O—O}\cdot \qquad (140)$$

$$C_6H_5\overset{CH_3}{\underset{CH_3}{C}}\text{—O—O}\cdot + C_6H_5\overset{CH_3}{\underset{CH_3}{C}}H \longrightarrow C_6H_5\overset{CH_3}{\underset{CH_3}{C}}OOH + C_6H_5\overset{CH_3}{\underset{CH_3}{C}}\cdot \qquad (141)$$

$$C_6H_5\overset{CH_3}{\underset{CH_3}{C}}OOH + Co^{3+} \longrightarrow C_6H_5\overset{CH_3}{\underset{CH_3}{C}}OO\cdot + Co^{2+} + H^+ \qquad (142)$$

$$C_6H_5\overset{CH_3}{\underset{CH_3}{C}}OOH + Co^{2+} \longrightarrow C_6H_5\overset{CH_3}{\underset{CH_3}{C}}O\cdot + Co^{3+} + OH^- \qquad (143)$$

$$C_6H_5\overset{CH_3}{\underset{CH_3}{C}}O\cdot \longrightarrow C_6H_5\overset{O}{\overset{\|}{C}}CH_3 + \cdot CH_3 \qquad (144)$$

given in Eqs. (139)–(144). Cumene hydroperoxide is produced from the free radical by reaction with oxygen followed by free-radical exchange with cumene. In the presence of cobalt(II), cumene hydroperoxide decomposes rapidly to acetophenone.

Natural, enzyme-catalyzed reactions involving reactive free-radical intermediates analogous to those described above have not been reported. Apparently nature prefers to carry out reactions smoothly and efficiently, without the formation of free radicals that are highly reactive, and would lead to many side reactions, and are difficult to control. The ionic mechanisms described above for partial insertion of oxygen provide what seem to be more reasonable kinetic pathways for biological systems.

4. *Synthetic Applications of Metal Ion-Catalyzed Hydroxylation Reactions*

A general method for the introduction of one hydroxy group (i.e., the insertion of one oxygen atom) into an organic substrate involves the utilization of metal ion-catalyzed hydroxylation reactions with Fenton's reagent, model peroxidase systems, or Udenfriend's system. Some of the organic compounds hydroxylated by these methods are presented in Table VII [151,153,157–162,165,181,189,190,202–212]. Hydroxylation in certain cases was effected by ionizing radiation but the method lacks general application since the yield of the hydroxylated product is very poor and in some cases decomposition of the substrate accompanies hydroxylation.

Hydroxylation by Fenton's reagent and model peroxidase systems is random in nature, characteristic of a free-radical reaction. About equal amounts of *ortho*-, *meta*-, and *para*-hydroxylated derivatives were obtained on the hydroxylation of benzoic acid, chlorobenzene, fluorobenzene, and nitrobenzene (Table VII). Hydroxylation of phenol and anisole mainly yield *ortho* and *para* isomers with small quantities of *meta* isomer. Hamilton *et al.* [149] reported a 3% yield of *m*-hydroxyanisole by a model peroxidase system consisting of catechol–Fe(II)–H_2O_2. The yield of hydroxylated products by Fenton's reagent and model peroxidase systems varies from 5 to 25% depending on the substrate.

The use of Udenfriend's system in aromatic hydroxylation results in the substitution of the hydroxyl group at the electronegative sites of the aromatic ring. The substituent group thus is electrophilic in nature, and the position of the OH group in the product can be easily predicted. Acetanilide, aniline, phenol, salicylic acid, and anthranilic acid yield mainly the *o*- and *p*-hydroxy isomers. A 9% yield of the *m*-hydroxy derivative of anisole has been reported by Norman and Rodda [190]. Hamilton and Friedman [157], however, reported no *meta* derivative in the hydroxylation of anisole by Udenfriend's system.

The yield of the hydroxylated products in Udenfriend's reaction varies from 3 to 30% depending on the substrate, the metal ion catalyst used, and the complexing agent. In the oxidation of benzoic acid, Nofre *et al.* [189] have reported the following order of catalytic activity of iron(II) complexes: Fe(II)–EDTA > Fe(II)–pyrophosphate > Fe(II)–aspartate > Fe(II)–stearate > Fe(II)–F > Fe(II)–glygly ≈ Fe(II)–ornithine ≈ Fe(II)–lysine > Fe(II)–mercaptoethylamine > Fe(II)–histidine > Fe(II)–trien > Fe(II)–CN. Fe(II)–EDTA and Fe(II)–CN thus occupy the two ends of the catalytic activity series. The activity of the EDTA complexes decreases in the order, Fe(II)–EDTA > Cu(II)–EDTA > Mn(II)–EDTA > Co(II)–EDTA > Sn(II)–EDTA. There seems to be no correlation between the stabilities of EDTA complexes and their relative catalytic activity.

One of the important synthetic applications of Udenfriend's reaction lies in the fact that the hydroxylation is relatively mild compared to other hydroxylation methods. As may be verified from Table VII, in the hydroxylation of tyramine [202] only the benzene ring is hydroxylated without oxidation of the side-chain amino group. In the oxidation of heterocyclic compounds

TABLE VII

HYDROXYLATION OF ORGANIC SUBSTRATES

	Compound	Structure	Products	Method[a]	References
1	Acetanilide	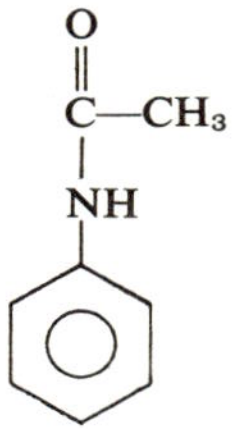	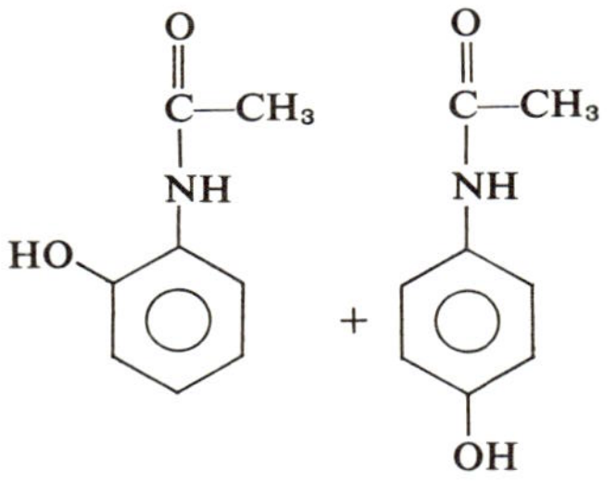	U.S.	202
2	*p*-Ethoxyacetanilide	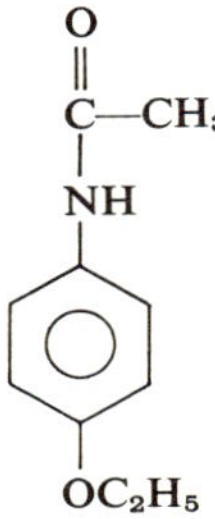	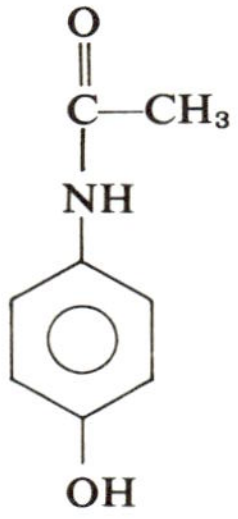 (no insertion)	U.S.	202
3	Aniline	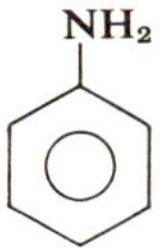	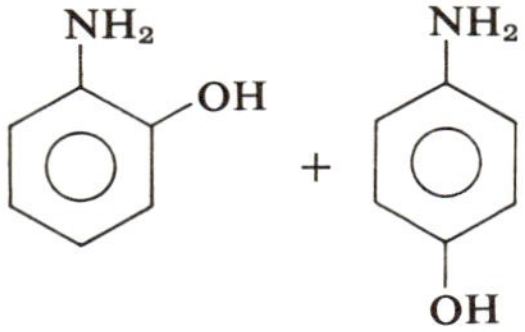	U.S.	202,203

4	Anthranilic acid	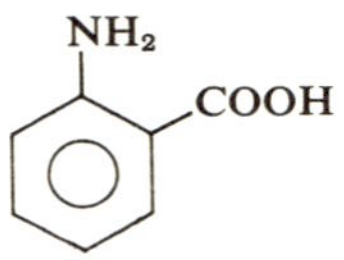	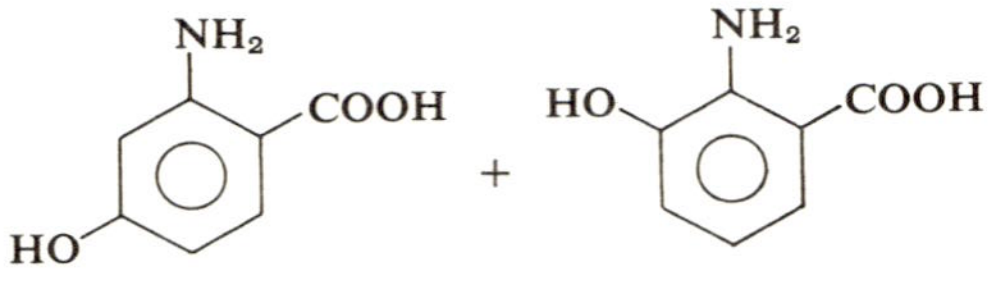	U.S.	204,205
5	Anisole	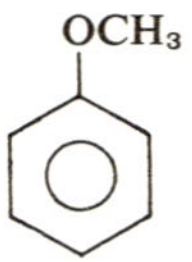	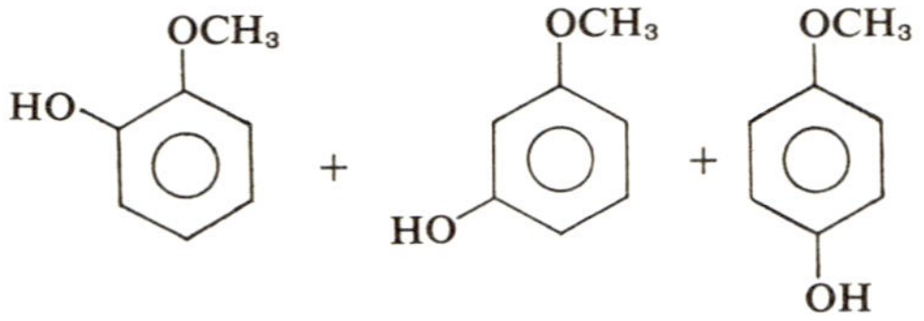	U.S. M.P. F.R.	157,190 157,190 157
6	Antipyrine	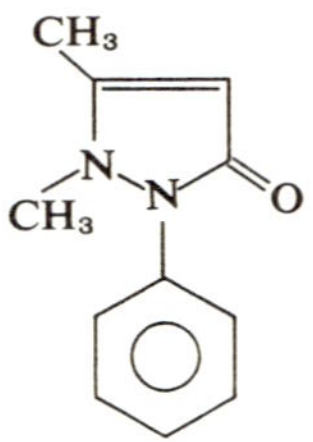	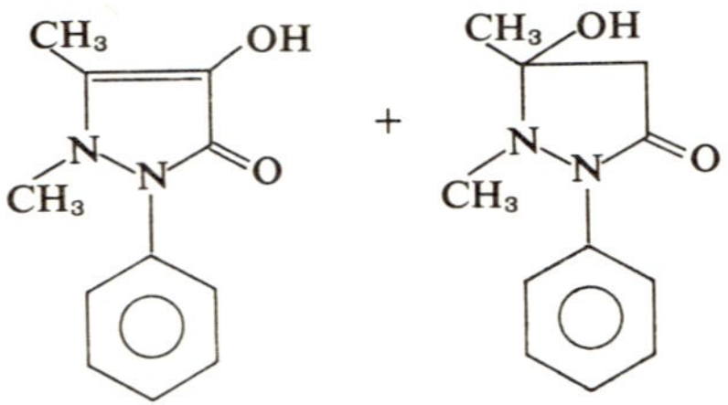	U.S.	202
7	Benzoic acid	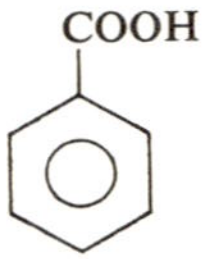	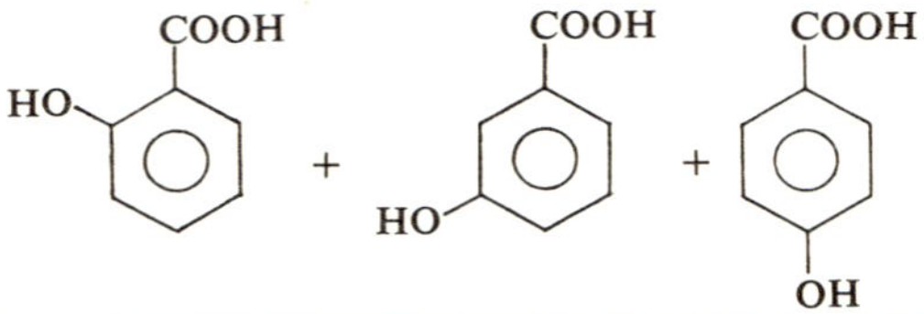	U.S. F.R. I U.S.[b]	151,206 160–162 160–162 189

TABLE VII *(continued)*

	Compound	Structure	Products	Method[a]	References
8	Chlorobenzene		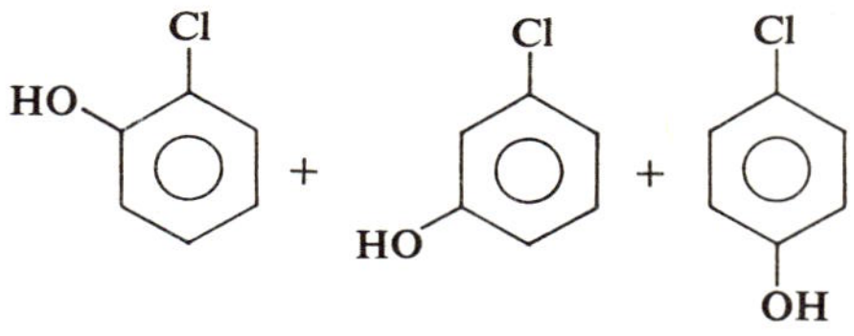	U.S. M.P. F.R.	190 159,190 159
9	Fluorobenzene	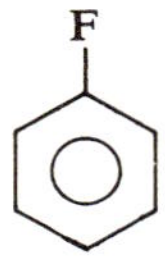	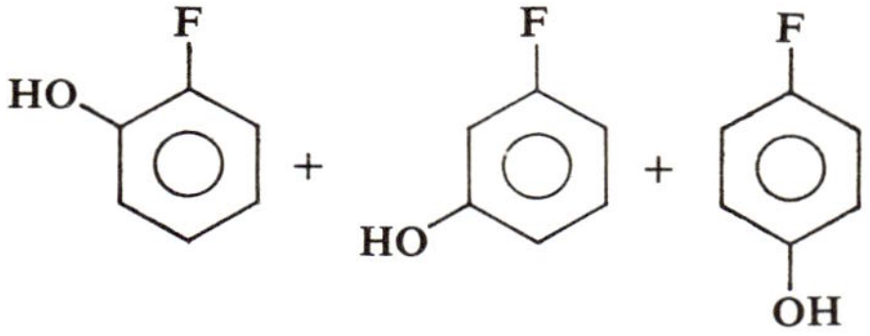	M.P.	190
10	Cyclohexane	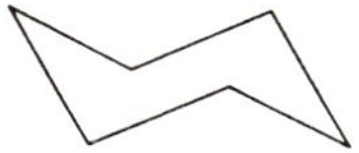	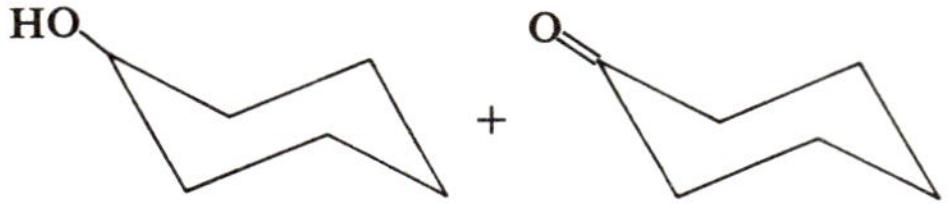	U.S. F.R.	158 158
11	Cyclohexene		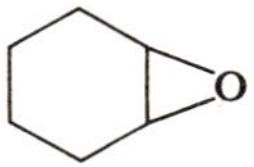	U.S. F.R.	158 158

12	2-Indolylacetic acid	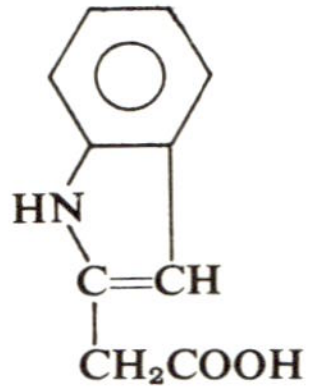	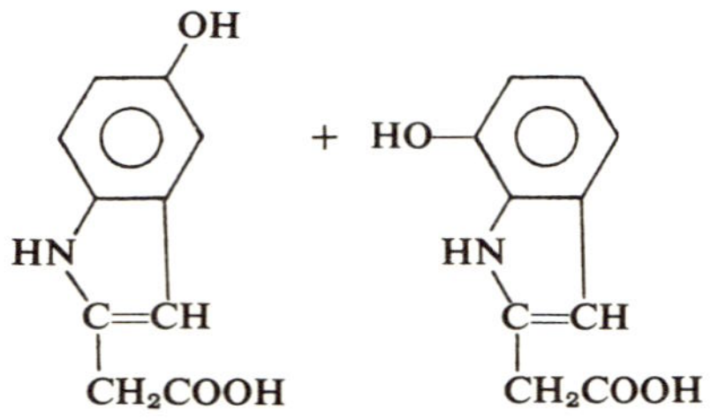	U.S.	205,206
13	Kyrunenine	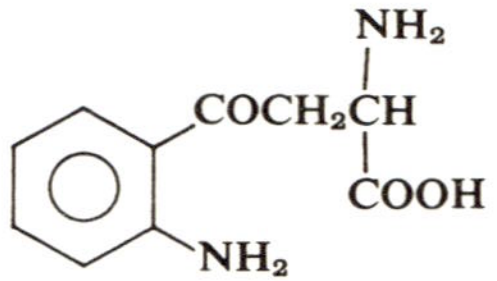	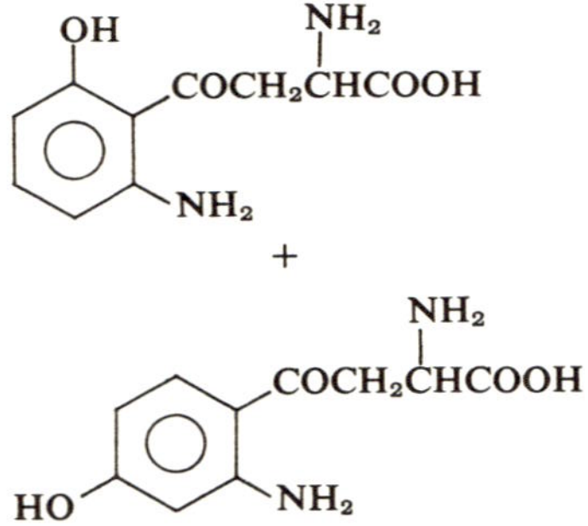	U.S.	205,206
14	Nitrobenzene		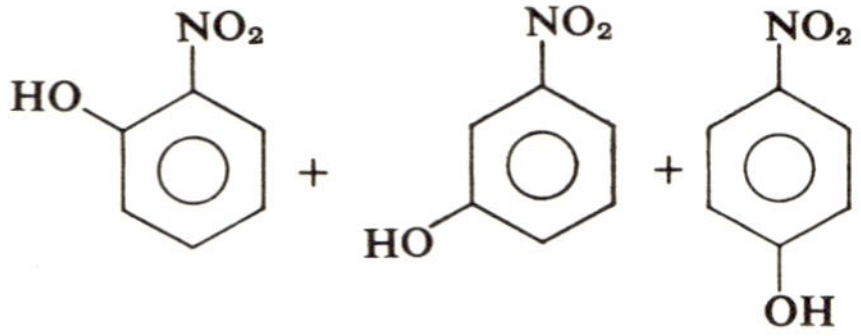	M.P. F.R. I	190 160–162 162

TABLE VII (*continued*)

Compound	Structure	Products	Method[a]	References
15 Phenol		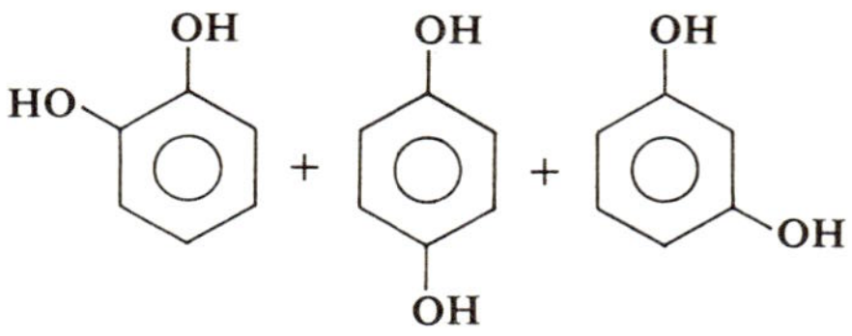	F.R.	165 207 208
16 Proline	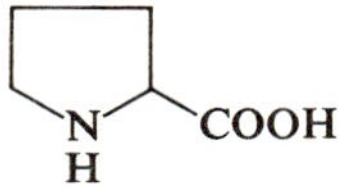	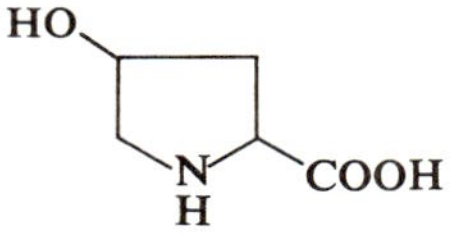	U.S. M.P.	209 209
17 Prolylglycine	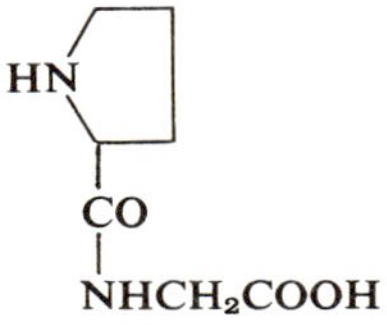	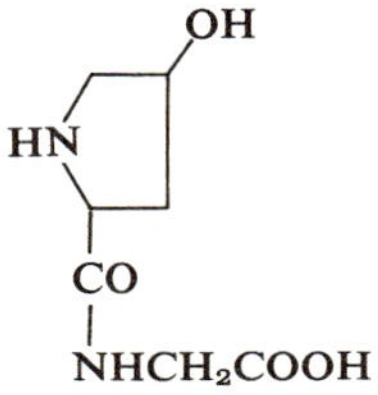	U.S. M.P.	209 209

18 Prolylglutamylglycine

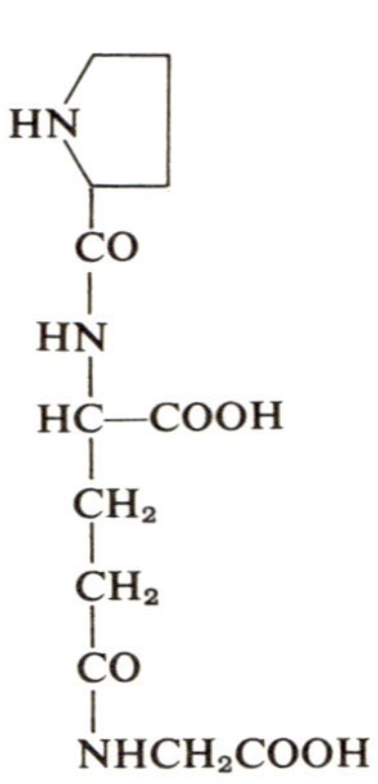

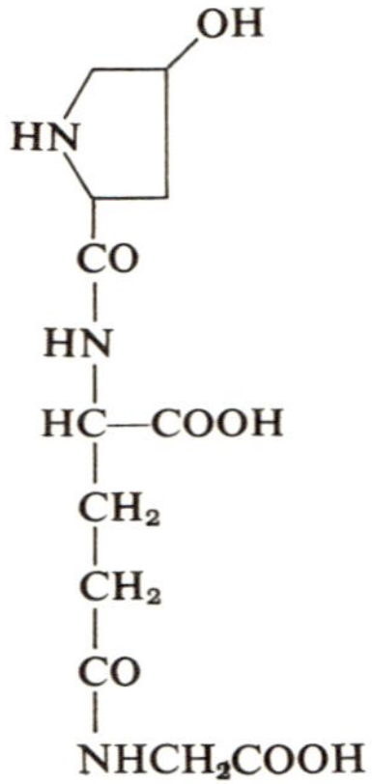

U.S. 209
M.P. 209

19 Pyrene

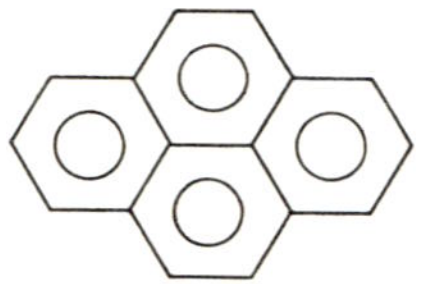

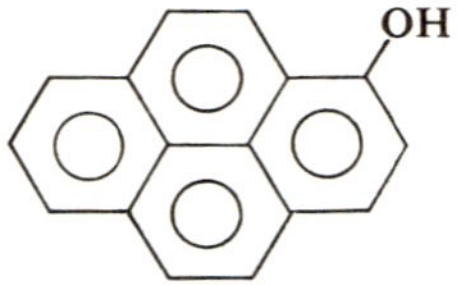

U.S. 210

20 Quinoline

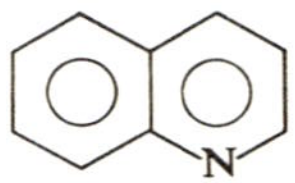

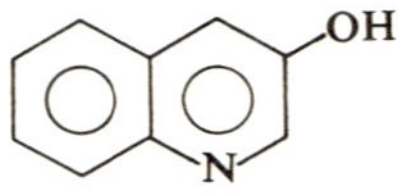

U.S. 153
F.R. 153
M.P. 153

TABLE VII (*continued*)

Compound	Structure	Products	Method[a]	References
21 Salicylic acid	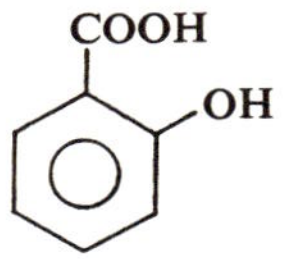	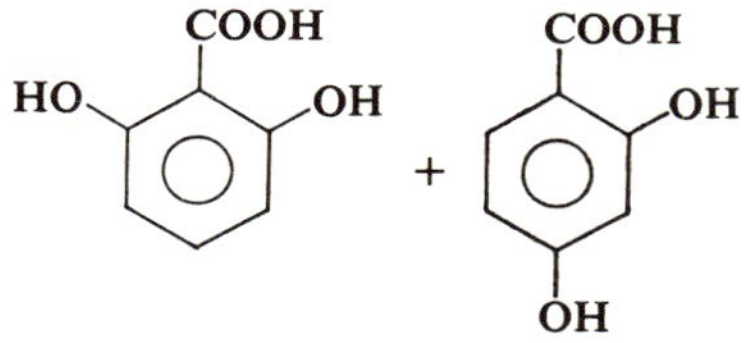	U.S. M.P.	181 181
22 Tyramine	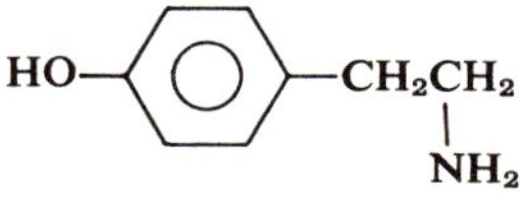	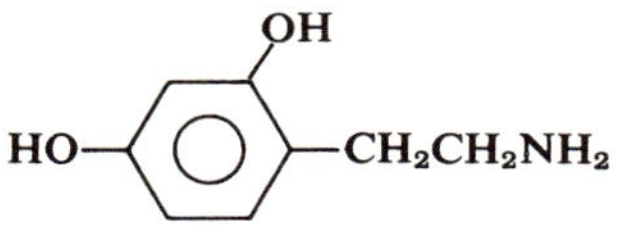	U.S.	204,205
23 Tryptophan	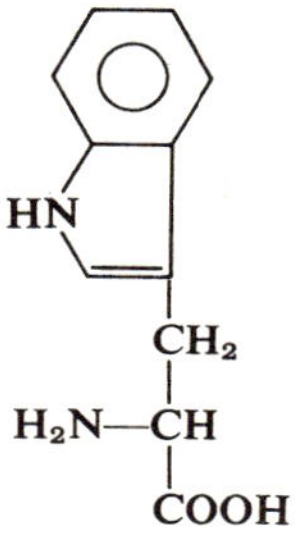	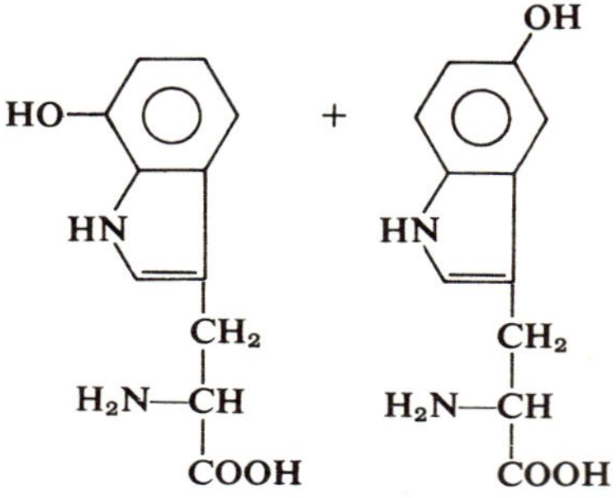	U.S.	204,205 211,212

24	1,3-Dimethylxanthine	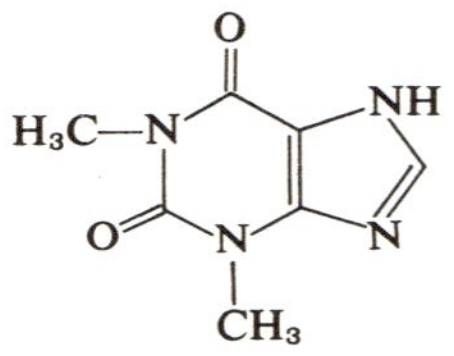	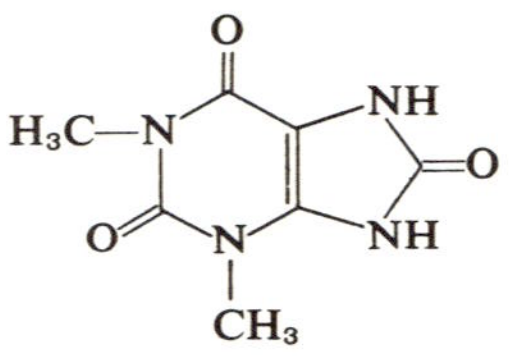	U.S.	202

[a] I = irradiation by X-rays; U.S. = Udenfriend's system, Fe(II)–EDTA + ascorbic acid + oxygen (unless otherwise specified); M.P. = model peroxidase system, Fe(II)–EDTA + ascorbic acid + H_2O_2; F.R. = Fenton's reagent, Fe(II) + H_2O_2.

[b] *Catalyst*

Fe(II)–EDTA	Fe(II)–pyrophosphate
Fe(II)–aspartate	Fe(II)–stearate
Fe(II)–fluoride	Fe(II)–glycylglycine
Fe(II)–ornithine	Fe(II)–lysine
Fe(II)–mercaptoethylamine	Fe(II)–histidine
Fe(II)–trien	Fe(II)–CN
Cu(II)–EDTA	Mn(II)–EDTA
Co(II)–EDTA	Sn(II)–EDTA
Cu(II)–ascorbic acid	Sn(II)–ascorbic acid
Fe(II)–ascorbic acid	Mn(II)–ascorbic acid
Co(II)–ascorbic acid	

like tryptophan and indolylacetic acid, the reagent hydroxylates only the benzene ring without attacking the reactive indole group.

Apart from synthetic applications, hydroxylation by Udenfriend's system may have considerable biological significance. In most of the cases the hydroxylated product obtained by the reaction of Udenfriend's system *in vitro* is identical with a substance formed from the same compound *in vivo*. An interesting example is the deethylation of *p*-ethoxyacetanilide which takes place both in the ascorbic acid system in the animal body to yield *N*-acetyl-*p*-aminophenol. Hydroxylation of proline [109] has considerable resemblance to the synthesis of collagen *in vivo*.

VIII. Total Insertion Reactions

Except for the specific reactions leading to peroxide formation, total insertion of oxygen is much less common, or in other words more difficult to accomplish, than partial or non-insertion reactions. This is probably due to the strong tendency of oxygen to be reduced to water, and the fact that catalysts that assist insertion generally polarize (and disproportionate) the oxygen molecule. In the following sections are a few examples of non-enzymatic and enzymatic total insertion reactions of oxygen.

A. Insertion without O—O Bond Cleavage

The combination of oxygen with an organic or organometallic compound sometimes occurs without cleavage of the O—O bond, resulting in the formation of a peroxide. Such compounds may be organoperoxides (symmetrical organoperoxides or organohydroperoxides) or metal organoperoxides. Although many such reactions are described in the literature, only a few representative examples will be given here. Organoperoxides may be formed by free-radical reactions with molecular oxygen. Initiation of free radicals may be readily accomplished with a metal ion, resulting in a chain reaction involving direct combination of oxygen with an organic free radical.

An example of this type of reaction is the formation of a bis(tetralin) peroxide from tetralin and molecular oxygen in the presence of alkanoates of manganese(II), copper(II), nickel(II), and iron(II) as catalysts [213,214]. Initiation of the reaction chain takes place by reaction of a trace of tetralin hydroperoxide with the metal ion [Eq. (145)]. The reaction then proceeds with a small steady concentration of tetralin hydroperoxide [Eqs. (146)–148)].

A similar free-radical reaction is the Cu(II)–phthalocyanine-catalyzed oxidation of cumene to cumene hydroperoxide by molecular oxygen. The mechanism shown in Eqs. (149)–(152) has been proposed by Kropf [215].

The function of the copper phthalocyanine–oxygen complex seems to be the generation of free radicals, with the formation of hydrogen peroxide. The main conversion to the organoperoxide is carried out by a chain reaction involving molecular oxygen.

Initiation

$$C_{10}H_{11}OOH \xrightarrow{M^{n+}} C_{10}H_{11}OO\cdot + M^{(n-1)+} + H^+ \qquad (145)$$

Propagation

$$C_{10}H_{12} + C_{10}H_{11}OO\cdot \longrightarrow C_{10}H_{11}\cdot + C_{10}H_{11}OOH \qquad (146)$$

$$C_{10}H_{11}\cdot + O_2 \longrightarrow C_{10}H_{11}OO\cdot \qquad (147)$$

Termination

$$2\,C_{10}H_{11}OO\cdot \longrightarrow C_{10}H_{11}OOC_{10}H_{11} + O_2 \qquad (148)$$

$$Cu(II){-}Pc + O_2 \rightleftharpoons (Cu(II){-}Pc)^{\delta+}O_2^{\delta-} \qquad (149)$$

$$[Cu(II){-}Pc]^{\delta+}O_2^{\delta-} + RH + ROOH \longrightarrow Cu(II){-}Pc + ROO\cdot + R\cdot + H_2O_2 \qquad (150)$$

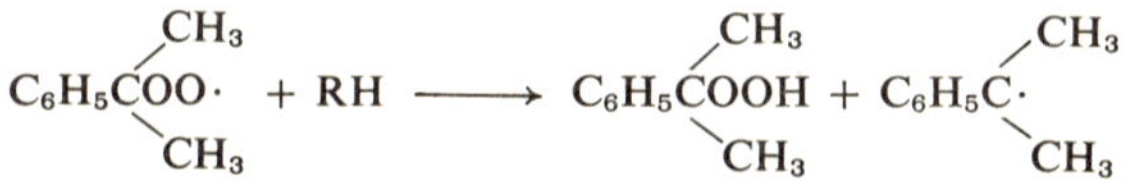

(151)

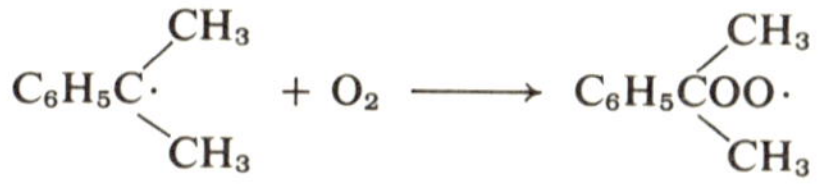

(152)

($RH = C_9H_{12}$)

a. Metal Organoperoxy Compounds. Organometallic compounds of B, Al, Zn, Cd, Mg, and Li react with molecular oxygen in ether solution to yield organoperoxymetallic compounds. A probable mechanism suggested [216] for the insertion of molecular oxygen into the organometallic compound is an initial nucleophilic attack of molecular oxygen on the metal ion, followed by a 1,3-rearrangement of the alkyl group from metal ion to oxygen in the manner depicted in Eq. (153). The two reaction steps shown may occur in a concerted fashion.

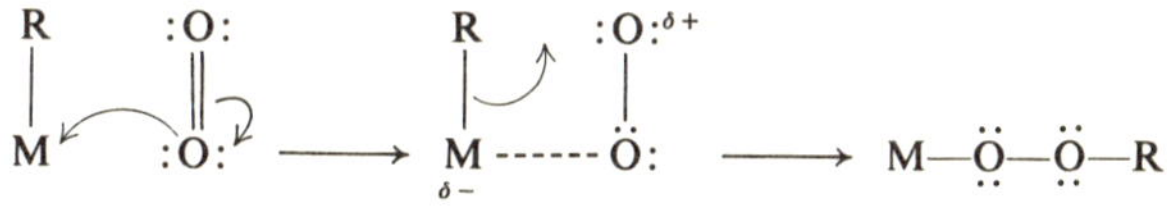

(153)

B. Insertion with O—O Bond Cleavage

1. *Enzymatic Systems*

Reactions in which both oxygen atoms become attached to the substrate may occur in two ways: (1) direct association of the oxygen with the substrate, followed by rearrangement to a more stable structure or (2) the breaking up of the oxygen to reactive fragments, followed by combination with the substrate. The latter type generally results in partial insertion of the oxygen because of the many side reactions that free radicals undergo. Let us consider first the direct insertion of the oxygen molecule, followed by rearrangement.

In biological systems total oxygen insertion is catalyzed by the dioxygenases, such as pyrocatechase, metapyrocatechase, and tryptophanpyrrolase

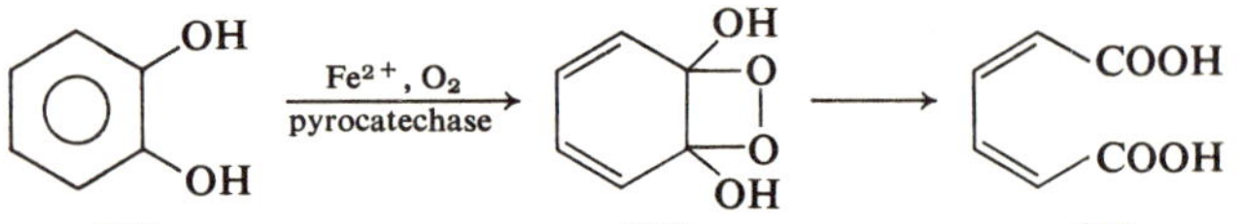

(154)

120 **121** **122**

[195–197,217]. Suggestions concerning the mechanism of the ferrous ion-catalyzed action of pyrocatechase were made by Hayaishi and Hashimoto [218] for the oxidation of catechol to *cis–cis* muconic acid. The reaction sequence proposed involves the formation of a hypothetical peroxy intermediate **121**, indicated in reaction (154). This mechanism is in conformity

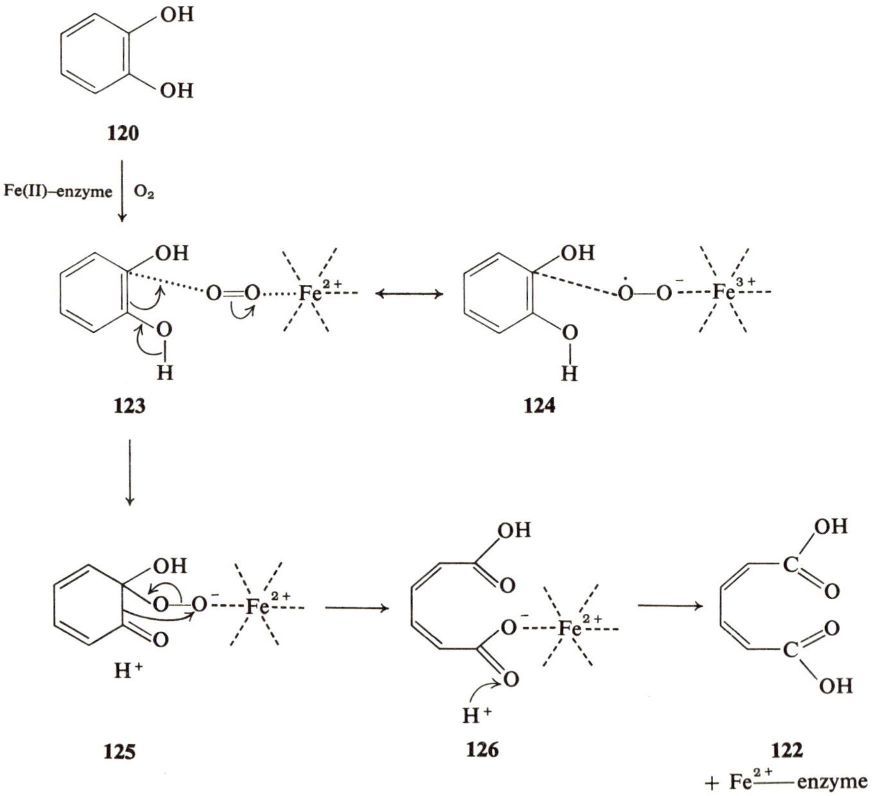

Fig. 26. Proposed mechanism for dioxygenase enzymes.

with the experimental observations that *o*-quinone and hydrogen peroxide were not detected as reaction intermediates, and that all of the oxygen is incorporated into the reaction product, as indicated by ^{18}O tracer studies. A more detailed mechanism, involving an intermediate iron(II)–oxygen complex, was proposed by Mason [212].

The peroxo-bridged intermediate proposed by Hayaishi has the disadvantage of requiring considerable energy in its formation, resulting from the loss of aromatic ring resonance and the strain of the four-membered peroxide ring. The mechanism illustrated in Fig. 26, which is analogous to a mechanism recently suggested by Hamilton [144] for tryptophan oxidation, is now proposed as a reasonable alternative. This mechanism is in accord with recent suggestions of ionic mechanisms of metal ion catalysis of various enzyme model reactions [144,219] and for the enzyme systems themselves [144]. Such ionic mechanisms seem to offer more reasonable explanations for the general experimental observations on the behavior of the oxygenases. In these new mechanisms, the metal ions are assigned more significant roles in the reaction mechanism. Also, the only substrates that show activity with the dioxygenase enzymes are those which can dissociate a proton in the initial step, as indicated above. The rearrangement of the organoperoxo intermediate has precedents in organic chemistry [7]. Similar mechanisms may be suggested for the oxidation of other substrates catalyzed by dioxygenases, involving total insertion of molecular oxygen.

2. *Model Systems*

A nonenzymatic reaction which seems to involve total insertion of oxygen and cleavage of the O—O bond is the Mn(III)/Mn(II)-catalyzed oxidation of acetophenone to benzoic acid [201]. The rate of oxidation was found to be independent of the metal ion concentration. Enolization of the ketone was

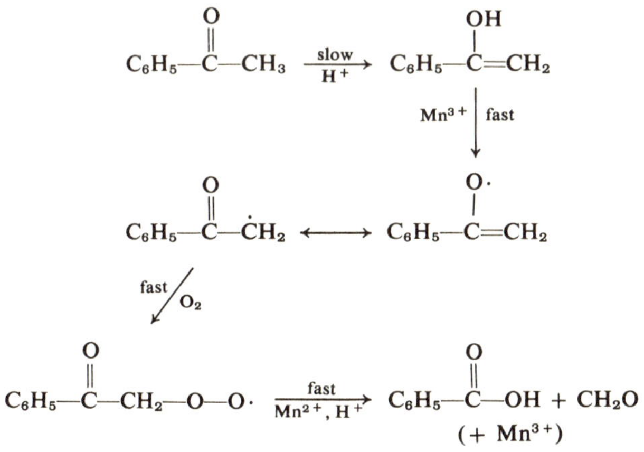

(155)

proposed as the rate-determining step. This is then followed by rapid free-radical reactions initiated and propagated by the oxidized and reduced forms of the metal ion, as indicated by reaction (155).

Comparison of reaction mechanism (155) with those suggested for the total insertion reactions of enzymes indicates that metal-catalyzed oxidation of acetophenone is not a satisfactory model for the dioxygenases. The main points of dissimilarity are the free-radical nature of the mechanisms and the fact that the oxygen atoms derived from the reacting oxygen molecule are found in two different reaction products. Model metal ion-catalyzed reactions in which both atoms of molecular oxygen enter the substrate molecule, become separated, and are found in a single oxidation product apparently have not yet been reported.

References

1. L. Pauling, "Nature of the Chemical Bond," pp. 351–353. Cornell Univ. Press, Ithaca, New York, 1960.
2. K. S. Pitzer, *J. Amer. Chem. Soc.* **70**, 2140 (1948).
3. A. D. Walsh, *J. Chem. Soc., London* pp. 331 and 398 (1948).
4. J. Chiriboga, *Arch. Biochem. Biophys.* **116**, 516 (1966).
5. K. B. Yatsimirskii, *Izv. Vyssh. Ucheb. Zaved., Khim. Khim. Tekhnol.* **2**, 480 (1959); *Chem. Abstr.* **54**, 6289h (1954).
6. W. H. Richardson, *J. Amer. Chem. Soc.* **87**, 247 and 1096 (1965).
7. E. S. Gould, "Mechanism and Structure in Organic Chemistry," p. 633. Holt, New York, 1959.
8. J. H. Wang, *in* "Oxygenases" (O. Hayaishi, ed.), pp. 469–516. Academic Press, New York, 1962.
9. F. Chirelli, *in* "Oxygenases" (O. Hayaishi, ed.), pp. 517–553. Academic Press, New York, 1962.
10. J. Weiss, *Nature (London)* **181**, 825 (1958).
11. E. Antonini, *in* "Oxygen in the Animal Metabolism" (F. Dickens and E. Neil, eds.), pp. 121–136. Macmillan, New York, 1964.
12. F. J. W. Roughton, *in* "Oxygen in the Animal Metabolism" (F. Dickens and E. Neil, eds.), pp. 5–28. Macmillan, New York, 1964.
13. R. R. Grinstead, *Biochemistry* **3**, 1308 (1964).
14. J. S. Griffith, *Proc. Roy. Soc., Ser. A* **235**, 23 (1946).
15. P. George, *Advan. Catal. Relat. Subj.* **4**, 367–428 (1952).
16. D. I. Crandall, *in* "Oxidases and Related Redox Systems" (T. F. King, H. S. Mason, and M. Morrison, eds.), p. 263. Wiley, New York, 1965.
17. P. George, *in* "Oxidases and Related Redox Systems" (T. F. King, H. S. Mason, and M. Morrison, eds.), pp. 3–33. Wiley, New York, 1965.
18. E. Frieden, S. Osaki, and H. Kobyashi, *Oxygen, Proc. Symp. N.Y. Heart Ass.* pp. 213–251 (1965).
19. F. J. W. Roughton, *Oxygen, Proc. Symp. N.Y. Heart Ass.*, pp. 105–124 (1965).
20. J. B. Wittenberg, *Oxygen, Proc. Symp. N.Y. Heart Ass.*, pp. 57–74 (1965).
21. R. Lemberg, J. W. Legge, and W. H. Lockwood, *Biochem. J.* **35**, 328, 339, and 363 (1941).

22. H. Vogt, Jr., H. M. Faigenbaum, and S. W. Wiberley, *Chem. Rev.* **63**, 269 (1963).
23. A. Vleck, *Trans. Faraday Soc.* **56**, 1137 (1960).
24. C. Brosset and N. G. Vannerberg, *Nature* (*London*) **190**, 714 (1961).
25. N. G. Vannerberg and C. Brosset, *Acta Crystallogr.* **16**, 247 (1968).
26. W. P. Schaefer and R. E. Marsh, *J. Amer. Chem. Soc.* **88**, 178 (1966).
27. E. Ebsworth and J. Weil, *J. Phys. Chem.* **63**, 1890 (1959).
28. A. E. Martell and M. Calvin, "Chemistry of the Metal Chelate Compounds." Prentice-Hall, Englewood Cliffs, New Jersey, 1952.
29. S. Fallab, *Chimia* **21**, 538 (1967).
30. S. Fallab, *Chimia* **13**, 177 (1969).
31. F. Miller and R. G. Wilkins, *J. Amer. Chem. Soc.* **92**, 2687 (1970).
32. F. Miller, J. Simplicio, and R. G. Wilkins, *J. Amer. Chem. Soc.* **91**, 1962 (1969).
33. A. G. Sykes and J. A. Weil, *Progr. Inorg. Chem.* **13**, 1 (1971).
34. R. Nakon and A. E. Martell, *J. Inorg. Nucl. Chem.* **34**, 1365 (1972).
35. R. Nakon and A. E. Martell, *J. Amer. Chem. Soc.* **94**, 3026 (1972).
36. R. Nakon and A. E. Martell, *Inorg. Chem.* **11**, 1002 (1972).
37. P. Biji and G. DeVries, *J. Chem. Soc.* (*Dalton*) p. 303 (1972).
38. H. K. J. Powell and G. H. Nancollas, *J. Amer. Chem. Soc.* **94**, 2664 (1972).
39. M. Woods, J. A. Weil, and J. K. Kinnaird, *Inorg. Chem.* **7**, 1713 (1972).
40. G. N. Schrauzer and L. Pin Lee, *J. Amer. Chem. Soc.* **92**, 1551 (1970).
41. M. Calligaris, G. Nardin, L. Randaccio, and A. Ripamonti, *J. Chem. Soc., A* p. 1069 (1970).
42. M. Calligaris, D. Ninchilli, G. Nardin, and L. Randaccio, *J. Chem. Soc., A* p. 2411 (1970).
43. L. Vaska, *Science* **140**, 809 (1963).
44. S. J. Laplaca and J. A. Ibers, *J. Amer. Chem. Soc.* **87**, 2581 (1965).
45. J. P. Collman and W. R. Roper, *Advan. Organometall. Chem.* **7**, 63 (1968).
46. L. Vaska and D. L. Catone, *J. Amer. Chem. Soc.* **88**, 532 (1966).
47. J. A. McGinnety, N. C. Payne, and J. A. Ibers, *J. Amer. Chem. Soc.* **91**, 6301 (1969).
48. J. J. Lavison and S. D. Robinson, *J. Chem. Soc. A* p. 762 (1971).
49. B. L. Shaw and R. E. Stainbank, *J. Chem. Soc.* (*Dalton*) p. 223 (1972).
50. R. W. Horn, E. Weissberger, and J. P. Collman, *Inorg. Chem.* **9**, 2367 (1970).
51. J. Valentine, D. Valentine, and J. P. Collman, *Inorg. Chem.* **10**, 219 (1971).
52. J. P. Collman and J. W. Kang, *J. Amer. Chem. Soc.* **89**, 844 (1967).
53. M. S. Weininger, I. F. Taylor, Jr., and E. L. Amma, *Chem. Commun.* p. 1172 (1971).
54. G. R. Clark, C. A. Reed, W. R. Roper, B. W. Skelton, and T. N. Waters, *Chem. Commun.* p. 758 (1971).
55. L. Vaska, L. S. Chen, and M. V. Miller, *J. Amer. Chem. Soc.* **93**, 6671 (1971).
56. S. A. Butler and J. Chatt, *J. Chem. Soc., A* p. 1411 (1970).
57. M. C. Baird, D. N. Lawson, J. T. Mague, J. A. Osborn, and G. Wilkinson, *Chem. Commun.* p. 129 (1967).
58. L. M. Haines, *Inorg. Chem.* **10**, 1685 (1971).
59. A. Nakamura, Y. Tatsuno, M. Yamamoto, and S. Otsuka, *J. Amer. Chem. Soc.* **93**, 6052 (1971).
60. M. J. Bennett and P. B. Donaldson, *J. Amer. Chem. Soc.* **93**, 3307 (1971).
61. T. E. Nappier, Jr. and D. W. Meek, *J. Amer. Chem. Soc.* **94**, 306 (1972).
62. N. W. Terry, III, E. L. Amma, and L. Vaska, *J. Amer. Chem. Soc.* **94**, 652 (1972).
63. M. M. Taqui Khan, R. K. Andal, and P. T. Manoharan, *Chem. Commun.* p. 561 (1971).
64. M. M. Taqui Khan and R. K. Andal, unpublished data.

65. B. W. Graham, K. R. Laing, C. J. O'Connor, and W. R. Roper, *Chem. Commun.* p. 1272 (1970).
66. B. E. Cavil, K. R. Grundy, and W. R. Roper, *Chem. Commun.* p. 60 (1972).
67. S. Takahashi, K. Sonogashira, and N. Hagihara, *Nippon Kagaku Zasshi* **87**, 610 (1966).
68. G. Wilke, H. Schott, and P. Hiembach, *Angew. Chem.* **79**, 62 (1967).
69. C. D. Cook and G. S. Jauhal, *Inorg. Nucl. Chem. Lett.* **3**, 31 (1967).
70. S. Otsuka, A. Nakamura, and Y. Tatsuno, *J. Amer. Chem. Soc.* **91**, 6994 (1969).
71. T. Kashiwagi, N. Yasuoka, N. Kasai, M. Kakudo, S. Takahashi, and N. Nagihara, *Chem. Commun.* p. 743 (1969).
72. M. C. Hall, B. T. Kilbourne, and K. A. Taylor, *J. Chem. Soc., A* p. 2539 (1970).
73. M. M. Taqui Khan and S. Vanchesan, unpublished data.
74. S. Cenini, A. Fusi, and G. Capparella, *Inorg. Nucl Chem. Lett.* **8**, 127 (1972).
75. M. M. Taqui Khan and R. K. Andal, *Proc. D.A.E. Symp.* (*India*) **1**, 253 (1970).
76. A. Nakamura, Y. Tatsuno, M. Yamamoto, and S. Otsuka, *J. Amer. Chem. Soc.* **93**, 6052 (1971).
77. J. A. McGinnety, R. J. Doedens, and J. A. Ibers, *Inorg. Chem.* **6**, 2243 (1967).
78. S. Otsuka, A. Nakamura, Y. Tatsuno, and M. Miki, *J. Amer. Chem. Soc.* **94**, 3761 (1972).
79. J. Halpern and A. L. Pickard, *Inorg. Chem.* **9**, 2798 (1970).
80. W. O. Seigel, S. J. Lapporte, and J. P. Collman, *Inorg. Chem.* **10**, 2158 (1971).
81. B. R. James and E. Ochiai, *Can. J. Chem.* **49**, 975 (1971).
82. B. R. James and F. T. J. Ng, *Chem. Commun.* p. 908 (1970).
83. A. G. Sykes, *Trans. Faraday Soc.* **59**, 1325 (1963).
84. A. G. Sykes, *Trans. Faraday Soc.* **59**, 1343 (1963).
85. A. G. Sykes, *Proc. Int. Conf. Coord. Chem., 7th*, p. 272 (1962).
86. J. A. Elvidge and A. B. P. Lever, *Proc. Chem. Soc., London* p. 195 (1959).
87. P. Przywarska-Boniecka, *Rocz. Chem.* **39**, 1377 (1965).
87a. M. Calvin, *Rev. Pure Appl. Chem.*, **15**, p.l. (1965).
88. D. Vonderschmitt, K. Berhauer, and S. Fallab, *Helv. Chim. Acta* **48**, 951 (1965).
88a. J. H. Weber and D. H. Busch, *Inorg. Chem.* **4**, 469 (1965).
89. M. T. Beck and S. Gorog, *Acta Chem. Acad. Sci. Hung.* **29**, 401 (1961).
90. R. Nakon, Ph.D. Dissertation, Texas A & M University, College Station, Texas (1971).
91. A. Weissberger, J. E. Luvalle, and D. S. Thomas, *J. Amer. Chem. Soc.* **65**, 1934 (1943).
92. H. Nord, *Acta Chem. Scand.* **9**, 430 (1955).
93. H. Nord, *Acta Chem. Scand.* **9**, 442 (1955).
94. R. R. Grinstead, *J. Amer. Chem. Soc.* **82**, 3464 (1960).
95. M. M. Taqui Khan and A. E. Martell, *J. Amer. Chem. Soc.* **89**, 4176 (1967).
96. M. M. Taqui Khan and A. E. Martell, *J. Amer. Chem. Soc.* **89**, 7014 (1967).
97. M. M. Taqui Khan and A. E. Martell, *J. Amer. Chem. Soc.* **90**, 6011 (1968).
98. C. R. Dawson, *in* "Biochemistry of Copper" (J. Peisach, P. Aisen, and W. E. Blumberg, eds.), p. 305. Academic Press, New York, 1966.
99. I. Yamasaki and L. J. Piette, *Biochim. Biophys. Acta* **50**, 62 (1961).
100. C. A. Tyson and A. E. Martell, *J. Amer. Chem. Soc.* **94**, 939 (1972).
101. D. Kertesz and R. Zito, *in* "Oxygenases" (O. Hayaishi, ed.), p. 307. Academic Press, New York, 1962.
102. H. S. Mason, *in* "Biochemistry of Copper" (J. Peisach, P. Aisen, and W. E. Blumberg, eds.), p. 307. Academic Press, New York, 1966.

103. D. W. Brooks and C. R. Dawson, *in* "Biochemistry of Copper" (J. Peisach, P. Aisen, and W. E. Blumberg, eds.), p. 339. Academic Press, New York, 1966.
104. D. Kertesz, *in* "Biochemistry of Copper" (J. Peisach, P. Aisen, and W. E. Blumberg, eds.), p. 359. Academic Press, New York, 1966.
105. H. S. Mason, *in* "Biological and Chemical Aspects of Oxygenases" (K. Block and O. Hayaishi, eds.), p. 287. Maruzen, Tokyo, 1966.
106. R. Zito and D. Kertesz, *in* "Biological and Chemical Aspects of Oxygenases" (K. Block and O. Hayaishi, eds.), p. 290. Maruzen, Tokyo, 1966.
107. H. S. Mason, E. Spencer, and I. Yamazaki, *Biochem. Biophys. Res. Commun.* **4**, 236 (1961).
108. P. George, *J. Chem. Soc., London* p. 434a (1954).
109. J. B. Ramsey, R. Sugimoto, and H. DeVorkin, *J. Amer. Chem. Soc.* **63**, 380 (1941).
110. J. W. Swinehart, *Inorg. Chem.* **4**, 1069 (1965).
111. M. Ardon and G. Stein, *J. Chem. Soc., London* p. 2095 (1956).
112. P. M. Henry, *Inorg. Chem.* **5**, 688 (1966).
113. A. M. Posner, *Trans. Faraday Soc.* **49**, 382 (1953).
114. M. Cher and N. Davidson, *J. Amer. Chem. Soc.* **77**, 793 (1955).
115. A. B. Lamb and L. W. Elder, *J. Amer. Chem. Soc.* **53**, 137 (1931).
116. E. Abel, *Monatsh. Chem.* **85**, 227 (1954).
117. R. E. Huffmann and N. Davidson, *J. Amer. Chem. Soc.* **78**, 4836 (1956).
118. T. W. Newton and F. B. Baker, *J. Phys. Chem.* **60**, 1417 (1956).
119. J. Halpern and J. G. Smith, *Can. J. Chem.* **34**, 1419 (1956).
120. G. A. Hamilton and A. Revesz, *J. Amer. Chem. Soc.* **88**, 2069 (1966).
121. J. M. Hill and P. J. G. Mann, *Biochem. J.* **99**, 454 (1966).
121a. M. Maruta and A. E. Martell, unpublished results.
122. A. E. Braunstein, *in* "The Enzymes" (P. D. Boyer, H. Lardy, and K. Myrbäck, eds.), 2nd rev. ed., Vol. 2, Part A, p. 113. Academic Press, New York, 1960.
123. F. Bufoni, *in* "Pyridoxal Catalysis" (E. E. Snell, *et al.* eds.), p. 363. Wiley (Interscience), New York, 1968.
124. H. Yamada, K. T. Yasunobu, Y. Yamano, and H. S. Mason, *Nature* (*London*) **198**, 1092 (1963).
125. A. Van Heuvelen, *Nature* (*London*) **208**, 888 (1965).
126. E. V. Goryachenkova, L. J. Stcherbatiuk, and C. I. Zamaraev, *in* "Pyridoxal Catalysis" (E. E. Snell, ed.), p. 391 Wiley (Interscience), New York, 1968.
127. H. R. Mahler, *in* "The Enzymes" (P. D. Boyer, H. Lardy, and K. Myrbäck, eds.), 2nd rev. ed., Vol. 8, p. 285. Academic Press, New York, 1963.
128. W. E. Blumberg, B. L. Horecker, F. Kelly-Falcoz, and J. Peisack, *Biochim. Biophys. Acta* **96**, 336 (1965).
129. F. C. Charalampous, *J. Biol. Chem.* **234**, 229 (1959).
130. F. C. Charalampous, *J. Biol. Chem.* **235**, 1286 (1960).
131. J. Chiriboga, *Arch. Biochem. Biophys.* **116**, 516 (1966).
132. G. Constantopoulos, A. Caroebter, P. A. Satch, and T. T. Chen, *Biochemistry* **5**, 1650 (1966).
133. H. S. Mason, E. Spencer, and I. Yamazaki, *Biochem. Biophys. Res. Commun.* **4**, 236 (1961).
134. W. G. Levine, *in* "Biochemistry of Copper" (J. Peisach, P. Aisen, and W. F. Blumberg, eds.), p. 371. Academic Press, New York, 1966.
135. T. Nakamura and Y. Ogura, *in* "Biochemistry of Copper" (J. Peisach, P. Aisen, and W. E. Blumberg, eds.), p. 389. Academic Press, New York, 1966.
136. W. W. Kaeding, R. O. Lindblom, and R. G. Temple, *Ind. Eng. Chem.* **53**, 805 (1961).

137. W. Brackmann and C. J. Baasbeek, *Rev. Trav. Chim. Pays-Bas* **85**, 242 (1966).
138. D. Huggins and W. C. Drinkard, *Advan. Chem. Ser.* **37**, 181 (1963).
139. J. E. Taylor, J. F. Yan, and J. L. Wang, *J. Amer. Chem. Soc.* **88**, 1663 (1966).
140. B. J. McCormick and G. Gorin, *Inorg. Chem.* **1**, 691 (1962).
141. D. L. Leussing and T. N. Tischer, *Advan. Chem. Ser.* **37**, 216 (1963).
142. E. Ochiai, *Tetrahedron* **20**, 1831 (1964).
143. W. I. Taylor and A. R. Battersby, "Oxidative Coupling of Phenols." Dekker, New York, 1967.
144. G. A. Hamilton, *Advan. Enzymol.* **24**, 54 (1969).
145. H. Siegel, *Angew. Chem., Int. Ed. Engl.* **8**, 167 (1969).
146. J. H. Wang, *J. Amer. Chem. Soc.* **77**, 4715 (1955).
147. R. C. Jarnigan and J. H. Wang, *J. Amer. Chem. Soc.* **80**, 6477 (1958).
148. P. Nichols and G. R. Schonbaum, *in* "The Enzymes" (P. D. Boyer, H. Lardy, and K. Myrbäck, eds.), 2nd rev. ed., Vol. 8, p. 147. Academic Press, New York, 1963.
149. G. A. Hamilton, J. P. Friedman, and P. M. Campbell, *J. Amer. Chem. Soc.* **88**, 5266 (1966).
150. G. A. Hamilton, J. W. Harrifin, Jr., and J. P. Friedman, *J. Amer. Chem. Soc.* **88**, 5269 (1966).
151. P. George, *in* "Currents in Biochemical Research" (D. E. Green, ed.), p. 338. Wiley (Interscience), New York, 1956.
152. B. Chance, *Arch. Biochem.* **21**, 416 (1949).
153. R. Breslow and L. N. Lukens, *J. Biol. Chem.* **235**, 292 (1960).
154. W. G. Barb, J. H. Baxendale, P. George, and K. R. Hargrave, *Nature* (*London*) **163**, 692 (1949).
155. W. G. Barb, J. H. Baxendale, P. George, and K. R. Hargrave, *Trans. Faraday Soc.* **47**, 462 (1951).
156. I. M. Kolthoff and A. I. Medalia, *J. Amer. Chem. Soc.* **71**, 3777 and 3784 (1949).
157. G. A. Hamilton and J. P. Friedman, *J. Amer. Chem. Soc.* **85**, 1008 (1963).
158. G. A. Hamilton, R. J. Workman, and L. Woo, *J. Amer. Chem. Soc.* **86**, 3390 (1964).
159. G. R. A. Johnson, G. Stein, and J. Weiss, *J. Chem. Soc., London* p. 3275 (1951).
160. H. Loebl, G. Stein, and J. Weiss, *J. Chem. Soc., London* p. 2074 (1949).
161. H. Loebl, G. Stein, and J. Weiss, *J. Chem. Soc. London* p. 2704 (1950).
162. H. Loebl, G. Stein, and J. Weiss, *J. Chem. Soc., London* p. 405 (1951).
163. G. Stein and J. Weiss, *Nature* (*London*) **161**, 650 (1948).
164. G. Stein and J. Weiss, *Nature* (*London*) **166**, 1140 (1950).
165. G. Stein and J. Weiss, *J. Chem. Soc., London* p. 3265 (1951).
166. J. H. Baxendale, M. G. Evans, and G. S. Park, *Trans. Faraday Soc.* **42**, 155 (1946).
167. F. Haber and J. Weiss, *Proc. Roy. Soc., Ser. A* **147**, 332 (1934).
168. J. H. Baxendale, *Advan. Catal. Relat. Subj.* **4**, 31 (1952).
169. M. G. Evans and N. Uri, *Trans. Faraday Soc.* **45**, 224 (1949).
170. M. G. Evans, P. George, and N. Uri, *Trans. Faraday Soc.* **45**, 230 (1949).
171. M. G. Evans, J. H. Baxendale, and N. Uri, *Trans. Faraday Soc.* **45**, 236 (1949).
172. W. C. Bray and M. H. Gorin, *J. Amer. Chem. Soc.* **54**, 2124 (1932).
173. J. Weiss, *Nature* (*London*) **181**, 825 (1958).
174. J. Weiss, *Trans. Faraday Soc.* **35**, 48 and 219 (1939).
175. J. Weiss, *Trans. Faraday Soc.* **37**, 463 (1941).
176. A. E. Cahill and H. J. Taube, *J. Amer. Chem. Soc.* **74**, 2313 (1952).
177. T. J. Conocchioli, E. J. Hamilton, Jr., and N. Sutin, *J. Amer. Chem. Soc.* **87**, 926 (1965).
178. N. K. King and M. E. Winfield, *Aust. J. Chem.* **12**, 138, 147 (1959).

179. I. M. Kolthoff and A. I. Medalia, *J. Amer. Chem. Soc.* **71**, 3784 (1949).
180. S. Kauffmann, *J. Biol. Chem.* **239**, 332 (1964).
181. R. R. Grinstead, *J. Amer. Chem. Soc.* **82**, 3472 (1960).
182. K. G. Paul, *in* "The Enzymes" (P. D. Boyer, H. Lardy, K. Myrbäck, eds.), 2nd rev. ed., Vol. 8, p. 227. Academic Press, New York, 1963.
183. B. C. Saunders, A. G. Holmes-Siedle, and B. P. Stark, "Peroxidases." Butterworth, London, 1964.
184. M. Mugdan and D. P. Young, *J. Chem. Soc., London* p. 2988 (1949).
185. N. Indictor and W. E. Bril, *J. Org. Chem.* **30**, 2074 (1965).
186. J. Axelrod, S. Udenfriend, and B. Brodie, *J. Pharmacol. Exp. Ther.* **111**, 176 (1954).
187. S. Udenfriend, C. T. Clark, J. Axelrod, and B. B. Brodie, *Fed. Proc., Fed. Amer. Soc. Exp. Biol.* **11**, 301 (1952).
188. S. Udenfriend, C. T. Clark, J. Axelrod, and B. B. Brodie, *J. Biol. Chem.* **208**, 731, and 741 (1954).
189. C. Nofre, A. Ceer, and A. Lefier, *Bull. Soc. Chim. Fr.* [5] p. 530 (1961), and references therein.
190. R. O. C. Norman and G. K. Rodda, *Proc. Chem. Soc., London* p. 130 (1962).
191. V. Ullrich and H. Standinger, *in* "Biological and Chemical Aspects of Oxygenases" (K. Block and O. Hayaishi, eds.), p. 235. Maruzen, Tokyo, 1966.
192. H. Standinger, B. Kerekjarto, V. Ullrich, and Z. Zubrzychi, *in* "Oxidases and Related Redox Systems" (T. E. King, H. S. Mason, and M. Morrison, eds.), p. 815. Wiley, New York, 1965.
193. R. O. C. Norman and J. R. L. Smith, *in* "Oxidases and Related Redox Systems" (T. E. King, H. S. Mason, and M. Morrison, eds.), p. 131. Wiley, New York, 1965.
194. A. Bobst and M. Viscontini, *Helv. Chim. Acta* **49**, 884 (1966).
195. H. Mason, *Annu. Rev. Biochem.* **34**, 595 (1965).
196. T. E. King, H. S. Mason, and M. Morrison, eds., "Oxidases and Related Redox Systems." Wiley, New York, 1965.
197. K. Block and O. Hayaishi, eds., "Biological and Chemical Aspects of Oxygenases." Maruzen, Tokyo, 1966.
198. R. Van Heldon, A. Bickel, and E. Kooyman, *Rec. Trav. Chim. Pays-Bas* **80**, 1237 and 1257 (1961).
199. G. Guroff, J. W. Daly, D. M. Jerina, J. Renson, B. Witkop, and S. Udenfriend, *Science* **157**, 1524 (1967).
200. N. H. Sloane and K. G. Untch, *Biochemistry* **3**, 1160 (1964).
201. R. Van Helden and E. C. Kooyman, *Rec. Trav. Chim. Pays-Bas* **80**, 230 (1961).
202. B. B. Brodie, J. Axelrod, P. S. Shore, and S. Udenfriend, *J. Biol. Chem.* **208**, 741 (1954).
203. J. H. Green, B. J. Ralph, and P. Schofield, *Nature (London)* **195**, 1309 (1962).
204. C. E. Dalgleish, *Biochem. J.* **58**, XLV (1954).
205. C. E. Dalgleish, *Arch. Biochem. Biophys.* **58**, 214 (1955).
206. R. H. Acheson and C. M. Hazelwood, *Biochim. Biophys. Acta* **42**, 49 (1960).
207. J. H. Merz and W. A. Waters, *J. Chem. Soc., London* p. 15 (1949).
208. J. H. Merz and W. A. Waters, *J. Chem. Soc., London* p. 2427 (1949).
209. M. Chvapil and J. Hurych, *Nature (London)* **184**, 1145 (1959).
210. F. Dewhurst and G. Calcutt, *Nature (London)* **191**, 808 (1961).
211. K. Torri and T. Moriyama, *J. Biochem. (Tokyo)* **42**, 193 (1955).
212. H. S. Mason, *Advan. Enzymol.* **19**, 79 (1957).
213. Y. Yamiya and K. U. Ingold, *Can. J. Chem.* **42**, 1027 (1964).
214. Y. Yamiya and K. U. Ingold, *Can. J. Chem.* **42**, 2424 (1964).

215. H. Kropf, *Justus Liebigs Ann. Chem.* **637**, 73 (1960).
216. A. G. Davies, D. G. Hare, and R. F. H. White, *J. Chem. Soc., London* p. 341 (1961).
217. O. Hayaishi, *Proc. Int. Congr. Biochem., Plen. Sess., 6th, 1964* IUB Vol. 33, p. 31 (1964).
218. O. Hayaishi and K. Hashimoto, *J. Biochem. (Tokyo)* **37**, 371 (1950).
219. A. E. Martell, *Proc. Symp. Coord. Chem., Plen. Sess., 3rd, 1970* Vol. 2, p. 125, Akademiai Kiado, Budapest (1971).

3

Activation of Molecular Nitrogen

I. Introduction

The nitrogen molecule is extraordinarily stable, with a dissociation energy of 225 kcal/mole. The N—N bond energy increases from 38 kcal/mole in single-bonded nitrogen (hydrazine) to 98 kcal/mole for double-bonded nitrogen in azo compounds. The increase in bond energy in passing from double- to triple-bonded nitrogen (in molecular N_2) is thus very large. The unusual stability of the nitrogen molecule is reflected in the difficulty of producing nitrogen compounds directly from molecular nitrogen, a process commonly referred to as the fixation of nitrogen. In the fixation of nitrogen by living organisms, enzymatic activation of molecular nitrogen takes place at ordinary temperature and pressure. Industrially, nitrogen activation is a process that is carried out under extreme reaction conditions. Enzymatic fixation of nitrogen is not only important in maintaining the fertility of the soil but also serves to keep a balance of free and fixed nitrogen in the geo- and biospheres.

A landmark in the chemistry of reactions of molecular nitrogen has been the recent studies on the preparation of complexes of molecular nitrogen with transition metal ions. After the first report of the complex $[Ru(NH_3)_5N_2]^{2+}$ by Allen and Senoff [1], a wide variety of nitrogen complexes of the transition metals have been reported by several investigators. The known nitrogen complexes now include those of ruthenium(II) [1–5], rhodium(II) [6], osmium(II) [7, 8], iridium(II) [9,10], cobalt(0) [11–13], and cobalt(I) [11,14].

Fixation of molecular nitrogen by reduction to ammonia in chemical systems has been investigated by Volpin and co-workers [15–21] and Van Tamelen and co-workers [22]. The best catalysts were found to be titanium(II) compounds in organic solvents. Fixation in aqueous solution has been studied with varying degrees of success by Haight and Scott [23], Yatsimirskii and Pavlova [24], and Taqui Khan and Martell [25]. Most of the work in the field of chemical fixation of nitrogen is of a qualitative nature and much remains to be done before detailed pictures of the mechanisms of these reactions can be clarified.

This chapter deals first with well-known biological systems for the activation of molecular nitrogen. This treatment is followed by a description of nitrogen complexes and chemical fixation. This field is developing at such a rapid pace that by the time this book appears the treatment given will inevitably be incomplete.

II. Activation by Microorganisms

A. Types of Organisms That Fix Nitrogen

Nitrogen is fixed in biological systems mainly by four types of micro organisms: (1) free-living bacteria, (2) symbiotic bacteria, (3) photosynthetic blue-green algae, and photosynthetic microorganisms. Each group is further subdivided into aerobic and anaerobic systems. A representative list of organisms for nitrogen fixation is given in Table I [26].

1. *Free-Living Bacteria*

The anaerobe *Clostridium pasteurianum* was the first organism reported to fix molecular nitrogen [27,28]. *Clostridium pasteurianum* is widely distributed in soils and has also been found in salt and fresh water.

The aerobic nitrogen-fixing organisms *Azotobacter agilis* and *Azotobacter chroococcum* were isolated from soil and canal water [29]. The aerobic *Azotobacter* species are widely distributed in soils to a depth of up to 50 cm. Acidity is favorable for the growth of *Azotobacter* species.

2. *Symbiotic Bacteria*

The symbiotic bacteria or nodule bacteria referred to as *Rhizobium* have been isolated from the plant nodules of legumes. The ability of *Rhizobium* to fix nitrogen depends on the strain of the bacteria, strain of the plant, and conditions of plant growth. Under the proper conditions, nitrogen fixation occurs as a symbiotic process, the plant supplying carbon intermediates for the metabolism of the bacteria and the bacteria in turn supplying fixed nitrogeneous material to the host plant. Without symbiosis the system does not function; the plant alone or the bacteria alone cannot fix nitrogen. The ability to fix nitrogen falls rapidly in excised or crushed nodules, although most symbiotic systems of this type are found in the nodules of legumes.

TABLE I

NITROGEN-FIXING MICROORGANISMS

Type	Aerobic	Anaerobic
(1) Free-living bacteria	*Azotobacter*	*Clostridium* *Aerobacter aerogenes* *Bacillus polymyxa* *Klebsiella* N_4—B
(2) Symbiotic bacteria	*Rhizobium* and leguminous plants	
(3) Photosynthetic blue-green algae	*Anabaena* *Anabaenopas* *Cylindrospermum* *Mastigoclaudus* *Nostoc* *Oscillatoria* *Tolypothrix*	
(4) Photosynthetic microorganisms		*Rhodospirillum* *Chromatium* *Rhodopseudomonas*

Bond [30] has reported the fixation of nitrogen in nonlegume root nodules of *Casuarina*.

3. *Photosynthetic Blue-Green Algae and Lichens*

Nitrogen-fixing species occur in the genera *Anabaena*, *Anabaenopsis*, *Aulosire*, *Calothrix*, *Cylindrospermum*, *Mastigocladus*, *Nostoc*, *Oscillatoria*, and *Tolypothrix* [26,31].

Symbiosis between algae and fungi frequently occurs in lichens. The symbiosis may or may not be of the nitrogen-fixing type, depending on the presence or absence of blue-green algae in the lichens. Most of the algae in lichens is usually the non-nitrogen-fixing green type, but some lichens do contain nitrogen-fixing blue-green algae.

4. *Photosynthetic Microorganisms*

Gest and co-workers [32,33] have discovered that the anaerobic photosynthetic purple nonsulfur bacterium *Rhodospirillum rubrum* is capable of fixing nitrogen. Fixation of nitrogen in the photosynthetic microorganisms *Chromatium* and *Rhodopseudomonas palustris* has been reported by Arnon *et al.* [34] and by Yamanaka and Kamen [35], respectively.

B. PRODUCTS OF NITROGEN FIXATION

An important achievement in the field of biological fixation of nitrogen has been the development of techniques for the preparation of cell-free extracts capable of reproducible nitrogen fixation. The methods used for the preparation of cell-free extracts in the case of the anaerobe *Clostridium pasteurianum* are ultrasonic irradiation of the whole cells with a Mullard 20 kcycle/second probe or rupture of the whole cells in a Hughes press at $-15°$ to $-55°C$ followed by centrifugation [36,37]. Sonic oscillation with a Raytheon 10 kcycle/second magnetostrictive oscillator has been used for the cell-free preparation of blue-green algae [31,38]. Nicholas *et al.* [39] prepared active extracts from *Azotobacter vinelandii* by either lysozyme treatment or ultrasonic irradiation by a Mullard 20 kcycle/second probe. In most cases the extent of fixation in cell-free extracts has been measured by the abundance of ^{15}N in the fixation products.

The fate of nitrogen in nitrogen fixation and the several hypothetical intermediates involved as suggested by Burris [40] are presented in Fig. 1. The oxidation states of nitrogen in the intermediates are indicated in parentheses for clarity. Burris [40] has defined a key intermediate as the stage at which fixation of nitrogen stops and assimilation begins. In other words, at the key intermediate stage, nitrogen goes from an inorganic to an organic compound. Schemes for nitrogen fixation have been advanced by two groups of investigators, Burris and co-workers [31,38] at Wisconsin, and Virtanen and co-workers [41,42] at Helsinki. The Burris group supports the formation of ammonia as the key intermediate in nitrogen fixation, whereas Virtanen's group advocates the formation of hydroxylamine as the key intermediate. Though details have been worked out by both the groups on the assimilation steps (formation of amino acids), there is no unequivocal evidence for the nature of the intermediate steps between nitrogen and ammonia shown in Fig. 1.

In the ammonia hypothesis of nitrogen fixation, ammonia and glutamic acid are the two main compounds through which fixed nitrogen enters the general metabolism of *Azotobacter* [40,43], *Rhodospirillum rubrum* [40], and *Clostridium* [44]. The ammonia is largely assimilated by conversion to glutamine and asparagine. In the aerobic *Azotobacter* species the excreted ammonia and glutamic and aspartic acids may be derived from the hydrolysis of glutamine and asparagine, respectively. Aspartic acid may also be formed by the transamination of oxaloacetic acid with glutamic acid or directly from oxaloacetic acid and ammonia by the reductive amination of the former in the presence of DPNH.

Based on the detection of the oxime of α-oxosuccinic acid in the excretory products of *Azotobacter*, hydroxylamine has been suggested as an intermediate in nitrogen fixation [41,42]. Aspartic acid was found to be the main excreted compound, though β-alanine and oximinosuccinic acid have also been detected. Oximes are finally reduced enzymatically to amino acids as shown in Fig. 1. The detection of glutamic acid in the excretions of nodulated

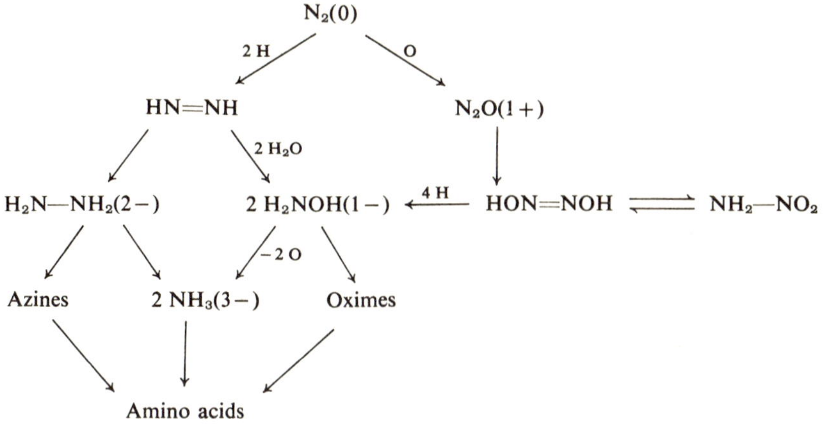

Fig. 1. Products of the fixation of nitrogen. After Burris [40].

plants led Virtanen *et al.* [41,42] to suggest that hydroxylamine is mainly reduced to ammonia, which condenses with α-ketoglutaric acid to give glutamic acid.

Another possible route suggested for the fixation of nitrogen involves conversion of nitrogen to nitrous oxide, followed by hydration to hyponitrous acid or nitramide and final reduction to hydroxylamine. Considerable experimental evidence has accumulated to indicate that this is not a significant pathway for nitrogen fixation, which appears to be entirely reductive. Allison and Burris [45] detected no labeled oxidized nitrogen compounds in *Azotobacter* supplied with $^{15}N_2$. The assimilation of $^{15}N_2O$ by nitrogen-fixing agents was found to be slower than that of molecular nitrogen [46]. Also, nitramide was found to decompose very rapidly [46], and its nitrogen was not utilized by the culture. According to McKee [26], the presence of oxidized nitrogen compounds in nitrogen-fixing organisms may be due to side reactions involving minor metabolic pathways.

Reductive fixation of nitrogen through the formation of diimide and hydrazine is also a possible pathway. The first step in such a mechanism is the formation of the hypothetical free radical $N_2H\cdot$ [47], analogous to $HO_2\cdot$, followed by reduction to diimide, N_2H_2. Diimide has also been proposed as an intermediate by Virtanen and Hakala [41]. Since it is very unstable and has not yet been isolated, it is presumed to undergo rapid further reduction to hydrazine. Reduction of nitrogen to hydrazine is partially supported by the work of Bach [48] in which labeled hydrazine with both nitrogen atoms labeled was supplied to *Azotobacter*. Bach [48] was able to isolate isotopic ^{15}N from the cell in the form of three azines, one of them being 2H,4H,5H-3-oxopyridazine-6-carboxylic acid **1**, obtained by the condensation of hydrazine with α-ketoglutaric acid. The same azines were isolated from *Azotobacter* grown on nitrogen with no hydrazine added. Azim and Roberts [49] reported inhibition by hydrazine at concentrations above 2×10^{-5} *M* but at lower concentrations hydrazine was found to stimulate fixation.

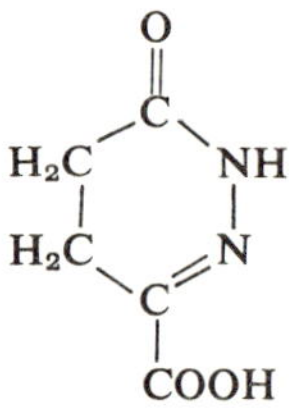

1

C. Requirements for Nitrogen Fixation

1. *Source of Electrons for the Reduction of Nitrogen*

If a reductive pathway is assumed for the fixation of nitrogen, then one of the essential requirements is a readily available source of electrons at the appropriate reduction potential. Extracts of *Clastridium pasteurianum* require a large concentration of pyruvate or 2-ketobutyrate for fixation [37]. The availability of pyruvate as a source of electrons for the fixation of N_2 has been supported by the work of Hardy *et al.* [50]. Pyruvate not only supplies electrons for the reduction of nitrogen, but also acts as a source of energy for the phosphoroclastic reaction (conversion of ADP to ATP) which accompanies fixation. Pyruvate as an electron donor can be replaced by other electron donors, such as KBH_4, H_2, NADH, or NADPH [50], if coupled with an ATP generator (acetyl phosphate or creatine phosphate + creatine kinase). Besides the requirement of an electron donor, nitrogen-fixing systems also require the low potential electron carrier, ferredoxin, a nonheme-iron protein.

2. *Metal Ion Requirement*

Fixation of nitrogen in microorganisms is directly affected by trace quantities of metal ions. A recent review by Nicholas [51] summarizes the metal ion requirements of various microorganisms. The metal ions that have direct bearing on the fixation of nitrogen are molybdenum(III,IV), tungsten-(III,IV), iron(II,III), and calcium(II). Direct involvement of other trace metal ions such as cobalt(II), manganese(II), and copper(II) in nitrogen-fixing metabolism has not yet been established with any degree of certainty.

Magee and Burris [46] have isolated a molybdenum-rich protein from *Azotobacter* cells. Since it was not possible to separate molybdenum from the protein by conventional dialysis, it was concluded accordingly that molybdenum is firmly bound to the *Azotobacter* protein, possibly as a molybdoprotein. A molybdenum-rich extract of *Azotobacter* has also been obtained by Keeler *et al.* [52]. The extract was later demonstrated to contain considerable nitrogenase activity [53]. The oxidation state of molybdenum in the molybdoprotein of *Azotobacter* is, however, not known with certainty. Though the oxidation states molybdenum(V) and molybdenum(III) have been suggested by Nicholas *et al.* [39] based on the electron paramagentic resonance (EPR) spectroscopy of the extract, it is nevertheless possible that the metal may possess entirely different oxidation states as part of the nitrogenase while fixing molecular nitrogen. Tungsten can replace molybdenum in *Azotobacter* and is incorporated into the same protein fraction as molybdenum [54]. The role of tungsten in nitrogen fixation is not, however, fully understood.

Iron is an essential element in the nitrogen-fixing metabolism of the large number of anaerobic, aerobic, photosynthetic, and symbiotic microorganisms that are known to date. The presence of iron in the nonheme-iron protein, ferredoxin, has been substantiated by the studies of Mortensen *et al.* [55] and later by Tagawa and Arnon [56]. Ferredoxin mainly acts as an electron carrier in the nitrogen-fixing mechanism of aerobic, anaerobic, and photosynthetic bacteria. The presence of hemoglobin in the red pigment of soybean nodules was first demonstrated by Kubo [57]. It was demonstrated by Ellfolk and Virtanen [58] that there is a close correlation between the nitrogen-fixing capacity of *Rhizobia* (root nodule bacteria) and the hemoglobin contents of the nodules. The red pigment has been characterized as a heme protein largely due to the work of Ellfolk and Virtanen [58], and its participation in nitrogen fixation mechanisms of symbiotic systems has been explained by Abel and co-workers [59,60].

Calcium(II) has been reported as an essential element in nitrogen fixation by *Azotobacter* [61]. It is believed [61] that calcium(II) is involved in the synthesis of ATP accompanying enzymatic nitrogen fixation. The details of the requirement of calcium(II) in the nitrogen fixation of *Azotobacter* are not known and more work is needed along these lines.

Various additional trace elements such as cobalt(II), copper(II), calcium(II), and manganese(II) have been reported in the metabolic processes of nitrogen-fixing microorganisms [62]. Their participation in nitrogen fixation mechanisms is, however, rather tenuous and controversial.

3. *Effect of the Concentration of Molecular Nitrogen*

The effect of the concentration of nitrogen on the rate of fixation of nitrogen is expressed by the Michaelis-Menton equation,

$$1/k = 1/k(\max) + K_M/[k(\max)p_{N_2}] \tag{1}$$

where k is the rate constant for the fixation of nitrogen, $k(\max)$ is the rate constant for maximum fixation, p_{N_2} is the partial pressure of nitrogen, and K_M is the Michaelis constant. The constant K_M is obtained by the plot of $1/k$ against $1/p_{N_2}$, whereby the intercept is $1/k(\max)$ and slope $K_M/k(\max)$. K_M is usually expressed in units of atmospheres and is equivalent to the dissociation constant of a hypothetical nitrogenase–nitrogen complex. The value of K_M varies from 0.02 to 0.03 atm for various nitrogen-fixing agents, including the aerobic, anaerobic, and symbiotic forms.

4. *Oxygen Dependence*

Oxygen completely destroys the hydrogenase activity of the anaerobe *C. pasteurianum*. For the aerobic bacterium *Azotobacter*, oxygen is required only for cell growth [63,64]. Bulen *et al.* [53] have reported that oxygen is not required for fixation of nitrogen by the cell-free extracts of *Azotobacter*. Also, fixation takes place in complete absence of oxygen with *Azotobacter* in the presence of hydrogenase extract from *C. pasteurianum*, ferredoxin, and a source of ATP [52].

5. *Effect of the Partial Pressure of Hydrogen*

Hydrogen acts as an electron donor for the fixation of nitrogen in cell-free extracts of *Azotobacter* when mixed with hydrogenase from *C. pasteurianum* and ferredoxin [53,64]. Molecular hydrogen can also act as electron donor in the nitrogen-fixing system of *C. pasteurianum* in the presence of a source of ATP. Thus in both the systems, aerobic and anaerobic, molecular hydrogen is not an inhibitor of nitrogen fixation as had been reported earlier [65]. Most of the nitrogen-fixing systems, anaerobic, aerobic, symbiotic, and photosynthetic, contain hydrogenases that may act as ready sources of electrons for the fixation of nitrogen. In the absence of nitrogen, the electrons from a suitable substrate such as pyruvate are used for the photoevolution of hydrogen in photosynthetic bacteria [32,33]. In the presence of nitrogen, the photoevolution of hydrogen is completely inhibited and the electrons from hydrogenase are used for the fixation of nitrogen.

D. Mechanism of Nitrogen Fixation

1. *Symbiotic Systems*

Hoch *et al.* [66,67] have posulated a working mechanism, depicted in Fig. 2, for the fixation of nitrogen in symbiotic systems. The nature of the enzymes involved in hydrogenase activity, path A, and nitrogenase activity, path B, are not yet well understood. The enzymes in question have sites capable of binding protons and molecular nitrogen with the subsequent transfer of two electrons to two protons (evolution of hydrogen) or to a nitrogen molecule (reduction). Bauer [68] has suggested that in the heme-iron protein, legoglobin [69], active in nitrogen fixation of the symbiotic system, the hydro-

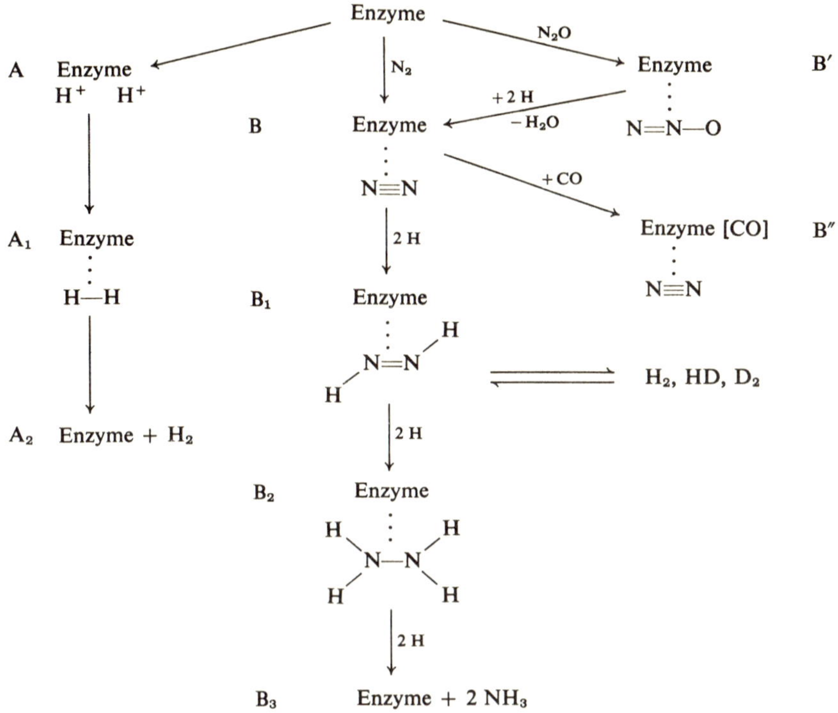

Fig. 2. Fixation of nitrogen in symbiotic systems. After Hoch *et al.* [67].

genase and the nitrogenase sites are either located near or on the same position in the protein. The fact that legoglobin iron atoms change their valence during nitrogen fixation suggests a key role of the heme-iron protein in the enzyme system of nitrogen fixation. Abel *et al.* [59,60] postulated the formation of a ferrolegoglobin–nitrogen complex in the nitrogen-fixing system of *rhizobia* and named the complex a ferroenzyme. Bauer [68] indicated the possibility of hydrogen evolution near such systems as reported by Hoch *et al.* [66,67]. The reduction of protons to molecular hydrogen results in a change of pH near the enzyme system and affects the binding characteristics of the ferroenzyme (ferrolegoglobin). The change of pH is also considered to bring about the appropriate configurational change in the ferro complex to bind molecular nitrogen in path B. The nature of the configurational change in the ferro complex and the factors responsible for nitrogen complex formation in path B are not clear [59]. In the proposed reaction two electrons are transferred to nitrogen and the ferro complex is changed to a ferri complex. Hydrogen atoms from the surrounding medium (path A) then attack the ferri-enzyme to produce diimide and the ferroenzyme. The ferroenzyme binds diimide and a further transfer of two electrons followed by attack of two hydrogen atoms (step B_2) results in the formation of hydrazine and the ferroenzyme. Ammonia is produced in an analogous manner in step B_3. Hoch *et al.* [67] suggested that the products of reduction are bound to the ferroenzyme (steps B_1, B_2, and B_3) until the final product, ammonia, is obtained.

In path B″, the enzyme is poisoned by carbon monoxide through the formation of a carbonyl complex of the ferroenzyme. Path B′ indicates inhibition by nitrous oxide in a similar fashion by formation of an iron complex. The complex can, however, be reduced to the nitrogen complex by hydrogen atoms produced in Path A.

2. *Free-Living Bacteria* (*Anaerobic Systems*)

Mortensen *et al.* [70,71] obtained two fractions from the cell-free extracts of *C. pasteurianum* by centrifugation under a hydrogen atmosphere. One of the fractions contained the nitrogenase activity of the system and the other all the hydrogenase and 80% of the pyruvic acid dehydrogenase activity. Neither of these two fractions was capable of fixing nitrogen, but a combination of the two with pyruvate and molecular nitrogen was highly active in nitrogen fixation. The fractions having a molecular weight range of 100,000–120,000 contain molybdenum and iron in a 1:12 molar ratio [72,73] and sulfide donor groups (S^{2-}). The other fraction, which is unstable in the absence of its counterpart (fraction 1), has nonheme iron and sulfide, but no molybdenum, and has a molecular weight in the range 40,000 to 60,000 [74]. Both fractions lose their activity in the presence of molecular oxygen.

The nitrogen-fixing systems were also found to contain an electron carrier that mediates the transfer of electrons from pyruvic dehydrogenase to dehydrogenase. This electron carrier was finally isolated from nitrogen-fixing bacteria *C. pasteurianum* [55,71,75] and given the name ferredoxin. Ferredoxin has also been isolated in crystalline form from photosynthetic bacteria *Rhodospirillum rubrum* and *Chromatium* by Tagawa and Arnon [56] and from *Rhodopseudomonas palustris* by Yamanaka and Kamen [35]. The term "ferredoxin" now applies to other iron–sulfur proteins involved in the photosynthetic process of chloroplasts. Investigation on clostridial ferredoxin [76] indicates that it is a nonheme-iron protein with a molecular weight of about 6000 and about 50 amino acid residues. The prosthetic group of ferredoxin is characterized [77] by the presence of eight iron, eight labile sulfide groups, and eight cysteine groups bound to iron. The presence of two types of iron complexes was suggested [78] on the basis of the EPR spectrum of reduced ferredoxin. This has been confirmed [79] recently by the X-ray structure of clostridal ferredoxin, which shows two identical clusters of Fe_4S_4 units separated by 12 Å (Fig. 3). Each cluster contains four iron and sulfur atoms

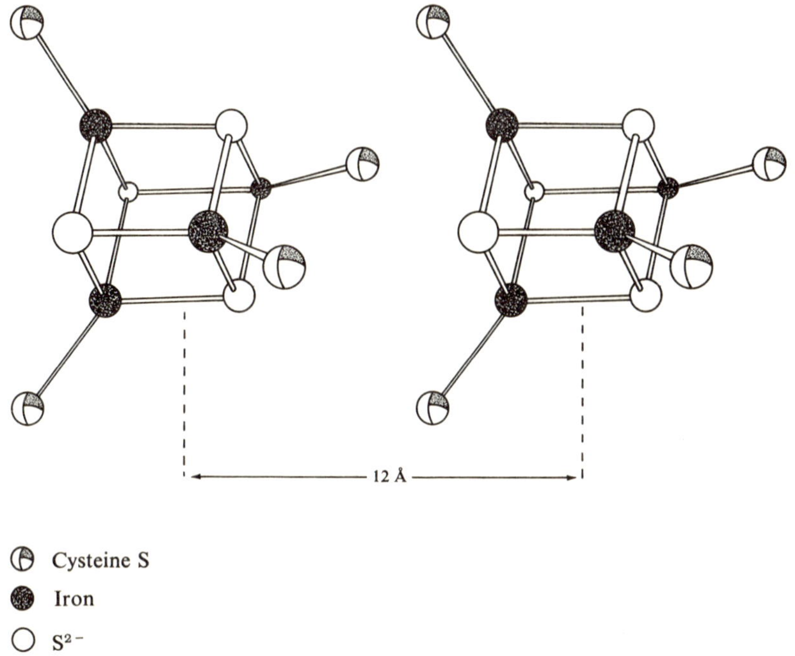

Fig. 3. Structure of clostridial ferredoxin. After Sieker *et al.* [79].

at the alternate corners of a cube with four cysteine sulfur atoms projecting from the iron atoms. In structural details the Fe_4S_4 cluster is similar to the Fe_4S_4 cluster [80] of high potential iron protein (HPIP) from *Chromatium*. High potential iron protein consists [81] of an 86-residue polypeptide chain with four cysteines, and an inorganic prosthetic group composed of four sulfides and four iron atoms bound to the apoprotein by the cysteine sulfur atoms (Fig. 4). Close structural analogs of ferredoxin and HPIP are the iron complexes $[(C_5H_5)FeS]_4$ [82] (see Fig. 5) and $Fe_4S_4(SR)_4$ [83] (R = Me, Et, $C_6H_5CH_2$). The properties of ferredoxin, HPIP, and $[C_5H_5FeS]_4$ and $[Fe_4S_4(RS)_4]$ are summarized in Table II [79,80,82–84].

A general mechanism for nitrogen fixation by anaerobic systems is presented in Fig. 6. The source of electrons for the reduction of ferredoxin is pyruvate through the phosphoroclastic reaction involving formation of ATP, ADP and phosphate. In nonenzymatic systems the source of electrons may well be nitrogen, NADH, or potassium borohydride coupled with acetyl phosphate and ATPase. A reductant, ATP, and a divalent metal cation, magnesium(II), are essential for activity. Other divalent cations such as manganese(II), iron(II), cobalt(II), or nickel(II) are also functional [74]. The

TABLE II

PROPERTIES OF Fe_4S_4 CLUSTER COMPOUNDS[a]

Property	Ferredoxin [79]	HPIP [80]	$[(C_5H_5)FeS]_4$ [82]	$[Fe_4S_4(SR)_4]$ [83]
Prosthetic group	Two clusters of Fe_4S_4 units each bound to four cysteines through iron and separated by 12 Å	One cluster of Fe_4S_4 unit bound through iron to four cysteines	$[(C_5H_5)FeS]_4$ cluster Fe and S at the corners of a distorted cube	$Fe_4S_4(SR)_4$ cluster a cubane structure
Fe—Fe distance (Å)	~3.06	3.06	2.650 3.365	2.776 2.732
Fe—S^{2-} distance (Å)	2.30	2.35	2.204 2.250	2.239 2.310
Fe—cysteine distance (Å)	~2.01	2.01	—	—
Molecular weight	6,000	10,000	—	—
Valence state of iron	3+[c]	2+[c]	3+	3+
Number of electrons transferred	1 or 2[b]	1[b]	—	1
Absorption maxima	390,280 nm (oxidized) 280 nm (reduced)	Broad band with shoulders at 450, 375, 325 nm (oxidized) 388 nm (reduced)	—	—

[a] References are indicated at the headings.
[b] Reference [83].
[c] Reference [84].

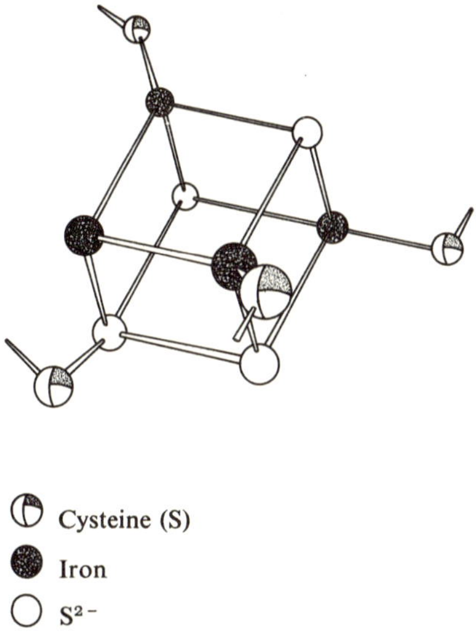

Fig. 4. Structure of $Fe_4S_4(cys)_4$ unit in HPIP from *Chromatium*. After Carter *et al.* [80].

steps involved in the transfer of electrons from reduced ferredoxin to hydrogenase (path A) or nitrogenase (path B) are not very well understood. Path A is operative in the absence of nitrogen fixation; under these conditions the evolution of hydrogen continues as the result of hydrogenase action. When the system fixes nitrogen, path A does not cease to operate and hydrogen is still evolved during the fixation [74]. The protons thus seem to compete successfully for a fraction of the electrons available from the reduced site of the enzyme.

Burris [38] suggests the involvement of ferredoxin and ATP at each step of the reduction of nitrogen to ammonia. The scheme also requires that the intermediates in the reduction of nitrogen to ammonia are bound to the enzyme throughout the reduction process. This is based upon the fact that the free intermediates were not detected in the system with ^{15}N as tracer [38], and also there was no isotopic [85] exchange reaction ($^{14}N_2 + {}^{15}N_2 \rightleftharpoons 2{}^{14}N^{15}N$). The nature of the nitrogen-binding enzyme site of nitrogenase and the metal ions involved in its activation is not clear. Gest [33] and Hardy and Burns [74] have proposed that the active site in nitrogenase may involve two adjacent metal ions (molybdenum or iron) that may be different or of the same type. This proposal is based on analogous systems for the activation of molecular oxygen by hemocyanin [copper(I) and copper(II)] sites and by hemerythrin [adjacent iron(II) sites] (Chapter 2, Table III) and on a model

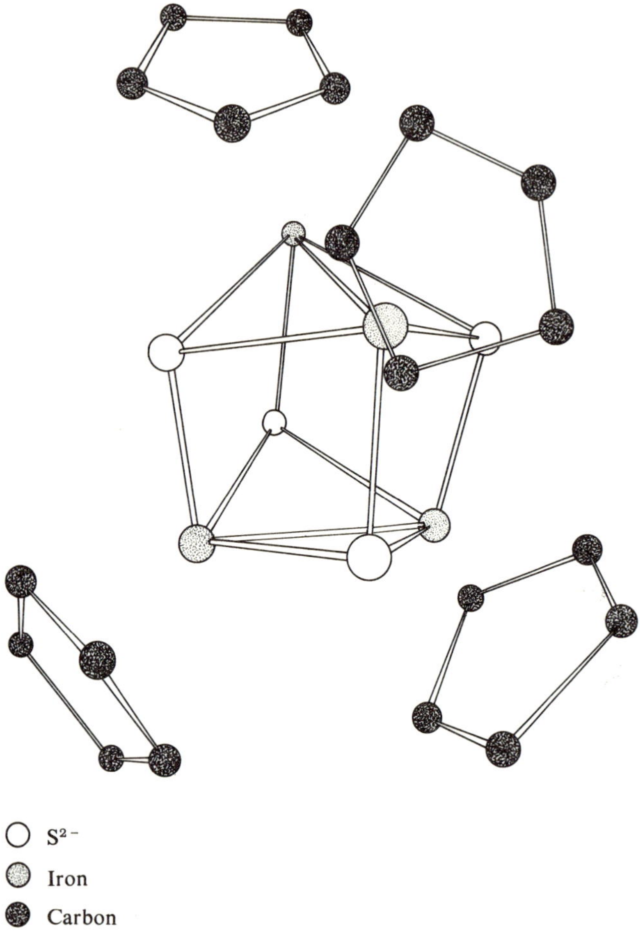

Fig. 5. Structure of (π-$C_5H_5FeS)_4$. After Wei *et al.* [82].

system for the activation of molecular hydrogen on two adjacent copper(I) sites [86].

3. *Free-Living Bacteria* (*Aerobic Systems*)

Nicholas *et al.* [39] reported fixation of nitrogen by a cell-free extract of *Azotobacter* in a medium in which the cells were grown. It was reported that the culture medium is a source of necessary supporting factors for nitrogen fixation [39]. A fresh culture medium was quite ineffective for the preparation of a nitrogen-fixing cell-free extract of *Azotobacter* [39]. The supernatant solution had very little nitrogen-fixing capacity, and the nitrogenase activity

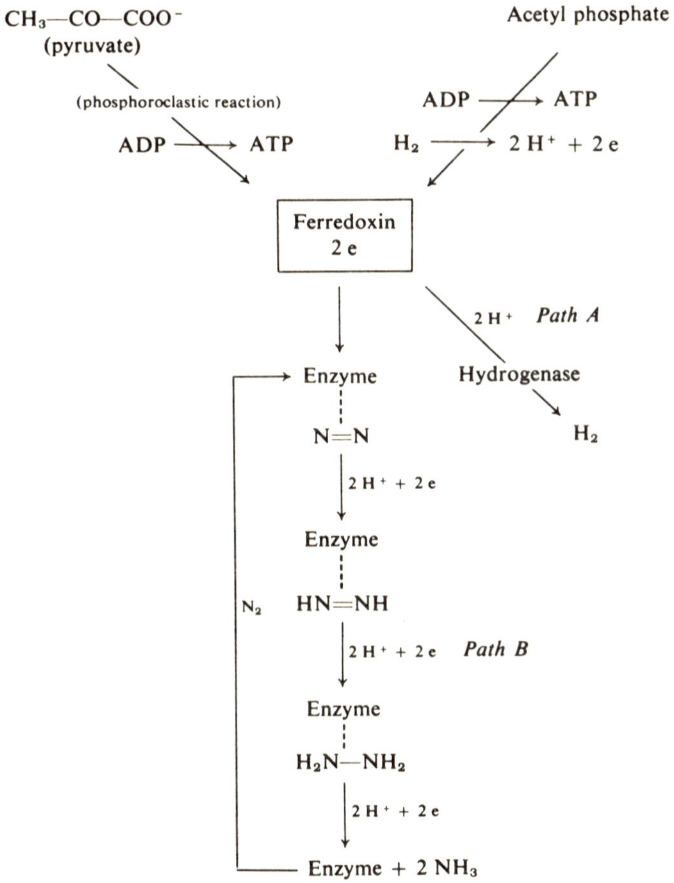

Fig. 6. Mechanism of nitrogen fixation by anaerobic systems. After Burris [38].

was concentrated in the molybdenum-rich particles. Fixation was observed when the particles were mixed with the clear extract. Bulen *et al.* [53] have prepared a nitrogenase-rich extract of *Azotobacter* cells by centrifugation. This extract was reported to possess 46% of the nitrogenase activity. Fixation was observed when this extract was mixed with potassium dihyrogen phosphate, ATP, creatine phosphate, creatine kinase, and the crude extract of *C. pasteurianum*, which supplies ferredoxin and hydrogenase. Burns *et al.* [87] have isolated the molybdenum-iron protein (nitrogenase) of *Azotobacter vinelandii* in crystalline form. The molecular weight of the protein is between 270,000 and 300,000 and it contains molybdenum, iron, cysteine, and labile sulfide in the ratio of 1:20:20:15. The protein was also isolated from other sources [73] and was shown to be essential for all typical nitrogenase reactions together with an additional iron protein (*Clostridium* extract) which serves as the electron-transfer system.

The similarity of fixation by anaerobic and aerobic bacterium *C. pasteurianum* and *A. vinelandii* has been studied in the respective cell-free extracts by Hardy *et al.* [50,88–90]. Besides nitrogen, the bacterial extracts reduced azide, nitrous oxide, hydrocyanic acid [88] and its analogs (alkyl nitriles), and acetylene and [88,91] and its homologs. Azide [88] and nitrous oxide [50,91] were reduced to molecular nitrogen and ammonia and to molecular nitrogen, respectively. The reduction of acetylene proceeded to the ethylene stage with no further reduction to ethane. The reduction of methyl nitrile to ethane and ammonia proceeded at a rate three orders of magnitude slower than that of hydrocyanic acid. Isonitriles [73,93] were reduced to methane, methylamine, ethylene, and ethane. According to Hardy and Burns [74]. Nitrogenase has higher affinity for molecular nitrogen than for any other substrate that has been studied. In all the reductions that were carried out, the number of electrons involved can be 2, 4, 6, 8, 10, or 12. In view of the reactions carried out, the authors suggested the nitrogen-fixing enzyme to be a stepwise reducing enzyme involving successive two-electron transfers [74] to the substrate.

The mechanism proposed for nitrogen fixation and for the reduction of other substrates is similar to that shown in Fig. 6 for anaerobic systems [50,88] and involves electron activation, substrate complexation, and reduction steps. The reductions catalyzed by the cell-free extracts of nitrogenase from *C. pasteurianum* and *A. vinelandii* require ATP and a reducing agent such as hydrogen, dithionate ($S_2O_4^{2-}$) or borohydride. All reductions are inhibited by carbon monoxide but carbon monoxide itself is not reduced by the system.

In the electron activation step (A) of the proposed mechanism [74] (Fig. 7), electrons obtained from hydrogen via hydrogenase or pyruvate via phosphoroclastic enzymes and ferredoxin reduce the oxidized enzyme of nitrogenase (X_{ox}) and produce an activated reduced form of the enzyme labeled X_{red}. The reaction is accompanied by the hydrolysis of ATP to ADP with ATPase. Hardy and Burns [74] proposed that the function of the ADP is to provide a better leaving group ADP or HPO_4 on X_{ox} and to facilitate its reduction to X_{red}. Although the sequence of these steps is not known, the electron activation step (formation of X_{red}) appears to be the rate-determining process in various redox reactions, including the fixation of nitrogen [74]. This interpretation is supported by the fact that the activation energy (12–14 kcal/mole) of the electron activation process is the same in the presence or absence of a reducible substrate (S). The metal ions involved in the nitrogenase enzyme X_{ox} are not known; they may be molybdenum, iron, or both. The electron activation site X_{ox} is distinguished from the substrate complexation and reduction site Y in that it is not affected by molecular hydrogen or the

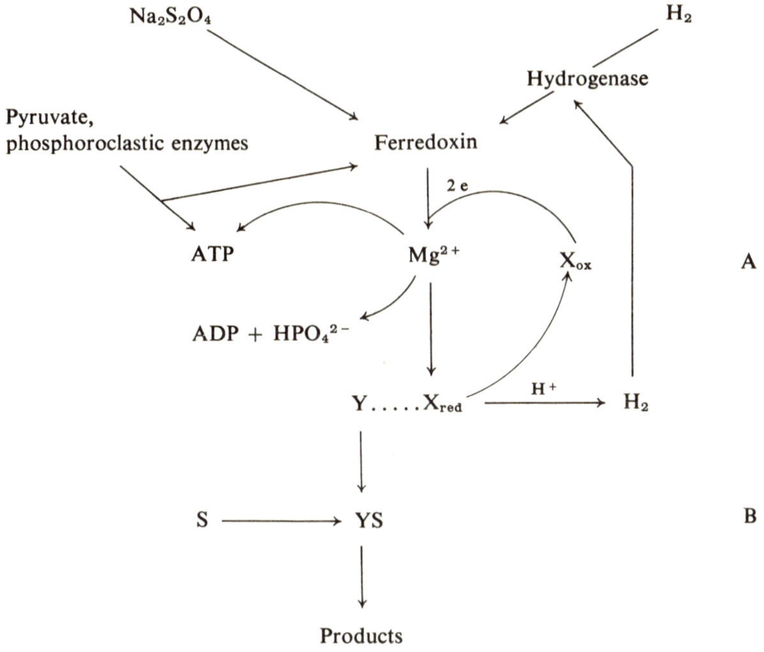

Reductions at Y:

$$N_2 + 6\,e \longrightarrow 2\,NH_3$$
$$N_2O + 2\,e \longrightarrow N_2 + H_2O$$
$$C_2H_2 + 2\,e \longrightarrow C_2H_4$$
$$N_3^- + 2\,e \longrightarrow N_2 + NH_3$$
$$HCN + 4\,e \longrightarrow CH_3NH_2$$
$$HCN + 6\,e \longrightarrow CH_4 + NH_3$$
$$CH_3NC + 6\,e \longrightarrow CH_4 + CH_3NH_2$$
$$CH_3NC + 10\,e \longrightarrow C_2H_4 + CH_2NH_2$$
$$CH_3NC + 12\,e \longrightarrow C_2H_6 + CH_3NH_2$$

Fig. 7. Nitrogen fixation and reduction reactions in aerobic and anaerobic systems.

inhibitor, carbon monoxide. The proposed reduction site in X_{red} is a hydride of the metal ion in X formed by the replacement of an ADP or phosphate ligand by hydride.

The substrate-binding site Y binds nitrogen or other substrates, and the bound substrate takes up electrons from the electron-activating site X_{red} for conversion to various reduced products including ammonia. Y is a site on a metal–protein complex and may involve molybdenum, nonheme iron, and labile sulfhydryl groups. The various substrates of nitrogenase are complexed by Y and held in position to be reduced by electrons from X_{red}. The positive charge on the metal ion at site Y may also help in attracting electrons from X_{red} and help in the reduction of the enzyme–substrate complex YS. All nitrogenase substrates have triple bonds and, with the exception of acetylene, have nonbonding electrons that may be involved in complexation with Y, thus assisting in the transfer of electrons to the end of the complexed substrate molecule and helping the protonation and reduction of the triple-bonded molecules.

Models of the active site Y have been developed that contain molybdenum [94], molybdenum and iron [95–97], and iron [98] in the presence of thiol groups and a reducing agent, borohydride, or dithionite. These systems, though much less effective than nitrogenase, can nevertheless reduce nitrogen [95], acetylene [94,96], cyanide [98], acetonitrile [98], and isocyanides [97] to the products shown in Fig. 7, indicating the similarity of the active site in nitrogenase to that of the model system.

The model developed by Schrauzer *et al.* [95] contains sodium molybdate, thioglycerol, sodium borohydride, and ferrous sulfate in aqueous medium. At room temperature and 200 atm of nitrogen a yield of ammonia corresponding to 15% of the total amount of nitrogen was obtained. It is interesting to note that no ammonia could be obtained with ferrous sulfate, thioglycerol, and sodium borohydride in the absence of molybdate. It was therefore argued that the active site in the model system may be a reduced molybdenum thiol species activated by iron.

For the reduction of isonitriles and acetylene, Schrauzer and Doemeny [96] and Schrauzer *et al.* [97] have used as catalyst a binuclear molybdenum(V) complex, **2**, of the composition $Na_2Mo_2O_4(cys)_2$. It was assumed that the complex undergoes hydrolysis to form a mononuclear species, **3**, which may combine with ATP to form the reactive intermediate **4**, containing molybdenum(IV). The conversion of **3** to **4** was presumed [96] to be facilitated by ATP through phosphorylation of the hydroxide ion in **3**, thus improving its leaving group properties. The activation energy of acetylene or isonitrile reduction is lowered by about 4–5 kcal/mole in the presence of ATP. It was suggested that the molybdenum(IV) complex **4** combines with the substrates acetylene and isonitriles to form complexes **5** and **6**, respectively, and by the series of steps illustrated for the isonitrile (**4** $\rightarrow$ **6** $\rightarrow$ **3**) is converted to the original molybdenum(V) complex and the products. It was further inferred [97] that hydrogen evolution by the molybdenum(V) complex occurs at the same site as acetylene or isonitrile reduction, through the formation of a labile molybdenum(IV) dihydride. Thus it appears that this catalytic system resembles the nitrogenase enzymatic system with respect to the nature of the substrate and the reduction products formed.

The idea of an exclusive molybdenum site for substrate binding and reduction has been disputed by Newton *et al.* [98], who reported on a model

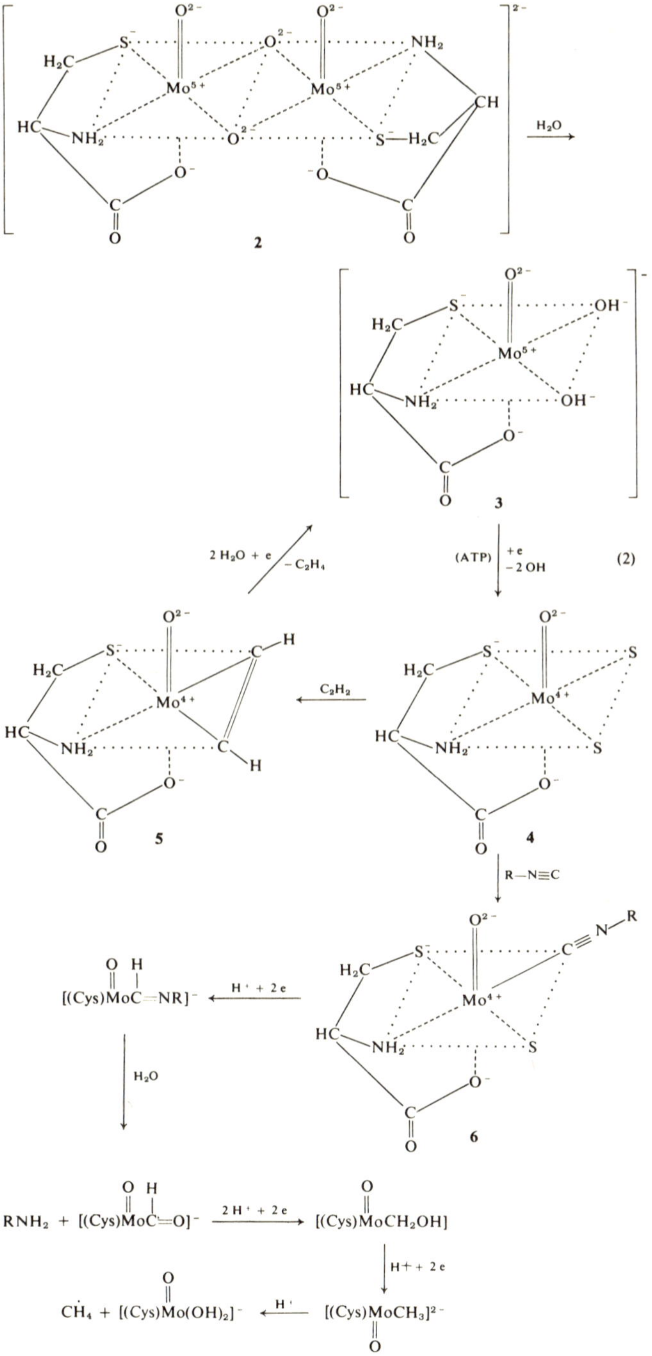
2
H_2O
3
(ATP)
+e
−2 OH
2 H_2O + e
− C_2H_4
C_2H_2
5
4
R—N≡C
6
H^+ + 2 e
H_2O
RNH_2 + [(Cys)MoC=O]$^-$
2 H^+ + 2 e
[(Cys)MoCH$_2$OH]
H^+ + 2 e
CH_4 + [(Cys)Mo(OH)$_2$]$^-$
H^+
[(Cys)MoCH$_3$]$^{2-}$

(2)

system containing iron and no molybdenum. With a catalyst containing dichlorobis (*N*-phenyl-*S*-methyl-2-aminoethanethiol)–Fe(II) and $NaBH_4$ or sodium dithionite in aqueous or ethanolic solution, a 5% yield of ammonia based on the weight of the catalyst was obtained. The system also reduced acetylene to ethylene, methylacetylene to propylene, cyanide ion to methane and ammonia, and acetonitrile to ethane and ammonia.

The model experiments that have appeared in the literature are far from conclusive, and at the present state it seems that sufficient evidence is not available to compare the action and properties of nitrogenase with model systems containing molybdenum, iron, and multimetal centers as the active catalytic site.

4. *Photosynthetic Systems*

Gest and co-workers [33] discovered considerable hydrogenase activity in *Rhodospirillum rubrum.* This photosynthetic microorganism evolves considerable quantities of molecular hydrogen during photosynthetic growth in a medium of glutamic acid. The evolution of hydrogen was completely suppressed in the presence of nitrogen, and nitrogen fixation took place with the formation of ammonia. On the basis of this observation, Kamen and Gest [99] postulated the presence of nitrogenase activity along with hydrogenase activity in *R. rubrum.* The fixation of molecular nitrogen by *R. rubrum* takes place only in light and under anaerobic conditions. The scheme in Fig. 8 has been proposed by Arnon *et al.* [34] for photosynthesis in green plants and bacteria. The sources of electrons for plants and bacterial photosynthesis, however, differ considerably. Plants utilize H_2O and OH^- as the electron donors (path A). The electrons are raised to the proper reduction level by cytochromes (aided by light) and are transferred to ferredoxin, which acts as the mediator for the transfer of electrons to TPN and the photoevolution of oxygen.

In bacteria, photophosphorylation takes place in a manner analogous to plant photosynthesis. Electrons from an internal donor, pyruvate or succinate (path B, Fig. 8), are raised to the proper reduction level by the accompanying photophosphorylation and travel through the cytochrome–ferredoxin chain to acceptors (nitrogenase, hydrogenase, or TPN). In the absence of nitrogen but in the presence of NH_4^+ salts, hydrogen is reduced by hydrogenase and dark-evolution of H_2 takes place. In the presence of nitrogen evolution of hydrogen stops and the electrons from ferredoxin are utilized to reduce N_2 in the presence of light.

III. Transition Metal Complexes of Molecular Nitrogen

Major advances in the complex chemistry of nitrogen are the recently reported syntheses of transition metal complexes containing molecular

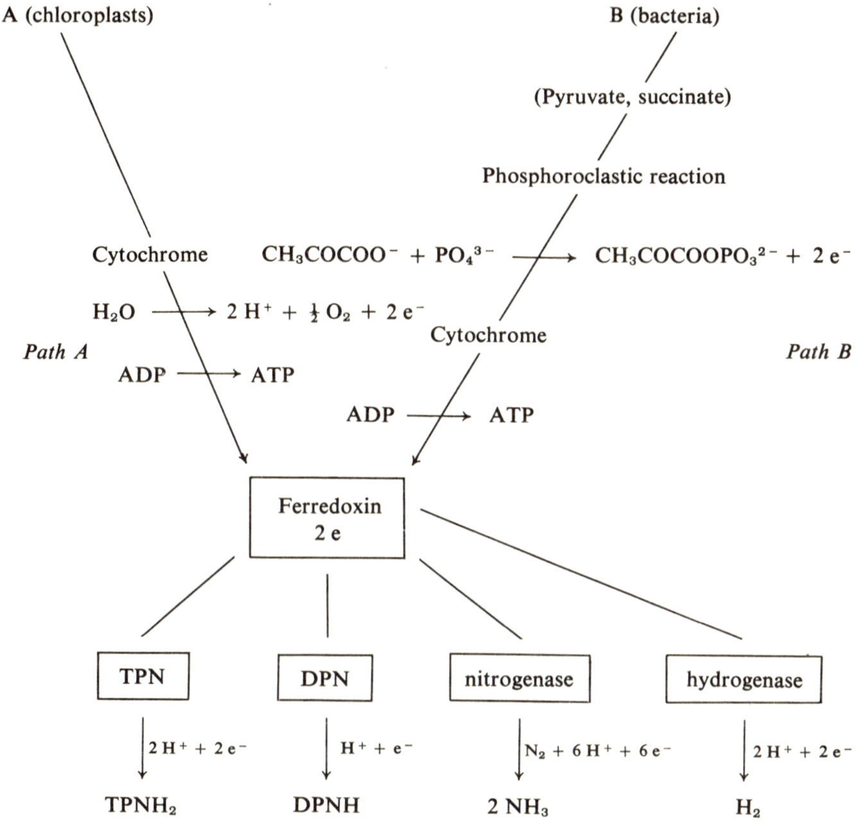

Fig. 8. Fixation in photosynthetic systems. TPN = triphosphopyridine nucleotide; $TPNH_2$ = reduced form of TPN; DPN = diphosphopyridine nucleotide. After Arnon *et al.* [34].

nitrogen as a ligand. These complexes have been prepared directly either by the displacement of coordinated water or other solvent molecules [3,100], by the displacement of hydride ion [12,14,101], or chloride ion [102–104] in a metal complex by molecular nitrogen, or by direct addition of molecular nitrogen to a coordinately unsaturated metal complex species [105,106]. Indirect preparations of molecular nitrogen complexes involve the generation of coordinated nitrogen within the coordination sphere of the metal ion from organic and inorganic azides [6,9,107–110], hydrazine [1,2,111,111a], coordinated nitrous oxide [112,113], coordinated ammonia [5,114], and from coordinated diazo compounds [7,115]. A study of the formation and stability of these complexes in solution may provide information that will be valuable in the design of efficient catalytic systems for nitrogen fixation. Some recent review papers [4, 116–121a] have surveyed the chemistry of dinitrogen complexes and nitrogen fixation from various points of view, depending on the interests of the reviewer. The emphasis in the present section will be based on the reactivity and bonding in metal–dinitrogen complexes.

A. Ruthenium(II) Complexes of Molecular Nitrogen

The first molecular nitrogen–metal complex, $[Ru(NH_3)_5(N_2)]X_2$ ($X = Br^-$, I^-, BF_4^-, PF_6^-)* was prepared by the action of hydrazine hydrate on an aqueous solution of ruthenium trichloride [1]. Besides ruthenium trichloride, aquopentammineruthenium(III) methanesulfonate, potassium pentachloroaquoruthenate(III), and ammonium hexachlororuthenate(IV) have also been employed [2] as starting materials for the preparation of the dinitrogenpentammineruthenium(II) ion. Action of azide ion on aquopentammineruthenium(III) methanesulfonate at pH 7 gave $[Ru(NH_3)_5N_2]X_2$ in 83% yield ($X = CH_3SO_3^-$).

The salts of $[Ru(NH_3)_5N_2]^{2+}$ are moderately stable in air [2] in the absence of water. Solutions of $[Ru(NH_3)_5N_2]X_2$ in dry dimethyl sulfoxide are stable at room temperature for periods up to 1 week, but solutions in water decompose slowly to polynuclear amine complexes of unknown composition [2]. Dinitrogenpentammineruthenium(II) ion salts are diamagnetic in solution. Conductivity measurements indicate that they are 2:1 electrolytes in dimethyl sulfoxide. The infrared spectra of the salts contain a sharp band in the region 2105–2170 cm^{-1} and a band at about 500 cm^{-1}, plus the usual bands characteristic of pentammine complexes of ruthenium(II). The bands assigned to the coordinated ammonia groups shift to lower frequencies in deuterium oxide and deuterohydrazine, but the band in the region 2105–2170 cm^{-1} shifts only by 5 cm^{-1}. Proton magnetic resonance (PMR) of the complex $[Ru(NH_3)_5N_2]^{2+}$ in both aqueous and dimethyl sulfoxide solutions indicates no high-field shift characteristic of hydrides. The PMR spectra show no peaks other than those attributed to coordinated ammonia groups. Thus the peak in the region 2105–2170 cm^{-1} has been assigned [2] to coordinated molecular nitrogen. The stretching frequency of molecular nitrogen is reduced by about 200 cm^{-1} in the complex as compared to that in free nitrogen (2331 cm^{-1}). The N—N band moves to higher frequencies with an increase of the size of the anion. Thus for the chloride, bromide, iodide, tetrafluoroborate, and hexafluorophosphate salts, the infrared N—N stretching frequencies are 2105, 2114, 2129, 2144, and 2167 cm^{-1}, respectively [2]. The metal–dinitrogen frequencies (bands about 500 cm^{-1}), however, decrease with the size of the anion. For the above-mentioned anions the

* In chemical formulas throughout this book, N_2 represents a dinitrogen group rather than two nitrides, unless specifically noted.

M—N frequencies are, respectively, 508, 499, 489, 487, and 474 cm^{-1}. The observed N—N and M—N frequency shifts with the size of the anion may be attributed to a lattice effect [117,122].

Coordinated dinitrogen is relatively labile in ruthenium(II) complexes and can be displaced by ammonia, chloride ion, and pyridine, yielding a variety of complexes containing the new ligand and no molecular nitrogen. Heating of the nitrogen complexes under anhydrous conditions results in evolution of more than 99% of the coordinated dinitrogen as gaseous nitrogen. It was originally reported by Allen and Senoff [1] that the coordinated dinitrogen in $[Ru(NH_3)_5N_2]^{2+}$ can be reduced to ammonia by sodium borohydride. Recently, however, Chatt *et al.* [122a] found by the use of $^{15}N_2$ that coordinated dinitrogen in $[Ru(NH_3)_5N_2]^{2+}$ cannot be reduced to ammonia, and the reduction observed in Allen's compound may be attributed to the presence of a coordinated hydrazine impurity in the sample. This fact has been more recently confirmed by Allen and Bottomley [123].

Dinitrogenpentammineruthenium(II) ion has been prepared directly from molecular nitrogen and $RuCl_3$ or $RuCl_3OH$ in tetrahydrofuran with zinc as the reducing agent [100]. The product was isolated from the THF solution by evaporation under vacuum. The infrared spectrum of the yellow complex contained a band at 2140 cm^{-1} which shifted to 2070 cm^{-1} when $^{15}N_2$ was used for the preparation of the complex. A shift of 70 cm^{-1} in the stretching frequency of the coordinated dinitrogen is in good agreement with the calculated value of the isotopic shift [100]. The product was identical in all respects to that obtained by the interaction of $RuCl_3OH$ or $RuCl_3$ with hydrazine in tetrahydrofuran in the absence of molecular nitrogen. Borodko *et al.* [124] obtained a ruthenium(II)–dinitrogen complex of the composition $RuN_2Cl_2(H_2O)C_4H_8O)$ in 80% yield by the reduction of ruthenium(III) chloride hydrate with zinc amalgam (1–1.5% Zn) in oxolan. The complex exhibited a ν_{N_2} stretching vibration at 2150 cm^{-1}. The coordinated water and solvent (C_4H_8O) in the complex may be replaced by more basic ligands such as ammonia and ethylenediamine to give a series of complexes of the composition $RuN_2Cl_2(NH_3)_2(H_2O)$, $RuN_2Cl_2(NH_3)_3$, and $RuN_2Cl_2(en)_2$ with ν_{N_2} vibrations in the infrared at 2090, 2080, and 2085 cm^{-1}, respectively. The ν_{N_2} stretching frequencies of these complexes decrease with an increase in the basicity of the donor, indicating enhanced ruthenium–dinitrogen bond strengths. This correlation is supported by the stabilities of the complexes $RuN_2Cl_2(H_2O)_2(C_4H_8O)$ and $RuN_2Cl_2(NH_3)_3$; the former complex dissociates completely at 50°C whereas the latter is stable up to 150°C.

Harrison and Taube [3] succeeded in the preparation of $[Ru(NH_3)_5N_2]^{2+}$ by the interaction of $[Ru(NH_3)_5H_2O]^{2+}$ with molecular nitrogen in aqueous solution of the ruthenium(II) salt at room temperature and 1 atm. A solution containing $[Ru(NH_3)_5H_2O]^{2+}$ was prepared by the reduction of $Ru(NH_3)_5Cl_3$

in 0.1 *M* H_2SO_4 by amalgamated zinc and was separated from the reducing agent. When nitrogen gas was bubbled through the solution, the absorption peak at 221 nm appeared, and a 50% conversion to $[Ru(NH_3)_5N_2]^{2+}$ took place in several hours. The coordinated dinitrogen may be displaced by oxidizing the complex with cerium(IV). It has been reported that between 0.5 to 0.6 mole of dinitrogen had been trapped for each mole of ruthenium(II) present.

Chatt and Ferguson [5] obtained a small quantity of $[Ru(NH_3)_5N_2]^{2+}$ in the course of an attempt to prepare pure $[Ru(NH_3)_6]Cl_2$ by the reduction of ruthenium trichloride in aqueous ammonia with zinc dust. The presence of the dinitrogen complex of ruthenium(II) was confirmed by the presence of an infrared band at 2120 cm^{-1} and by evolution of nitrogen when the complex was treated with hydrochloric acid. The source of nitrogen in the coordinated dinitrogen complex seems to be the oxidation of a part of the ammonia present in the system by ruthenium(III). This conclusion is based on the fact that the dinitrogen complex of ruthenium(II) is formed even when the reduction of ammonia with zinc dust is conducted *in vacuo* or under argon with rigorous exclusion of atmospheric nitrogen. The maximum yield of $[Ru(NH_3)_5N_2]Cl_2$ (10–15%) is obtained when the reduction of ruthenium trichloride with zinc is conducted in anhydrous ammonia at its boiling point. No gas is evolved during the reduction stage. When the reduction is conducted in aqueous ammonia both hydrogen and nitrogen are evolved. It is suggested [5] that intermediate species such as $[RuH_x(NH_3)_y]^{n+}$ or $[Ru(NH_3)_5(H_2O)]^{n+}$ are first formed in the system. The complexes then pick up nitrogen gas formed by the oxidation of ammonia. It is not quite clear how ammonia could be oxidized to nitrogen by ruthenium(III) under strongly reducing conditions. It is nevertheless possible that the strong thermodynamic stability of the Ru(II)–N_2 complex may cause part of the ammonia to be oxidized to dinitrogen on the ruthenium ion.

Pell and Armor [114] have reported a novel preparation of $[Ru(NH_3)_5N_2]^{2+}$ by saturating an alkaline solution of ruthenium(III) hexammine, $[Ru(NH_3)_6]^{3+}$, with nitric oxide. The yield of the dinitrogen complex was quantitative at pH 8.5–9. It was suggested that the dinitrogen complex might have formed by the nucleophilic attack of nitric oxide on a coordinated amide group of the complex $[Ru(NH_3)_5(NH_2)]^{2+}$ formed by the dissociation of $[Ru(NH_3)_6]^{3+}$ in alkaline solution. The reaction is extremely simple since one of the reactants nitric oxide also helps in the reduction of ruthenium(III) to ruthenium(II) without the necessity of an external reducing agent.

A detailed X-ray determination of the crystal structure (Fig. 9) of $[Ru(NH_3)_5N_2]Cl_2$ [125] led to the conclusion that the system Ru—N—N is linear and that the bonding is therefore analogous to that in metal carbonyls. The crystal of $[Ru(NH_3)_5N_2]Cl_2$ is disordered, the nitrogen molecule ran-

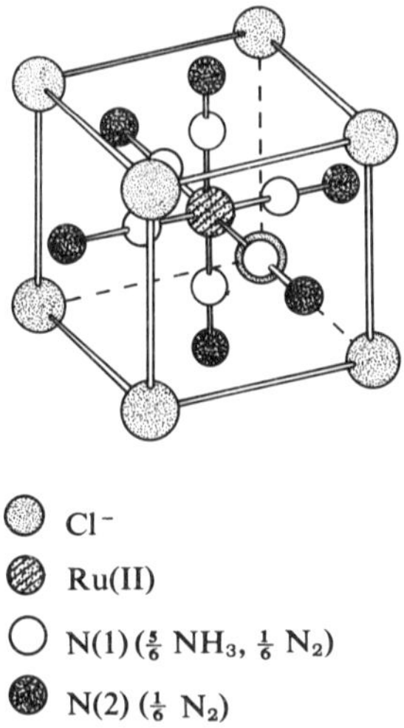

Fig. 9. Disordered crystal structure of $[Ru(NH_3)_5N_2]Cl_2$. According to Bottomley and Nyburg [125].

domly occupying any of the six octahedral positions around ruthenium(II). The space-averaged result of such a situation is six inner nitrogen atoms [five ammonia nitrogens and the inner nitrogen, N(1), of the coordinated nitrogen molecule] and six outer nitrogens [each nitrogen being one-sixth of the coordinated nitrogen atom, N(2)]. The distance between the N(1) and the N(2) nitrogens is 1.12 Å. Perpendicular coordination of nitrogen to ruthenium(II) in a manner analogous to the olefin complexes has been eliminated on the basis of the fact that the distance of N(2) from ruthenium, 3.24 Å, is greater than N(1), 2.12 Å. Since the molecular nitrogen complex is analogous in structure to those of metal carbonyls it is concluded that back donation of electrons from the filled t_{2g} nonbonding orbitals of the metal ion to antibonding molecular orbitals mainly located on the nitrogen ($t_{2g}{}^*\pi$ or $t_{2u}{}^*\pi$) is important in determining the stability and strength of metal–dinitrogen bonding.

A binuclear ruthenium(II)–μ-dinitrogen complex with an absorption at 260 nm was synthesized by Harrison and Taube [3] by the reaction of molecular nitrogen with aquopentammineruthenium(II) in aqueous solution. Harrison *et al.* [126] have recently characterized the complex on the basis of its infrared spectrum. In place of the strong sharp peak at about 2100 cm^{-1} characteristic of the N—N stretch in dinitrogenpentamineruthenium(II) cation, there was a broad weak absorption at 2060 cm^{-1}. Though the absorption at 2060 cm^{-1} has not been definitely assigned to N—N stretch in the binuclear species, the absence of strong absorption at 2100 cm^{-1} nevertheless implies that nitrogen is symmetrically bonded in the binuclear ion. The

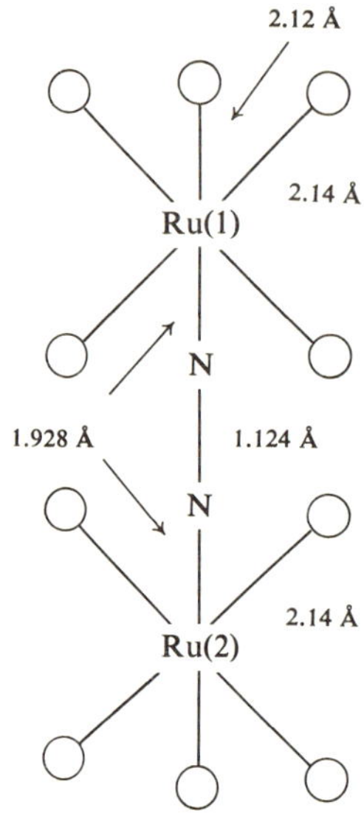

Fig. 10. X-Ray structure of $\{[Ru(NH_3)_5]_2N_2\}^{4+}$ [127].

binuclear μ-dinitrogen complex is remarkably stable to hydrolytic dissociation and is slowly formed in almost quantitative yield by mixing equimolar solutions of dinitrogenpentammineruthenium(II) and aquopentammineruthenium(II) species. When $[Ru(NH_3)_5N_2]^{2+}$ and $[Ru(NH_3)_5H_2O]^{2+}$ are mixed in solution, the absorption at 221 nm due to the dinitrogenruthenium(II) complex diminishes while that at 262 nm characteristic of the binuclear μ-dinitrogen complex increases. The stoichiometric composition of the binuclear μ-dinitrogen complex was established by Job's method [126].

The X-ray structure of μ-dinitrogendecaamminediruthenium(II) cation has been investigated by Trietel *et al.* [127] and is presented in Fig. 10. The geometry of the complex conforms to D_4h symmetry with the Ru(1) and Ru(2) atoms and the axially bound ammonia groups in a straight line, and the remaining coordinated ammonias in an eclipsed conformation, as is indicated in Fig. 10. The Ru—NH_3 distance in the plane of the two ruthenium atoms is 2.14 Å, which is slightly longer than the axial Ru—NH_3 distance (2.12 Å). The average Ru—NH_3 distance for all the ten Ru—NH_3 bonds is 2.12 Å. The distance Ru—N_2 is 1.928 Å and the N—N bond length 1.124 ± 0.015 Å. Thus the N—N distance in the binuclear complex is not appreciably different from that of the mononuclear complex (1.12Å) and is slightly longer than that in free nitrogen (1.098 Å). The Raman spectrum of $[Ru(NH_3)_5N_2(NH_3)_5Ru]^{4+}$, recently reported by Chatt *et al.* [103], has a strong Raman band at 2100 ± 2 cm^{-1} which is shifted to 2030 ± 2 cm^{-1} in the $^{15}N_2$ analog and has been assigned to coordinated N_2 stretch. On the basis of the frequency of the Raman band, it was independently concluded [128] that the binuclear μ-dinitrogen complex has a linear symmetrical structure $[(NH_3)_5RuN{=}NRu(NH_3)_5]^{4+}$ with a delocalized electronic system and with the metal ion formally assigned to the 2+ oxidation state.

Thus it is seen that the structure of the complex, as determined by X-ray crystal analysis (Fig. 10), is in accord with the symmetrical coordinated N—N stretching vibration, observed as a weak IR band in the 2050–2100 cm^{-1} range, and as a strong Raman band at 2100 ± 2 cm^{-1}.

Allen and Bottomley [129] have obtained the complexes $[Ru(NH_3)_5N_2]^{2+}$ and $[Ru(NH_3)_5N_2(NH_3)_5Ru]^{4+}$ from aquopentammineruthenium(II) by using air instead of molecular nitrogen. A solution of chloropentammineruthenium(III) chloride (0.1 *M*) in 0.1 *M* sulfuric acid was reduced with amalgamated zinc under argon until the solution became neutral. A stream of air was then passed through the solution until a pink color developed, possibly due to the formation of $[Ru(NH_3)_5OH]^{2+}$. At this stage air was replaced by argon, a fresh portion of sulfuric acid added, and the solution reduced with zinc amalgam. Argon was then replaced by air until the pink color reappeared. The sequence of oxidation and reduction of the ruthenium solution was continued till no further pink color developed on passing air through the solution. At this stage, the solution was separated from the reducing agent, and the nitrogen complexes were precipitated by the addition of sodium fluoroborate. Both the dinitrogenpentammineruthenium(II) complex and Taube's binuclear μ-dinitrogen complex were prêcipitated as the fluoroborates. The experiment bears a resemblance to the aerobic fixation of nitrogen and is significant in pointing out that molecular nitrogen can combine with ruthenium(II) even in the presence of molecular oxygen.

A convenient route for the synthesis of $[Ru(NH_3)_5N_2]^{2+}$, with only a very small percentage of the dimer as impurity, is the reduction of chloropentaammineruthenium(III) with amalgamated zinc under nitrous oxide [112]. In about an hour an 81% yield of the mononuclear and 7.4% of the binuclear complexes were obtained. It has been suggested by Diamantis and Sparrow [112] that a nitrous oxide complex of ruthenium(II) may be formed in the first step and is subsequently reduced to $[Ru(NH_3)_5N_2]^{2+}$ by amalgamated zinc. Direct evidence for the formation of nitrous oxide–pentaamineruthenium(II) species in solution has been provided by Armor and Taube [113]. Addition of nitrous oxide to an aqueous solution of aquopentaamineruthenium(II) at 6.8°C gave rise to a maximum at 238 nm attributed to $[N_2ORu(NH_3)_5]^{2+}$ species. The maximum at 238 nm disappears when argon is passed through the solution. The rate constants [Eq. (3)] for the formation, k_f, and dissociation, k_r of the nitrous oxide complex of ruthenium(II) in solution have been measured spectrophotometrically as 9.5×10^{-3}

M^{-1} sec^{-1}, and 1.35×10^{-3} sec^{-1}, respectively. From these rate constants the equilibrium constant corresponding to the formation of the nitrous oxide complex according to the equilibrium (3) is calculated as 7.0 M^{-1}.

$$[(NH_3)_5RuOH_2]^{2+} + N_2O \underset{k_r}{\overset{k_f}{\rightleftharpoons}} [(NH_3)_5RuN_2O]^{2+} + H_2O \qquad (3)$$

The nitrous oxide complex is slowly reduced in solution by the aquo complex of ruthenium(II) and after a few days $[Ru(NH_3)_5H_2O]^{3+}$, $[(NH_3)_5RuCl]^+$, $[(NH_3)_5RuN_2]^{2+}$, and the binuclear complex $\{[Ru(NH_3)_5]_2N_2\}^{4+}$ are formed. In the presence of chromium(II), the nitrous oxide complex is reduced to the dinitrogen complex at a rate which is equal to the rate of formation of the nitrous oxide complex from $[(NH_3)_5RuH_2O]^{2+}$ and nitrous oxide (k_f). The reduction follows the rate law, Eq. (4).

$$\frac{d[(NH_3)_5RuN_2]^{2+}}{dt} = k[(NH_3)_5RuH_2O^{2+}][N_2O] \qquad (4)$$

The value of k at 6.8°C is 10.1×10^{-3} M^{-1} sec^{-1}.

The nitrous oxide complex $[Ru(NH_3)_5(N_2O)](BF_4)_2$ has recently been obtained in the solid state from a solution of $[Ru(NH_3)_5H_2O]^{2+}$ and oxygen-free nitrous oxide at 36–40 atm in the presence of $NaBF_4$. The compound is diamagnetic and stable in dry air and in high vacuum at room temperature for 2–3 days. The observed IR bands of medium intensity at 2275 cm^{-1} and a strong band at 1210 cm^{-1} have been attributed to the ν_3 and ν_1 stretching vibrations of nitrous oxide. Molecular oxygen, iron(III), and cerium(IV) oxidize the compound with the evolution of nitrous oxide. Thermal decomposition of the compound takes place at 135°C in high vacuum, giving nitrogen, nitrous oxide, water, $[Ru(NH_3)_5N_2]^{2+}$, and ruthenium red [$2\ RuCl_2(OH) \cdot 7\ NH_3 \cdot 3\ H_2O$] as the products of decomposition.

Bottomley and Crawford [131] have prepared $[Ru(NH_3)_5(N_2O)]X_2$ ($X = Cl^-$, I^-) by the reaction of $[Ru(NH_3)_5(NO)]X_3$ with hydroxylamine, hydrazine hydrate, and ammonia. Hydroxylamine reacted with an aqueous solution of $[Ru(NH_3)_5(NO)]X_3$ to give a good yield of the Ru(II)–N_2O complex, characterized by a strong IR band at 1175 cm^{-1} (I^- salt) and a weak band at 2250 cm^{-1} (I^- salt) attributed to the ν_1 (symmetric) and ν_3 (asymmetric) stretching vibrations of nitrous oxide, respectively. In the case of the reaction of $[Ru(NH_3)_5(NO)]X_3$ with hydrazine hydrate, the Ru(II)–N_2O and $[Ru(NH_3)_5(N_2)]X_2$ complexes were formed, the latter through the possible formation of a $[Ru(NH_3)_5(N_3)]X_2$ intermediate. Reaction of $[Ru(NH_3)_5(NO)]X_3$ with ammonia yields a mixture of products containing $[Ru(NH_3)_4(OH)(NO)]X_2$, $[Ru(NH_3)_5(N_2O)]X_2$, and $[Ru(NH_3)_5N_2]X_2$.

Armor and Taube [132] have isolated a μ-N_2O dimer of the composition $[Ru(NH_3)_5]_2N_2O^{4+}$ by the reaction of $[Ru(NH_3)_5H_2O]$ with nitrous oxide at 1 atm pressure. The bromide compound has an IR band of medium intensity at 2110 cm^{-1} and a strong band at 1160 cm^{-1}. These bands were assigned to ν_3 and ν_1 stretching vibrations of nitrous oxide. The characteristic bands at 3450, 3240, 1625, 1290, 1275, and 790 cm^{-1}, which are due to coordinated ammonia and water, have also been observed. The amount of nitrous oxide obtained on decomposition is 43–51% of that expected for a 1:1 Ru:N_2O ratio, indicating that the compound is an N_2O-bridged dimer.

The experiment by Armor and Taube [132] has an interesting analogy to the biological fixation of nitrogen. In the chemical pathway it is possible that the reduction of nitrous oxide proceeds with the formation of nitramide and hydroxylamine as intermediates, even though these substances were not detected. A useful application of the reaction would be to devise mild reducing agents to stop the reduction of nitrous oxide at the hydroxylamine stage. Such reactions would probably have considerable synthetic potential, as well as interest for comparison with biological reactions involving hydroxylamine.

Kane-Maguire *et al.* [108] have prepared a number of azido Ru(II)–dinitrogen complexes by the acid cleavage of the diazide complexes which decompose by the cleavage of the N—N bond in one of the azide ligands.

$$cis\text{-}[Ru(en)_2(N_3)_2]^+ \longrightarrow cis\text{-}[Ru(en)_2N_2N_3]^+ + \tfrac{1}{2}N_2 \quad (5)$$

$$cis\text{-}[Ru(trien)(N_3)_2]^+ \longrightarrow cis\text{-}[Ru(trien)N_2N_3]^+ + \tfrac{1}{2}N_2 \quad (6)$$

The complex *cis*-$[Ru(en)_2N_2(N_3)]PF_6$ shows absorption maxima in acetone solution at 465 nm ($\varepsilon = 1100$) and $\nu(N_2)$ at 2130 and 2050 cm^{-1}. For the complex *cis*-$[Ru(trien)N_2(N_3)]^+$ in acetone solution two electronic absorption maxima were recorded at 500 nm ($\varepsilon = 2800$) and 375 nm ($\varepsilon = 2080$). Infrared N—N stretch absorptions were observed at 2030 and 2050 cm^{-1}. The complexes are brown, diamagnetic, and lose molecular nitrogen at room temperature. They can, however, be stored in the dark at −15°C for several weeks without any significant change. They are soluble in water and dimethyl sulfoxide, in which they are 1:1 electrolytes. Reaction of *cis*-$[Ru(en)_2(N_2)(N_3)]^+$ with cerium(IV) releases molecular nitrogen both from coordinated nitrogen and from coordinated azide ion.

The crystal and molecular structure of the complex $[Ru(en)_2(N_3)(N_2)](PF_6)]$ has been determined by Davis and Ibers [133]. The material is X-ray-sensitive and it was necessary to use four different crystals for data collection. The structural details are presented in Fig. 11. The central metal ion is coordinated octahedrally to six nitrogen atoms. The N—N bond length of coordinated dinitrogen is 1.106 Å and the Ru—N—N bond angle is 179.3°. The Ru—N_2 bond length of 1.894 Å is significantly shorter than a metal–nitrogen single

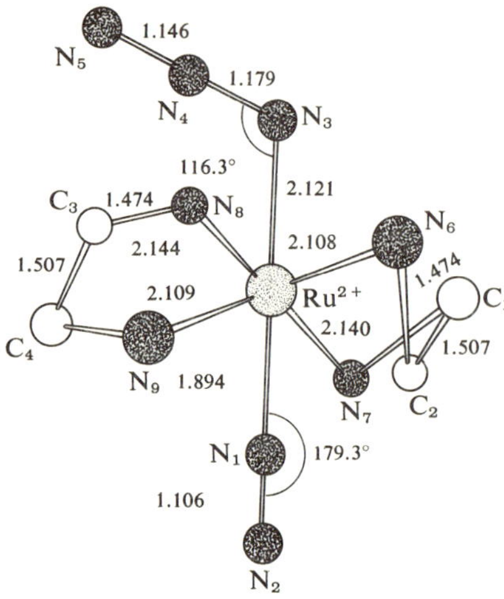

Fig. 11. Structure of azidodinitrogenbis(ethylenediamine)ruthenium(II). After Davis and Ibers [133].

bond (~2.19 Å) in metal–ammine complexes but slightly longer than that of a metal–nitrogen double bond (1.69–1.17 Å) in metal–nitrosyl complexes [133]. The ethylenediamine rings are in the gauche conformation with a mean C—N distance of 1.474 Å and a mean C—C distance of 1.507 Å. The N—N bond lengths in the coordinated azide are those of an ionic azide, their average being 1.162 Å.

1. *Rate and Equilibrium Studies of Ruthenium(II)–Dinitrogen Complexes*

Itzkovitch and Page [134] have measured the rate of formation, k_f, of $[(NH_3)]_5RuN_2]^{2+}$ from molecular nitrogen and $[(NH_3)_5RuH_2O]^{2+}$ at 25°C and pH 2.6 as $7.1 \times 10^{-2}\ M^{-1}\ sec^{-1}$. Armor and Taube [135] obtained the value of k_f at 25°C ($\mu = 0.1$) as $7.3 \times 10^{-2}\ M^{-1}\ sec^{-1}$, in close agreement with that reported by Itzkovitch and Page [134]. The enthalpy and entropy of activation corresponding to k_f have been reported as 18.3 ± 0.3 kcal/mole and -2 ± 1 eu, respectively [135].

The rate of decomposition of the dinitrogen complex, k_r, was obtained by Armor and Taube [113,135] by passing argon through a solution of the dinitrogen complex and measuring the concentration of aquopentaammine-ruthenium(II) complex by the isonicotinamide reaction [135] (absorption at 748 nm) and the concentration of the binuclear species $\{[(NH_3)_5Ru]_2N_2\}^{4+}$ by absorption at 260 nm. The equilibrium constant for the formation of the

$$[Ru(NH_3)_5(H_2O)]^{2+} + N_2(aq) \underset{k_r}{\overset{k_f}{\rightleftharpoons}} [Ru(NH_3)_5N_2]^{2+} \tag{7}$$

Ru(II)–N_2 complex was determined by a static method in which $[Ru(NH_3)_5N_2]^{2+}$ was allowed to reach equilibrium at fixed nitrogen activity (1.0 atm) and constant hydrogen ion concentration. The equilibrium constant obtained at 25°C ($\mu = 0.1$) is $3.3 \times 10^4\ M^{-1}$ and ΔH° and ΔS° were found to be -10.1 ± 1.4 kcal/mole and -13 ± 5 eu, respectively.

The rate constant for the formation of the binuclear species $[(NH_3)_5Ru]_2 \cdot N_2]^{4+}$ [Eq. (8)] from aquopentaammineruthenium(II) and dinitrogenpentaammineruthenium(II) has been reported as $4.2 \times 10^{-2}\ M^{-1}\ sec^{-1}$ [134]. The equilibrium constant K_d for the formation of the binuclear species has been reported as $10^4\ M^{-1}$ by Itzkovitch and Page [134] and more recently [135] as $7.3 \times 10^3\ M^{-1}$ at 25°C ($\mu = 0.1$). The standard entropy and enthalpy changes corresponding to K_d have been calculated as -11.2×1.4 kcal/mole and -20 ± 5 eu, respectively [135]

$$[Ru(NH_3)_5H_2O]^{2+} + [Ru(NH_3)_5N_2]^{2+} \overset{K_d}{\rightleftharpoons} [Ru(NH_3)_5]_2N_2^{4+} + H_2O \qquad (8)$$

The enthalpy change for reaction (9) has been reported as -22 ± 2 kcal/mole at 25°C [136]. The rate of decay of the binuclear species calculated from K_d and from the rate of the forward reaction is $4.2 \times 10^{-6}\ sec^{-1}$. Thus the binuclear species decomposes to the mononuclear complex at about the same rate as the mononuclear complex does to the aquo complex and molecular nitrogen.

$$N_2(aq) + 2\,[Ru(NH_3)_5H_2O]^{2+} \rightleftharpoons [Ru(NH_3)_5]_2N_2^{4+} + 2\,H_2O \qquad (9)$$

Shilova and Shilov [100] measured the rate of isotopic exchange of complexed nitrogen in $RuCl_2L_3N_2$ (L = NH_3, H_2O, THF) with $^{15}N_2$ in tetrahydrofuran (THF) [Eqs. (10)–(12)] and concluded that the exchange and solvolysis of $RuCl_2L_3N_2$ proceed through a five-coordinate intermediate $RuCl_2L_3$ by an S_N1 mechanism. The value of k_1 at 20°C has been reported as

$$RuCl_2L_3N_2 \xrightarrow{k_1} RuCl_2L_3 + N_2 \qquad (10)$$

$$RuCl_2L_3 + {}^{15}N_2 \xrightarrow{k_2} RuCl_2L_3{}^{15}N_2 \qquad (11)$$

$$RuCl_2L_3 + L' \xrightarrow{k_3} RuCl_2L_3L' \qquad (12)$$

$3.3 \times 10^{-4}\ sec^{-1}$. From the temperature dependence of k_1, the activation energy has been computed as 23 kcal/mole [100]. When the ligand L′ is water or tetrahydrofuran, the values of the ratio k_2/k_3 are, respectively, 3.1 and 175. Thus it appears that for the above reactions dinitrogen is about three times more reactive than water as a ligand, the order of reactivity being $N_2 > H_2O \gg$ THF.

Although a mechanism for the formation of $[(NH_3)_5RuN_2]^{2+}$ from $[(NH_3)_5RuH_2O]^{2+}$ has not yet been proposed, it is possible that the formation of $[(NH_3)_5RuN_2]^{2+}$ also takes place through a five-coordinate intermediate, $[Ru(NH_3)_5]^{2+}$, formed by the dissociation of $[(NH_3)_5RuH_2O]^{2+}$. The reaction sequence may be represented as

$$[Ru(NH_3)_5H_2O]^{2+} \underset{k_2}{\overset{k_1}{\rightleftharpoons}} [Ru(NH_3)_5]^{2+} + H_2O \xrightarrow[N_2]{k_3} [(NH_3)_5RuN_2]^{2+} \quad (13)$$

Assuming a steady-state approximation for the concentration of $[Ru(NH_3)_5]^{2+}$, the rate of formation of the dinitrogen complex may be expressed as

$$\frac{d[Ru(NH_3)_5N_2{}^{2+}]}{dt} = \frac{k_1k_3[Ru(NH_3)_5H_2O^{2+}][N_2]}{k_2 + k_3[N_2]} \quad (14)$$

If k_2 is much greater than $k_3[N_2]$, the rate of formation of the dinitrogen complex will depend on the concentrations of the aquo complex and molecular nitrogen. If $k_3[N_2] \gg k_2$, first-order kinetics will be observed, the rate of formation of the dinitrogen complex depending in this case on the rate of dissociation of the aquo species to the five-coordinate intermediate. Studies of the rate of exchange of coordinated nitrogen in $[(NH_3)_5RuN_2]^{2+}$ with $^{15}N_2$ would be very helpful for testing this mechanism. Kinetics and mechanistic studies on the formation of nitrogen complexes by the displacement of hydride ion or chloride ion would also assist the understanding of the reactivity patterns of dinitrogen complexes.

2. *Mechanism of the Formation of Ruthenium(II)–Dinitrogen Complexes*

Two mechanisms have been proposed for the formation of the ruthenium(II)–dinitrogen complexes, one involving the decomposition of the coordinated azide in acid solution and the other the reaction of hydrazine with $[Ru(NH_3)_6]^{3+}$ [111,111a].

In their investigation of the acid-catalyzed decomposition of azide, Kane-Maguire *et al.* [108] proposed a protonated nitrene intermediate, **7**, which reacts either with the original azide to form the ruthenium(II)–dinitrogen complex **8**, or with itself to give the dinitrogen-bridged binuclear ion **9** [Eqs. (15)–(17)].

The possible formation of a four-nitrogen intermediate in the above reactions may be related to the recent findings of Douglas *et al.* [110] that the electrophilic attack of nitrosyl ion on a Ru(II)–$^{15}NN_2$ complex results in the formation of a Ru(II)–^{15}NN complex with ^{15}N attached to ruthenium(II) [Eq. (18)]. This reaction probably involves the formation of $[(das)_2ClRu^{15}NNNNO]^+$ as an intermediate.

According to Bottomley [111, 111a] the reaction of hydrazine hydrate with $RuCl_3$, $[RuCl_5(H_2O)]^{2-}$, or $RuCl_6{}^{3-}$ results in the formation of $[Ru(NH_3)_6]^{3+}$, **10** (having an absorption at 267 nm). The ruthenium(III) hexammine undergoes the reactions shown in Eq. (20), to give ultimately the ruthenium(II)–dinitrogen complex **13**.

$$[(NH_3)_5RuN_3]^{2+} + H^+ \rightleftharpoons [(NH_3)_5Ru{-}\overset{\displaystyle H}{\overset{|}{N}}{\cdots}N{\cdots}N]^{3+} \longrightarrow \underset{\mathbf{7}}{[(NH_3)_5Ru{=}NH]^{3+}} + N_2 \qquad (15)$$

$$\underset{\mathbf{7}}{[(NH_3)_5Ru{=}NH]^{3+}} + [N_3{-}Ru(NH_3)_5]^{2+} \longrightarrow [(NH_3)_5Ru{-}\overset{\displaystyle H}{\overset{|}{N}}{\cdots}N{\cdots}N{\cdots}N{-}Ru(NH_3)_5]^{5+} \longrightarrow 2\,\underset{\mathbf{8}}{[(NH_3)_5RuN_2]^{2+}} + H^+ \qquad (16)$$

$$2\,[(NH_3)_5Ru{=}NH]^{3+} \longrightarrow \underset{\mathbf{9}}{[(NH_3)_5Ru{-}N{=}N{-}Ru(NH_3)_5]^{4+}} + 2\,H^+ \qquad (17)$$

$$[Ru(^{15}NN_2)Cl(das)_2] + NO^+ \longrightarrow [Ru(^{15}N{-}^{14}N)Cl(das)_2]^+ + N_2O \qquad (18)$$

$$N_2H_4{-}H_2O + \left\{\begin{array}{l} RuCl_3 \\ RuCl_5(H_2O)^{2-} \\ RuCl_6^{3-} \end{array}\right. \longrightarrow [Ru(NH_3)_6]^{3+} \qquad (19)$$

$$\underset{\mathbf{10}}{[Ru(NH_3)_6]^{3+}} + N_2H_4 \longrightarrow \underset{\mathbf{11}}{[Ru(NH_3)_5N_2H_4]^{3+}} + NH_3 \xrightarrow{-H_2} \underset{\mathbf{12}}{[Ru(NH_3)_5(NH{=}NH)]^{3+}} \xrightarrow{O_2} \underset{\mathbf{13}}{[Ru(NH_3)_5N_2]^{2+}} \qquad (20)$$

The mechanism predicts hydrazine and diimide complexes **11** and **12** as intermediates in the formation of the dinitrogen complex **13**. An interesting feature of the mechanism is the involvement of molecular oxygen in the last step. Since nitrogen fixation by microorganisms takes place in the presence of oxygen the above mechanism seems to have a biological parallel and merits further investigation.

3. *Ruthenium(II)–Dinitrogen Complexes with Tertiary Phosphines and Arsines*

Yamamoto *et al.* [138] have prepared a solution of a true Ru(II)–dinitrogen carrier in which molecular nitrogen is loosely and reversibly bound to ruthenium(II) and may be displaced by passing a current of an inert gas such as argon through the solution. The Ru(II)–dinitrogen carrier was prepared in tetrahydrofuran or benzene solution of ruthenium trichloride or of tris(acetylacetonato)ruthenium(III) by reduction with triethylaluminum in the presence of triphenylphosphine in an atmosphere of molecular nitrogen at room temperature. Initially, a red color developed in the solution, corresponding to the formation of a hydrido complex of ruthenium(II) with the probable formula $H_2Ru(PPh_3)_4$. When nitrogen was continuously bubbled through the benzene solution of the red hydrido complex of ruthenium(II), the red color turned brownish and a sharp infrared band appeared at 2143 cm^{-1}. This band disappeared on passing argon through the solution, and the color changed from brown to red. The cycle of absorption and resorption of molecular nitrogen was repeated several times. The band at 2143 cm^{-1} in the infrared spectrum of the brown solution has been assigned to the coordinated nitrogen molecule. The assignment was verified by an isotopic shift from 2143 to 2110 cm^{-1} through the use of ^{15}N in place of one of the two nitrogen atoms of the N_2 molecule. Coordinated nitrogen in the ruthenium(II)–dinitrogen carrier may also be reversibly displaced when hydrogen or ammonia is used in place of argon. It was not possible to isolate the nitrogen carrier complex of ruthenium(II) in a crystalline form. A large excess of triphenylphosphine interferes with the combination of molecular nitrogen with ruthenium(II). A similar triphenylphosphinerhodium(I) complex fails to combine with molecular nitrogen.

A ruthenium–dinitrogen complex of the composition $Ru(H)_2N_2(PPh_3)_3$ **14**, has recently been prepared by Knoth [139] through the interaction of hydridochlorotris(triphenylphosphine)ruthenium(II) with triethylaluminum and molecular nitrogen in ether. The dinitrogen complex has recently been obtained by the reaction of nitrogen with solid dihydride $[Ru(H_2)(PPh_3)_3]$, prepared by passing molecular hydrogen through a solution of the corresponding dichloride $[RuCl_2(PPh_3)_3]$ in benzene containing a small quantity of triethylamine [140]. Also, Takashi *et al.* [141] have prepared the hydride $[RuH_2(PPh_3)_3]$ by the reaction of ruthenium trichloride or tris(acetylacetonato)ruthenium(III), triphenylphosphine, and triethylaluminum in THF. Reversible combination of $[RhH_2(PPh_3)_3]$ with nitrogen in THF gave the dinitrogen complex [Eq. (21)].

$$RuH_2(PPh_3)_3 + N_2 \rightleftharpoons \underset{\mathbf{14}}{Ru(H)_2N_2(PPh)_3} \qquad (21)$$

The compound is air-sensitive and almost colorless when recrystallized from hexane–benzene in a nitrogen atmosphere (m.p. 185°C). The infrared spectrum of the compound shows a strong sharp band at 2147 cm^{-1} assigned to N—N stretch, and bands of moderate intensity at 1947 and 1917 cm^{-1} assigned to Ru—H stretch. The complex readily loses molecular nitrogen when treated with excess of triphenylphosphine, which was the basis for an earlier attempt [138] to isolate this complex. The coordinated nitrogen in the complex is displaced reversibly by ammonia and hydrogen with the formation of the complexes $(PPh_3)_3Ru(NH_3)_2$ and $(PPh_3)_3RuH_4$, respectively. Other reactions [142] involving the displacement of dinitrogen from $Ru(H)_2N_2(PPh_3)_3$ are as indicated in Eqs. (22)–(25).

$$\text{(1 or 2)}Ru(H)_2N_2(PPh_3)_3 \ \textbf{14} \xrightarrow{PPh_3} Ru(H)_2(PPh_3)_4 + N_2 \quad (22)$$

$$\xrightarrow{CO} Ru(H)_2(CO)(PPh_3)_3 + N_2 \quad (23)$$

$$\xrightarrow{NCC_6H_5} Ru(H)_2(NCC_6H_5)(PPh_3)_3 + N_2 \quad (24)$$

$$\xrightarrow{B_{10}H_8(N_2)_2} \underset{\textbf{14a}}{[Ru(H)_2(PPh_3)_3]_2B_{10}H_8(N_2)_2} + 2\,N_2 \quad (25)$$

The interaction of $Ru(H)_2N_2(PPh_3)_3$ with inner diazonium salts of $B_{10}H_{10}{}^{2-}$ such as $B_{10}H_8(N_2)_2$ [Eq. (25)] also seems to result in the displacement of dinitrogen. In this case the $B_{10}H_8$ moiety combines with two complex molecules. The product **14a** of reaction (25) has been shown to contain bridging dinitrogens between boron and ruthenium(II), probably derived from the diazonium groups.

The compound $Ru(H)_2N_2(PPh_3)_3$, **14**, is a *cis*-dihydride as indicated by its NMR spectrum (peaks at 18.1 τ and 22.5 τ). The H—P coupling constants indicate a meridional arrangement of phosphine ligands in the complex.

Douglas *et al.* [109,110] obtained the first dinitrogen complex of Ru(II) containing coordinated arsine, **15**, by the reaction [Eq. (26)] of the corresponding azide complex with nitrosyl fluorophosphate, $NOPF_6$, in methanol. The azide complex $[RuN_3Cl(das)_2]$ was obtained by reactions (27) and (28).

$$[RuN_3Cl(das)_2] + NOPF_6 \longrightarrow \underset{\textbf{15}}{[Ru(N_2)Cl(das)_2]^+PF_6^-} + N_2O \quad (26)$$

[das = *o*-phenylenebis(dimethylarsine)]

$$RuNOCl_3 + 2\,das \xrightarrow{\text{methanol}} [RuCl(NO)(das)_2]Cl_2 \quad (27)$$

$$[RuCl(NO)(das)_2]Cl_2 + 3\,N_2H_4 \xrightarrow{\text{methanol}} [RuN_3Cl(das)_2] + 2\,N_2H_5Cl + H_2O \quad (28)$$

With an azide coordinated through ^{15}N to Ru(II) in the complex $[Ru(^{15}NN_2)Cl(das)_2]$ it was found that reaction with $NOPF_6$ gave a ruthenium(II)–dinitrogen complex with ^{15}N still linked to the metal, indicating that the Ru(II)—^{15}N bond is not ruptured in the azide complex during attack by NO^+. This conclusion is supported by the fact that no linkage isomerism [135] was observed in reaction (29).

$$[Ru(^{15}NN_2)Cl(das)_2] \xrightarrow{NOPF_6} [Ru(^{15}N—^{14}N)Cl(das)_2]^+PF_6^- + N_2O \quad (29)$$

The complex $[Ru(N_2)Cl(das)_2](PF_6)$ shows an IR band at 2130 cm^{-1} corresponding to the nitrogen–nitrogen stretching frequency and one of medium intensity at 489 cm^{-1} assigned to the metal–nitrogen bond. The NMR spectrum of the compound in dimethyl sulfoxide (d_6) shows sharp singlets at 1.88 and 1.84 ppm for the coordinated das groups, indicating a *trans* configuration for the complex.

Bancroft *et al.* [102] have reported an impure ruthenium(II)–dinitrogen complex of the composition *trans*-$[RuH(N_2)(depe)_2]BPh_4$ [depe = 1,2-bis-(diethylphosphino)ethane]. The complex decomposes in the solid state and could not be obtained in pure form. In chloroform solution the complex exhibits an intense band at 2175 cm^{-1}, assigned to the N—N stretching vibration. The NMR hydride peak was observed at 25.87 τ.

Taqui Khan and Andal [143] have obtained a brown dinitrogen complex, $RuHCl(N_2)(AsPh_3)_3$, by passing molecular nitrogen through a solution of $RuHCl(AsPh_3)_3$ containing dimethylformamide. The dinitrogen complex $RuHCl(N_2)(AsPh_3)_3$ exhibits strong peaks at 2050 and 1950 cm^{-1} assigned to $\nu(N_2)$ and $\nu(H)$ stretching vibrations, respectively.

An unusual mixed-ligand dinitrogen dioxygen complex of composition $RuCl_2(AsMePh_2)_2(N_2)(O_2)$, **16**, has been obtained by Taqui Khan and Andal [144] by the reaction of the nitrosyl complex $RuCl_2(AsMePh_2)_2(O_2)(NO)$, **17**, with anhydrous hydrazine in methanol. The coordinated dioxygen in complex **17** is quite stable towards reduction, and hydrazine attacks only the coordinated nitrosyl group to form the dinitrogen complex. Complex **16** is characterized by intense infrared peaks at 2070 and 855 cm^{-1}, assigned to coordinated dinitrogen and dioxygen, respectively. The compound is quite stable at room temperature but decomposes on heating with the loss of both molecular nitrogen and oxygen. Attempts to combine molecular nitrogen and oxygen within the coordination sphere of the metal ion to produce nitric oxide have not yet been successful.

4. *Ruthenium(II)–Bis(dinitrogen) Complexes*

An unstable bis(dinitrogen)–ruthenium(II) complex with a probable *cis* arrangement of dinitrogen ligands **18** has been obtained [115] by the reaction of nitrous acid with *cis*- dinitrogen–bis(ethylenediamine)azidoruthenium(II)

$$cis\text{-}[Ru(en)_2N_3N_2]^+ + HNO_2 + H^+ \xrightarrow{\text{cold}} \underset{\mathbf{18}}{cis\text{-}[Ru(en)_2(N_2)_2]^{2+}} + N_2O + H_2O \qquad (30)$$

at 0°C. The addition of a cold solution of sodium tetraphenylborate to the reaction mixture resulted in the precipitation of the yellow tetraphenylborate salts of both the aquodinitrogen and bis(dinitrogen) complexes. The infrared spectrum of the *cis*-aquodinitrogen–bis(ethylenediamine)ruthenium(II) complex shows the N—N stretch at 2130 cm^{-1}. The bands at 2220 and 2190

cm^{-1} in the infrared spectrum of the *cis*-bis(dinitrogen) complex have been assigned to the N—N stretching vibrations of the two *cis* dinitrogen groups in the complex. These bands (at 2220 and 2190 cm^{-1}) disappear in about 30 minutes at 0°C in solution and in the solid state at room temperature, as the bis(dinitrogen) complex decomposes completely to form the *cis*-aquodinitrogen–bis(ethylenediamine)ruthenium(II) complex.

A tetrammine–bis(dinitrogen) complex of the composition $[Ru(NH_3)_4 \cdot (N_2)_2]^{2+}$ has been obtained [145] by the treatment of chloropentammine-ruthenium(III), $[Ru(NH_3)_5Cl]^{2+}$, or *cis*-dichlorotetrammineruthenium(III), *cis*-$[Ru(NH_3)_4Cl_2]^+$, with hydrazine hydrate at −23°C. An oily compound of an indefinite composition separated and decomposed to the dinitrogen and the bis(dinitrogen) complexes of ruthenium(II) at room temperature. The bis(dinitrogen) complex is very similar to the ethylenediamine analog, has infrared absorption bands at 2220 and 2185 cm^{-1}, and decomposes completely to the dinitrogen complex within 3 days at room temperature.

B. Osmium(II)–Dinitrogen Complexes

Dinitrogenpentammineosmium(II) complexes of the general formula $[Os(NH_3)_5N_2]X_2$ [X = Cl^-, Br^-, I^-, ClO_4^-, BF_4^-, and $B(C_6H_5)_4^-$] have been obtained [7,146] by refluxing a solution of hexachloroosmate(IV) or osmium tetrachloride and hydrazine hydrate followed by precipitation of the nitrogenpentammineosmium(II) salts of the anions indicated above. The chloride salt, $[Os(NH_3)_5N_2]Cl_2$, is a pale yellow diamagnetic solid, stable in air for extended periods [7], but decomposing at elevated temperatures under reduced pressure. The complex decomposes slowly in neutral aqueous solution but is stable in strong halogen acids even under reflux. The infrared spectra of these complexes have bands between 2101 and 2061 cm^{-1} corresponding to the N—N stretching vibration of coordinated nitrogen. The frequency of the band in the solid complexes increases somewhat with the size of the anion in the order Cl < Br < I < ClO_4 < BF_4 < BPh_4 due to a crystal lattice effect [102,146]. The chloride complex has a coordinated nitrogen infrared frequency of 2010 cm^{-1}, which is about 100 cm^{-1} less than the corresponding ruthenium(II)–dinitrogen complex, indicating a stronger metal–nitrogen bond for the osmium(II) complex. The osmium(II)–nitrogen stretching mode is observed between 546 and 520 cm^{-1}.

The crystal and molecular structure of the complex $[Os(NH_3)_5N_2]Cl_2$ has been determined by Fergusson *et al.* [147]. The structural details are presented in Fig. 12. The osmium(II) ion is the center of the distorted octahedron coordinated to six nitrogen atoms, five of which are ammonia ligands and one of which is part of a dinitrogen ligand. The dinitrogen N—N bond distance is 1.12 Å and the Os—N—N bond angle was found to be 179°.

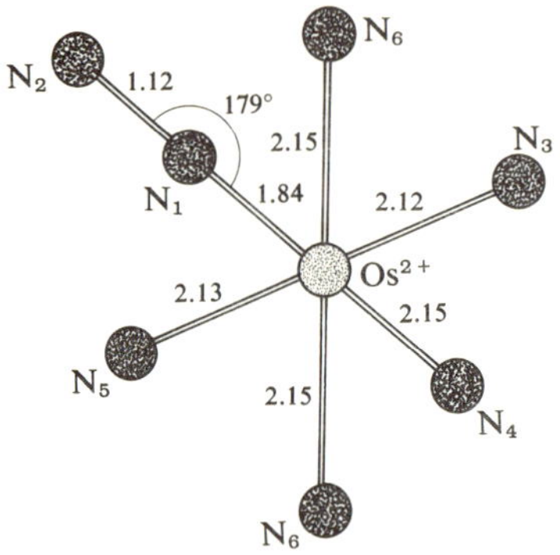

Fig. 12. Structure of dinitrogenpentammineosmium(II). According to Fergusson *et al* [147].

The osmium–ammonia–nitrogen bond lengths range from 2.12 to 2.15 Å, and the osmium–dinitrogen bond distance is 1.84 Å, which is similar to the osmium(II)–carbon bond length in carbonyls and osmium(II)–nitrogen bond length in nitrosyls, indicative of multiple bonding between the osmium and the adjacent atom of the coordinated ligand.

The osmium(II)–dinitrogen complex undergoes decomposition with the evolution of nitrogen in the presence of copper(I) and silver(I) [148]. Binuclear complexes of the type $[(NH_3)_5OsN_2M]^{3+}$ [M = silver(I), copper(I)] have been proposed as intermediates in the decomposition of the nitrogen complex [146]. The intermediate $[(NH_3)_5OsN_2M]^{3+}$ is thought to decompose by a one-electron oxidation of osmium(II) by M(I) to form a transient $[Os(NH_3)_5N_2]^{3+}$ species, which has been found to have an absorption at 2140 cm^{-1} corresponding to the N—N stretching vibration [146]. Oxidation of $[Os(NH_3)_5N_2]^{2+}$ with I_2 gave a mixture of products with a strong N—N stretching band at 2130 cm^{-1}, attributed to the formation of $[Os(NH_3)_5N_2]^{3+}$ [149]. Electrochemical oxidation of $[Os(NH_3)_5N_2]^{2+}$ in aqueous solution also produces the oxidized osmium(II)–dinitrogen complex that decomposes to $[Os(NH_3)_5H_2O]^{3+}$ and N_2 with a rate constant of 0.02 sec^{-1} [150].

Thermal dissociation of dinitrogen from $[Os(NH_3)_5N_2]^{2+}$ in aqueous solution has been investigated by Myagkov *et al.* [151] in the range 150°–185°C in a sealed tube. The reaction was first order with respect to the complex with a rate constant $k = 4 \times 10^{13} \exp(-3500/RT)$ sec^{-1}. It was concluded that the rate-determining step is the dissociation of $[(Os)(NH_3)_5N_2]^{2+}$ to the pentammine species $[Os(NH_3)_5]^{2+}$ which is rapidly aquated to $[Os(NH_3)_5H_2O]^{2+}$.

Molecular nitrogen complexes of osmium(II) of incompletely known composition were obtained by Borodko *et al.* [8,152] by the reduction of

K_2OsCl_6, $OsCl_4$, OsO_4, or $Os(OH)Cl_3$ with hydrazine in aqueous or tetrahydrofuran solutions. The presence of molecular nitrogen in the coordination sphere of osmium(II) has been confirmed by the infrared spectra of the complexes. Bands due to coordinated dinitrogen (2010, 2095, and 2168 cm^{-1}) shifted to lower frequencies (60–80 cm^{-1}) when $^{15}N_2H_4$ was employed in the synthesis of the complexes. These dinitrogen–osmium(II) complexes are stable in air up to 200°C and in aqueous solution, but are decomposed by sulfuric acid with the evolution of nitrogen.

Chatt *et al.* [103,153] have obtained a series of osmium(II)–dinitrogen complexes of the type $[OsX_2(N_2)(MPhR_2)_3]$ (where M = P or As, X = Cl or Br, R = alkyl) by the reaction of amalgamated zinc with *mer*-$[OsX_3(MPhR_2)_3]$ in dry tetrahydrofuran under nitrogen at 1 atm pressure. Analogs containing ^{15}N have also been prepared to establish the coordination of molecular dinitrogen in the complexes. The solid complexes are stable in air but in solution undergo oxidative halogen addition and loss of dinitrogen to form the parent compound, $[OsX_3(MPhR_2)_3]$. The complexes are diamagnetic and the proton magnetic resonance spectra show that they possess a meridional configuration of the phosphine ligands with the halogen *trans* to nitrogen. The complexes have very sharp and intense N—N stretching bands indicating linear coordination of nitrogen.

Chatt *et al.* [154] have prepared halo hydrido dinitrogen complexes of the type *mer*-$[OsXY(N_2)(PR_3)_3]$ (X = Cl or Br, Y = H or D) by the reduction of $[OsX_2(N_2)(PR_3)_3]$ with sodium borohydride or lithium aluminum hydride in methanol or THF. The compounds have been obtained as crystalline diamagnetic solids quite stable in air. As in the case of *mer*-$[OsX_2N_2(PPhR_2)_3]$, the proton magnetic resonance spectra of the compounds show that they possess a meridional configuration of the phosphine ligands with the halogen *trans* to nitrogen. The compound *mer*-$[OsHClN_2(PEt_2Ph)_3]$ reacts with anhydrous HCl in toluene to give $[OsCl_2(N_2)(PEt_2Ph)_3]$.

A cationic hydride complex of the composition $[OsHN_2(Et_2PCH_2CH_2\cdot PEt_2)]^+$ has been obtained [104] by passing nitrogen through an acetone solution of $[OsHCl(Et_2PCH_2CH_2PEt_2)]$. The compound obtained in chloroform solution has an N—N stretching frequency at 2144 cm^{-1} and a proton magnetic resonance absorption at 20.47 τ. It decomposed in the solid state and could not be obtained in pure crystalline form.

Maples *et al.* [155] studied the rate of nitrogen substitution by phosphine ligands in complexes of the type *mer*-$OsN_2X_2L_3$ (**19**) (L = PMe_2Ph, Et_2PhP; X = Cl, Br) and *mer*-$OsN_2X_2(PMe_2Ph)_2[PPh(OMe)_2]$ in toluene solution by following the rate of disappearence of the N—N stretching infrared absorption band of the complexes. The rate of replacement of nitrogen in *mer*-$OsN_2X_2(PR_2Ph)_3$ with PR_2Ph was first order with respect to complex concentration and independent of ligand concentration. The ranges of activation parameters for these reactions have been reported as $\Delta H\dagger = 27\text{–}34$ kcal/mole and $\Delta S\dagger = 9\text{–}19$ eu. Reaction sequence (31) and (32) was proposed [155] for the substitution reaction.

$$\underset{\mathbf{19}}{Os(N_2)X_2[PR_2Ph]_3} \xrightarrow{\text{slow}} OsX_2[PR_2Ph]_3 + N_2 \quad (31)$$

$$OsX_2[PR_2Ph]_3 + L \xrightarrow{\text{fast}} OsX_2[PR_2Ph]_3L \quad (32)$$

The rate of the reaction decreases, in the order of the increasing basicity of the phosphine ligands, $P(OCH_3)_2(Ph) \gg P(CH_3)_2Ph > P(C_2H_5)_2Ph$ in accord with the above dissociative mechanism. As the electron density increases on the metal ion, the metal–nitrogen bond becomes stronger by back donation of electrons into π^* orbitals of the nitrogen, and dissociation of the nitrogen becomes increasingly more difficult.

1. *Osmium(II)–Bis(dinitrogen) Complexes*

The high stability of osmium(II) dinitrogen complexes is reflected in the recent preparation of the first bis(dinitrogen) complex of osmium(II). This complex was obtained by the diazotization of a coordinated ammonia ligand in the complex dinitrogenpentammineosmium(II) chloride. The reaction proceeded at room temperature with the formation of the diamagnetic bis(dinitrogen)tetrammineosmium(II) complex in 34% yield. The formation of the complex was followed by the disappearance of the absorption peak due to the dinitrogen complex at 208 nm and the growth of the peak due to the bis(dinitrogen) complex at 221 nm. Chloride, bromide, and iodide salts of $[Os(NH_3)_4(N_2)_2]^{2+}$ have been prepared [156]. The complex $[Os(NH_3)_4(N_2)_2]^{2+}$ has infrared peaks at 2120 and 2175 cm^{-1}, higher than those of the dinitrogen complex (2010–2060 cm^{-1}). The higher infrared frequency of the bis(dinitrogen) complex is in accord with the concept of the sharing of the π-donor capacity of metal ion between the two coordinating nitrogen donors. On the basis of the splitting of infrared stretching mode of coordinated nitrogen, it was suggested that the two coordinated nitrogen molecules occupy *cis* positions in the complex [156]. The thermal stability of the bis(dinitrogen) complex is much less than that of the dinitrogen complex. The former decomposes in solution at about 50°C to form dark brown solutions containing unidentified products.

2. *μ-Dinitrogen Complexes of Ruthenium(II) and Osmium(II)*

Nitrogen bridge complexes containing two ruthenium(II) ions or one ruthenium(II) and one osmium(II) have been reported recently [121,151,158]. They are listed in Table III (entries 2–8). In most of these complexes the N—N stretching frequency is weakly infrared-active. Raman spectra have been measured only for $[Ru(NH_3)_5]_2N_2^{4+}$ where a strong band appears at about 2100 cm^{-1}. As in the case of $[Ru(NH_3)_5]_2N_2^{4+}$, it is expected that in the μ-N_2 complexes the metal—N—N—metal group will frequently be approximately linear with the N—N bond distance, lengthening as the N—N bond weakens and the M—N bond strengthens. It may readily be seen from Table III (entries 2 and 5) that when two ruthenium atoms are replaced by one ruthenium and one osmium atom, the N—N stretching frequency is lowered by about 30 cm^{-1}. Nitrogen–bridged compounds with two osmiums would be expected to have still weaker N—N bonds. In the compound $[(NH_3)_5RuN_2Os(NH_3)_4CO]$, the nitrogen rather than the carbon monoxide is the bridging group. Here the N—N stretching frequency increases by about 60 wave numbers over that of $[(NH_3)_5RuN_2Os(NH_3)_5]^{4+}$ because of electronic interaction with carbonyl group.

An interesting trimeric compound of the type *cis*-$[(NH_3)_5Ru—N_2—$ $—Os(NH_3)_4—N_2—Ru(NH_3)_5]^{6+}$ has been described by Allen [121]. It is interesting to note that the compound exhibits only one weak N—N stretching absorption at 2097 cm^{-1}, although one would have expected more than one band due to the *cis* arrangement of the two μ-N_2 groups. The Raman spectrum of this compound would probably help to resolve these structural questions.

The second-order rate constants and the activation energies for the reactions of the nucleophiles *cis*-$[Ru(NH_3)_4(H_2O)(N_2)]^{2+}$ and $[Os(NH_3)_5N_2]^{2+}$ with $[Ru(NH_3)_4(H_2O)_2]^{2+}$ to form the μ-dinitrogen complexes $[(NH_3)_4(H_2O)\cdot RuN_2Ru(H_2O)(NH_3)_4]^{4+}$ and $[(NH_3)_4(H_2O)RuN_2Os(NH_3)_5]^{4+}$, respectively, at 25°C are, respectively, 0.72×10^{-1} M^{-1} sec^{-1} and 17.1 ± 0.2 kcal/mole [157], and 1.2×10^{-1} M^{-1} sec^{-1} and 16.8 ± 0.1 kcal/mole [158]. The second-order rate constant and the activation energy for the formation of $[(NH_3)_5RuN_2Ru(NH_3)_4(H_2O)]^{4+}$ from the nucleophile $[Ru(NH_3)_5N_2]^{2+}$ and $[Ru(NH_3)_4(H_2O)_2]^{2+}$ have been reported as 0.68×10^{-1} M^{-1} sec^{-1} and 18.2 ± 0.1 kcal/mole [157]. Thus the rate of reaction of the osmium nucleophile $[Os(NH_3)_5N_2]^{2+}$ with $[Ru(NH_3)_4(H_2O)_2]^{2+}$ is about twice those of the ruthenium nucleophiles $[Ru(NH_3)_5N_2]^{2+}$ and [*cis*-$Ru(NH_3)_4(H_2O)\cdot(N_2)]^{2+}$. The increased nucleophilicity of dinitrogen bound to osmium may be explained on the basis of a greater π-donating ability of osmium(II) as compared to ruthenium(II).

Electrolytic oxidation of the $[(NH_3)_4(H_2O)RuN_2Os(NH_3)_5]^{4+}$ at 1.0 V gave a stable binuclear complex of 5+ ionic charge [158]. Further oxidation of the latter complex at +3.0 V resulted in the oxidative decomposition of the complex to $[(NH_3)_5Ru(H_2O)]^{3+}$ and $[(NH_3)_5OsN_2]^{3+}$. The decomposition obeyed first-order kinetics with $k_r = 1.5 \times 10^{-5}$ sec^{-1} at 25°C. The complex of 5+ ionic charge has a mixed-valence, charge-transfer band at 755 nm, characteristic of the mixed-valence oxidation state of osmium and ruthenium. An ESR study of this compound would be very interesting as a possible indication of the delocalized charge density on the metal ions.

C. Cobalt–Dinitrogen Complexes

Dinitrogentris(triphenylphosphine)cobalt(0), **20**, has been prepared by Yamamoto *et al.* [11] by bubbling dinitrogen through a toluene or ether solution of tris(2,4-pentanediono)cobalt(III) containing diethylaluminum-monoethoxide and triphenylphosphine. Yellow crystals of $[(PPh_3)_3CoN_2]$ (**20**) obtained from the reaction mixture contained one molecule of dinitrogen per cobalt atom. The complex is moderately stable in the absence of air. Thermal decomposition above 80°C liberated stoichiometric amounts of dinitrogen as indicated by the mass spectrum of the complex. The infrared spectrum of the complex shows a strong band at 2088 cm^{-1}, which was assigned to the N—N stretching frequency of the coordinated dinitrogen moiety.

Dinitrogentris(triphenylphosphine)cobalt(0) was also obtained by Misono *et al.* [13] from tris(triphenylphosphine)cobalt dihydride and molecular nitrogen. The complex $(Ph_3P)_3CoH_2$ was prepared by the reaction of tris(2,4-pentanediono)cobalt(III) with tris(isobutyl)aluminum in ether at 10°C in the presence of triphenylphosphine and hydrogen bubbling through the reaction mixture. The dihydride crystallizes as an air-sensitive yellow solid, decomposing at 77°C in vacuum with the evolution of hydrogen and benzene. A benzene solution of the dihydride complex when kept under nitrogen changes its color from brown to red and, on addition of petroleum ether, orange crystals are obtained. The infrared spectrum of the orange complex shows an absorption band at 2088 cm^{-1} characteristic of coordinated dinitrogen. The bands at 1755, 1900, and 1935 cm^{-1} characteristic of the hydride complex $(PPh_3)_3CoH_2$, are absent in the dinitrogen complex, and the elemental analysis agrees with the formula $(Ph_3P)_3CoN_2$, reported earlier by Yamamoto *et al.* [11].

When dinitrogentris(triphenylphosphine)cobalt(0) in solution is swept with molecular hydrogen [11], the red color of the dinitrogen complex changes to yellow with the loss of dinitrogen, as shown by mass spectrometry, and the dihydrido complex $(Ph_3P)_3CoH_2$ is obtained. Molecular nitrogen is also displaced from $(Ph_3P)_3CoN_2$ on treatment with ammonia or ethylene. Nitrogen could not be displaced from the complex by passing an inert gas such as argon through the solution, indicating that coordinated dinitrogen is firmly bound to cobalt and could only be displaced by a more strongly coordinated ligand.

Speier and Marko [159] have characterized the compounds $CoN_2(PEt_3)_3$ (**21**) and $CoN_2(PPh_3)_3$ (**20**) as yellow paramagnetic species with oxidation state zero for the cobalt atom. Strong reducing agents such as *n*-butyllithium or potassium diphenyl phosphide react with these cobalt(0) complexes in tetrahydrofuran solution, resulting in cleavage of the phosphorus–carbon bond and formation of phosphide, PR_2^-, ligands [160]. These reactions are outlined in Eqs. (33) and (34).

$$\underset{\substack{\mathbf{21} \\ [\nu(N_2) = 2059\ cm^{-1}]}}{Co(N_2)(PEt_3)_3} + n\text{-BuLi} \longrightarrow \underset{\substack{\mathbf{22} \\ [\nu(N_2) = 2014\ cm^{-1}]}}{Li^+[Co(N_2)(PEt_3)_2(PEt_2)]^-} + C_6H_{14} \quad (33)$$

$$\underset{\substack{\mathbf{20} \\ [\nu(N_2) = 2093\ cm^{-1}]}}{Co(N_2)(PPh_3)_3} + n\text{-BuLi} \longrightarrow \underset{\substack{\mathbf{23} \\ [\nu(N_2) = 2016\ cm^{-1}]}}{Li^+[Co(N_2)(PPh_3)_2(PPh_2)]^-} + C_4H_9{-}C_6H_5$$

$$\xrightarrow{n\text{-BuLi}} \underset{\substack{\mathbf{24} \\ [\nu(N_2) = 1904\ cm^{-1}]}}{2\ Li^+[Co(N_2)(PPh_3)(PPh_2)_2]^{2-}} + C_4H_9{-}C_6H_5 \quad (34)$$

The result of the presence of negatively charged strongly π-donating phosphide groups as ligands is to increase the negative charge density on the metal ion and consequently strengthen the cobalt–dinitrogen π bonding. The anionic dinitrogen complexes exhibit the lowest N—N stretching frequencies recorded thus far.

Hydridodinitrogentris(triphenylphosphine)cobalt(I) (**26**) was obtained along with dinitrogentris(triphenylphosphine)cobalt(0) (**20**) [14] by reduction of a toluene solution of tris(2,4-pentanediono)cobalt(III) with tris(isobutyl)-aluminum in the presence of triphenylphosphine under a nitrogen atmosphere at $-5°$ to $-10°C$. The complex $(PPh_3)_3CoH(N_2)$ (**26**) was also obtained by Sacco and Rossi [12] by the reaction of molecular nitrogen at room temperature and atmospheric pressure with trihydridotris(triphenylphosphine)cobalt(III) (**25**) in benzene solution, as shown in Eq. (35).

$$\underset{\mathbf{25}}{(Ph_3P)_3CoH_3} + N_2 \rightleftharpoons \underset{\mathbf{26}}{(Ph_3P)_3CoH(N_2)} + H_2 \quad (35)$$

Equilibrium (35) is displaced far to the right in the presence of molecular nitrogen and is displaced in the reverse direction under a hydrogen atmosphere. The presence of coordinated dinitrogen along with hydrogen in the coordination sphere of cobalt was demonstrated by the thermal decomposition of the dinitrogen complex, which resulted in the formation of a gaseous mixture containing 66.6% nitrogen and 33.3% hydrogen. The hydridodinitrogencobalt(I) complex is diamagnetic, in agreement with the expected behavior of a five-coordinated cobalt(I) complex [86]. Though the complex has a strong N—N absorption band in the infrared at 2080–2084 cm^{-1}, it is surprising that no bands corresponding to the Co—H stretching vibration are observed in the infrared spectrum.

Rossi and Sacco [106] have recently obtained a very good yield of **26** by the interaction, shown in Eq. (36), of molecular nitrogen with the coordinately unsaturated species $CoHL_3$ (L = PPh_3, $PMePh_2$, $PEtPh_2$, $PBuPh_2$, PEt_2Ph, PBu_3) in benzene solution. The hydrido tris phosphine complexes of

$$CoHL_3 + N_2 \rightleftharpoons CoHN_2L_3 \tag{36}$$

cobalt(I) have been prepared by the reaction of σ-cyclooctenyl-π-cycloocta-1,5-dienecobalt(I), **27**, with the phosphine ligand in benzene solution in the presence of argon. The formation of $CoHL_3$ is presumed to have taken place according to the steps given in Eqs. (37) and (38). When nitrogen is passed

$$\underset{\mathbf{27}}{Co(C_8H_{13})(C_8H_{12})} + 2\,L \longrightarrow Co(C_8H_{13})L_2 + C_8H_{12} \tag{37}$$

$$Co(C_8H_{13})L_2 + L \longrightarrow [Co(C_8H_{13})L_3] \longrightarrow CoHL_3 + C_8H_{12} \tag{38}$$

through a benzene solution of **27** and an excess of the phosphine ligand, the cobalt(I)–dinitrogen complexes were obtained in very good yields.

The ease of formation of the Co(I)–dinitrogen complex from a σ-bonded cyclooctenyl group in reaction (38) lends support to the idea that an alkyl–cobalt complex may be an intermediate in the formation of **26** from tris(acetylacetonato)cobalt(III) and triisobutylaluminum in the presence of triphenylphosphine. Recently, a complex of the composition $[Co(C_2H_4)\cdot(PPh_3)_3]$ has been isolated as dark brown crystals by the reaction of $(CH_3)_2AlOEt$ with tris(acetylacetonato)cobalt(III) and triphenylphosphine in an argon atmosphere [11]. In the presence of nitrogen, however, $HCo(N_2)(PPh_3)_3$ (**26**) is obtained. Accordingly, it was proposed [11] that an intermediate alkyl–cobalt complex of the type $[Co(C_2H_5)(PPh_3)_3]$ is formed in the first step of the reaction. In the absence of molecular nitrogen, the cobalt–alkyl bond is cleaved homolytically to form an ethyl radical $\cdot C_2H_5$ that disproportionates to ethane and ethylene, the latter combining with the cobalt complex to form $[Co(C_2H_4)(PPh_3)_3]$. In the presence of molecular nitrogen a hydride is abstracted from the σ-bonded alkyl group to give a Co—H bond and subsequent formation of $Co(H)(N_2)(PPh_3)_3$ with the release of ethylene.

Enemark *et al.* [161] have prepared the monoetherate of **26**, $(PPh_3)_3\cdot CoH(N_2)Et_2O$, and have determined the structure of the compound by X-ray analysis. The complex was found to be mononuclear with the Co—N—N linkage very nearly linear, as was the case of the Ru—N—N linkage in the dinitrogenpentammineruthenium(II) ion [125] (Fig. 9). The coordination sphere

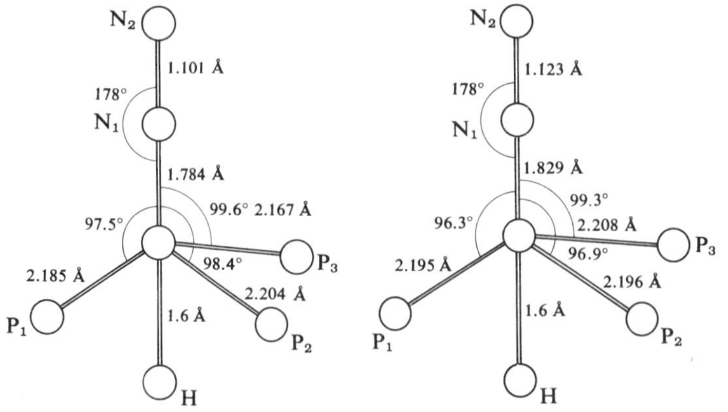

Fig. 13. The structure of $Co(H)(N_2)(PPh_3)_3$. After Davis *et al.* [112,163].

of cobalt is very nearly trigonal bipyramidal, with the cobalt atom 0.3 Å above the equatorial plane of the phosphorus atoms and with one apical site apparently vacant. Though the presence of a hydrogen atom on the apical site *trans* to coordinated nitrogen cannot be established on the basis of X-ray analysis, indirect stereochemical evidence very strongly favors this possibility. The phenyl hydrogen atoms are not within 3.0 Å of the cobalt and there is no possibility of the blocking by a phenyl hydrogen of the apical site *trans* to the coordinated dinitrogen molecule. Thus there is enough room for a coordinated hydrogen atom on cobalt; however, Sacco and Rossi did not identify a metal–hydrogen stretching frequency [12]. Although the previous workers failed to resolve the hydride and nitrogen stretching bands, Enemark *et al.* [161] found peaks at 2085 and 2105 cm^{-1} and suggested that they correspond to Co—H and N—N stretching frequencies, respectively. Several metal–hydrogen stretching frequencies have been observed in the 2100 cm^{-1} region, and it is plausible that the presence of a *trans*-coordinated dinitrogen ligand reduces the M—H stretching to about 2085 cm^{-1}.

Davis *et al.* [162,163] have recently determined the structure of **26** by the X-ray analysis of good single crystals of the substance based on 3312 independent reflections. The structure consists of two discrete but nearly identical molecules of $Co(H)(N_2)(PPh_3)_3$ (**26**) as shown in Fig. 13. The coordination about each cobalt atom is that of a trigonal bipyramid with the three phosphorus atoms in the equatorial planes and with hydrogen and nitrogen occupying apical positions. The Co—N—N bond is not exactly linear but at an angle of 178°. The position parameters of the hydride ligands have been derived from an interpolation of electronic density peaks in the Fourier maps and a distance of 1.61 ± 0.01 Å was computed. This distance is the sum of the covalent radii of cobalt(I) and H^-. The remaining structural parameters are given in Fig. 13.

The proton NMR spectrum of the complex in tetrahydrofuran solution has a single quartet at τ 29 ($J = 50$ cycle/second) with intensity ratios 1:3:3:1 for the four equally spaced bands [164]. The NMR spectrum thus supports the equivalence of the three coordinated phosphines and implies the presence of a coordinated hydride anion at an apical site in the trigonal bipyramid structure of hydridodinitrogentris(triphenylphosphine)cobalt(I).

1. *Reactions of* $H(N_2)Co(PPh_3)_3$

When a stream of carbon dioxide was passed through a benzene solution of hydridodinitrogentris(triphenylphosphine)cobalt(I) (**26**) at room temperature, the original red color of the solution changed to brownish green with the loss of molecular nitrogen from the complex. The resulting solution gave formatotris(triphenylphosphine)cobalt(I) in 45% yield [11,165]. The presence of a coordinated formato group in the complex was indicated by the infrared bands at 1620 and 1300 cm^{-1} characteristic of a coordinated $HCOO^-$ group. The formatotris(triphenylphosphine)cobalt(I) complex has been obtained directly by the displacement of hydrogen and nitrogen from **26** by formic acid. The insertion of carbon dioxide with the formation of the

$$H(N_2)Co(PPh_3)_3 + CO_2 \longrightarrow \underset{\mathbf{28}}{HCOOCo(PPh_3)_3} + N_2 \tag{39}$$

$$H(N_2)Co(PPh_3)_3 + HCOOH \longrightarrow \underset{\mathbf{28}}{HCOOCo(PPh_3)_3} + N_2 + H_2 \tag{40}$$

formatotris(triphenylphosphine)cobalt(I) complex **28** has also been reported by Misono *et al.* [166].

When carbon monoxide was bubbled through a benzene solution of **26**, coordinated hydrogen and nitrogen were displaced, the original red color of the solution changed rapidly to yellow-brown, and after several hours a complex of the composition $[Co(CO)_3(PPh_3)]_2$ was precipitated [166]. Tricarbonyltris(triphenylphosphine)cobalt(0) was identified by elemental analysis and infrared spectra. The addition of light petroleum to the residual yellow-brown solution precipitated two crystalline complexes, $CoH(CO)\cdot(PPh_3)_3$ and $CoH(CO)_2(PPh_3)_2$, identified by elemental analysis. These reactions indicate that carbon monoxide forms more stable complexes than does nitrogen with cobalt(I).

The hydrogen–deuterium exchange of a benzene solution of **26** with deuterium at 25°C was studied by Parshall [167]. The extent of exchange obtained corresponded to about 19 hydrogen atoms per mole of the cobalt complex. It has been proposed [167] that besides the coordinated hydrogen, six aryl hydrogens per phosphine ligand exchange with deuterium. It was confirmed

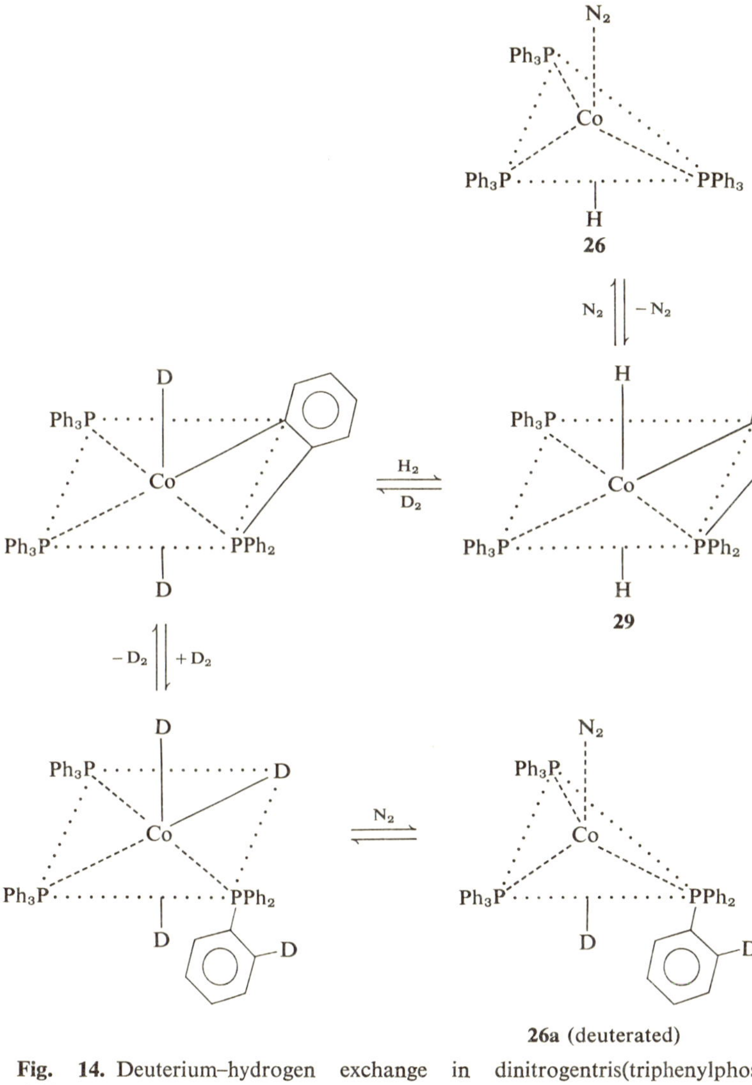

(41)

26a (deuterated)

Fig. 14. Deuterium–hydrogen exchange in dinitrogentris(triphenylphosphine)-hydridocobalt (I).

by NMR that the deuterium introduced into triphenylphosphine ligands is in a position *ortho* to the phosphorus atom. It was suggested that **26** is in equilibrium with a six-coordinate compound, **29**, involving bonding between the *ortho* carbon atoms of one of the phenyl rings and cobalt [167]. The mechanism proposed [167] to explain the hydrogen–deuterium exchange is illustrated in Fig. 14.

Nitrous oxide reacts with the dinitrogen complex **26** in benzene solution at room temperature with the evolution of dinitrogen and the disappearance of the band at 2088 cm^{-1} (N—N stretch) [168]. The original red color of the complex changes to yellow-green and then to pale blue. The chemical basis for these color changes and the complexes formed in solution have not yet been determined. In the presence of an excess of triphenylphosphine, oxidation to triphenylphosphine oxide has been observed. It has accordingly been postulated that nitrous oxide expels coordinated dinitrogen from **26** to form a nitrous oxide complex which is subsequently reduced to the original dinitrogen complex and oxygen. The oxygen is trapped by triphenylphosphine with the formation of its oxide (P—O stretch at 1200 cm^{-1}). Mass spectrometric analysis indicated the presence of both hydrogen and nitrogen in the gas phase. The presence of hydrogen in the gas phase indicates a more complicated degradation of the original dinitrogen complex than was assumed by the investigators.

In a similar reaction between nitrous oxide and tris(triphenylphosphine)-trihydridocobalt, **25**, $CoH_3(PPh_3)_3$, in benzene solution, evolution of nitrogen and hydrogen was observed and the formation of the dinitrogen complex **26** as an intermediate was indicated by the appearance of a peak at 2080 cm^{-1}. In this case also, the catalytic reduction of nitrous oxide to nitrogen [169] was accompanied by simultaneous oxidation of triphenylphosphine to the phosphine oxide.

The dinitrogen complex **26** catalyzes the hydrogenation of ethylene to ethane at room temperature and at atmospheric pressure. In this case the reaction probably proceeds through formation of an intermediate ethylene complex, $CoH(C_2H_4)(PPh_3)_3$, which rearranges to a σ-bonded ethyl complex, $Co(C_2H_5)(PPh_3)_3$. Reaction of the latter with hydrogen releases ethane and forms the hydridocobalt complex. The hydridodinitrogen complex **26** also catalyzes the isomerization of 1-butene to 2-butene and the polymerization of olefins [169].

Ammonia reacts reversibly with **26** in benzene solution to form the complex $CoH(NH_3)(PPh_3)_3$, **30**, and molecular nitrogen. The reversible reactions of **26**, **25**, and **30**, involving nitrogen, hydrogen, and ammonia are summarized in Eq. (42) [169a].

Reaction of **26** with hydrogen chloride or iodine gave 1 mole of nitrogen, some hydrogen, and $CoCl_2(PPh_3)_2$ or $CoI_2(PPh_3)_3$.

When $CoH(N_2)(PPh_3)_3$ reacts in benzene solution with tributylphosphine, successive replacement of triphenylphosphine by tributylphosphine takes place with the concomitant shift of the N—N stretching frequency from 2088 to 2075, 2061, and 2048 cm^{-1}. These frequencies correspond to the displacement of 1, 2, and 3 moles of triphenylphosphine by tributylphosphine. With a large excess of tributylphosphine the absorption band at 2048 cm^{-1}

completely disappears because of the formation of tetrakis(tributylphosphine) hydridocobalt, $CoH(PBu_3)_4$.

Aresta *et al.* [170] have prepared binuclear complexes of cobalt(0) having the composition $(L_3Co)_2N_2$, where L = PPh_3, PEt_2Ph, by the interaction of $CoClL_3$ in toluene (L = PPh_3) or tetrahydrofuran (L = PEt_2Ph) with sodium metal under nitrogen (Co/Na ~ 1). The complexes $[(PPh_3)_3Co]_2N_2$ and $[(PEt_2Ph)_3Co]_2$ have been isolated as dark red and dark brown crystalline

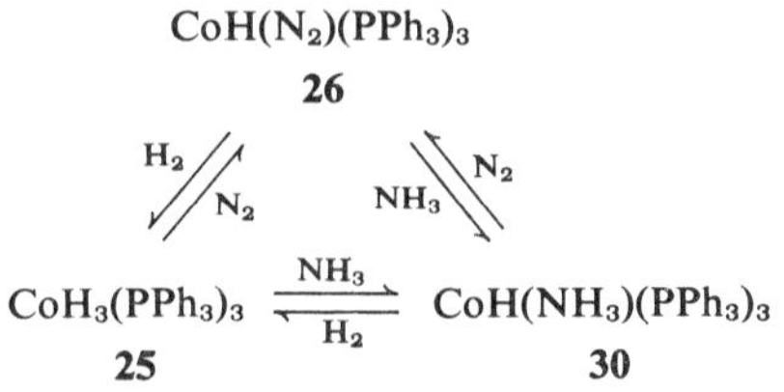

(42)

solids, respectively, which are unstable in air and soluble in nonpolar solvents. The infrared spectra of these complexes do not have absorptions corresponding to N—N stretching vibrations. Thermal dissociation at 100°–120°C or treatment with iodine at 60°C liberated 0.50 mole of dinitrogen per mole of cobalt, as expected from the binuclear composition of the complexes.

The reaction of $CoClL_3$ (L = PPh_3, PEt_2Ph) in toluene or tetrahydrofuran with an excess of sodium under nitrogen gave the mononuclear cobalt (1−) complexes of the composition $Na(CoN_2L_3)$, obtained as black (L = PEt_2Ph) or red (L = PPh_3) diamagnetic crystals, respectively. Infrared spectra of the complexes in Nujol mull show a strong absorption band at 1875 cm^{-1} (ν_{N_2}) for $Na[Co(N_2)(PEt_2Ph)_3]$ and at 1845 cm^{-1} (ν_{N_2}) for $Na[Co(N_2)(PPh_3)_3]$. On treatment with water the cobalt complexes decomposed with formation of the cobalt(I) complexes $CoH(N_2)L_3$ and a strongly alkaline solution (due to formation of sodium hydroxide).

D. Rhodium(I)–Dinitrogen Complexes

Reaction of chlorocarbonylbis(diphenylphosphine)rhodium(I) with excess propionyl azide in toluene at −20°C gave, on addition of hexane at −70°C, a precipitate of chlorodinitrogenbis(triphenylphosphine)rhodium(I) [6]. The reaction was conducted in an atmosphere of argon at < −60°C. The product was reprecipitated from toluene solution by the addition of methanol in the form of a yellow solid stable in an inert atmosphere and moderately stable at room temperature. A solution of the rhodium(I)–dinitrogen complex in an organic solvent such as chloroform, benzene, or toluene immediately liberated molecular nitrogen at room temperature. The complex is insoluble in methanol, ethanol, and hexane. The presence of coordinated nitrogen in the rhodium(I) complex has been confirmed [6] by the presence of a stretching band at 2152 cm^{-1} in the infrared spectrum of the rhodium(I) complex.

A μ-dinitrogenrhodium(III) complex of the composition $2\,NH_4^+$, $\{[Rh(OH)(NO_2)_3(NH_3)_2]_2N_2\}^{2-}$ has been reported by Volkova *et al.* [171]. The potassium salt of the complex was obtained as colorless crystals when an aqueous solution of $K[Rh(NO_2)_4(NH_3)_2]$ and ammonium sulfate at pH 1–2 was heated on a water bath for 2 hours. The complex is reasonably stable in air and has a Raman ν_N- stretching frequency at 2070 cm^{-1} corresponding to a linear Rh(III)—N≡N—Rh(III) group, with no absorption in the infrared. When ammonium sulfate labeled with ^{15}N was used, two peaks were observed at about 2045 cm^{-1} corresponding to the presence of both μ-$^{14}N_2$ and μ-$^{14}N^{15}N$. An attempt to reduce the complex with sodium dithionate was unsuccessful.

E. Iridium(I)–Dinitrogen Complexes

Chlorocarbonylbis(diphenylphosphine)iridium(I) (**31**) reacts with a series of aroyl azides in a medium containing ethanol or water to form chlorodinitrogenbis(triphenylphosphine)iridium(I) [9, 10], **34**. Collman *et al.* [10] proposed the mechanism given in Fig. 15 for the formation of the iridium(I)–dinitrogen complexes by this reaction. When complex **31** is mixed with an aroyl azide in chloroform solution containing a nucleophile such as water or ethanol, the carbonyl peak of the complex **31** at 1970 cm^{-1} and the azide peak at 2145 cm^{-1} decrease in intensity with the appearance of a new peak at 2105 cm^{-1} corresponding to the complex **32** and the carbamide, **36**, peaks at 1780 and 1710 cm^{-1}. In the absence of a nucleophile, in dry chloroform the peaks due to the isocyanate **33** and the dinitrogen complex **34** appear at 2250 and 2105 cm^{-1}, respectively. Compounds **33** and **34** then react with the rapid disappearance of the peak at 2250 cm^{-1} and enhancement of the peaks at 1720 and 1890 cm^{-1} characteristic of the acyl isocyanate complex, **35**. The reaction sequence proposed [10] involves an initial attack of the acyl or aryl azide on **31** to form a labile intermediate **32** which collapses to the dinitrogen complex **34** and an acyl or aryl isocyanate **33**. If an alcohol is present a carbamate, **36**, is formed. The hypothetical intermediate **35** is proposed as a π-bonded complex, where the metal is coordinated to the N=C π bond and also forms a partial σ bond with the isocyanate nitrogen.

F. Iron(II)–Dinitrogen Complexes

Reaction of Fe(II) chloride dihydrate and ethyldiphenylphosphine with sodium borohydride in ethanol under hydrogen or argon gave a yellow crystalline compound, $FeH_2(PEtPh_2)_3$ [101,172]. The formula of this complex was determined, and the compound was characterized by elemental

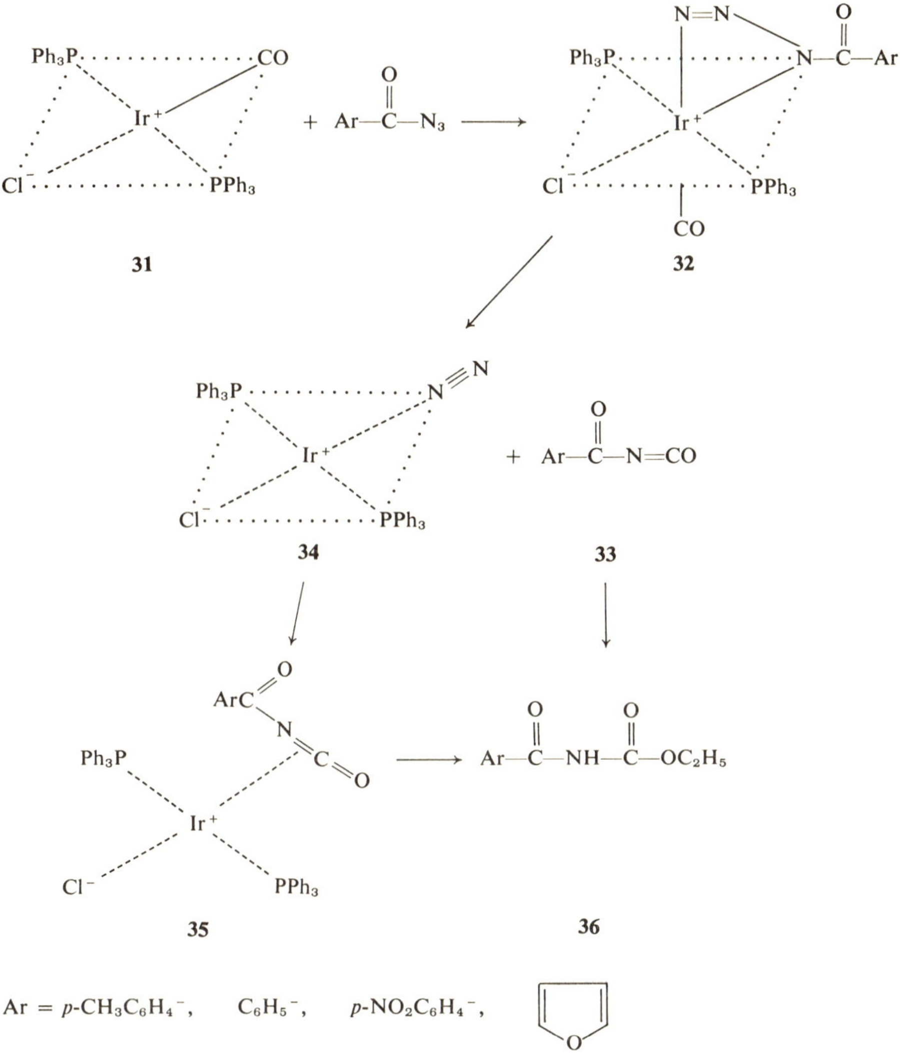

Fig. 15. Formation of iridium(I) dinitrogen complexes.

analysis and infrared bands of medium intensity at 1922 and 1860 cm^{-1} (Fe—H stretch). Molecular nitrogen reacts with this complex in the solid state or in solution to form a yellow crystalline diamagnetic dinitrogen complex, $FeH_2N_2(PEtPh_2)_3$ [101], **37**. The dihydridodinitrogeniron(II) complex shows absorption due to the coordinated dinitrogen (N—N stretch) at 2055–2060 cm^{-1} and absorptions due to Fe—H stretch at 1950–1960 (w) and 1855–1863 (m) cm^{-1}. The complex reacts quantitatively with iodine and with hydrochloric acid according to Eqs. (43) and (44). The dinitrogen complex also undergoes a reversible reaction in which a mole of hydrogen is lost from a mole of the complex, and rapid migration of hydrogen from ligand to iron(II) takes place. The complex formed, **38**, is considered to

$$\underset{\mathbf{37}}{FeH_2(N_2)(PEtPh_2)_3} + I_2 \longrightarrow H_2 + N_2 + FeI_2(PEtPh_2)_2 + PEtPh_2 \quad (43)$$

$$\underset{\mathbf{37}}{FeH_2(N_2)(PEtPh_2)_3} + 2\,HCl \longrightarrow 2\,H_2 + N_2 + FeCl_2(PEtPh_2)_2 + PEtPh_2 \quad (44)$$

possibly have a bond between one of the *ortho* carbon atoms of a phenyl ring and iron(II) in a manner analogous to that of **29**, described above.

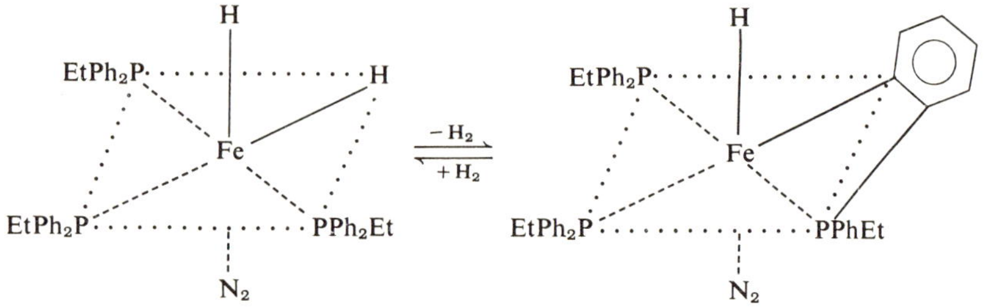

37 **38** (45)

The dihydridodinitrogentris(ethyldiphenylphosphine)iron(II) complex, **37**, loses a mole of ethyldiphenylphosphine when it reacts with triethylaluminum in benzene solution, as shown in Eq. (46). The five-coordinate dihydro

$$\underset{\mathbf{37}}{FeH_2(N_2)(PEtPh_2)_3} + AlEt_3 \longrightarrow \underset{\mathbf{39}}{FeH_2(N_2)(PEtPh_2)_2} + Et_3AlPEtPh_2 \quad (46)$$

Fe(II)–dinitrogen complex, **39**, has an N—N stretching vibration at 1989 cm^{-1}. When the six-coordinate dinitrogen complex, $FeH_2(N_2)(PEtPh_2)_3$, **37**, reacts with carbon monoxide at room temperature and pressure, the coordinated dinitrogen is displaced.

$$FeH_2(N_2)(PEtPh_2)_3 + CO \longrightarrow FeH_2CO(PEtPh_2)_3 + N_2 \quad (47)$$

Bancroft *et al.* [102,104] prepared *trans*-hydridodinitrogendi[bis(diethylphosphino)ethane]iron(II) by the displacement of a chloride ion by molecular nitrogen from an acetone solution of *trans*-hydridochlorodi[bis(diethylphosphino)ethane]iron(II) in the presence of sodium tetraphenylborate. The complex was obtained as orange needles that are stable in dry air and melt with darkening and decomposition at 135°–145°C. It was characterized by its sharp infrared band at 2090 cm^{-1}, assigned to coordinated nitrogen, and a high-field quintet in its proton NMR spectrum with a J_{PH} value of 49 Hz.

An analogous iron(II)–dinitrogen complex with bis(dimethylphosphino)-ethane has also been reported with ν (N—N) at 2093 cm^{-1}. These synthetic reactions are of special interest since they involve the displacement of a fairly strongly bound chloride ion by molecular nitrogen under mild conditions. The complexes are very reactive in solution, nitrogen being readily lost in an atmosphere of hydrogen or argon and easily replaced by carbon monoxide.

Silverthorn [172] has reported the synthesis of a binuclear μ-dinitrogen compound of iron(II), $\{[\pi\text{-}C_5H_5Fe(dmpe)]_2N_2\}(BF_4)_2$, obtained as an orange crystalline material by the interaction of $\pi\text{-}C_5H_5Fe(dmpe)$, where dmpe = $(CH_3)_2PCH_2CH_2P(CH_3)_2$, in acetone with thallous fluoroborate, $TlBF_4$ at 0°C under nitrogen. The compound does not show an absorption in the infrared but the Raman spectrum of the solid has a very intense band at 2054 cm^{-1} assigned to a symmetric Fe—N=N—Fe stretching frequency. The complex reacts readily with carbon monoxide in acetone solution and with lithium aluminum hydride in tetrahydrofuran to yield the complexes $[\pi\text{-}C_5H_5Fe(dmpe)CO]^+BF_4^-$ and $\pi\text{-}C_5H_5Fe(dmpe)H$, respectively.

G. Rhenium(I)–Dinitrogen Complexes

The reaction of benzoylazo complexes of rhenium(III) of the general formula $Re(L)Cl_2(C_6H_5CON_2)$ (where L = a tertiary monophosphine or half-tertiary diphosphine) with an excess of the phosphine in methanol resulted in the formation of paramagnetic complexes of rhenium(I) with the composition $[ReN_2Cl(L)_4]$ [173]. The complexes are pale yellow and are thermally quite stable. From the proton magnetic resonance spectra it has been concluded that chloride ion and dinitrogen occupy mutually *trans* positions in the complexes. In chloroform solution the nitrogen complexes undergo oxidative halogen addition to rhenium(III) or rhenium(IV) complexes, depending on the phosphine, with quantitative evolution of molecular nitrogen. It is of interest to note that steric factors play an important role in the formation of these complexes and only the less sterically hindered phosphines carry the reaction to completion. In line with the other complexes of molecular nitrogen, it has not been possible to reduce the coordinated nitrogen to ammonia. Direct reaction of nitrogen with trichlorotris(dimethylphenylphosphine)rhenium(III) $ReCl_3(PMe_2Ph)_3$, in dry tetrahydrofuran and amalgated zinc also gave the compound $ReCl(N_2)(PMe_2Ph)_4$, **40**, in good yield.

Chlorodinitrogentetrakis(dimethylphenylphosphine)rhenium(I), **40**, reacts with a variety of electron acceptor molecules in tetrahydrofuran to form μ-dinitrogen polynuclear complexes [174]. These polynuclear complexes are formed by the attack of acceptor metal compounds on the dinitrogen group of the complex $ReCl(N_2)(PMe_2Ph)_4$, **40**, by reaction (48).

The N—N stretching frequency of $ReCl(N_2)(PMe_2Ph)_4$ is reduced by about 200–300 cm^{-1} by combination with a transition metal with empty or nearly empty *d* orbitals [titanium(III), zirconium(III), niobium(V), tantalum(V)] and by about 30–70 cm^{-1} in the case of d^2–d^7 metal ions, chromium(III), rhenium(V), iron(II), and cobalt(II). The dark violet complex obtained by the reaction of **40** with the THF solvate of chromium trichloride, $CrCl_3 \cdot (THF)_3$,

$$\underset{\mathbf{40}}{ReCl(N_2)(PMe_2Ph)_4} + MX \longrightarrow (PMe_2Ph)_4ClReN_2MX \quad (48)$$

[MX = $TiCl_2(THF)_3$, $TiCl_3$, $ZrCl_4$, $HfCl_4$, $NbCl_5$, $TaCl_5$, $MoCl_4(THF)_2$, $MoCl_4(Et_2O)_2$, $ReOCl_3(PPh_3)_2$, $FeCl_2(THF)$, $CoCl_2(THF)$, $AlCl_3$]

in dichloromethane has been fully characterized [175] and has the composition $[(PMe_2Ph)_4ClReN_2CrCl_3(THF)_2]$, **41**. This complex has an intense N—N stretching band at 1875 cm^{-1} with $^{15}N_2$ at 1815 cm^{-1} and a magnetic moment corresponding to three unpaired spins. In this complex, rhenium(I) is in a low-spin d^6 configuration, so that the paramagnetism corresponds to that of the d^3 chromium(III) moiety. Exposure to air in solution or in the solid state results in oxidative rupture of the compound to $[ReCl(N_2)(PMe_2Ph)_4]^+$ and $CrCl_3(THF)_3$. The oxygen probably attacks the chromium atom and the μ-N_2 bridge in the Re⋯N⋯N⋯Cr group and provides a pathway for rapid electron transfer from rhenium(I) to molecular oxygen. The μ-dinitrogen complex **41** has a strong charge-transfer band at 537 nm characteristic of a Re⋯N⋯N⋯Cr bridge and may be assigned to the charge transfer from rhenium(I) to chromium(III) states [175].

Oxidation of chlorodinitrogendi[bis(diphenylphosphino)ethane]rhenium(I), $[ReCl(N_2)(Ph_2PCH_2CH_2PPh_2)_2]$, **42**, with silver(I), copper(II), or iron(III) salts or halogens results in conversion of the compound to a purple cationic species, $[ReCl(N_2)(Ph_2PCH_2CH_2PPh_2)]^+$, characterized by an N—N stretching frequency at 2060 cm^{-1}, which is about 60 cm^{-1} higher than that of the parent compound [149]. The increase in the N—N stretching frequency with an increase in the oxidation state of the metal ion is consistent with the expected decrease in electron transfer from the metal to π orbitals or the dinitrogen ligand. There was no indication of the formation of a nitrogen-bridged polynuclear species in the reaction of **42** with silver(I), copper(II), or iron(III) salts.

The structure of $ReCl(N_2)(PMe_2Ph)_4$, **40**, determined by Davis and Ibers [176], and some of its structural parameters are indicated in Fig. 16. The phosphorus atoms P_1 and P_3, and P_2 and P_4 form two pairs that lie, respectively, above and below a plane passing through rhenium at right angles to the bonds of the remaining two groups. The N—N distance of 1.055 Å is somewhat shorter than expected from the low N—N stretching frequency of 1925 cm^{-1}. This is the shortest N—N distance recorded for any dinitrogen

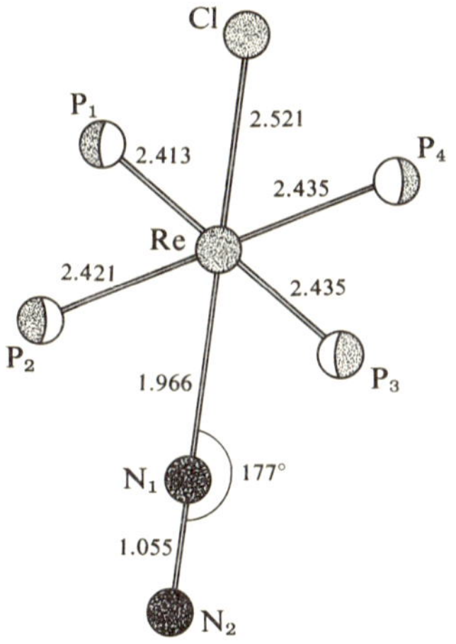

40

Fig. 16. Structure of chlorodinitrogentetrakis(dimethylphenylphosphine)rhenium. According to Davis and Ibers [176].

complex so far, but the uncertainty in the N—N distance is very large and firm conclusions cannot yet be made.

Chatt *et al.* [177] prepared a mixed carbonyl dinitrogen complex of the type $[ReCl(CO)_2(N_2)(PPh_3)_2]$, **44**, by reacting carbon monoxide with the benzoylazo complex $ReCl_2(PPh_3)_2(C_6H_5CON_2)$, **43**, in refluxing benzene–methanol, as in Eq. (49). The product is a yellow, diamagnetic air-stable solid

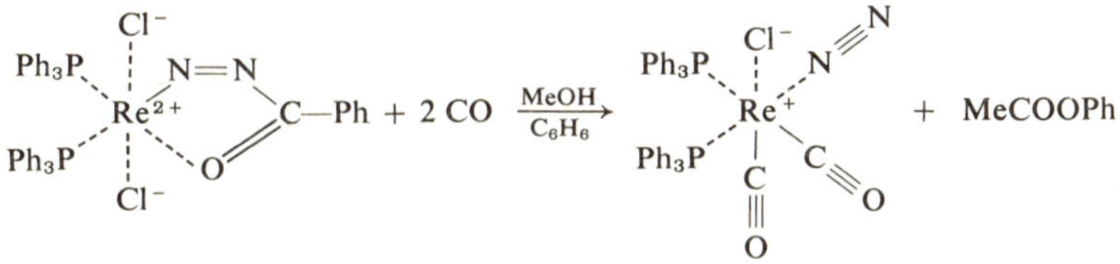

(49)

43 **44**

with an N—N stretching frequency of 1930 cm^{-1} and carbonyl stretching frequencies at 2105 and 2020 cm^{-1}. The nitrogen and carbonyl vibrations are probably coupled. Oxidation of the compound with chlorine gives 2 moles of carbon monoxide and 1 mole of nitrogen. Reaction of **43** with bis(diphenylphosphino)ethane, $Ph_2PCH_2CH_2PPh_2$, gave $[ReCl(CO)(Ph_2PCH_2CH_2PPh_2)_2]$, **44**, and two equivalents of gaseous products, consisting of nitrogen and carbon monoxide.

The reaction of *cis*-$[Re(CO)_4Br(PMe_2Ph)]$ and $[Re(CO)_3Cl(PMe_2Ph)_2]$ with hydrazine in benzene solution resulted [178] in the formation of air-sensitive yellowish brown dinitrogen complexes of rhenium(I), $[Re(CO)_3\cdot$

$(NH_2)(N_2)(PMe_2Ph)_2]$, **45**, and $[Re(CO)_2(NH_2)N_2(PMe_2Ph)_2]$, **46**. The presence of coordinated dinitrogen in these complexes has been confirmed by sharp peaks at 2220 and 2225 cm^{-1}, respectively, the highest frequencies observed thus far for molecular nitrogen complexes. The arsenic analog of **45** is also an air-sensitive solid with ν(N—N) at 2240 cm^{-1}. The presence of a coordinated σ-NH_2 group in these complexes has been confirmed by a shift in the ν(N—H) band on deuteration and also by ^1H NMR spectra of the complexes. Compounds **44**, **45**, and **46** are the first examples of dinitrogen complexes containing both carbonyl and dinitrogen groups coordinated to the same metal ion. The high value of the N—N stretching vibration frequency in these complexes may be due to the electron-withdrawing effect or competition of π-bonding by the carbon monoxide. Another point of considerable interest is the conversion (disproportionation) of the hydrazine moiety to coordinated dinitrogen and an amido group. The metal–nitrogen stretching vibration was observed at 520–530 cm^{-1} in these complexes.

Sellman [179] has reported a rhenium(I)–dinitrogen complex having the composition π-$C_5H_5Re(CO)_2N_2$, **48**. It was obtained by the copper(II)-catalyzed oxidation of the hydrazine complex π-$C_5H_5Re(CO)_2(N_2H_4)$, **47**, with hydrogen peroxide in tetrahydrofuran at −20°C, as indicated by Eq. (50).

$$\underset{\mathbf{47}}{\pi\text{-}C_5H_5Re(CO)_2N_2H_4} + 2\,H_2O_2 \xrightarrow[\text{THF}]{Cu^{2+}/-20^\circ C} \underset{\mathbf{48}}{\pi\text{-}C_5H_5Re(CO)_2N_2} \tag{50}$$

The complex π-$C_5H_5Re(CO)_2(N_2)H_4$, **47**, was obtained from π-$C_5H_5Re(CO)_3$, **49**, by reactions (51) and (52). The infrared spectrum of the dinitrogen com-

$$\underset{\mathbf{49}}{\pi\text{-}C_5H_5Re(CO)_3} \xrightarrow[\text{THF}/10^\circ C]{H_n} \pi\text{-}C_5H_5Re(CO)_2THF + CO \tag{51}$$

$$\pi\text{-}C_5H_5Re(CO)_2THF + N_2H_4 \xrightarrow{\text{THF}/0^\circ C} \pi\text{-}C_5H_5Re(CO)_2N_2H_4 + THF \tag{52}$$

plex, **48**, has an absorption band at 2141 cm^{-1}, assigned to the ν_{N_2} stretching frequency. There are two additional infrared bands at 1970 and 1915 cm^{-1}, corresponding to ν_{CO} stretching frequencies of the two carbonyl groups.

H. Molybdenum(0)–Dinitrogen Complexes

An orange molybdenum(0)–dinitrogen complex of the composition $Mo(N_2)(PPh_3)_2(PhMe)$ has been obtained by passing nitrogen through a toluene solution containing molybdenum(III) acetylacetonate, a tenfold excess of triphenylphosphine, and triethylaluminum or triisobutylaluminum for several days [180]. The complex is moderately air-stable and sparingly soluble in toluene, acetone, and tetrahydrofuran. Thermal decomposition at 120°C gave nitrogen, toluene, and traces of hydrogen. The IR spectrum of the complex shows a band at 2005 cm^{-1} assigned to coordinated N—N stretch [180,181].

Reduction of molybdenum(III) tris(acetylacetonate) in the presence of triethylaluminum and bidentate phosphines of the general formula $Ph_2P(CH_2)_n \cdot PPh_2$ ($n = 1, 2, 3$) in toluene solution under nitrogen gave [181] a series of dinitrogen complexes of the composition *trans*-$Mo(N_2)_2[Ph_2P(CH_2)_nPPh_2]_2$. Bidentate phosphines with $n > 3$ and 1,2-bis(dimethylphosphino)ethane failed to give analogous bis(dinitrogen) complexes of molybdenum(0). The complex $Mo(N_2)_2(Ph_2PCH_2PPh_2)_2$, **50**, could not be isolated as a crystalline solid. *trans*-$Mo(N_2)_2(Ph_2PCH_2CH_2PPh_2)_2$, **51**, is an orange-yellow air-stable crystalline solid soluble in toluene but insoluble in petroleum ether. *trans*-$Mo(N_2)_2[Ph_2P(CH_2)_3PPh_2]_2$, **52**, is an orange solid, which is less soluble in toluene than complex **50**.

Complexes **50**, **51**, and **52** each have a weak absorption near 2010 cm^{-1} and a very strong absorption in the range 1925–1995 cm^{-1}. The weak band at 2010 cm^{-1} and the strong bonds (1925–1995 cm^{-1}) have been assigned to an N—N symmetric stretching (A_{1g}, Raman-active) and N—N asymmetric stretching (A_{2u}, infrared-active) modes, as expected for a *trans*-bis(dinitrogen) structure. The strong vibrations around 1925–1995 cm^{-1} shift to lower frequencies with an increase in the methylene chain length of the ditertiary phosphine. Thus for compounds **50**, **51**, and **52** these vibrations were observed at 1995, 1970, and 1925 cm^{-1}, respectively.

Complex **51** reacts reversibly with hydrogen in toluene or tetrahydrofuran solution to give *trans*-$Mo(H)_2(Ph_2PCH_2CH_2PPh_2)_2$, **53**, which is a yellow crystalline solid fairly stable in air ($\nu_{M\text{-}H}$ 1745 cm^{-1}). In toluene solution the complex obtained was a μ-(diphos), where diphos = $Ph_2PCH_2CH_2PPh_2$, hydrido complex of the composition [*trans*-$Mo(H)_2$(diphos)]$_2$-μ-(diphos), **54**, in which two trigonal bipyramids [*trans*-$Mo(H)_2$(diphos)] are bridged by the diphos ligand.

When complex **51** was allowed to react with $CoH_3(PPh_3)_3$ in benzene or toluene, the hydrido complexes **53** or **54** were obtained, accompanied by the formation of $CoH(N_2)(PPh_3)_3$, **26**, in accordance with Eq. (53). Complex **26** was detected only in solution and could not be isolated from the reaction mixture.

$$\underset{\mathbf{51}}{trans\text{-}Mo(N_2)_2(diphos)_2} + CoH_3(PPh_3)_3 \xrightarrow{C_6H_6} \underset{\mathbf{53}}{trans\text{-}MoH_2(diphos)_2} + \underset{\mathbf{26}}{CoH(N_2)(PPh_3)_3} + N_2$$

$$\xrightarrow{C_6H_5CH_3} \underset{\mathbf{54}}{[trans\text{-}Mo(H)_2(diphos)]_2\text{-}\mu\text{-}diphos} + \underset{\mathbf{26}}{CoH(N_2)(PPh_3)_3} + N_2 \quad (53)$$

I. Molybdenum(I)–Dinitrogen Complexes

A yellow molybdenum(I)–dinitrogen complex of the composition $[MoCl(N_2)(PPh_2CH_2CH_2PPh_2)_2]$, **55**, has been obtained by Atkinson *et al.* [182] by the reduction of $[MoOCl_2(PPh_2CH_2CH_2PPh_2)_2(THF)]$ with zinc dust (Mo:Zn ratio 1:2) in tetrahydrofuran for 20 hours in the presence of molecular nitrogen. A longer reduction time or excess of zinc gave the orange molybdenum(0) complex, *trans*-$Mo(N_2)_2(PPh_2CH_2CH_2PPh_2)_2$, **51**, previously reported by Hidai *et al.* [183]. The yellow complex is air-stable and in dichloromethane solution has an intense N—N absorption band at 1975 cm^{-1} (Nujol 1970 cm^{-1}). The compound reacts with $[MoCl_4(THF)_2]$ in a 1:1 ratio to give a blue solid of undetermined composition with a greatly reduced N—N stretching frequency of 1770 cm^{-1} (in Nujol).

J. Tungsten(0)–Dinitrogen Complexes

Bell *et al.* [184,185] have synthesized tungsten(0) complexes of the compositions *cis*-$[W(N_2)_2(PMe_2Ph)_4]$, **56**, and *trans*-$[W(N_2)_2(Ph_2PCH_2CH_2PPh_2)_2]$, **57**, by the reduction of tungsten(IV) complexes of the type $[WCl_4(PR_3)_2]$ with 2% sodium amalgam in anhydrous tetrahydrofuran in the presence of an excess of the free phosphine ligand. The complexes *cis*-$[W(N_2)_2(PMe_2Ph)_4]$, and *trans*-$[W(N_2)_2(Ph_2PCH_2CH_2PPh_2]$ (**57**) are yellow and orange solids, respectively, with N—N stretching absorption bands at 1931 and 1998 cm^{-1} for the former and at 1953 cm^{-1} for the latter. The compounds deteriorate in the presence of air after a few days at room temperature.

K. Nickel(0)–Dinitrogen Complexes

Jolly and Jonas [105] synthesized a μ-dinitrogen complex of nickel(0) of the composition $N_2\{Ni[P(C_6H_{11})_3]_2\}_2$ by the reduction of nickelbis(acetylacetonate) with trimethylaluminum in the presence of tricyclohexylphosphine and nitrogen, as shown in Eq. (54). The complex **58** was obtained in a 40%

$$2\ Ni(acac)_2 + 4\ (C_6H_{11})_3P + 4\ Al(CH_3)_3 + N_2 \longrightarrow \underset{\mathbf{58}}{N_2\{Ni[P(C_6H_{11})_3]_2\}_2} + 2\ C_2H_6 + 4\ (CH_3)_2Al(acac) \quad (54)$$

yield as dark red crystals. It dissociates in benzene to give a monomer having an IR absorption at 2028 cm^{-1}, assigned to the N—N stretching vibration. Nitrogen is lost when argon is passed through a solution of the complex, the color of the solution changing from red to yellow. The change is reversible and the red color is readily recovered when nitrogen is passed through the yellow solution.

Jolly *et al.* [186] have determined the structure of $\{[(C_6H_{11})_3P]_2Ni\}_2N_2$,

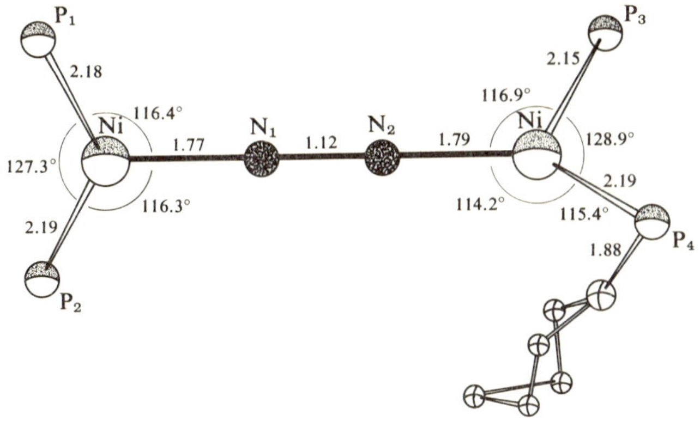

Fig. 17. The structure of $\{[(C_6H_{11})_3P]_2Ni\}_2N_2$. (Only one cyclohexane ring is shown for clarity.) After Jolly *et al.* [186].

58, by X-ray crystallography. The structural parameters are indicated in Fig. 17. The molecule contains a linear Ni—N≡N—Ni system in which the nitrogen molecule is enclosed in a cage formed by four of the cyclohexyl rings. All the cyclohexyl rings are in a chair conformation. The N—N bond distance is 1.12 Å, which compares well with distances in other μ-N_2 complexes (1.12–1.16 Å), indicating that there is no extra weakening of the N≡N bond due to the μ coordination. The Ni—N_1 and Ni—N_2 bond distances are identical (reported as 1.77 and 1.79 Å, respectively). The nickel atoms have an exactly planar trigonal–bipyramidal geometry with slightly distorted bonding angles to the phosphorus atoms due to steric repulsion of the bulky $P(C_6H_{11})_3$ ligands. The phosphorus atoms of the ligands form distorted tetrahedra with average C—P—Ni angles of 114°–115°, and average P—C distances of 1.88 Å.

L. Manganese(II)–Dinitrogen Complexes

Johnson and Beveridge [187] reported the formation of a 1:1 adduct of dinitrogen with the *N,N′*-disalicylal-1,3-propanediiminemanganese(II) ion in benzene solution as well as in the solid state. Molecular nitrogen is bonded reversibly to the manganese(II) chelate and is given off on boiling the benzene solution. After correcting for the solubility of molecular nitrogen in benzene, the uptake of nitrogen by the manganese(II) complex amounted to the formation of a 1:1 adduct. The infrared spectrum of the dinitrogen complex did not show the infrared band characteristic of the N—N stretch (about 2000 cm^{-1}). This indicates that the complex is different from all the other nitrogen complexes prepared thus far, in which the M—N—N bonding is considered to be linear. It was tentatively proposed by Johnson and Beveridge [187] that in this complex nitrogen is coordinated to manganese in a perpendicular manner (infrared-inactive complex of C_{2v} symmetry). This proposal may be tested by examining the Raman spectrum of the complex.

M. Manganese(I)–Dinitrogen Complexes

Taqui Khan and Anwaruddin have recently obtained [188] the first stable manganese complex by the reduction of dichlorobis[1,2-bis(diphenylphosphino)ethane]manganese(II) in benzene–alcohol solution with sodium borohydride in the presence of molecular hydrogen and then passing nitrogen through the solution. It was suggested that $Mn(H)(Cl)(diphos)_2$ might have been formed as an intermediate, and that the nitrogen complex is formed by displacement of the coordinated hydride. The nitrogen complex has been obtained as a buff-colored solid and analyzed for $Mn(N_2)Cl(diphos)_2$ [diphos = 1,2-bis(diphenylphosphino)ethane]. The presence of coordinated dinitrogen in the complex was confirmed by the presence of an IR band at 2100 cm^{-1}. The compound is moderately stable in dry air, is sparingly soluble in benzene, alcohol, and chloroform, and decomposes without melting.

A manganese(I)–dinitrogen complex of the composition π-$C_5H_5Mn \cdot (CO)_2N_2$, **60**, has been obtained by Sellman [189] in 40% yield by the oxidation of π-$C_5H_5Mn(CO)_2N_2H_4$, **59**, with H_2O_2/Cu^{2+} in tetrahydrofuran at $-40°C$, as indicated in Eq. (55). After removal of the solvent and sublimation,

$$\underset{\mathbf{59}}{\pi\text{-}C_5H_5Mn(CO)_2N_2H_4} + 2\,H_2O_2 \xrightarrow[\text{THF}/-40°C]{Cu^{2+}} \underset{\mathbf{60}}{\pi\text{-}C_5H_5Mn(CO)_2N_2} + 4\,H_2O \quad (55)$$

reddish brown air-stable crystals of π-$C_5H_5Mn(CO)_2N_2$, **60**, were obtained. The compound melts at 60°C with evolution of gas. The infrared spectrum of the compound in *n*-hexane has an intense absorption band at 2169 (ν_{N-N}) and two additional intense bands at 1980 and 1923 cm^{-1}, assigned to the stretching frequencies of the two carbonyl groups. The PMR spectrum of the compound in acetone-d_6 shows a sharp singlet at τ 5.16 due to C_5H_5 protons. The singlet is shifted upfield by 12 Hz with respect to the signal of $C_5H_5Mn(CO)_3$. This shift has been attributed [190] to the greater σ-donor capacity of dinitrogen in π-$C_5H_5Mn(CO)_2N_2$ as compared to CO in π-$C_5H_5 \cdot Mn(CO)_3$.

N. Titanium(III)–Dinitrogen Complexes

Tueben and de Liefde Meijer [191] reported a well-characterized series of Ti(III)–dinitrogen complexes of the composition $(Cp_2TiR)_2N_2$, **61**, (R = phenyl, benzyl, 2-methylphenyl, 3-methylphenyl, 4-methylphenyl, 2,6-dimethylphenyl, and pentafluorophenyl) which was obtained by the reaction of bis(cyclopentadienyl) aryl or aralkyl titanium complexes, Cp_2TiR, in toluene with molecular nitrogen at $-20°C$ [Eq. (56)]. The compounds Cp_2TiR were synthesized with the appropriate Grignard reagent, as indicated by Eq. (57). When alkyl Grignard reagents were employed, the products,

$$2\,Cp_2TiR + N_2 \xrightarrow[-20°C]{\text{toluene}} \underset{\mathbf{61}}{(Cp_2TiR)_2N_2} \tag{56}$$

$$Cp_2TiCl + RMgBr \xrightarrow[-20°C]{\text{ether}} Cp_2TiR + MgClBr \tag{57}$$

Cp_2TiR (R = alkyl) could not be isolated. The reaction of molecular nitrogen with a mixture of Cp_2TiCl and alkylmagnesium bromide at low temperatures, however, gave the deep blue nitrogen complexes *in situ*.

The dinitrogen complexes $(Cp_2TiR)_2N_2$ (R = aryl) were isolated as deep blue diamagnetic crystalline compounds that were characterized on the basis of their elemental analysis, spectra, and physical properties. Toluene solutions of the dinitrogen complexes exhibited intense absorptions at 590–600 nm ($\varepsilon = 4 \times 10^4$ liter mole^{-1} cm^{-1}). Linear μ coordination of dinitrogen to two titanium atoms was proposed on the basis of the absence of the N—N stretching frequency in the infrared spectra of the complexes.

Reactions of the dinitrogen complexes, **61**, with hydrogen chloride and bromide yielded Cp_2TiCl and Cp_2TiBr_2, respectively, with the liberation of 1 mole of dinitrogen per mole of complex. The compounds with R = phenyl and benzyl are more stable than the others and decomposed above 70°C to give a quantitative yield of nitrogen together with dark unidentified products. Reduction of the complexes with sodium naphthalide in tetrahydrofuran gave hydrazine and ammonia, the yield depending on the ratio of the complex to sodium naphthalide. Other dinitrogen complexes of titanium(III) that have not been well characterized [192–194] are listed in Table III (entries 95, 99, and 100).

O. Platinum(III)– and Platinum(IV)–Dinitrogen Complexes

A μ-dinitrogen platinum(III) compound of the composition $K_2\{[Pt(OH)_2 \cdot (NO_2)_2(NH_3)]_2N_2\}$, **62**, was obtained [195] as fine pale yellow crystals from the reaction between aqueous $K_2[Pt(NO_2)_4]$ and ammonium sulfate at 60°C and pH 1–2. Acidification with perchloric acid in the presence of ammonium perchlorate gave $K_2\{[Pt(ClO_4)(NO_2)_3(NH_3)]_2N_2\}$, **63**. Compounds **62** and **63** are diamagnetic and extremely unstable in air. The symmetrical ν_{N_2} frequencies of the Raman spectra of these complexes are 2034 and 2065 cm^{-1}, respectively. The compounds are not well characterized and the oxidation state (III) seems quite unusual for platinum.

King [196] has reported a platinum(IV)–dinitrogen complex, $Pt(CH_3)_3 \cdot$

$(N_2)[H_2B(C_3H_3N_2)_2]$, **64**, that was obtained as colorless crystals by the interaction of iodotrimethylplatinum(IV) and the potassium salt of bis(pyrazolyl)-dihydridoborate in ethanol–tetrahydrofuran under nitrogen. The infrared spectrum of this compound has a ν_{N_2} stretching frequency at 2054 cm^{-1}.

$(CH_3)_3PtI + K[H_2B(C_3H_3N_2)_2] + N_2$

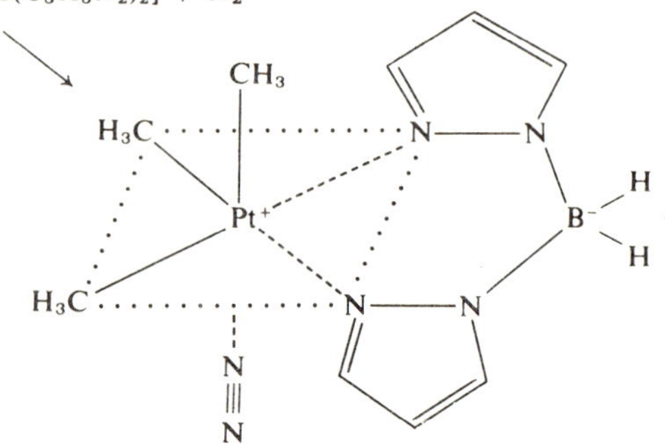

+ KI (58)

64

P. Summary of Dinitrogen Complexes

A summary of the properties of dinitrogen complexes that have been synthesized up to the present time is given in Table III.

Q. Metal–Dinitrogen Surface Complexes

Because of the relevance of the surface complexes of dinitrogen to the complex species that are formed in solution, a brief account of the work done along these lines will be presented without considering in detail the mechanism of chemisorption of molecular nitrogen on metal surfaces. A review of the nature of chemisorption of dinitrogen on metal surfaces has been given by Emmett [197].

An interesting observation indicating the formation of a metal–dinitrogen surface complex is the report by Eischens and Jackson [198] of an IR band at 2246 cm^{-1} for molecular nitrogen absorbed on metallic nickel. Since the vibrational spectrum of molecular nitrogen contains only a single Raman-active band at 3300 cm^{-1}, the IR band at 2246 cm^{-1} was attributed to molecular nitrogen bound in an end-on manner to the metal surface. The IR band at 2246 cm^{-1} was further confirmed by substitution of isotopic $^{15}N_2$, which resulted in a shift towards lower frequency, 2128 cm^{-1}. The Ni—N═N surface complex was found to be sensitive to hydrogen [198,199], the frequency of the band due to N—N stretch being shifted to higher values, 2260 and 2275 cm^{-1}. This shift was interpreted as being due to a change in bonding

TABLE III
PROPERTIES OF METAL–DINITROGEN COMPLEXES

	Complex	Color	Solvent	Stability	N—N IR bands[a] (cm^{-1})	References
1	$[Ru(NH_3)_5N_2]^{2+}$	Yellow	Water	Stable in the absence of water and when pure	2105–2170	1–5 100 111 121
2	$[(NH_3)_5RuN_2Ru(NH_3)_5]^{4+}$	Yellow	Water	Stable in the absence of water and when pure	Broad band at 2060	4 126
3	$[(H_2O)_5RuN_2Ru(H_2O)_5]^{4+}$	—	Water	Stable in the absence of water and when pure	2080	121
4	$[(trienBrRu)_2N_2]^{2+}$	—	Water	Stable in the absence of water and when pure	2060	121
5	$[(NH_3)_5RuN_2Os(NH_3)_5]^{4+}$	—	Water	Stable in the absence of water	2047	158
6	$[(NH_3)_5RuN_2Os(NH_3)_5]^{5+}$	—	Water	Stable in the absence of water	2034	158
7	$[(NH_3)_5RuN_2Os(NH_3)_4CO]^{4+}$	—	Water	Stable in the absence of water	2105	121
8	$[Os(NH_3)_4(N_2)_2[(Ru(NH_3)_5]_2]^{6+}$	—	Water	Stable in the absence of water	2097	121
9	$[Ru(H_2O)_5N_2]^{2+}$	—	Water	Stable in the absence of water	2125	121
10	$[Ru(en)_2N_2(H_2O)]^{2+}$	Yellow	Water	Stable in the absence of water	2130	115
11	*cis*-$[Ru(en)_2(N_2)_2]^{2+}$	Yellow	Water	Very unstable above 0°C	2190 2220	115

TABLE III (*continued*)

	Complex	Color	Solvent	Stability	N—N IR bands[a] (cm^{-1})	References
12	*cis*-$[Ru(trien)(N_2)(N_3)]^+$	Yellow	Water	Very unstable above 0°C	2120	108
13	*cis*-$[Ru(en)_2(N_2)(N_3)]^+$	Yellow	Water	Very unstable above 0°C	2130	108
14	$Ru(N_2)Cl_2(H_2O)_2(C_4H_8O)$	Yellow	Water	Stable up to 50°C	2090	124
15	$Ru(N_2)Cl_2(NH_3)_3$	Yellow	Water	Stable up to 150°C	2080	124
16	$Ru(N_2)Cl_2(en)_2$	Yellow	Water	Stable up to 150°C	2085	124
17	$[Ru(NH_3)_4(N_2)_2]^{2+}$	Yellow	Water	Unstable at room temperature	2220 2185	145
18	Ru(III)–N_2 carrier in presence of triphenylphosphine	Red solution	C_6H_6 or THF	Stable only in solution	2143	138
19	$[RuCl(N_2)(das)_2]^+$	Colorless	C_6H_6 or THF	Thermally stable up to 120°C (stable in air)	2130	109 110
20	$RuClH(N_2)(AsPh_3)_3$	Brown	C_6H_6	Stable in air	2050	143
21	$RuCl_2(N_2)(O_2)(AsMePh_2)_2$	Brown	C_6H_6	Stable in air	2070	144
22	$(PPh_3)_3Ru(N_2)H_2$	Colorless	C_6H_6	Air-sensitive	2147	139 140 141
23	$[RuH(N_2)(diphos)_2]^+$	Colorless	$CHCl_3$	Air-sensitive	2163–2175	104
24	$[Os(NH_3)_5N_2]^{2+}$	Yellow	Water	Stable in air and solution	2010–2060	7
25	$[Os(NH_3)_4(CO)N_2]^{2+}$	—	—	—	2180	121
26	$[Os(NH_3)_5N_2]^{3+}$	Yellow	Water	Unstable, decomposes to $[Os(NH_3)_5H_2O]^{3+}$ and N_2	2140	149 150
27	*cis*-$[Os(NH_3)_4(N_2)_2]^{2+}$	Deep yellow	Water	Stable below 50°C	2120	156
28	$OsCl_2(N_2)(PMe_2Ph)_3$	Colorless	C_6H_6	Stable in the solid state	2082	103 153

29	$OsBr_2(N_2)(PMe_2Ph)_3$	Colorless	C_6H_6	Stable in the solid state	2090	103, 153
30	$OsCl_2(N_2)(PEt_2Ph)_3$	Colorless	C_6H_6	Stable in the solid state	2063	103, 153
31	$OsCl_2(^{15}N_2)(PEt_2Ph)_3$	Faintly pink	C_6H_6	Stable in the solid state	1198–2010	103, 153
32	$OsBr_2(N_2)(PEt_2Ph)_3$	Colorless	C_6H_6	Stable in the solid state	2070–2087	103, 153
33	$OsCl_2(N_2)(PPr_2Ph)_3$	Colorless	C_6H_6	Stable in the solid state	2062	103, 153
34	$OsCl_2(N_2)(PBu_2Ph)_3$	Colorless	C_6H_6	Stable in the solid state	2060–2080	103, 153
35	$[OsH(N_2)(diphos)_2]^+$	Colorless	$CHCl_3$	Stable in the solid state	2136	104
36	$OsCl_2(N_2)(PEtPh_2)_3$	Colorless	C_6H_6	Stable in the solid state	2090	103
37	$OsCl_2(N_2)(PMePh_2)_3$	Colorless	C_6H_6	Stable in the solid state	2092	153
38	$OsCl_2(N_2)(PEt_3)_3$	Faintly pink	C_6H_6	Stable in the solid state	2064–2069	103, 153
39	$OsCl_2(N_2)(AsMe_2Ph)_3$	Colorless	C_6H_6	Stable in the solid state	2063–2073	153
40	$OsCl_2(N_2)(AsEt_2Ph)_3$	Colorless	C_6H_6	Stable in the solid state	2063–2070	153
41	$[OsH(N_2)(Et_2PCH_2CH_2PEt_2)_2]^+$	Colorless	$CHCl_3$	Stable in the solid state	2136–2144	104
42	$OsHCl(N_2)(PMe_2Ph)_3$	Colorless	Toluene	Stable in the solid state	2059	154
43	$OsHBr(N_2)(PMe_2Ph)_3$	Colorless	Toluene	Stable in the solid state	2065	154
44	$OsHCl(N_2)(PEt_2Ph)_3$	Colorless	Toluene	Stable in the solid state	2050	154
45	$OsHCl(N_2)(PEtPh_2)_3$	Colorless	Toluene	Stable in the solid state	2070	154
46	$OsDCl(N_2)(PMe_2Ph)_3$	Colorless	Toluene	Stable in the solid state	2062	154
47	$[Co(N_2)(PEt_2Ph)_3]^-$	Black	Toluene	Stable in the absence of air	1875	170
48	$[Co(N_2)(PPh_3)_3]^-$	Red	Toluene	Stable in the absence of air	1845	170
49	$Co(N_2)(PPh_3)_3$	Yellow	Toluene ether	Stable in the absence of air, decomposes 100°–150°C	2088	11, 13, 159

TABLE III *(continued)*

	Complex	Color	Solvent	Stability	N—N IR bands[a] (cm^{-1})	References
50	$[(PPh_3)_3Co]_2N_2$	Dark red	Toluene ether	Stable in the absence of air decomposes 100°–150°C		170
51	$[(PEt_2Ph)_3Co]_2N_2$	Dark brown	Toluene ether	Stable in the absence of air decomposes 100°–150°C		170
52	$Co(N_2)(PEt_3)_3$	Yellow	Toluene ether	Stable in the absence of air, decomposes 100°–150°C	2059	159
53	$Co(H)(N_2)(PPh_3)_3$	Brown	Toluene	Stable in the absence of air, decomposes at 80°C	2080–2094	12 14 164
54	$Co(H)(N_2)(PMePh_2)_3$	Brown	Toluene	Stable in the absence of air	2077	106
55	$Co(H)(N_2)(PEtPh_2)_3$	Brown	Toluene	Stable in the absence of air	2060	106
56	$Co(H)(N_2)(PBuPh_2)_3$	Brown	Toluene	Stable in the absence of air	2070	106
57	$Co(H)(N_2)(PEt_2Ph)_3$	Brown	Toluene	Stable in the absence of air	2060	106
58	$Co(H)(N_2)(PBu_2Ph)_3$	Brown	Toluene	Stable in the absence of air	2060	106
59	$Co(H)(N_2)(PEt_3)_3$	Brown	Toluene	Stable in the absence of air	2033	190
60	$Co(H)(N_2)(PBu_3)_3$	Brown	Toluene	Stable in the absence of air	2032	190
61	$Rh(N_2)Cl(PPh_3)_2$	Yellow	Toluene	Stable in an inert atmosphere, moderately stable in air at low temperature	2105	6 141

62	$\{[Rh(OH)(NO_2)_3(NH_3)]_2N_2\}^{2-}$	Colorless	Water	Stable in air	R-2070	171
63	$Ir(N_2)Cl(PPh_3)_2$	Yellow	Benzene	Stable in air, decomposes in vacuum at 80°C	2105	9 10
64	$Ir(N_2)Br(PPh_3)_2$	Yellow	Benzene	Stable in air	2107	107
65	$Ir(N_2)N_3(PPh_3)_2$	Yellow	Benzene	Stable in air	2112	107
66	$Fe(H)_2(N_2)(PPh_3)_3$	Yellow	Benzene	Stable in dry air	2074	101
67	$Fe(H)_2(N_2)(PEtPh_2)_2$	Yellow	Benzene	Stable in dry air	2055–2060	101
68	$Fe(H)_2(N_2)(PEtPh_2)_2$	Yellow	Benzene	Stable in dry air	1989	101
69	$[Fe(H)(N_2)(Et_2PCH_2CH_2PEt_2)_2]^+$	Orange	Nitromethane	Stable in dry air	2090	102
70	$\{[\pi\text{-}C_5H_5Fe(Me_2PCH_2CH_2PMe_2)]_2N_2\}^{2+}$	Orange	Acetone	Stable in dry air	R-2054	172
71	$[FeH(N_2)(Me_2PCH_2CH_2PMe_2)_2]^+$	Orange	Nitromethane	Stable in dry air	2093	102
72	$ReCl(N_2)(Ph_2PCH_2CH_2PPh_2)_2$	Pale yellow	$CHCl_3$	Quite stable	1980	173
73	$ReCl(N_2)(PMe_2Ph)_4$	Pale yellow	$CHCl_3$	Quite stable	1925	173
74	$(PMe_2Ph)_4ClReN_2CrCl_3^-(THF)_2$	Violet (dark)	CH_2Cl_2	Quite stable	1875	175
75	$ReCl(N_2)(Ph_2PCH_2CH_2PPh_2)_2$	Pale yellow	$CHCl_3$	Quite stable	1980	173
76	$ReCl(N_2)(Ph_2PCH_2CH_2PPh_2)_2^+$	Pale yellow	$CHCl_3$	Quite stable	2060	149
77	$ReBr(N_2)(Ph_2PCH_2CH_2PPh_2)_2$	Pale yellow	$CHCl_3$	Quite stable	1980	173
78	$ReCl(N_2)(PMePh_2)_4$	Pale yellow	$CHCl_3$	Quite stable	1922	173
79	$Re(CO)_3(NH_2)(N_2)(PMePh_2)_2$	Pale yellow	Benzene	Stable in the absence of air	2220	178
80	$Re(CO)_3(NH_2)(N_2)(AsMePh_2)_2$	Pale yellow	Ethanol	Stable in the absence of air	2240	178
81	$ReCl(CO)_2(N_2)(PPh_3)_3$	Brown	Ethanol	Stable in the absence of air	1930	177
82	$\pi\text{-}C_5H_5Re(CO)_2(N_2)$	Yellow	C_6H_6, THF	Moderately air-stable	2141	179
83	*trans*-$[Re(CO)_2(NH_2)N_2(PMePh_2)_2$	Pale yellow-brown	Benzene	Stable in the absence of air	2225	178
84	$Mo(N_2)(PPh_3)_2(PPhMe_2)$	Orange-yellow	Sparingly soluble in C_6H_6, THF, or acetone	Moderately air-stable	2005	180, 181

TABLE III *(continued)*

	Complex	Color	Solvent	Stability	N—N IR bands[a] (cm^{-1})	References
85	*trans*-[$Mo(N_2)_2(Ph_2PCH_2PPh_2)_2$]	Orange	Sparingly soluble in C_6H_6, THF, or acetone	Moderately air-stable	2020 W 1995 S	181
86	*trans*-[$Mo(N_2)_2(Ph_2PCH_2CH_2PPh_2)_2$]	Orange-yellow	Sparingly soluble in C_6H_5, THF, or acetone	Moderately air-stable	2020 W 1970 V.S.	183
87	*trans*-[$Mo(N_2)_2(Ph_2PCH_2CH_2CH_2PPh_2)_2$]	Orange	Sparingly soluble in C_6H_6, THF, or acetone	Moderately air-stable	2020 W 1925 V.S.	181
88	$Mo(Cl)(N_2)(Ph_2PCH_2CH_2PPh_2)_2$	Yellow	$CHCl_3$	Air-stable	1975	182
89	*trans*-[$W(N_2)_2(Ph_2PCH_2CH_2PPh_2)_2$]	Orange-yellow	C_6H_6	Moderately air-stable	1953	185
90	*cis*-[$W(N_2)_2(PMe_2Ph)_4$]	Yellow	C_6H_6	Moderately air-stable	1931 & 1998	185
91	*N,N*-Disalicylaldehyde-1,3-propanediimine-Mn(II) nitrogen carrier	Yellow	Benzene	Stable in the solid state	—	187
92	$Mn(N_2)Cl(Ph_2PCH_2CH_2PPh_2)_2$	Buff colored	Sparingly soluble in alcohol, acetone C_6H_6	Moderately air-stable	2100	188
93	π-$C_5H_5Mn(CO)_2(N_2)$	Reddish brown	THF	Stable in air	2169	189

94	$N_2[Ni(P(C_6H_{11})_3)_2]$	Red	Soluble in C_6H_6 with dissociation to the monomer	Moderately air-stable	2028	105 186
95	$[(C_5H_5)_2Ti]_2N_2$	Dark blue	Toluene-hexane	Solid stable at room temperature, N_2 atmosphere		192
96	$\{[Pt(OH)_2(NO_2)_2(NH_3)]_2N_2\}^{2-}$	Yellowish		Very unstable	R-2034	195
97	$\{[Pt(ClO_4)(NO_2)_3(NH_3)]_2N_2\}^{2-}$	Colorless		Very unstable	R-2065	195
98	$Pt(CH_3)_2(N_2)[H_2B(C_3H_3N_2)_2]$	Colorless	Soluble in C_6H_6, THF	Stable	2054	196
99	$[(C_5(CH_3)_5)_2Ti]_2N_2$	Dark blue	Toluene hexane	Exists only in solution at −80°C		193
100	$[(C_5H_5)_2TiN_2]_2$	Dark blue	Benzene	Exists only in solution in the range 20°–25°C	1960	194
101	$[(C_5H_5)TiR]_2N_2$ R = phenyl, 2-methylphenyl, 3-methylphenyl, 4-methylphenyl, 2,6-dimethylphenyl, pentafluorophenyl, benzyl	Deep blue crystalline compounds	Toluene	Stable up to 70°C		191

[a] R, Raman; W, weak; S, strong; V.S., very strong.

(hydride formation) at the surface of nickel and a consequent weakening in the strength of the nickel–nitrogen bond.

Infrared absorption bands at 2260 and at 2230 cm^{-1} of dinitrogen chemisorbed on palladium and platinum, respectively, have been reported by Van Hardeveld and Van Montfoort [200]. It is of interest that for the Ni—N_2, Pd—N_2, and Pt—N_2 surface complexes the bands due to N—N stretch have high extinction coefficients and the integrated absorption intensity is approximately the same as for other stable dinitrogen complexes [119], indicating an end-on coordination of molecular nitrogen. The enthalpy of formation of the Ni—N_2 surface complex varies from -5 to -9.2 k cal/mole depending on the surface coverage [199]. Measurement of the heat of chemisorption in the case of Pd and Pt surface complexes indicates that ΔH varies in the range -2 and -20 kcal/mole [119]. The polarity of the metal–dinitrogen bond in the surface complexes depends on the nature of the metal atom and the relative contribution of the donor and acceptor metal–dinitrogen bonds [119]. In this respect the surface complexes of molecular nitrogen closely resemble the metal ion–dinitrogen complexes. The importance of surface complexes in the synthesis of ammonia by the Haber and Bosch process has been discussed in detail by Murray and Smith [116] and will not be considered here.

R. Binding of Dinitrogen by Nonmetallic Systems

One of the significant advances in the chemistry of the fixation of nitrogen is the recent finding [119] that nonmetallic diazo compounds do not differ significantly in bonding from metal–dinitrogen complexes. Borodko and Shilov [119] have shown that the force constants of the Ru—N_2 and N—N bonds in $[(NH_3)_5RuN_2]^+$ are close to the corresponding force constants of the N—N bonds in diazomethane, $CH_2{=}N^+{=}N:^-$. In some cases diazo compounds can be directly formed from molecular nitrogen. Thus Borodko *et al.* [201] have prepared labeled diazomethane by the radiolysis of diazomethane, ketene, diazirine, or phenylcyclopropene in the presence of $^{15}N_2$. Of the compounds investigated, those giving rise to unsubstituted carbene as an intermediate were the most reactive for the binding of molecular nitrogen [202].

An interesting observation in the binding of molecular nitrogen by organic species has been the formation of the dinitrogen adduct of the benzenesulfenium cation, $C_6H_5S^+N_2$ [203]. The compound $(C_6H_6SN_2)^+$ has not been obtained in pure form and therefore has not been completely characterized, but the presence of the dinitrogen in the compound was confirmed by IR, UV, and mass spectrometry. The compound undergoes the usual diazonium coupling reaction to form $C_6H_5SN{=}NAr$.

A spirocyclic bisdinitrogen "complex" of molecular nitrogen has been reported by Ellermann *et al.* [204]. Reaction of a 1:4 molar mixture of tetrakis[(phenylsodiophosphino)methyl]methane, **65**, and carbon disulfide in tetrahydrofuran gave the phosphorane thiolate **66**, which absorbs molecular nitrogen at room temperature to give the nitrogen complex **67** in quantitative yield. The structure of the nitrogen complex of the spiro compound is not yet

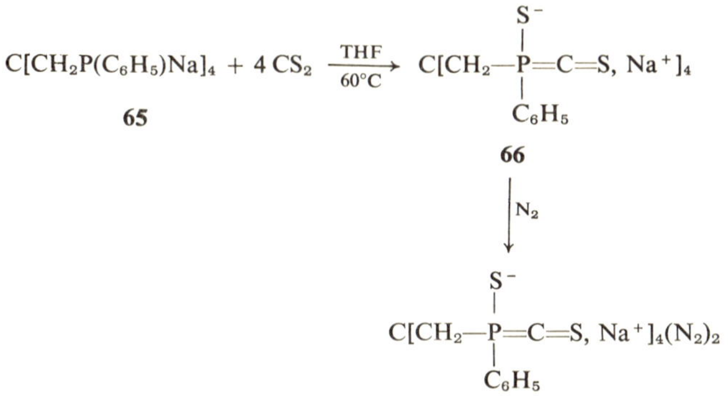

65 **66** **67** (59)

known. The IR spectrum of **67** gave a band at 2090 cm^{-1}, assigned to the N—N stretch of chemically bound nitrogen. The intermediate compound **66** also absorbs carbon monoxide to form a carbonyl compound that gives a characteristic carbonyl absorption at 1708 cm^{-1}. The formation of the benzene sulfenium cation and the spirocyclic compound with molecular nitrogen is interesting in that it suggests the possibility of nitrogen fixation on enzymes without metal coordination.

S. Structure and Bonding in Metal–Dinitrogen Complexes

Molecular nitrogen is a very stable molecule with a bond dissociation energy of 225 kcal/mole. The relative inertness of molecular nitrogen as compared to acetylene, with a C—C triple bond, may be due to the energy of dissociation of its first bond, which is about 130 kcal/mole, very much higher than the dissociation of the first bond in acetylene (53 kcal/mole). The second and third bond dissociation energies of nitrogen in diazo compounds and hydrazine, respectively, are much less than the first and correspond to 100 and 32 kcal/mole, respectively. The nonlinearity of bond dissociation energy with bond order in molecular nitrogen has been attributed by Pauling [205] to relatively lower repulsion energy of the nonbonding pairs of electrons in molecular nitrogen.

The molecular orbital diagram of molecular nitrogen is presented in Fig. 18. The distribution of electrons in the various orbitals in the order of

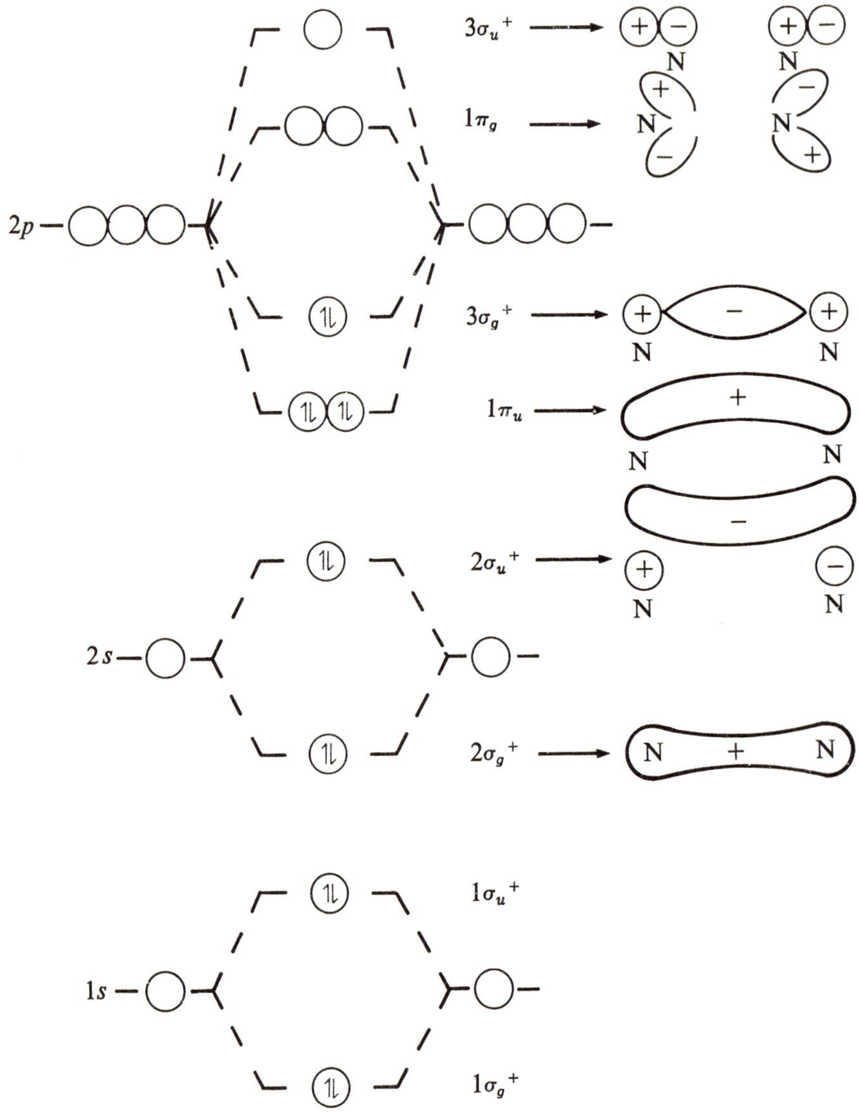

Fig. 18. Molecular orbitals for nitrogen.

increasing energy is $1\sigma_g^+$, $1\sigma_u^+$, $2\sigma_g^+$, $2\sigma_u^+$, $1\pi_u$, $3\sigma_g^+$ [206–208]. According to Chatt [209,210] the highest filled molecular orbital in nitrogen ($3\sigma_g^+$) is at a lower energy (−15.6 eV) than the donor π orbital of acetylene (−11.4 eV), and the lowest empty $1\pi_g$ orbital is at about −7 eV. The ionization potential of nitrogen corresponding to the removal of an electron from $3\sigma_g^+$, $1\pi_u$, and $2\sigma_u^+$ levels, as obtained from the photoelectron spectrum of gaseous nitrogen, corresponds, respectively to 367 kcal/mole, 390 kcal/mole, and 430 kcal/mole [211]. Thus the three levels $3\sigma_g^+$, $1\pi_u$, and $2\sigma_u^+$ are relatively closer to each other and small changes in the interatomic distance of molecular nitrogen can alter this sequence of levels [212]. Borodko and Shilov [119] have calculated an electron affinity of about +84 kcal/mole (i.e., energy of repulsion) for molecular nitrogen. The very high values of electron affinity and ionization potential of molecular nitrogen thus reflect the difficulty of reduction and oxidation of the nitrogen molecule.

The possibility of molecular nitrogen acting as a ligand was first suggested by Orgel [213]. Molecular nitrogen may be visualized as coordinated to a metal ion either with an end-on linkage M—N≡N or in an edge-on manner.

Other less important but interesting possibilities have been pointed out by Borodko and Shilov [119]. X-Ray structures of the nitrogen complexes studied thus far, $[Ru(NH_3)_5N_2]^{2+}$, $[Ru(NH_3)_5]_2N_2^{4+}$ [127], $[Ru(en)_2(N_3)(N_2)]^+$ [133], $ReCl_2(PMe_2Ph)_4(N_2)$ [176], and $CoH(N_2)(PPh_3)_3$ [162,163], $[Os(NH_3)_5N_2]^{2+}$ [148], and $\{[(C_6H_{11})_3P]_2Ni\}_2N_2$ [185] indicate that molecular nitrogen is coordinated in a linear manner to the metal ion. Although other complexes of molecular nitrogen have not been analyzed by X-ray crystallographic methods, similarities of the frequencies and intensities of bands in the 1900–2200 cm^{-1} range corresponding to the N—N stretching vibrations indicates that molecular nitrogen is coordinated in an end-on manner in all these complexes. The high intensity of the N—N band stretch is due to the polar character of coordinated nitrogen resulting from linear coordination of nitrogen to the metal ion. The integrated intensities for ν(N—N) in metal–nitrogen complexes vary from 2.7×10^{-4} to 8.5×10^{-4} liter $mole^{-1}$ cm^{-2} [119]. If the coordination of molecular nitrogen is analogous to that of metal–olefin or metal–acetylene complexes (coordinated in an "edge-on" manner), one would observe much lower values of integrated absorption intensities. To date there has been no substantiated report of an edge-on coordination of nitrogen in metal–nitrogen complexes. Such complexes with local symmetry of C_{2v}, if obtained, would give N—N stretching absorption bands only in the Raman spectra.

Quantum mechanical calculation of hypothetical Fe—N═N and

$$\begin{array}{c} \quad N \\ Fe—\| . \\ \quad N \end{array}$$

species has indicated that the linear arrangement Fe—N═N is more stable than the edge-on structure [119]. For the end-on arrangement of dinitrogen (Fig. 19) the formation of a metal–dinitrogen complex, in this case a ruthenium(II) complex, may be conceived to have taken place by the donation of a pair of electrons from the $3\sigma_g{}^+$ level of nitrogen (bonding orbital on nitrogen) to a suitable σ-bonding orbital on the metal ion (d_{z^2}, s, or p_z). The orbitals on nitrogen giving rise to this σ bond are of very low energy, so that the resulting bonding would be very weak. The major contribution to the stability of the $M—N_2$ bond is the back donation of electrons from filled metal orbitals (d_{xz}, d_{yz}) to the lowest empty antibonding orbitals $1\pi_g$ on molecular nitrogen. In Fig. 19 the bonding scheme in $[Ru(NH_3)_5N_2]^{2+}$ [end-on coordination of molecular nitrogen to Ru(II)] has been depicted. If the plane of the metal ion is taken as the xy plane, then both d_{xz} and d_{yz} orbitals on Ru^{2+} can overlap with the degenerate $e(1\pi_g)$ orbitals on nitrogen to form bonds with an average of four π electrons per metal ion.

In the dinuclear complex, $[Ru(NH_3)_5]_2N_2{}^{4+}$ (Fig. 20), two σ bonds are formed by the overlap of $3\sigma_g{}^+$ orbital on nitrogen with $(4d_{z^2} + 5s + 5p)$ hybrid orbitals on the two ruthenium atoms ($Ru—N_2—Ru$ arrangement being linear). In the formation of π bonds, the d_{xz} and the d_{yz} orbitals on both the ruthenium atoms overlap with the degenerate $e(1\pi_g)$ orbitals on nitrogen with an average of four π electrons for both the ruthenium(II) ions. Thus there is not as much drift of electrons to molecular nitrogen in the binuclear complex as occurs in the mononuclear complex. This may be readily justified by almost identical N—N distance in the mononuclear and the dinuclear complexes, 1.120 and 1.124 Å, respectively.

The bonding scheme for an edge-on arrangement of dinitrogen in a metal–dinitrogen complex is depicted in Fig. 21. As in the complexes of molecular nitrogen with a linear arrangement of the metal–nitrogen bond, a σ bond is formed in this case by the overlap of a $3\sigma_g{}^+$ orbital on nitrogen with a suitable hybrid orbital on the metal ion. Of the d_{yz} and the d_{xz} orbitals, the former can form only a δ bond and the latter a π bond with the $1\pi_g$ orbitals of dinitrogen. Since the combination of a $(\delta + \pi)$ bond in this case is much weaker than two π bonds in the end-on complexes, the edge-on complexes of molecular nitrogen are thus expected to be much less stable than the end-on complexes. The possibility of the formation of edge-on complexes as reactive intermediates in nitrogen fixation cannot, however, be ruled out at the present stage.

Back donation of electrons to the highest empty antibonding orbitals $(1\pi_g)$

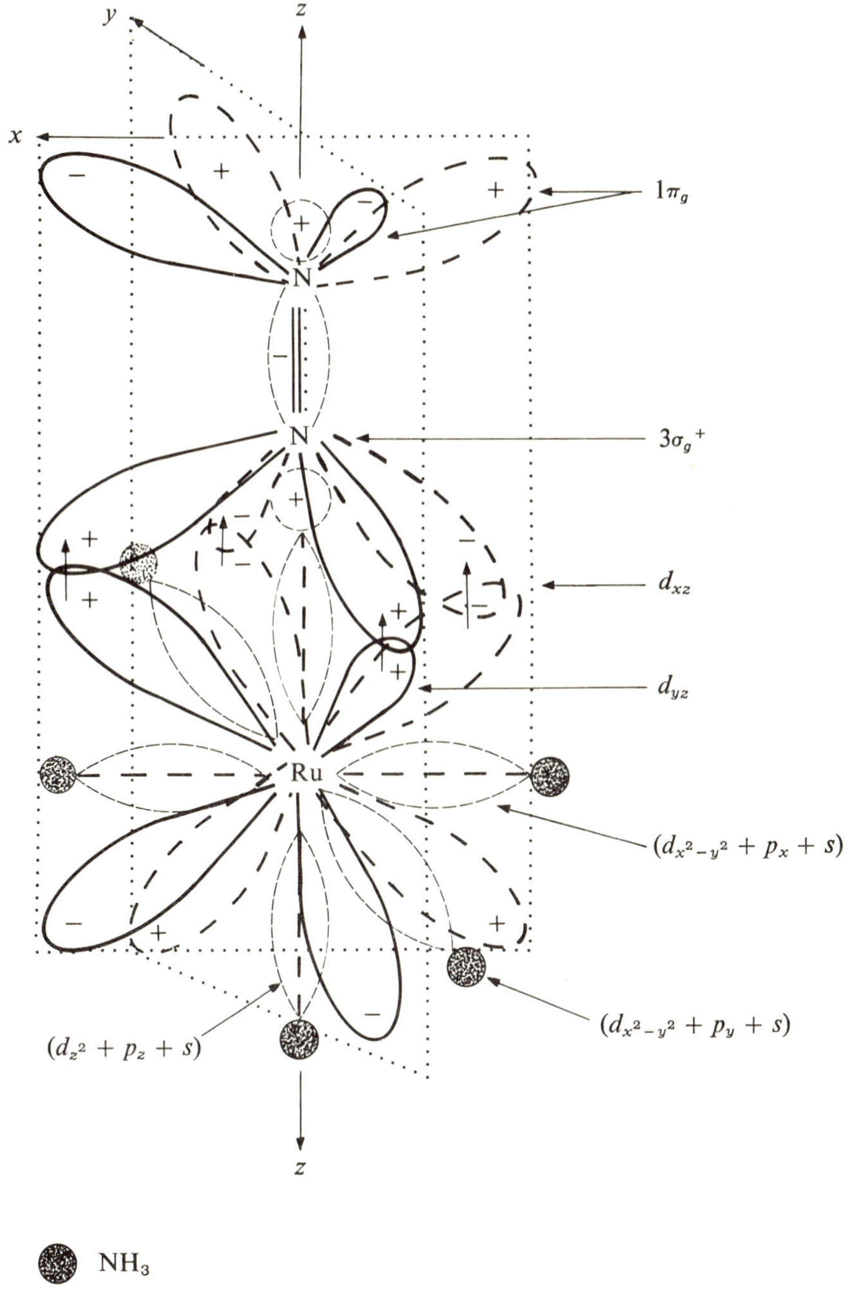

Fig. 19. Bonding in $[Ru(NH_3)_5N_2]^{2+}$.

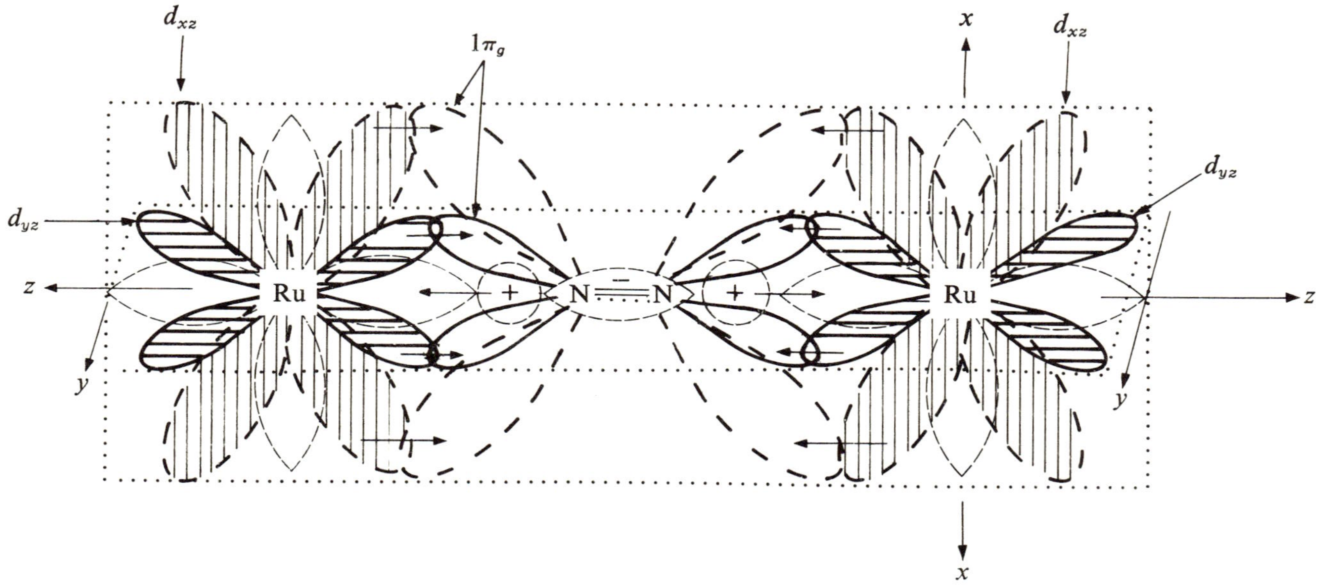

Fig. 20. Bonding in $[Ru(NH_3)_5]_2N_2^{4+}$.

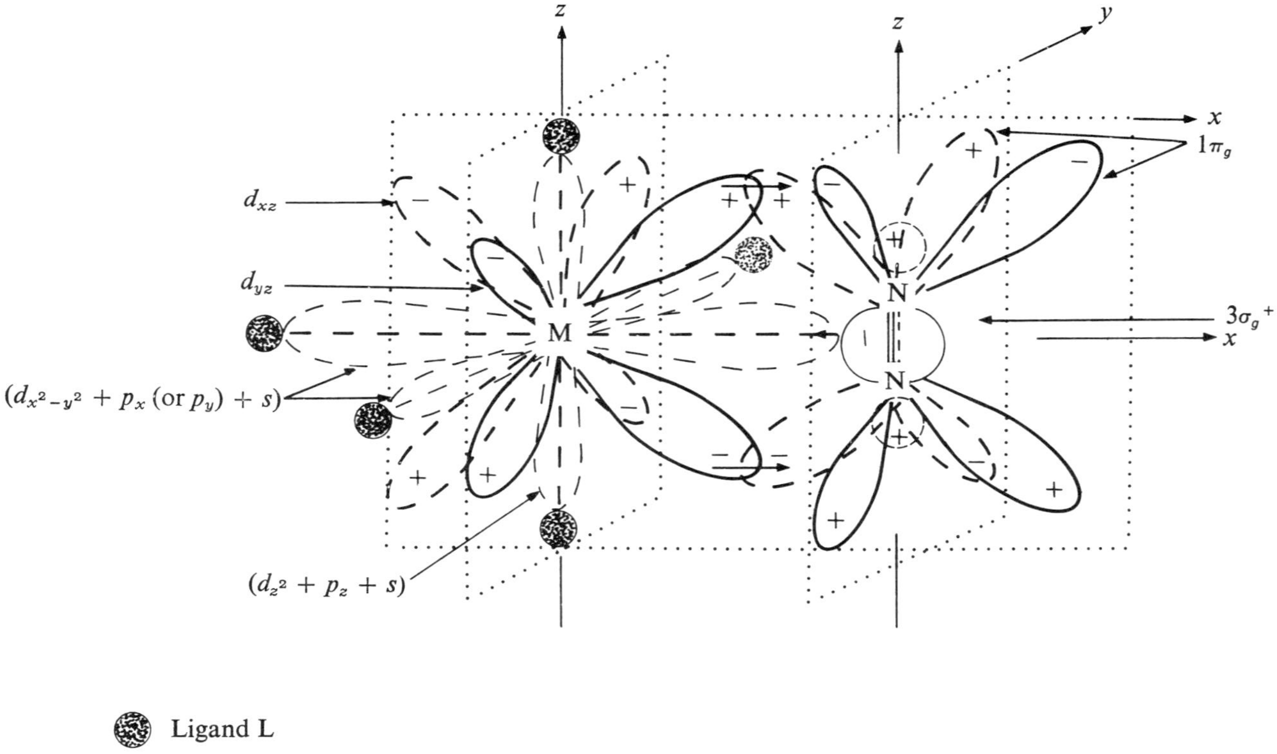

Fig. 21. Bonding in $[ML_5N_2]^{x+}$ complex (with an edge-on bonding of nitrogen).

on nitrogen requires a high ligand-field separation, Δ, on the metal complex. For 4*d* and 5*d* transition elements, nitrogen complexes can be formed with ligands of low or high ligand-field strengths but for 3*d* elements—manganese(I), iron(II), cobalt(I), nickel(II)—only ligands of high ligand-field strength (e.g., tertiary phosphines) seem to produce stable complexes. Thus for effective back bonding the value of Δ (high enough for spin pairing) seems to be very critical. Even small changes in the nature of the other donor groups attached to the metal ion can drastically change the stability of a dinitrogen complex.

The stability of a given dinitrogen complex may also be due partly to a steric effect as suggested by Misono *et al.* [190]. The presence of bulky phosphine groups on $Co(H)(N_2)(L_3)$, where L = phosphine, prevents the approach of a fourth such group near cobalt(I) and leaves the fifth coordination position vacant for molecular nitrogen. The relative contribution of steric and electronic effects to the stability of metal–dinitrogen complexes cannot, however, be quantitatively assessed at the present stage. It may, nevertheless, be broadly stated that the stability of metal–dinitrogen complexes increases as one goes down in the transition series, $3d < 4d < 5d$, as would be expected since this is also the order of ability of the metal ions to form π bonds by back donation. This is very well supported by the data in Table IV, where the N—N and M—N frequencies are compared along with the bond dissociation energies, D(M—N), for ruthenium(II) and osmium(II) dinitrogen complexes. A relatively large metal ion can form a stronger d_π–p_π bond with molecular nitrogen as evidenced by a decrease of the N—N stretch and an increase in the M—N stretch [119].

It may, however, be added here that in the series of complexes $[M(H)(N_2)(Et_2PCH_2CH_2PEt_2)_2]^+$ [M = iron(II), ruthenium(II), osmium(II)]

TABLE IV

VIBRATIONAL FREQUENCIES AND STABILITIES OF SOME METAL–NITROGEN COMPLEXES[a]

Complex	(N—N) (cm^{-1})	(M—N) (cm^{-1})	D(M—N)[b] (kcal/mole)	References
$[Ru(NH_3)_5N_2]Cl_2$	2105	500	25	2,119
$[Os(NH_3)_5N_2]Cl_2$	2012	546	39	7,119
$[Ru(NH_3)_5N_2]Br_2$	2114	499	—	2
$[Os(NH_3)_5N_2]Br_2$	2038	540	—	7
$RhN_2Cl(PPh_3)_2$	2150	—	—	152
$IrN_2Cl(PPh_3)_2$	2095	—	—	9

[a] Both donation and back donation of electrons in molecular nitrogen complexes weaken the N—N bond, as evidenced by an increase in the N—N bond length in metal–dinitrogen complexes to values greater than that of molecular nitrogen (1.098 Å).

[b] Dissociation energy of metal–nitrogen bond.

the stabilities follow the order iron(II) ≫ osmium(II) > ruthenium(II). The osmium(II) and ruthenium(II) complexes are very unstable, the solids decomposing spontaneously with the evolution of nitrogen. The reason for the higher stability of the iron(II) complex compared to the ruthenium(II) and osmium(II) complexes in this case is not clear. It would be highly desirable to synthesize stable analogous dinitrogen complexes of iron(II), ruthenium(II), and osmium(II) and to compare their stabilities and bond type by comparing the vibrational frequencies of their N—N and M—N bonds.

Since the major contribution to the strength of metal–nitrogen bond is the back donation of electrons into the empty $1\pi_g$ orbital of molecular nitrogen, any ligand that increases the electron density at the metal ion (decreased π acidity, i.e., π-electron donor activity, or increased π basicity, i.e., π activity), should increase the metal–nitrogen bond strength by increasing back donation, resulting in a corresponding decrease in the N—N bond strength. This has been observed in a series of complexes as indicated in Table V. Increasing the basicity of one of the ligands attached to the metal ion while keeping other factors constant causes a decrease in the N—N stretching vibrational frequency. In the iridium(I) complexes (1–3, Table V) a change from I^- to Cl^- causes a decrease of about 8 cm^{-1} in the N—N stretch. The effect is more significant in iridium(I) (4–6) and osmium(II) (14–19) complexes where the aryl phosphine ligand is replaced by more basic aralkyl phosphines. A similar effect is observed in complexes 7–9, Table V, where water is replaced by ammonia. The reduction in N—N stretching frequency is also observed in cobalt(I) complexes [106] (10–13) by using alkyl phosphines in place of aryl phosphines. It should be mentioned here that apart from a general trend in the weakening of the N—N bond with increased basicity of the donor group attached to the metal ion, a quantitative correlation of vibrational frequency with basicity is not possible. Coupling of the various degenerate vibrations may also play a prominent role in the lowering of the frequency of the N—N stretching vibration, a factor that cannot be fully evaluated at the present time.

The effect of decreasing the π acidity (π-acceptor activity) of the ligands coordinated to the metal may be seen in the osmium(II) complexes 15–19 of Table V. As the ligands of the osmium(II) complexes change from PMe_2Ph to $AsMe_2Ph$ (17,18) and from PEt_2Ph to $AsEt_2Ph$ (15,19), changes of 15–21 cm^{-1} are observed in the stretching frequency corresponding to lower π acidity of the arsine ligand as compared to that of the phosphine.

Oxidation of the metal ion causes a marked increase in the N—N stretching vibration of coordinated nitrogen, and a decrease in the strength of metal–dinitrogen coordination [215]. In both the rhenium (20,21) and osmium complexes (22,23) a one unit increase in the oxidation state of the metal ion increases the N—N frequency by about 80–130 cm^{-1}. This is in line with

TABLE V

CHANGES IN THE N—N STRETCHING FREQUENCY IN MOLECULAR NITROGEN COMPLEXES WITH CHANGING BASICITY OF THE LIGAND DONOR GROUPS

	Complex	N—N (cm^{-1})	References
1	$IrI(N_2)(PPh_3)_2$	2113	107
2	$IrBr(N_2)(PPh_3)_2$	2107	107
3	$IrCl(N_2)(PPh_3)_2$	2105	9
4	$IrCl(N_2)(PPh_2Et)_2$	2093	107
5	$IrCl(N_2)(PPhEt_2)_2$	2088	107
6	$IrCl(N_2)(PPhMe_2)_2$	2051	107
7	$RuCl_2(N_2)(H_2O)(NH_3)_2$	2163	119
8	$RuCl_2(N_2)(H_2O)_3$	2090	119
9	$RuCl_2(N_2)(NH_3)_3$	2080	119
10	$CoH(N_2)(PPh_3)_3$	2088	11,14,215
11	$CoH(N_2)(PPh_2Me)_3$	2077	106
12	$CoH(N_2)(PPh_2Et)_3$	2060	106
13	$CoH(N_2)(PBu_3)_3$	2050	106
14	$OsCl_2(N_2)(PEtPh_2)_3$	2090	153
15	$OsCl_2(N_2)(PEt_2Ph)_3$	2080	153
16	$OsCl_2(N_2)(PEt_3)_3$	2064	153
17	$OsCl_2(N_2)(PMe_2Ph)_3$	2078	153
18	$OsCl_2(N_2)(AsMe_2Ph)_3$	2063	153
19	$OsCl_2(N_2)(AsEt_2Ph)_3$	2059	153
20	$ReCl(N_2)(Ph_2PCH_2CH_2PPh_2)_2$	1980	173
21	$[ReCl(N_2)(Ph_2PCH_2CH_2PPh_2)_2]^+$	2060	149
22	$Os(NH_3)_5N_2]^{2+}$	2010	7
23	$[Os(NH_3)_5N_2]^{3+}$	2140	150

decreased charge transfer to π orbitals of molecular nitrogen from a metal ion in a higher oxidation state. A decrease in the oxidation state of the metal ion decreases the N—N stretching frequency and increases the ν_{N_2} absolute intensity [214]. Since the infrared intensity of the dinitrogen stretching vibration is largely dependent on the extent of π-electronic charge transfer from the transition metal to dinitrogen, the intensity should increase with a decrease in the nuclear charge on the metal ion. In rhenium(I), tungsten(0), and molybdenum(0) dinitrogen complexes the intensities are in the range 9–11 $\times$ 10^4 M^{-1} cm^{-2} (maximum observed intensities). The electron density is transferred to dinitrogen to such an extent that these complexes are easily protonated in solution.

It is of considerable interest at this stage to compare the bonding in dinitrogen complexes with that of the isoelectronic carbonyl complexes. Jaffe and Orchin [216] have compared the complex-forming ability of dinitrogen and carbon monoxide, and have come to the conclusion that the bonding $5\sigma_g{}^+$ orbital in carbon monoxide is better suited than the $3\sigma_g{}^+$ orbital of molecular nitrogen to the formation of σ bonds with metal ions. The $5\sigma_g{}^+$ orbital in carbon monoxide is of the nonbonding (slightly antibonding) type [217] and is located mostly on the carbon atom. Because of the combination of the energetically favored $2p(\sigma)$ orbital of oxygen with the $2s$ orbital on carbon, the $5\sigma_g{}^+$ orbital on carbon monoxide has more p character than the $3\sigma_g{}^+$ orbital on molecular nitrogen, and thus possesses better directional properties to form a σ bond. The $3\sigma_g{}^+$ orbital on molecular nitrogen is of the bonding type, has about 25% s character, and is about 32 kcal/mole lower in energy than the $5\sigma_g{}^+$ orbital in carbon monoxide [208]. The donation of a pair of electrons to a metal ion is thus energetically unfavorable in metal–dinitrogen complexes compared to those of metal carbonyls. The N—N bond in metal–dinitrogen complexes is thus weakened by donation of electrons from $3\sigma_g{}^+$ orbitals, whereas the formation of a σ bond by the donation of electrons from the slightly antibonding $5\sigma_g{}^+$ orbital in carbon monoxide enhances the strength of the C—O bond. Back donation of electrons from the metal ion to the antibonding $1\pi_g$ levels in dinitrogen and in carbon monoxide results in the weakening of the N—N and C—O π bonds.

The $1\pi_g$ level in carbon monoxide is a better π-electron acceptor than the $1\pi_g$ level of dinitrogen. This is best borne out by the infrared measurements of carbonyl and dinitrogen stretching frequencies in the analogous dinitrogen and carbonyl complexes such as $Ir(L)(PPh_3)_2Cl$ and $Rh(L)(PPh_3)_2Cl$ (L = CO or N_2). The integrated absorption intensity of the M—C bond in the carbonyls is about three times that of the corresponding M—N bond in metal–dinitrogen complexes [218]. Since integrated absorption intensities are highly dependent on the π-electron density in the M—L bond, it may be concluded that the π-acceptor ability of the $1\pi_g$ orbital of carbon monoxide is much higher than that of $1\pi_g$ orbital of molecular nitrogen. This is further confirmed by a lowering of the τ values of the hydrido ligand to the extent of 5τ to 6τ in the hydrido carbonyl complexes $[OsH(CO)(PR_3)_3]$, $CoH(CO)(PR_3)_3$, and *trans*-$[FeH(CO)(diphos)_2]^+$ as compared to the corresponding dinitrogen complexes [104,154]. Since the chemical shift of the hydride resonance has been shown to be dependent upon the d-orbital electron density on the metal ion, the carbonyl complexes should have their hydride resonances at lower field. This is in line with a lower d-orbital population on the metal ion in carbonyls due to the higher π-acceptor ability of carbon monoxide over nitrogen.

Because of the complex interplay of effects listed above, the assessment of the overall strength of the C—O and N—N bonds in the coordinated ligands should be judged with caution and should be considered on the basis of combined σ and π effects.

The higher σ-donor and π-acceptor capacity of carbon monoxide as compared to dinitrogen has been demonstrated further in terms of the center

Mossbauer shifts (CS shift) of the complexes *trans*-[Fe(H)(N_2)(diphos)$^+$] and *trans*-[Fe(H)(CO)(diphos)$_2$]$^+$ +0.16 and −0.04 mm/second, respectively. The much lower center shift in the carbonyl compound indicates that carbon monoxide is more strongly bound ($\sigma + \pi$) than dinitrogen [219,220]. Recent studies [220] on the quadrupole (QS) splitting of the complexes *trans*-[Fe(H)(diphos)$_2$L] (L = CO or N_2) indicate that the dinitrogen complex (L = N_2) has the more positive QS than carbon monoxide. The QS of a complex becomes more positive as the π-acceptor function of the ligand increases and more negative as the σ-donor function increases. (In other words QS is a measure of $\pi - \sigma$.) The higher positive value of QS for dinitrogen indicates that it is a better π ligand. The positive value of center shift indicates that it is a poorer ($\pi + \delta$) ligand as compared to carbon monoxide. These results show that π-acceptor bonding relative to σ-donor bonding is more important in the coordination of dinitrogen than in the formation of carbon monoxide complexes. This is also true in $[Ru(NH_3)_5L]^{2+}$ (L = CO, N_2), as indicated by the frequencies of the Ru—C and Ru—N stretching vibrations, which are, respectively, 585 and 500 cm^{-1} [137]. This difference in stability is also the reason for the fact that carbon monoxide forms metal polycarbonyls, whereas in the metal–polydinitrogen complexes, coordination of more than one nitrogen molecule drastically reduces bond strength [115,145,156] to the point where more than two dinitrogens coordinated per metal ion are never found.

A tentative schematic molecular orbital diagram for $[Ru(NH_3)_5N_2]^{2+}$ is presented in Fig. 22. The complex molecule belongs to the point group C_{4v} and the splitting of the d levels is that expected for a field of tetragonal symmetry. The σ-bonding orbitals of the complex are composed of the five σ_b orbitals of ammonia and one $3\sigma_g{}^+$ orbital of molecular nitrogen. It is assumed that the three $a_1(\sigma_b)$ molecular orbitals of the complex are composed of the ($4d_{z^2} + 5p_z + 5s$) orbitals of ruthenium(II), σ_b orbitals on ammonia and $3\sigma_g{}^+$ orbitals of dinitrogen. If dinitrogen is bonded to the metal ion end-on along the z axis, then the $e(\sigma_b)$ and $b_1(\sigma_b)$ molecular orbitals of the complex may be constructed, respectively, from the ($5p_x$, $5p_y$) and $4d_{x^2-y^2}$ orbitals on the metal ion, and σ_b orbitals from ammonia. The bonding π levels $e(\pi_b)$ are mostly $4d_{xz}$ and $4d_{yz}$ orbitals of ruthenium(II) and the $1\pi_g$ orbital of molecular nitrogen. The $4d_{xy}(b_2)$ level remains nonbonding on the metal ion. The intense band at 221 nm may be assigned to the allowed metal–ligand charge transfer from $e(\pi_b) \rightarrow e(\pi^*)$ or $b_2 \rightarrow e(\pi^*)$.

A tentative molecular orbital diagram of the dinuclear complex $[Ru(NH_3)_5]_2N_2{}^{4+}$ is schematically depicted in Fig. 23. The molecule belongs to the point group D_{4h}, with the two ruthenium atoms and molecular nitrogen collinear and the ammonia groups in an eclipsed conformation [127]. The σ-bonding molecular orbitals a_{1g} may be constructed from the σ_b orbitals of

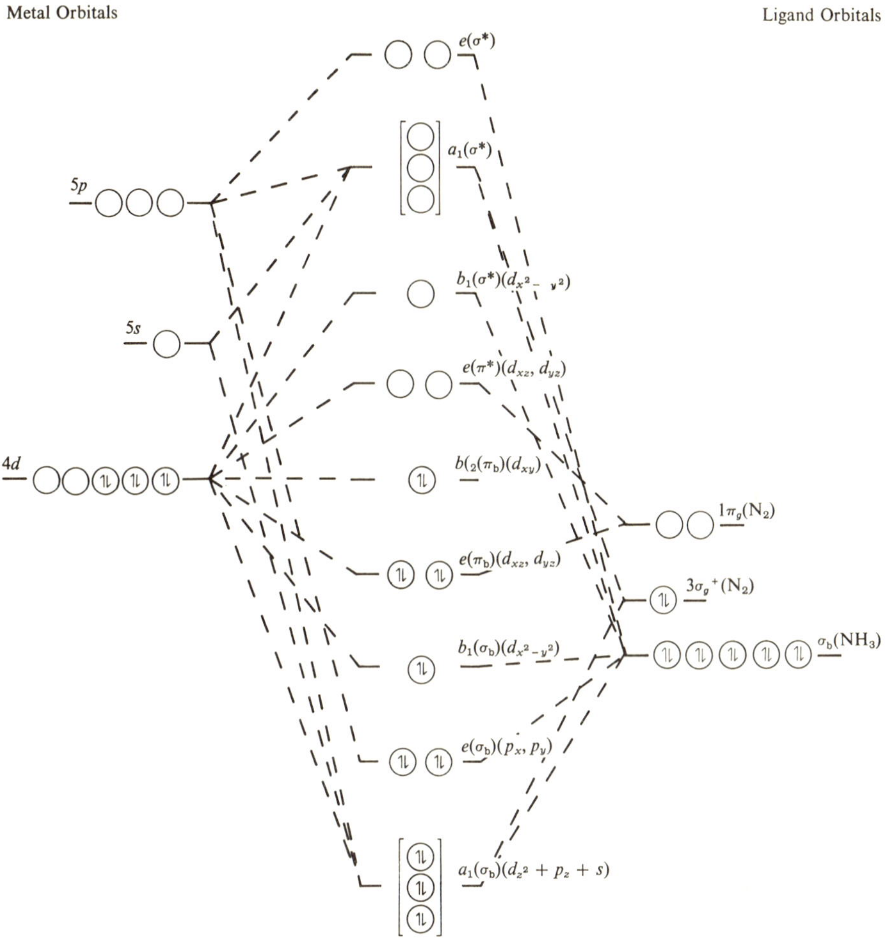

Fig. 22. Molecular orbital diagram of $[Ru(NH_3)_5N_2]^{2+}$.

ammonia and the $3\sigma_g^+$ orbital of molecular nitrogen, with hybrid ($4d_{z^2}$ + $5p_z + 5s$) orbitals on Ru_1 and Ru_2. The e_u and b_{1g} orbitals may be constructed, respectively, from $5p_x$, $5p_y$, and $4d_{x^2-y^2}$ orbitals on the metal ion, and the σ_{bg} orbitals of ammonia. The relative ordering of the σ_g levels, $a_{1g}(\sigma_b)$, $e_u(\sigma_b)$, and $b_{1g}(\sigma_b)$ are only schematic and not to scale. The d orbitals of the metal ions ($Ru_1 + Ru_2$) are split into $e_g(\pi_b)$, $b_{2g}(\pi)$, $b_{1u}(\pi)$, and $e_u(\pi)$. According to Trietel *et al.* [127], these molecular orbitals may be arranged in terms of increasing energy as $e_g(\pi_b)(d_{xz_1} + d_{xz_2}, d_{yz_1} + d_{yz_2}) < b_{2g}(\pi)(d_{xy_1} + d_{xy_2}) < b_{1u}(\pi)(d_{xy_1} - d_{xy_2}) < e_u(\pi)(d_{xz_1} - d_{xz_2}, d_{yz_1} - d_{yz_2}) < e_g(\pi^* N_2)$. In the ground state of the molecule the 12 d electrons from the two ruthenium(II) ions occupy the above molecular levels in the order $(e_g)^4$, $(b_{2g})^2$, $(b_{1u})^2$, $(e_u)^4$.

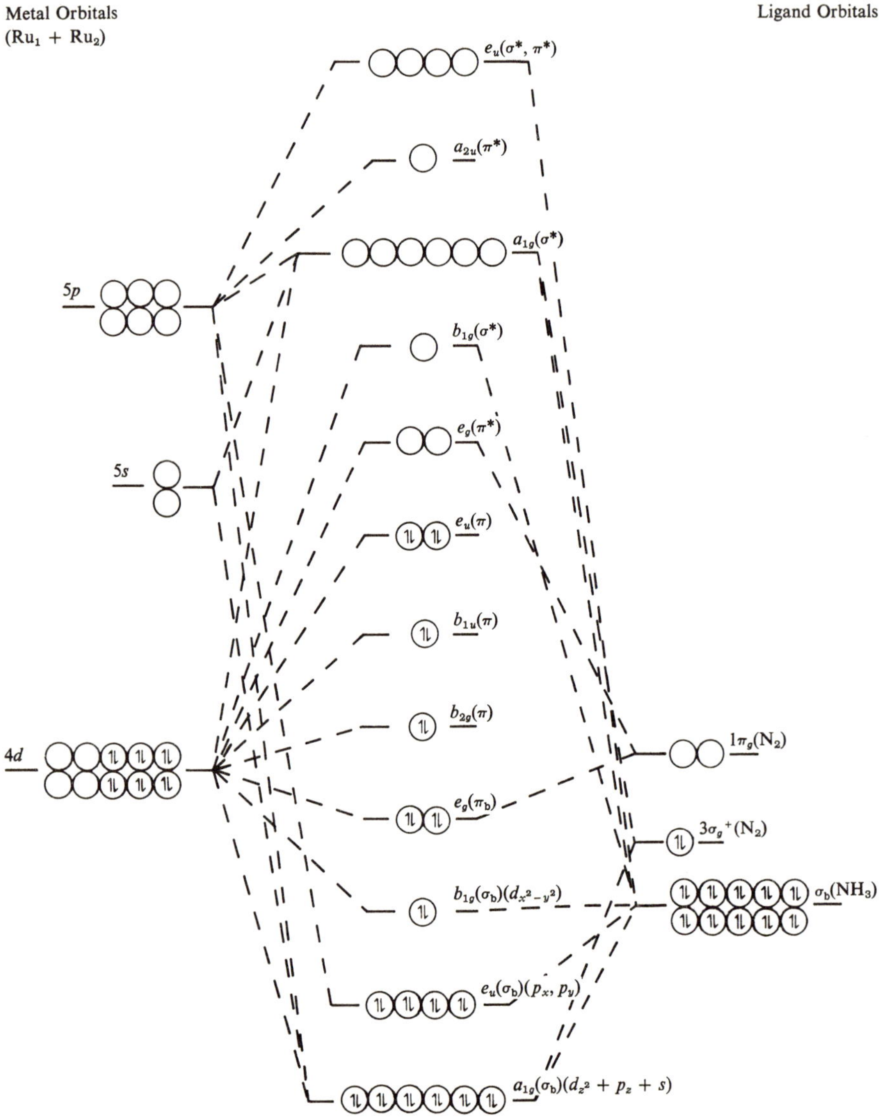

Fig. 23. Molecular orbital diagram of $[Ru(NH_3)_5]_2N_2^{4+}$.

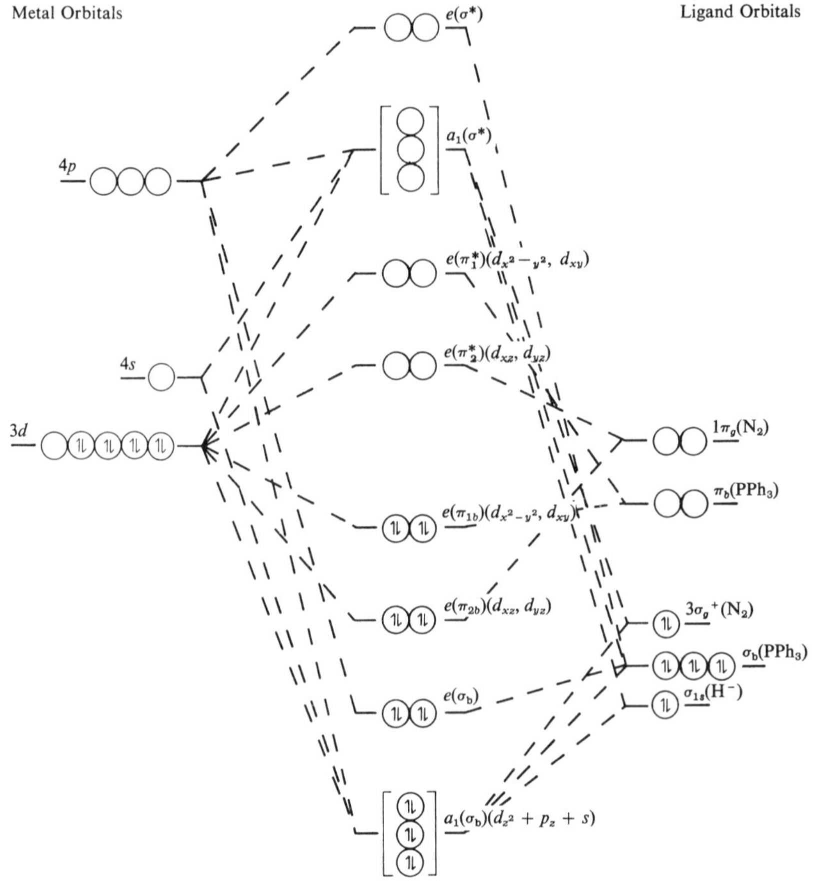

Fig. 24. Molecular orbital diagram of $Co(PPh_3)_3H(N_2)$.

The π bonding from the two ruthenium centers to molecular nitrogen is provided by the four electrons in the $e_g(\pi)$ level, giving an average of one π bond per ruthenium(II) ion. It has been suggested that the intense absorption band at 263 nm in the complex is due to the excitation of the $e_u(\pi)$ electrons to the $e_g(\pi^*)$ level [127]. It is noted that at this stage the position of one of the antibonding orbitals [designated $a_{2u}(\pi^*)$] is highly uncertain.

A suggested schematic molecular orbital diagram for $CoH(N_2)(PPh_3)_3$ is presented in Fig. 24. The molecule has a distorted trigonal bipyramidal structure [162,163] and therefore belongs to the symmetry group C_{3v}. It is assumed that the σ_b levels on the ligands may be arranged in the order $\sigma_{1s}(H^-) < \sigma_b(PPh_3) < 3\sigma_g^+(N_2)$. The relative ordering of these levels may only be determined by actual calculations. The bonding $a_1(\sigma_b)$ orbitals in the complex may be formed by the combination of ($3d_{z^2}$ + $4p_z$ + $4s$) orbitals on the metal ion with $\sigma_{1s}(H^-)$, $\sigma_b(PPh_3)$, and $3\sigma_g^+(N_2)$ orbitals on the ligands. The $e(\sigma_b)$ orbitals are mostly composed of the metal $4p_x$ and $4p_y$ orbitals and $\sigma_b(PPh_3)$ ligand orbitals. The two degenerate sets of $e(\pi_b)$ levels may be built from $3d_{xz}$, $3d_{yz}$, and $3d_{x^2-y^2}$ $3d_{xy}$ pairs on the metal ion and π_b (PPh_3) and $1\pi_g(N_2)$ orbitals on the ligands, and may be very much closer in energy. Only two π-bonding orbitals from the three phosphine ligands can take part in the formation of the bonding $e(\pi_b)$ levels. Though an electronic spectrum has not been reported for the complex, one can expect an intense UV band due to $e(\pi) \rightarrow e(\pi^*)$ transition. Such information is needed to establish the relative positions of the $e(\pi_1^*)$ and $e(\pi_2^*)$ levels, which may be the reverse of that shown in Fig. 24.

IV. Fixation of Molecular Nitrogen by Metal Ions and Metal Complexes

A. Catalytic Systems in Homogeneous Solution

Studies of the chemical fixation of nitrogen by the use of metal ions and metal complexes in solution are of recent origin. A knowledge of the formation of metal–dinitrogen complexes discussed in Section III is very helpful in the selection of suitable catalysts for the fixation of nitrogen, mainly in its conversion to ammonia. The search for suitable catalysts for the fixation of nitrogen at room temperature and pressure is in progress throughout the world and is a major goal of many laboratories. The data in Table VI [15–20,22,23,25,221] summarize the results reported on the fixation of nitrogen up to the present time.

Haight and Scott [23] employed molybdenum(VI) and tungsten(VI) as catalysts for the fixation of nitrogen in the presence of metallic zinc, stannous chloride in hydrochloric acid, and platinum or porous graphite cathodes, as reducing agents. Dinitrogen was bubbled through a porous graphite cathode or around a platinum cathode in a medium containing molybdate(VI) in 3 *M* hydrochloric acid. When tungstate was used as a catalyst, the dinitrogen was bubbled through a solution of tungstate (WO_4^{2-}) in 12 *M* hydrochloric acid in the presence of stannous chloride as a reducing agent. Kjeldahl analysis, followed by a spectroscopic test with Nessler's reagent, was employed for the detection of ammonia.

Before the beginning of all experiments, all the reagents including dinitrogen were determined to be free of ammonium ion and precautions were taken to prevent introduction of ammonium ion from any source except reduction of gaseous nitrogen. The amount of ammonia formed in various experiments is presented in Table VI. After a period of 24 hours during which all molybdenum(VI) was reduced to molybdenum(III), no further reduction of dinitrogen was detected. It was [23] postulated that molybdenum(V) or molybdenum(IV) forms an activated complex with nitrogen followed by a homolytic fission of the N—N bond. Molybdenum and tungsten nitrides are proposed [23] as intermediates which on hydrolysis give ammonia and hydroxo complexes of the metal ions.

Volpin and Shur [15,19] employed halides of chromium(III), molybdenum(VI), tungsten(VI), and titanium(IV) in the presence of the reducing agents, lithium aluminum hydride, ethylmagnesium bromide, and triisobutylaluminum, in a reaction medium of ether or heptane at pressures of nitrogen from 1 to 100 atm. The yield of ammonia depended on the nature of the solvent, the nature of the reducing agent, and the pressure of dinitrogen. Ether was found to be a better solvent than tetrahydrofuran (THF) for the reduction of dinitrogen. The yield of ammonia at 150 atm was found to be higher than that obtained at 1 or 7 atm. Ferric chloride was inactive at 1 atm of dinitrogen, but activates dinitrogen at 150 atm. The transition metal chlorides of nickel(II), cobalt(II), copper(I), and palladium(II) were inactive even at 150 atm. The highest yield of ammonia was obtained with dicyclopentadienyltitanium in diethyl ether in the presence of ethylmagnesium bromide [16].

The fixation of molecular nitrogen by the chromium(III) chloride–ethylmagnesium bromide system in ether has been demonstrated [17] by using ^{15}N. The labeled dinitrogen under a pressure of 14 atm reacted with the $CrCl_3$–EtMgBr mixture in ether at room temperature. The ammonia which formed over a period of 11 hours amounted to 0.048 mole per mole of chromic chloride and was analyzed for ^{15}N by conversion to molecular nitrogen, followed by mass spectrometric determination of its isotopic composition.

Volpin *et al.* [18] employed a mixture of dinitrogen and hydrogen for the synthesis of ammonia. With tetraethoxytitanium(IV) in the presence of ethylmagnesium bromide in ether at a dinitrogen pressure of 38–100 atm and hydrogen of 32–50 atm, the yield of ammonia obtained in about 9 hours was 0.50 mole per mole of titanium compound. Use of tris(acetylacetonato)oxovanadium(IV) under the same conditions gave the same yield of ammonia as did the ethoxytitanium complex. Acetylacetone complexes of chromium(III), manganese(III), and molybdenum(VI) were less active than vanadium(IV) and titanium(II). For the acetylacetonato complexes, the yield of ammonia decreased in the order, vanadium(IV) $>$ chromium(III) $>$ manganese(III) $>$ molybdenum(VI) $>$ iron(III) $\approx$ nickel(II) $\approx$ copper(II).

Triphenylphosphine complexes of titanium(IV), iron(III), cobalt(II), nickel(II), palladium(II), and platinum(II), in the presence of ethylmagnesium bromide in ether, *n*-butyllithium in heptane, and tris(isobutyl)aluminum in *n*-heptane, have been used as catalysts [20] for the fixation of nitrogen. All the experiments were conducted at room temperature. At 150 atm of dinitrogen,

Conversion of N_2 to NH_3 on Metallic Centers

No.	Reaction time (hours)	Catalyst[a]	Reducing agent	Solvent	Pressure N_2 (atm)	Yield of NH_3 (mole/mole catalyst)	References
1	24	MoO_4^{2-}	Pt cathode	3 *M* HCl	1	0.01	23
2	24	MoO_4^{2-}	Porous graphite cathode	3 *M* HCl	1	0.01	23
3	24	MoO_4^{2-}	Porous graphite cathode	3 *M* HCl	1	< 0.001	23
4	24	MoO_4^{2-}	Zn	3 *M* HCl	1	0.001	23
5	24	WO_4^{2-}	$SnCl_2$	12 *M* HCl	1	0.1	23
6	10–11	$CrCl_3$	$LiAlH_4$	Ether	7	0.07	15
7	10–11	$CrCl_3$	$LiAlH_4$	Ether	1	0.02	15
8	10–11	$CrCl_3$	EtMgBr	Ether	150	0.17	17, 19
9	10–11	WCl_6	EtMgBr	Ether	150	0.16	15, 19
10	10–11	$MoCl_5$	EtMgBr	Ether	150	0.08	15, 19
11	10–11	$TiCl_4$	EtMgBr	Ether	150	0.10	15, 19
12	10–11	$TiCl_4$	$(iso\text{-}Bu)_3Al$	Heptane	150	0.25	15, 19
13	10–11	$FeCl_3$	EtMgBr	Ether	150	0.09	15, 19
14	10–11	$(C_5H_5)_2Ti$	EtMgBr	Ether	150	0.84	16
15	31	$(C_5H_5)_2Ti$	EtMgBr	Ether	150	0.91	16
16	8	$(C_5H_5)_2Ti$	EtMgBr	Ether	1	0.70	15
17	7	$(C_5H_5)_2Ti$	EtMgBr	Toluene	150	0.22	16
18	11	$(C_5H_5)_2Ti$	EtMgBr	THF	150	0.40	16
19	10–11	$Ti(OC_2H_5)_4$	EtMgBr	Ether	38–100 + H_2(32–50 atm)	0.5	18
20	10–11	$VO(acac)_2$	EtMgBr	Ether	38–100 + H_2(32–50 atm)	0.5	18
21	10–11	$Cr(acac)_3$	EtMgBr	Ether	38–100 + H_2(32–50 atm)	< 0.5	18

22	10–11	Mn(acac)$_3$	EtMgBr	Ether	38–100 + H_2(32–50 atm)	< 0.5	18
23	10–11	MoO_2(acac)$_2$	EtMgBr	Ether	38–100 + H_2(32–50 atm)	< 0.5	18
24	10–11	Fe(acac)$_3$	EtMgBr	Ether	38–100 + H_2(32–50 atm)	< 0.5	18
25	10–11	Ni(acac)$_2$	EtMgBr	Ether	38–100 + H_2(32–50 atm)	< 0.5	18
26	10–11	Co(acac)$_3$	EtMgBr	Ether	38–100 + H_2(32–50 atm)	0.0	18
27	10–11	$(PPh_3)_2TiCl_4$	Al(*i*-C_4H_9)$_3$	Ether	120	0.12	20
28	10–11	$(PPh_3)_2TiCl_4$	EtMgBr	Ether	120	0.10	20
29	10–11	$(PPh_3)_2FeCl_3$	EtMgBr	Ether	120	0.03	20
30	½	VCl_3	Lithium naphthalide metal ion (3:1)	THF	120	0.9	221
31	½	VCl_3	5:1	THF	120	1.2	221
32	½	VCl_3	7:1	THF	120	2.0	221
33	½	VCl_3	10:1	THF	120	1.2	221
34	½	$CrCl_3$	10:1	THF	120	1.3	221
35	½	$CrCl_3$	10:1	THF	1	0.4	221
36	½	$TiCl_4$	10:1	THF	120	1.3	221
37	Several weeks	Ti(II)	K metal, K-*t*-butoxide	Ether	1	0.1 0.15	22
38	24	Ru(II)	$TiCl_3$	3 *M* HCl	1	0.4	25
39	24	Ru(III)	$SnCl_2$	3 *M* HCl	1	0.07	25
40	24	Ru(III)	$(CH_3)_2SnCl_2$	3 *M* HCl	1	0.05	25
41	24	Ru(III)	N_2H_4	3 *M* HCl	1	0.06	25
42	24	Rh(III)	$TiCl_3$	3 *M* HCl	1	0.36	25
43	24	Rh(III)	$SnCl_2$	3 *M* HCl	1	0.40	25
44	24	Ir(III)	$SnCl_2$	3 *M* HCl	1	0.10	25
45	24	Ir(III)	$TiCl_3$	3 *M* HCl	1	0.20	25
46	24	Os(II)	$TiCl_3$	3 *M* HCl	1	0.10	25
47	24	Os(III)	$SnCl_2$	3 *M* HCl	1	0.12	25

[a]acac = 2,4-pentanedione; THF = tetrahydrofuran.

only bis(triphenylphosphine)titanium tetrachloride and bis(triphenylphosphine)iron trichloride have been found to be active catalysts for the fixation of nitrogen. The yield of ammonia varied from 0.12 to 0.03 mole per mole of catalyst. The complexes $(PPh_3)_2NiCl_2$, $(PPh_3)_2CoCl_2$, $(PPh_3)_2PdCl_2$, and $(PPh_3)_2PtCl_2$ were found to be inactive as catalysts in the presence of reducing agents such as lithium aluminum hydride in ether and sodium borohydride in aqueous ethanol. A review of Volpin and Shur's work on the fixation of nitrogen has appeared recently [21,222].

Volpin *et al.* [223] have described a titanium(II) catalyst of the composition $C_6H_6 \cdot TiX_2 \cdot 2\,AlX_3$ ($X = Cl^-, Br^-$) obtained by heating titanium tetrachloride with a mixture of aluminum and aluminum halide in benzene. A mixture of the catalytic titanium(II) species, metallic aluminum, and aluminum halide in the ratio (1:150:250) when heated at 130°C reduces dinitrogen to ammonia at 100 atm and 130°C. A total yield of 115 moles of ammonia per mole of the catalyst was obtained after hydrolysis of the reaction mixture. The yield of ammonia was increased by increasing the proportion of the reducing agents (aluminum and aluminum halide). A maximum of 284 moles of ammonia per mole of the titanium(II) catalyst was obtained when the catalyst, aluminum, and aluminum halide were employed in the ratio 1:600:1000. In the absence of aluminum and aluminum halide, the catalytic titanium(II) species reacts with dinitrogen at 130°C to form a compound with the composition $C_6H_6(TiX_2 \cdot 2\,AlX_3)_3N$. The nitride on hydrolysis gave a stoichiometric yield of ammonia.

Yatsimirskii and Pavlova [24] have used solutions containing reduced forms of titanium, vanadium, niobium, molybdenum, tungsten, and rhenium in 3 *M* hydrochloric or sulfuric acid solution as catalysts for the fixation of molecular nitrogen. Dinitrogen was passed through acidified solutions of these metal ions for 6 hours and the amount of dinitrogen fixed, detected colorimetrically, was in the range 0.00–0.04 mole per mole of the catalyst. Solutions of zirconium, hafnium, and tantalum were found to be inactive as catalysts for the fixation of nitrogen under these conditions.

Molecular nitrogen may be converted to ammonia [221] by complexes of vanadium, chromium, and titanium in low oxidation states, obtained by reduction of the corresponding transition metal salt with lithium naphthalide in tetrahydrofuran. Solutions of vanadium trichloride, chromium trichloride, or titanium trichloride were added to a solution of lithium naphthalide in tetrahydrofuran under 1 atm of dinitrogen. At higher pressures of dinitrogen, an ampoule containing the salt solution was broken in a tetrahydrofuran solution of lithium naphthalide in an autoclave at the desired pressure. The yield of ammonia increased with increasing concentration of lithium naphthalide and in some cases (Table VI) resulted in the fixation of 1 mole of dinitrogen per mole of the metal complex. It was suggested that in the presence of an excess of lithium naphthalide, the transition metals under consideration behaved as highly reactive metal ion species in the zero oxidation state and the reduction of dinitrogen to ammonia takes place within the coordination sphere of the metal ion.

Van Tamelen *et al.* [22] employed a titanium(II) catalyst obtained by the reduction of dialkoxydichlorotitanium(IV) in ether with potassium metal (1 equiv) at room temperature. Reduction of titanium(IV) species in solution was indicated by the change in color of the solution from off-white to intense black, and by the disappearance of metallic potassium within 12–24 hours. Purified dinitrogen was bubbled through the solution and into a trap that contained a dilute aqueous solution of sulfuric acid or boric acid. In about 48 hours ammonia was detected in the exit traps and was determined by vapor-phase chromatography, mass spectrometry, and by titration. When ammonia production ceased after several weeks further addition of potassium metal to the solution produced more ammonia. The yield of ammonia was about 0.10 to 0.15 mole per mole of titanium catalyst. Substitution of argon for dinitrogen stopped the formation of ammonia in the above reaction. Alkali metal, titanium tetrachloride, and alkali metal alkoxides were all necessary ingredients for the synthesis of ammonia. Hydrogen needed for the synthesis of ammonia most probably comes from the solvent [22]. Replacement of ether by *o*-xylene markedly reduced the yield of ammonia.

Van Tamelen *et al.* [224] have developed a cyclic process of fixation of nitrogen involving the use of titanium(IV) isopropoxide, isopropyl alcohol, and sodium naphthalide in an ether solvent. Sodium naphthalide is the reducing agent and isopropyl alcohol acts as the source of hydrogen. In the operation of the cyclic process, the isopropyl alcohol was added after the absorption of dinitrogen had been completed, and the ammonia formed was evolved in gaseous form. The titanium (Ti^{n+}) catalyst was regenerated either by the addition of sodium naphthalide or metallic sodium. The treatment with metallic sodium or sodium naphthalide also regenerates the naphthalide radical anion **68** formed from naphthalene in the nitrogen reduction stage. In the course of five cycles, a yield of about 3.4 moles of ammonia per mole

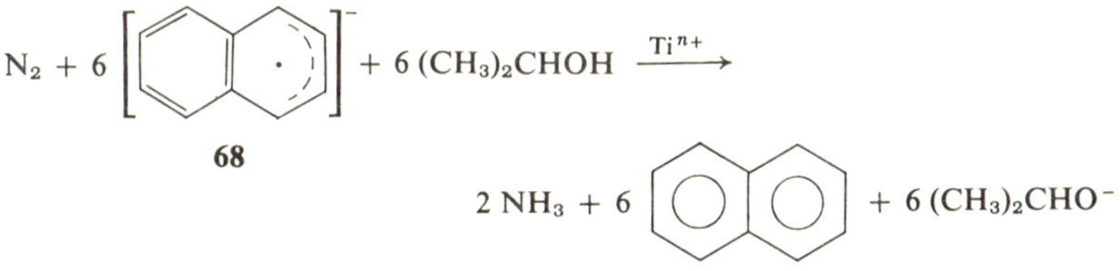

68

(60)

of titanium was obtained. The net reaction may be represented by Eq. (60). The oxidation state of the catalytic titanium species is not known with certainty. Synthesis of ammonia was also attempted with air as the source of nitrogen. In this case the yield of ammonia was considerably reduced because of oxidation of the titanium catalyst and side reactions of molecular oxygen with the naphthalide free radical [224].

A recently reported cyclic process for the fixation of nitrogen consists of the reduction of dinitrogen with titanium(IV) isopropoxide–sodium naphthalide catalyst in benzene or tetrahydrofuran at 20°–25°C [225]. Titanium(II) has been proposed as the active species for the reduction of dinitrogen and is most probably produced in solution by the reduction of titanium(IV) with sodium naphthalide [225]. The maximum yield of ammonia was realized when titanium(IV) and sodium naphthalide were employed in the ratio of 1:6. That the solvent acts as a source of protons in the fixation of nitrogen was confirmed by conducting the reaction in perdeuterotetrahydrofuran, whereby ammonia with 10% deuterium was obtained during a period of 6 weeks. A nitride intermediate that eventually becomes hydrolyzed to ammonia was proposed, the titanium(II) catalyst being regenerated by the addition of fresh sodium naphthalide [225].

An improvement over the above procedure which also establishes the formation of a metal nitride in the reduction of nitrogen utilizes an electrolytic cell with an aluminum anode, a nichrome cathode, and an electrolyte consisting of titanium(IV) isopropoxide, naphthalene, aluminum isopropoxide, and tetramethylammonium chloride in 1,2-dimethoxyethane [226]. Nitrogen gas was bubbled through the solution for 11 days while the solution was electrolyzed. When the resulting solution was treated with 8 *M* sodium hydroxide a 6.1-fold yield of ammonia, based on the titanium employed, was obtained [226]. No ammonia was obtained when aluminum isopropoxide was removed from the solution. Removal of naphthalene which probably acts as an electron carrier reduced the yield of ammonia. It has been proposed that aluminum isopropoxide forms aluminum nitride from a titanium nitride species, thus recycling titanium(II) in solution to form more titanium–nitrogen complex, followed by its reduction to titanium nitride. The aluminum nitride is then hydrolyzed to ammonia.

Sodium naphthalide reduction of biscyclopentadienyl complexes of niobium(V), zirconium(IV), and the alkoxide complexes of tungsten(IV) and tungsten(V) in tetrahydrofuran solution resulted in the fixation of nitrogen to a small extent ($<3\%$) [227]. Thus the heavy metal ions of Groups IV, V, and VI seem to be much less efficient as catalysts than is titanium(IV) in the fixation of nitrogen.

Lithium naphthalide reduction of a ferric chloride solution in tetrahydrofuran resulted in a stoichiometric (1:1) yield of ammonia per mole of the iron catalyst [228]. For the optimum yield of ammonia 4 moles of naphthalide ion per mole of iron(III) is required, which indicates that the valence of the catalytic iron species may be zero. The concentration of the naphthalide anion was always kept at the required level by the use of an excess of lithium.

Van Tamelen *et al.* [229] have obtained both hydrazine and ammonia by the hydrolysis of the species obtained after reduction of molecular nitrogen with titanium(IV) isopropoxide–sodium naphthalide in tetrahydrofuran at room temperature. The yield of ammonia and hydrazine depends on the ratio of titanium(IV) to sodium naphthalide. The yield of hydrazine increased to 15–19% [based on titanium(IV)] at a Ti(IV): sodium napthalide ratio of 1.5:6, while the NH_3/N_2H_4 ratio varied in the range 3.3–5.0. No hydrazine was obtained when titanium(IV) was replaced by cobalt(II), molybdenum(V), tungsten(VI), and chromium(III). In the presence of aluminum(III), the nitrogen bound to titanium is transferred to aluminum with the formation of a nitride, probably AlN. No hydrazine was obtained from such a system. The formation of hydrazine in molecular nitrogen reduction is an important step and if properly controlled and developed may lead to the synthesis of substituted hydrazines and other organic intermediates.

Taqui Khan and Martell [25] have used ruthenium(II), ruthenium(III), rhodium(III), osmium(II), osmium(III), and iridium(III) catalysts for the reduction of dinitrogen to ammonia. A mixture of one volume of nitrogen and three volumes of hydrogen at 1 atm pressure and room temperature was circulated through the catalyst solution (10^{-3}–10^{-4} M) in 3 M hydrochloric acid. The catalyst solutions were obtained by adding a suitable reducing agent (titanium trichloride, stannous chloride, dimethyltin dichloride, or hydrazine) to the appropriate transition metal ion in a 10:1 ratio. The resulting solution has a characteristic color of the transition metal ion. At the end of 24 hours, a sample of the solution was withdrawn from the reaction mixture and the amount of ammonia estimated with a micro-Kjeldahl apparatus. The yield of ammonia varied from 0.05 to 0.4 mole per mole of the catalyst. The best catalysts were found to be Ru(II) + $TiCl_3$ and Rh(III) + $SnCl_2$. Titanium trichloride and stannous chloride alone were inactive in the fixation of nitrogen under the same experimental conditions. Fixation of nitrogen from the reaction mixture ($N_2 + 3\ H_2$) was established by circulating isotopic nitrogen ($^{15}N_2$, 25% enrichment) and hydrogen in the ratio of 1:3 through the catalyst solution containing Ru(II) + $TiCl_3$ or Rh(III) + $SnCl_2$. The liberated ammonia was absorbed in the Kjeldahl apparatus and the resulting NH_4NO_3 and $^{15}NH_4NO_3$ solutions were carefully dried and examined with a mass spectrograph. Prominent peaks for nitrous oxide at mass numbers 44 and 45 (corresponding to $^{14}N_2O$ and $^{14}N^{15}NO$) established the fixation of nitrogen in the experimental reaction mixture.

B. Mechanism of Nitrogen Fixation by Chemical Systems

1. *The Volpin-Shur Reaction*

The Volpin-Shur catalyst for the fixation of nitrogen, consisting of dicyclopentadienyltitanium dichloride and ethylmagnesium bromide [$(C_5H_5)_2TiCl_2$

and C_2H_5MgBr] dissolved in ether, has attracted much interest as a model system for the fixation of nitrogen. At 1 atm pressure the yield of ammonia obtained with this catalyst is nearly 1 mole of ammonia per mole of catalyst. The mixing of Grignard reagent and the titanium complex resulted in the evolution of gaseous products consisting mainly of ethylene and ethane in the first phase of the reaction prior to nitrogen fixation. Any delay in the addition of nitrogen resulted in a lower yield of ammonia. Hydrolysis at this stage resulted in the formation of hydrogen, indicating that metal hydrides are being formed during the reaction.

The mechanism of fixation of nitrogen with the Volpin-Shur catalyst was studied by Nechiporenko *et al.* [230] by the use of deuterated Grignard reagent and deuterated ether. The use of D_2O for the hydrolysis of reaction products gave HD. If C_2D_5MgBr was used as a reducing agent, C_2D_6 was obtained, indicating that the source of protons in the reaction is not the solvent. This was confirmed by using C_2D_5MgBr and $(C_5H_5)_2TiCl_2$ as reagents in $(C_2D_5)_2O$. After fixation and hydrolysis deuterated ammonia was not obtained. These results indicate that protonation of nitrogen does not take place during reduction, so that the reduced species is probably a metal nitride which ultimately gives ammonia on hydrolysis. By measuring the rate of fixation of nitrogen as a function of the partial pressure of dinitrogen, Maskill and Pratt [231] have also arrived at the conclusion that the reduced nitrogen species in the fixation of nitrogen is a nitride, possibly of zero-valent titanium.

The presence of hydride intermediate in the Volpin-Shur system (prior to fixation) has been confirmed by Brintzinger [232], who carried out an EPR study on the catalyst [$(C_5H_5)_2TiCl_2$ and C_2H_5MgX] in tetrahydrofuran. On mixing these two reagents in equivalent amounts, a greenish brown solution was formed with an absorption maximum at 710 nm and an EPR signal at $g = 1.979$. Excess of Grignard reagent resulted in the disappearance of the original EPR signal; a new signal appeared at $g = 2.00$ and the solution turned very dark brown. When a tenfold excess of ethylmagnesium bromide was added to $(C_5H_5)_2TiCl_2$, the change to brown color was completed in an hour at room temperature, and ethylene and ethane were evolved. The resulting solution gave an EPR signal at $g = 1.993$ consisting of a hyperfine triplet with a 1:2:1 intensity ratio. When hydrochloric acid was added to the final reaction product, red $(C_5H_5)_2TiCl_2$ was formed. The triplet was obtained with reduced intensity by the use of CD_3CH_2MgX, indicating that it is due to the presence of two equivalent hydrogen atoms in a paramagnetic complex formed from dicyclopentadienyltitanium dichloride, $(C_5H_5)_2TiCl_2$, **69**, and ethylmagnesium bromide in equivalent proportions. The original structure suggested for this complex [232] was di-μ-hydrido-bis[dicyclopentadienyltitanium(III)], **70**. Brintzinger [233] has, however, more recently assigned the triplet to a dihydrido mononuclear anion, $[\pi\text{-}(C_5H_5)_2TiH_2]^-$, **71**.

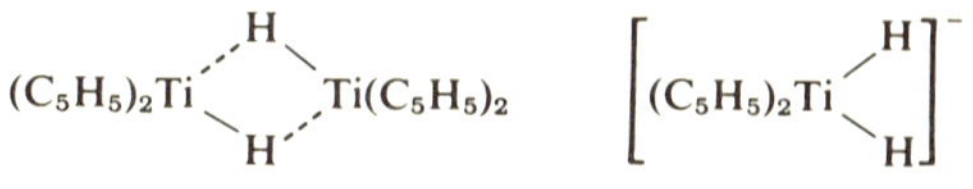

70 **71**

A solution of 0.080 *M* **69** and 0.80 *M* ethylmagnesium bromide in dry tetrahydrofuran at room temperature and 150 atm of dinitrogen gave 1 mole of ammonia per mole of catalyst in 15 days. The final solution before hydrolysis exhibited weak and complicated EPR signals [234]. When the resulting solution was acidified, however, a quintuplet with a *g* value centered at 1.987 (spaced by 2.3 gauss) appeared in the EPR spectrum. Brintzinger [232] first interpreted this result by postulating an intermediate with two equivalent nitrogen nuclei, but later suggested [233] that the hyperfine structure is due to four methylene protons of $[\pi\text{-}(C_5H_5)_2Ti(C_2H_5)_2]^-$, formed by the action of excess of Grignard reagent on the halide species resulting from acidification of the nitrogen-containing complex.

Henrici-Olive and Olive [235] have measured the ESR spectrum of **69** and ethylmagnesium bromide in the ratio of 1:12 in benzene solution at 20°C and obtained a 1:2:1 triplet at $g = 1.9938$. The triplet, which undergoes further splitting on interaction with ring protons, was assigned to the monomeric dihydride complex **71** in confirmation of Brintzinger's conclusions.

Reduction of **69** with lithium naphthalide in the Li/Ti molar ratio of 3:5 in tetrahydrofuran gave a strong ESR signal consisting of a triplet in which each line is further split into a quartet due to interaction with ^7Li ($I = 3/2$) [118]. The ESR signal disappears at Li/Ti ratios of 4 and above. If nitrogen was passed through the solution there was no change in the ESR signal even when one nitrogen atom per titanium was converted in solution to reduced intermediates. This result indicated [118] that hydride hydrogen is not involved in the reduction of dinitrogen. The amount of nitrogen fixed did not change even after 17 hours at molar NH_3/Ti ratios equal to or higher than unity, which supports the view [118] that two atoms of titanium are involved in the reduction of one molecule of nitrogen. Based on the experimental results Henrici-Olive and Olive [118] have proposed the mechanism given in Fig. 25 for the formation of active reducing agents in the fixation of nitrogen. It is suggested that the formation of the Ti(III)-dihydrido species **70** in solution takes place by the reduction of a $[Cp_2Ti]^-$ species, **73**. Two such $[Cp_2Ti]^-$ species extract two protons from the solvent, THF, to form the titanium hydrido dimer **70**. One-electron reduction of **70** then results in the formation of a mixed-valence dinuclear hydride, **74**, which on further

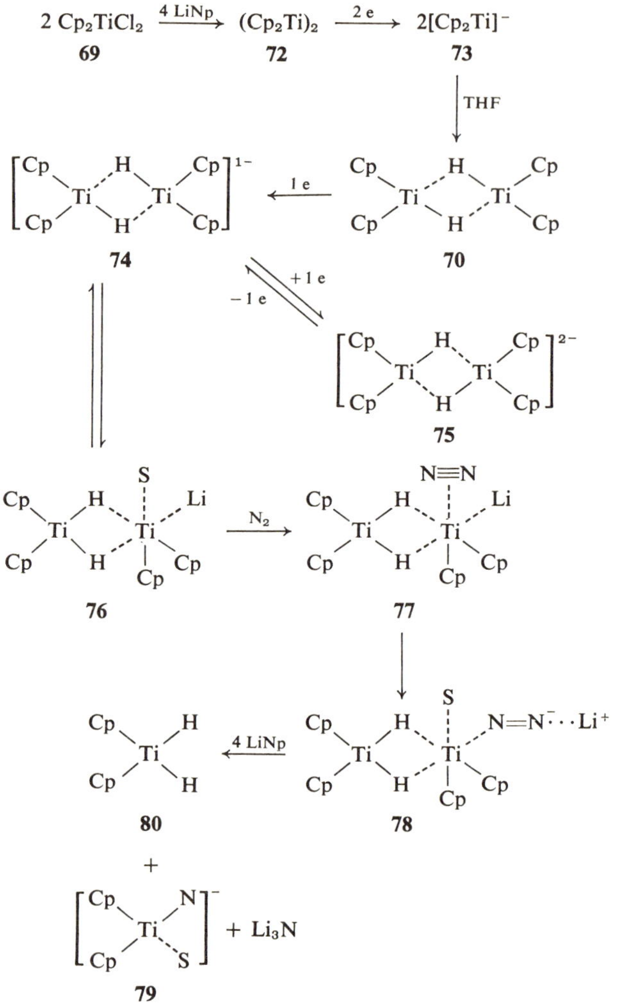

Fig. 25. Reduction of nitrogen with lithium naphthalide and bis(cyclopentadienyl)-titanium dichloride.

reduction gives the ESR-inactive species **75**. The most reactive species in solution is assumed to be the mixed-valence titanium hydride **74**, which is coordinately unsaturated and may have lithium in one of the coordination positions, giving rise to **76**, as evidenced by splitting of each line in the ESR triplet of **74** to give a quartet. It is suggested that the species **76** binds dinitrogen to form **77**. *cis* Insertion of dinitrogen in the Ti—Li bond then gives the species **78**, which is reduced by lithium naphthalide to a stable nitride species **79** and lithium nitride, which on hydrolysis gives ammonia. The titanium(III)-

dihydrido species **80** which is also formed in solution along with the nitride Cp_2TiN gives the ESR triplet even after fixation of nitrogen. It has been suggested [118] that the mechanism depicted in Fig. 25 may be related to biochemical fixation, in that a binuclear azo nitrogen complex (species **78**) is an intermediate in the reaction. The complex **78** is thought to be readily reduced to a nitride while molecular nitrogen requires very high energy for reduction.

Van Tamelen [194,236] has questioned the mechanism suggested by Henrici-Olive and Olive [118] on the basis of the fact that the nitrogen-fixing species **74** (a three-electron reduction product of the titanocene) is not able to fix nitrogen when obtained directly from titanocene by a three-electron naphthalide reduction. Further, deprotonation of tetrahydrofuran should result in the formation of an anion which should immediately become tritiated on reaction with tritium oxide. Treatment of the completed titanocene-dinitrogen fixation product with tritium oxide, however, did not lead to isotopic labeling of THF [194], indicating the absence of THF anion in the fixation process. Starting with a titanium dinitrogen complex, Van Tamelen [194] obtained, in a rapid reaction, 2 moles of ammonia per gram atom of titanium rather than the 1:1 ratio expected with Olive's [118] mechanism. The mechanism given in Eq. (61) was proposed by Van Tamelen [194] to explain his results.

$$\begin{array}{ccccc} 2\,Cp_2TiCl_2 & \xrightarrow{4\,e} & [Cp_2Ti]_2 & \xrightarrow{N_2} & 2\,Cp_2TiN_2 \\ \mathbf{69} & & \mathbf{72} & & \mathbf{81} \\ & & & & \Big\updownarrow \\ & & 2\,N^{3-} & \xleftarrow{e} & [Cp_2TiN_2]_2 \\ & & & & \mathbf{82} \end{array} \qquad (61)$$

Sodium naphthalide reduction of the titanocene **69** gives the dimer **72**, which reacts with nitrogen to form the monomeric species **81**, in equilibrium with the dimer **82**. The dimer **82** has been substantiated by molecular weight determination in benzene solution and an IR band at 1960 cm^{-1}, assigned to $\nu(N_2)$ in the complex. Reduction of **82** yields the nitride, which on hydrolysis gives ammonia.

Brintzinger and Bartell [237] provided evidence for the formation of a highly reactive titanocene species that behaves in a manner similar to a carbene in solution. The transient species bis(cyclopentadienyl)titanium $(C_5H_5)_2Ti$, **83**, is obtained in solution [192] by suspending in ether the gray-green hydride $[\pi(C_5H_5)Ti(\sigma\text{-}C_5H_4)H]_2$ **84**, which decomposes by the evolution of 0.5 mole of hydrogen per mole of titanium to form the mononuclear titanocene $(C_5H_5)_2Ti$, **83**. The presence of **83** has been confirmed by the stoichiometry of the decomposition of **84** and by the infrared spectrum of the resulting solution, which shows strong bands at 790 and 1010 cm^{-1}, corresponding to a simple metallocene. A toluene solution of **83** turns bright green at 100°C due to the formation of a dimer (**72**) which deactivates [237] via a ring to metal hydride shift to form **84** $[\pi(C_5H_5)Ti(\sigma\text{-}C_5H_4)H]_2$. An ether solution of **83** reacts with carbon monoxide at room temperature and nitrogen at −80°C to form the well-characterized dicarbonyl and binuclear dinitrogen complexes $[(C_5H_5)_2Ti(CO)]_2$, **85**, and $[(C_5H_5)_2Ti]_2(N_2)$, **86**. The μ-dinitrogen complex **86** is dark blue in the solid state and quite stable at room temperature when stored under dinitrogen. It decomposes, however, in toluene solution with the evolution of 1 mole of dinitrogen per 2 moles of titanium. It is therefore different from the dimer **82**, in Van Tamelen's mechanism that is expected to give 1 mole of dinitrogen per 2 moles of titanium.

Recently Bercaw and Brintzinger [193] have been able to obtain the methyl-substituted reactive species $[C_5(CH_3)_5]_2Ti$, **87**, that reacts with carbon monoxide and dinitrogen in a manner analogous to that of **83** to form the analogous carbonyl and dinitrogen complexes, $[(C_5(CH_3)_5)_2Ti]_2(CO)_2$, **88**, and $[(C_5(CH_3)_5)_2Ti]_2N_2$, **89**, in solution. In contrast to the dinitrogen complex obtained from $[(C_5H_5)_2Ti]_2N_2$, **86**, the methyl-substituted complex $[C_5(CH_3)_5]_2Ti_2N_2$, **89**, is stable in toluene solution only at low temperature.

Volpin *et al.* [222] suggested a mechanism for the activation of nitrogen which is similar to that proposed by Van Tamelen [236]. According to Volpin *et al.* [222] the most important prerequisite for the fixation of nitrogen is the creation of conditions conducive to the splitting of the metal nitride bond in the transition metal nitride derivative formed, and the subsequent regeneration of the active transition metal compound. For such purposes an aprotic acid such as aluminum bromide may be used. The steps in Eqs. (62)–(67) have been proposed for the catalytic reduction of nitrogen [222] by a titanium tetrahalide, aluminum, and aluminum bromide in benzene solution at 130°C and 100 atm of dinitrogen pressure.

The active catalyst **90**, a titanium(II) species, activates molecular nitrogen to form the μ-dinitrogen complex, **91**. This complex then undergoes reduction by electrons from titanium(II) or aluminum metal to form possible hydrazine intermediates **92** and **93** that are converted to the nitrides **94**, **95**, and **96** in subsequent reduction steps. Splitting of the titanium(III)–nitride bonds by aluminum bromide and reduction of titanium(III) to titanium(II) may be concerted steps which ultimately lead to the regeneration of the catalyst. The end product of reduction is an aluminum nitride compound which gives ammonia on hydrolysis. The fact that aluminum bromide cannot be replaced by aluminum chloride is considered to reflect the noncatalytic nature of the reaction, possibly by producing a nitride product which is very stable. This has been verified in the case of a mixture of titanium tetrachloride, aluminum, and aluminum chloride which gives the active compound $C_6H_6 \cdot TiCl_2 \cdot 2\,AlCl_3$. The latter compound reacts quantitatively with dinitrogen to form a stable nitride corresponding to the formula $C_6H_6(TiCl_2 \cdot 2\,AlCl_3)_3N$, **97**, which hydrolyzes to give a quantitative yield of ammonia. The use of aluminum compounds by Van Tamelen in a cyclic process for the fixation of nitrogen strongly supports [226] Volpin's view about the importance of an aprotic acid in catalyst regeneration.

$$TiX_4 \xrightarrow{Al\,+\,AlBr_3} \underset{\mathbf{90}}{TiX_2 \cdot nAlBr_3} \qquad (62)$$

$$2\,TiX_2 \cdot nAlBr_3 + N_2 \longrightarrow \underset{\mathbf{91}}{N_2(TiX_2 \cdot nAlBr_3)_2} \qquad (63)$$

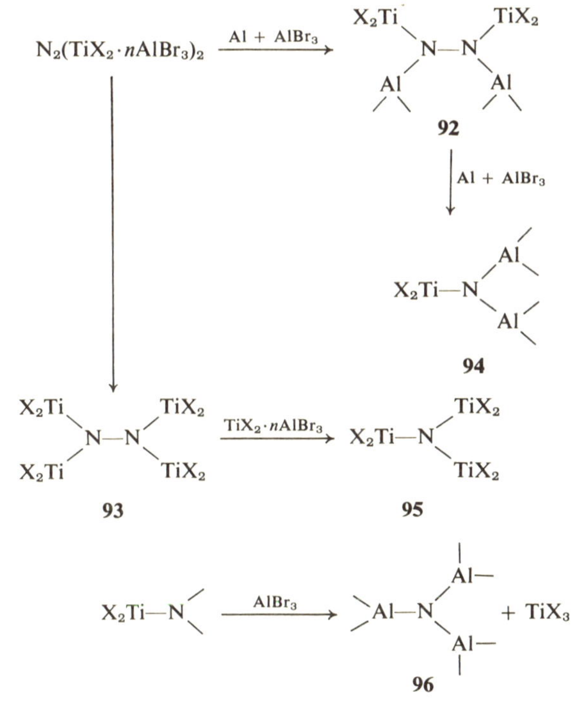

(64)

(65)

(66)

$$TiX_3 + Al + AlBr_3 \longrightarrow \underset{\mathbf{90}}{TiX_2 \cdot nAlBr_3} \qquad (67)$$

In the examples cited above the formation of a metal nitride as the ultimate product of the reduction of nitrogen is supported only by indirect evidence. Direct evidence for the existence of nitrides in systems that reduce molecular nitrogen has recently been provided [238] by the preparation of titanium nitride of the composition $[TiN \cdot Mg_2Cl_2 \cdot THF]$, **98**, which was obtained by the reduction of $TiCl_3 \cdot 3\,THF$ with magnesium in tetrahydrofuran at room temperature and 1 atm pressure of dinitrogen. The complex is a black air-sensitive solid which on hydrolysis produces 1 mole of ammonia per mole of titanium. The overall reaction for the formation of the complex may be written as Eq. (68). The infrared spectrum of the complex gave no evidence

$$TiCl_3 \cdot 3\,THF + \tfrac{5}{2}\,Mg + \tfrac{1}{2}\,N_2 \longrightarrow \underset{\mathbf{98}}{[TiN \cdot Mg_2Cl_2 \cdot THF]} + \tfrac{1}{2}\,MgCl_2 \cdot 2\,THF \qquad (68)$$

for an N—N stretching vibration. The presence of nitrogen in a reduced state (nitride form) is supported by its reaction with protic solvents. Also, magnesium can be completely removed from the complex by the addition of benzoyl chloride.

2. *Formation of Azo and Diimine Complexes*

An interesting example of the reduction of molecular nitrogen through the intermediate formation of an azo metal complex has recently been reported by Volpin *et al.* [239]. On bubbling dinitrogen through a mixture of bis(cyclopentadienyl)titanium(II) chloride or bis(cyclopentadienyl)titanium(III)(diphenyl) with a fivefold excess of phenylithium in ether solution at room temperature and pressure, 0.03 mole of aniline and 0.17 mole of ammonia per mole of the catalyst were formed after hydrolysis of the reaction product. Increase in the dinitrogen pressure to 100 atm increased the yield of aniline and ammonia to 0.15 and 0.65 mole, respectively, per mole of Ti(II). Other titanium catalysts such as titanium tetrachloride, cyclopentadienyltitanium trichloride, and titanium tetrabutoxide also produced aniline and ammonia on bubbling dinitrogen through an ether solution of the catalyst with phenyllithium as the reducing agent. It was proposed that the intermediate is an aryltitanium(II) compound which adds dinitrogen to form a dinitrogen complex [239], **99**, as indicated in Eq. (69). Insertion of molecular nitrogen in the titanium–carbon bond is then considered to take place with the formation of an aryl azo derivative, which on hydrolysis gave aryl amine and ammonia [239], in accordance with the reaction sequence of Eq. (69).

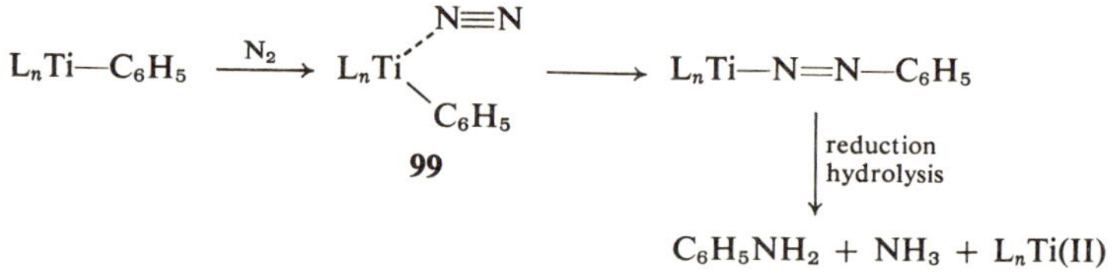

(69)

In the reaction of bis(cyclopentadienyl)titanium dichloride, **69**, $(C_5H_5)_2$-$TiCl_2$, with alkylmagnesium bromides and butyllithium in ether, aliphatic amines were obtained in much lower yields (<0.002 mole). The stability of the titanium–carbon bond also decreases in the order Ti–aryl > Ti–aralkyl > Ti–alkyl, supporting the concept of nitrogen insertion in the Ti–carbon bond as the mechanism for the reaction.

An interesting reaction that incorporates dinitrogen into organic carbonyl compounds to give a good yield of primary and secondary amines and nitriles has recently been reported by Van Tamelen [236]. Magnesium powder is added to a solution of $(C_5H_5)_2TiCl_2$ in tetrahydrofuran under dry dinitrogen at 23°C. After the absorption of dinitrogen is completed, an excess of ketone or aldehyde is added to the black solution and the resulting mixtured stirred at room temperature for 5 days. After work-up the solution gave a 50% yield of primary and secondary amines. Aroyl chlorides gave aryl nitriles in good yield. The overall reactions may be represented as in Eqs. (70) and (71).

$$R_2CO + Ti^{2+} + N^{3-} + 3\,HX \longrightarrow R_2CHNH_2 + X_2TiO + X^- \qquad (70)$$

$$RCOCl + Ti^{2+} + N^{3-} \longrightarrow RCN + TiO + Cl^- \qquad (71)$$

Volpin *et al.* [240] have recently effected the insertion of molecular nitrogen in a metal–carbon bond through the reaction of molecular nitrogen with a mixture of naphthalene and sodium in THF in the presence of titanium tetrachloride. Hydrolysis of the mixture gave ammonia together with aromatic amines, α-naphthylamine, 5,8-dihydro-α-naphthylamine, 5,6,7,8-tetrahydro-α-naphthylamine, and β-naphthylamine. The higher the sodium to naphthalene ratio in the mixture, the higher the yield of amines. No amines or ammonia were obtained in the absence of the titanium compound. The use of lithium gave a higher yield of amines as compared to sodium. With a $TiCl_4/C_{10}H_8/Li$ ratio of 1:6:10, a total of 2.7 moles of amines per mole of titanium was obtained.

Reduction of an aromatic diazonium salt with a metal hydride, which may be considered a model for reduction by metal–enzyme complexes of molecular nitrogen in biological systems, has been reported by Parshall [241]. An extremely rapid reaction (see Fig. 26) takes place between the platinum hydride **100** and the diazonium salt **101** in aqueous ethanol at 25°C to give an aryl diimide complex, **102**. The formation of the diimide complex, **102**, may be analogous to the addition of two hydrogen atoms to an enzyme–dinitrogen complex in biological systems. The diimide complex was isolated as the tetrafluoroborate salt. Reduction of **102** with $Na_2S_2O_4$ or by hydrogenation with a platinum catalyst gives the phenylhydrazine complex **103**. Further hydrogenation of the phenylhydrazine complex for several hours results in the cleavage of the Pt—N and N—N bonds, with the formation of an ammonium salt and aniline. Parshall's [241] model system (Fig. 26) for nitrogenase is in some respects similar to the biological fixation of nitrogen. It operates at room temperature and atmospheric pressure and is influenced by the same inhibitors (carbon monoxide and hydrogen) that affect biological

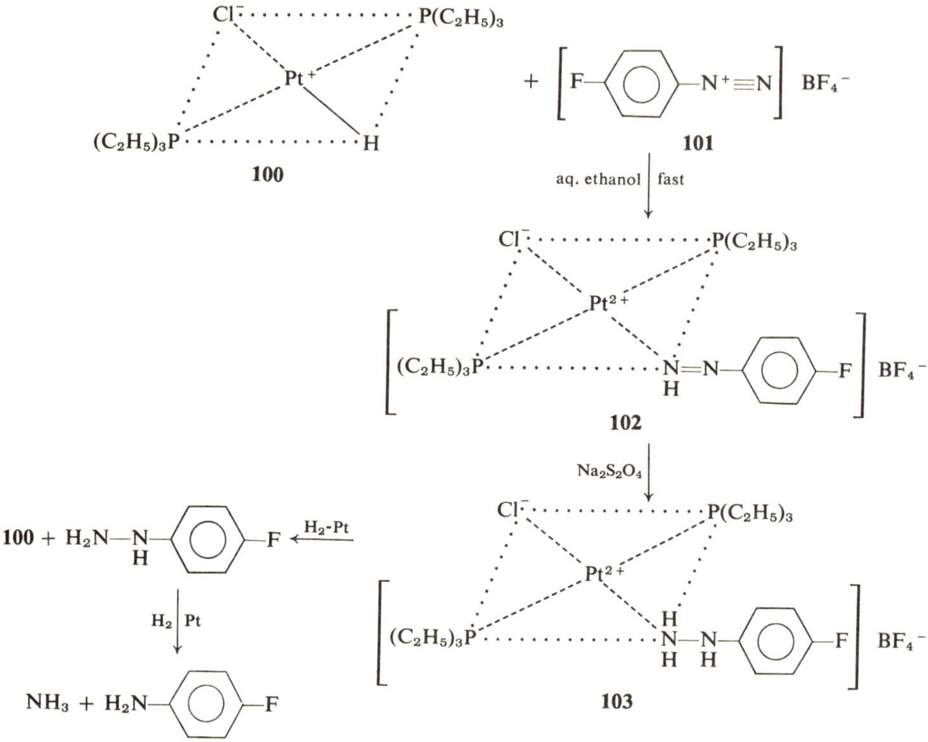

Fig. 26. Parshall's nitrogenase model.

systems. The yield of $(Et_3P)_2PtHCl$ from $(Et_3P)_2PtCl_2$ is low but may be improved by the replacement of chloride with better leaving groups such as phosphate or tosylate. In biological systems adenosine triphosphate possibly may be the leaving group for the formation of an enzyme–hydride complex [241]. The intermediates in Parshall's mechanism are stable to air and neutral water but are very susceptible, as are the biological intermediates, to acids and nucleophiles.

3. *Protonation of Dinitrogen Complexes and Reduction of Dinitrogen to Hydrazine*

Protonation of coordinated dinitrogen and the subsequent reduction of the metal–dinitrogen complex to hydrazine is a very recent and significant development [242–247] in the field of nitrogen fixation. Protonation of coordinated dinitrogen has been achieved in binuclear μ-dinitrogen complexes [242–245] or mononuclear dinitrogen complexes of low-valent metal ions [246,247].

Shilov *et al.* [242] have obtained a dinitrogen complex of titanium(III) of the composition $[(Cp)_2Pr^iTi]_2N_2$, **104**, by the reaction of bis(dicyclopentadienyl titanium(III) chloride and isopropylmagnesium chloride in ether at −80° to −110°C under nitrogen. Complex **104** is a dark blue solid stable at −110°C but decomposing at −70°C with evolution of nitrogen. Prolonged treatment of **104** with isopropylmagnesium chloride at −60°C gave a black compound, **105**, of the composition $[(Cp)_2Ti]_2N_2MgCl$ [242a]. Complex **105** hydrolyzes to give a quantitative yield of hydrazine. An appreciable quantity of hydrazine was also obtained when diethyltitanium dichloride was reduced by sodium naphthalide in oxolan in the presence of molecular nitrogen. Complexes of the type **106** and **107** were proposed [243] as intermediates in the formation of hydrazine by a rearrangement reaction such as illustrated in reaction (72).

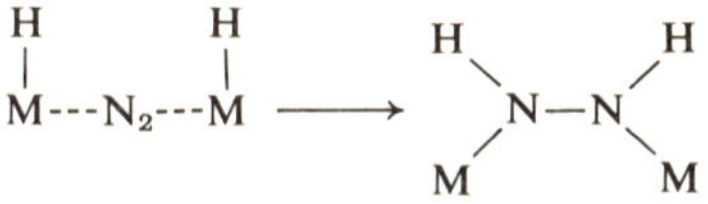

106 **107** (72)

Denisor *et al.* [244] have isolated an unstable iron complex of the composition $[(PPh_3)_2FePr^i(H)N_2Pr^iFe(PPh_3)_2]$, **108**, by the interaction of $(PPh_3)_2FeCl_3$ with isopropylmagnesium chloride in ether at −50°C. Decomposition of **108** with hydrochloric acid resulted in the formation of hydrazine. The mechanism illustrated by Eq. (73) was proposed [244]. The reaction, though quantitative, is noncatalytic with respect to the production of hydrazine.

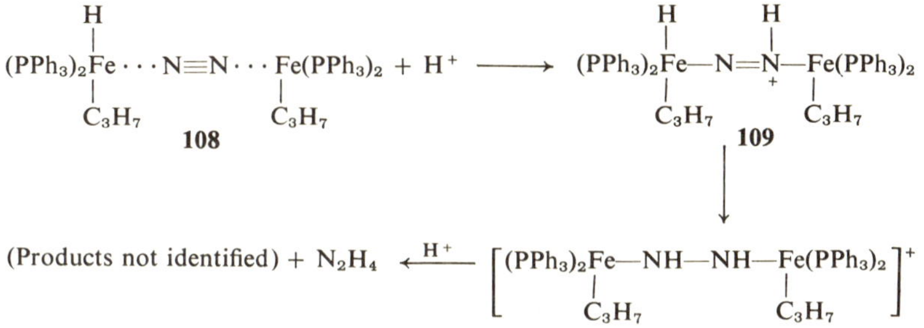

108 **109** **110** (73)

The reduction of molecular nitrogen to hydrazine in the presence of low-valent molybdenum and titanium compounds in aqueous or aqueous–alcoholic media was reported by Denisov *et al.* [244]. The reaction simulates biological nitrogen fixation by the fact that nitrogen is activated by complexes of molybdenum in a low oxidation state [probably [244] molybdenum(III)]. In this system titanium(III) probably plays the role of an electron-transferring agent. Attempts to employ $FeCl_3$, $RuCl_3$, WCl_6, and $KReO_4$ instead of molybdenum compounds were unsuccessful in the reduction of nitrogen by titanium(III). Vanadium(IV) or vanadium(III) were found [244] to be good catalysts for hydrazine formation.

The use of vanadium(II) compounds [244,245] in the presence of magnesium(II) ions in alkaline media under 10 atm of dinitrogen resulted in quantitative formation of hydrazine according to Eq. (74).

$$4\,V(OH)_2 + N_2 + 4\,H_2O \longrightarrow N_2H_4 + 4\,V(OH)_3 \tag{74}$$

The yield of hydrazine was found to increase with the concentration of alkali or magnesium(II). At a hydroxide concentration of 1.0 *M* and at a molar $[Mg^{2+}]/[V^{2+}]$ ratio of about 30, hydrazine formation goes to completion in seconds at room temperature and atmospheric pressure, the yield of hydrazine relative to the reducing agent reaching about 20%.

Hydrazine formation was catalytic [245] at 85°C under 100 atm of nitrogen both for molybdenum–titanium and vanadium(II) complexes. Magnesium(II) was found [245] to strongly activate the reduction of nitrogen in both these systems. The maximum activity of the titanium–molybdenum system was reached when the ratio of magnesium(II) to titanium(III) was 0.5. The mechanism of these catalyzed reactions is speculative at this stage and requires further investigation.

Chatt [246] has reported the protonation reactions of bis(dinitrogen)-tungsten(0) complexes of the type $W(diphos)_2(N_2)_2$, **57**, in 12 *M* hydrochloric acid, resulting in the formation of the diimine complex, $W(diphos)_2(N_2H_2)_2$, **111**. Although **111** has not been fully characterized, the diimine moiety is considered [246] to be coordinated through one of the nitrogen atoms, as indicated. Compound **111** thus seems to be an oxidative addition product of

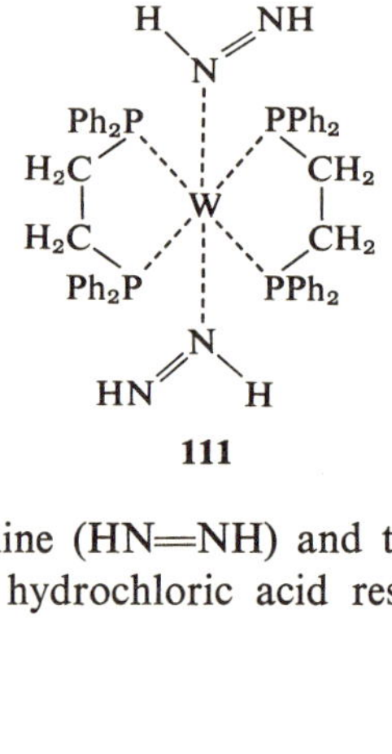

111

the reaction between diimine (HN=NH) and tungsten(0). Reaction of **90** with acetyl chloride and hydrochloric acid resulted in the formation of (diphos)$_2$W(N$_2$HCOCH$_3$)$_2$, **112**, which is an acyl derivative of **111**, with a proton on the coordinated diimine species

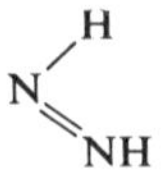

replaced by an acetyl group.

According to Chatt [246] the same types of products are obtained by the protonation of Mo(diphos)$_2$(N$_2$)$_2$, **51**, as are described above for tungsten–dinitrogen complexes.

V. Summary

Fixation of nitrogen to this date has been achieved only in highly reactive systems capable of reducing nitrogen to nitrides. Metal–nitrogen complexes thus far have defied nearly all attempts to reduce them to nitrogen compounds. The transfer of charge to one or both nitrogen atoms in these complexes apparently is not sufficient for protonation and subsequent cleavage of the N—N bond under mild conditions. In nature this has been achieved through the use of binuclear complexes in which the same or different metal ions are coordinated to a bridging nitrogen molecule. Much work needs to be done before the mechanism and energetics of such biological reductions can be understood and fully exploited. It does seem, however, that a logical route for nonbiological nitrogen fixation would be through binuclear complexes in which nitrogen is symmetrically coordinated.

It may or may not be possible eventually to produce ammonia at a price competitive with that of ammonia presently obtained by the Haber-Bosch and other methods, but the experience gained in nitrogen coordination chemistry will certainly reveal new approaches to the synthesis of organic nitrogen compounds such as amines and hydrazines. Insertion of molecular nitrogen in metal–carbon, metal–oxygen, and metal–metal bonds may provide new and challenging synthetic routes for the production of nitrogen compounds. It is perhaps in this direction—the synthesis of new organic nitrogen compounds—that the development of metal ion-catalyzed nitrogen fixation will prove to be the most fruitful.

References

1. A. D. Allen and C. V. Senoff, *Chem. Commun.* p. 621 (1965).
2. A. D. Allen, F. Bottomley, R. O. Harris, V. P. Reinsalu, and C. V. Senoff, *J. Amer. Chem. Soc.* **89**, 5595 (1967).

3. D. R. Harrison and H. Taube, *J. Amer. Chem. Soc.* **89**, 5706 (1967).
4. A. D. Allen and F. Bottomley, *Accounts Chem. Res.* **1**, 360 (1968).
5. J. Chatt and J. E. Fergusson, *Chem. Commun.* p. 126 (1968).
6. L. Yu. Ukhin, Yu. A. Shvetsov, and M. L. Khidekel, *Bull. Acad. Sci. USSR, Div. Chem. Sci.* **7**, 934 (1967).
7. A. D. Allen and J. R. Stevens, *Chem. Commun.* p. 1147 (1967).
8. Yu, G. Borodko, V. S. Bukreev, G. I. Kozub, M. L. Khidekel, and A. E. Shilov, *Zh. Strukt. Khim.* **8**, 542 (1967); *Chem. Abstr.* **67**, 96413b (1967).
9. J. P. Collman and J. W. Kang, *J. Amer. Chem. Soc.* **88**, 3459 (1966).
10. J. P. Collman, M. Kubota, J. Y. Sun, and E. Vastine, *J. Amer. Chem. Soc.* **89**, 169 (1967).
11. A. Yamamoto, S. Kitazume, L. S. Pu, and S. Ikeda, *Chem. Commun.* p. 841 (1969).
12. A. Sacco and M. Rossi, *Chem. Commun.* p. 316 (1967).
13. A. Misono, Y. Uchida, T. Saito, and K. M. Song, *Chem. Commun.* p. 419 (1967).
14. A. Misono, Y. Uchida, and T. Saito, *Bull. Chem. Soc. Jap.* **40**, 799 (1967).
15. M. E. Volpin and V. B. Shur, *Dokl. Akad. Nauk SSSR* **156**, 1102 (1964); *Chem. Abstr.* **61**, 8933a (1964).
16. M. E. Volpin, V. B. Shur, and M. A. Illatsovskaya, *Bull. Acad. Sci. USSR, Div. Chem. Sci.* **4**, 1644 (1964).
17. M. E. Volpin, V. B. Shur, and L. P. Bichin, *Bull. Acad. Sci. USSR, Div. Chem. Sci.* **5**, 698 (1965).
18. M. E. Volpin, M. A. Illatovskaya, E. I. Larikov, M. L. Khidekel, Yu. A. Shvetsov, and V. B. Shur, *Dokl. Akad. Nauk SSSR* **164**, 331 (1965); *Doklady Chemistry*, **164**, 861 (1965).
19. M. E. Volpin and V. B. Shur, *Nature* (*London*) **290**, 1236 (1966).
20. M. E. Volpin, N. K. Chapovskaya, and V. B. Shur, *Bull. Akad. Sci. USSR, Div. Chem. Sci.* **6**, 1033 (1966).
21. M. E. Volpin and V. B. Shur, *Zh. Vses. Khim. Obshchest.* **12**, 31 (1967); *Chem. Abstr.* **66**, 101119p (1967).
22. E. E. Van Tamelen, G. Boche, S. W. Ela, and R. B. Fechter, *J. Amer. Chem. Soc.* **89**, 5707 (1967).
23. G. P. Haight, Jr. and R. Scott, *J. Amer. Chem. Soc.* **86**, 743 (1964).
24. K. B. Yatsimirskii and V. K. Pavlova, *Dokl. Akad. Nauk SSSR* **165**, 130 (1965); *Chem. Abstr.* **64**, 7645d (1965).
25. M. M. Taqui Khan and A. E. Martell, unpublished results.
26. H. S. McKee, "Nitrogen Metabolism in Plants." Oxford Univ. Press, London and New York, 1962.
27. S. Winogradsky, *C. R. Acad. Sci.* **116**, 1385 (1893).
28. S. Winogradsky, *C. R. Acad. Sci.* **118**, 353 (1894).
29. M. W. Beijerinck, *Zentralbl. Bakteriol. Parasitenk., Infektionskr. Hyg., Abt. 2*, **7**, 561 (1901).
30. G. Bond, *Nature* (*London*) **204**, 600 (1964).
31. K. C. Schneider, C. Bradbeer, R. N. Singh, L. C. Wang, P. W. Wilson, and R. H. Burris, *Proc. Nat. Acad. Sci. U.S.* **46**, 726 (1960).
32. H. Gest and M. D. Kamen, *Science* **109**, 558 (1949).
33. H. Gest, J. Judis, and H. D. Peck, Jr., in "Inorganic Nitrogen Metabolism" (W. D. McElroy and B. Glass, eds.), p. 298. Johns Hopkins Press, Baltimore, Maryland, 1956.
34. D. I. Arnon, M. Losada, M. Nozaki, and K. Tagawa, *Nature* (*London*) **190**, 601 (1961).

35. T. Yamanaka and M. D. Kamen, *Biochem. Biophys. Res. Commun.* **18**, 611 (1965).
36. J. E. Carnahan, L. E. Mortensen, H. E. Mower, and J. E. Castle, *Biochim. Biophys. Acta* **38**, 188 (1960); **44**, 520 (1960).
37. J. E. Carnahan and J. E. Castle, *Annu. Rev. Plant Physiol.* **15**, 125 (1964).
38. R. H. Burris, *Annu. Rev. Plant Physiol.* **17**, 155 (1966).
39. D. J. D. Nicholas, D. J. Fischer, W. J. Redmond, and W. A. Wright, *J. Gen. Microbiol.* **22**, 191 (1960).
40. R. H. Burris, in "Inorganic Nitrogen Metabolism" (W. D. McElroy and B. Glass, eds.), p. 316. Johns Hopkins Press, Baltimore, Maryland, 1956.
41. A. I. Virtanen and M. Hakala, *Acta Chem. Scand.* **3**, 1044 (1949).
42. A. I. Virtanen, A. Kemppii, and E. L. Salmenoja, *Acta Chem. Scand.* **8**, 1729 (1954).
43. R. H. Burris and P. W. Wilson, *J. Bacteriol.* **52**, 505 (1946).
44. I. Zelitch, E. D. Rosenblum, R. H. Burris, and P. W. Wilson, *J. Biol. Chem.* **191**, 295 (1951).
45. R. M. Allison and R. H. Burris, *J. Biol. Chem.* **224**, 351 (1957).
46. W. E. Magee and R. H. Burris, *Plant Physiol.* **29**, 199 (1954).
47. N. Bauer, *Nature (London)* **188**, 471 (1960).
48. M. K. Bach, *Biochim. Biophys. Acta* **26**, 104 (1957).
49. M. B. Azim and E. R. Roberts, *Biochim. Biophys. Acta* **21**, 308 (1956).
50. R. W. F. Hardy, E. Knight, Jr., and A. J. D'Eustachio, *Biochem. Biophys. Res. Commun.* **23**, 409 (1966).
51. D. J. D. Nicholas, *Annu. Rev. Plant. Physiol.* **12**, 63 (1961).
52. R. F. Keeler, W. A. Bulen, and J. E. Varner, *J. Bacteriol.* **72**, 394 (1956).
53. W. A. Bulen, R. C. Burns, and J. R. Lecomte, *Biochem. Biophys. Res. Commun.* **17**, 265 (1964).
54. R. F. Keeler and J. E. Varner, *Arch. Biochem. Biophys.* **70**, 585 (1957).
55. L. E. Mortensen, R. C. Valentine, and J. E. Carnahan, *Biochem. Biophys. Res. Commun.* **7**, 448 (1962).
56. K. Tagawa and D. I. Arnon, *Nature (London)* **195**, 537 (1962).
57. H. Kubo, *Acta Phytochim.* **11**, 195 (1939).
58. N. Ellfolk and A. I. Virtanen, *Acta Chem. Scand.* **6**, 411 (1952).
59. K. Abel, *Phytochemistry* **2**, 429 (1963).
60. K. Abel, N. Bauer, and T. Spence, *Arch. Biochem. Biophys.* **100**, 339 (1963).
61. R. G. Esposito and P. W. Wilson, *Proc. Soc. Exp. Biol. Med.* **93**, 564 (1956).
62. J. Peive, *Agrokhimiya* **3**, 18 (1964).
63. C. A. Parker and P. B. Scutt, *Biochim. Biophys. Acta* **29**, 662 (1958).
64. C. A. Parker and P. B. Scutt, *Biochim. Biophys. Acta* **38**, 230 (1960).
65. A. L. Shug, P. W. Wilson, D. E. Green, and H. R. Mahore, *J. Amer. Chem. Soc.* **76**, 3355 (1954).
66. G. E. Hoch, H. N. Little, and R. H. Burris, *Nature (London)* **179**, 430 (1957).
67. G. E. Hoch, K. C. Schneider, and R. H. Burris, *Biochim. Biophys. Acta* **37**, 273 (1960).
68. N. Bauer, *J. Phys. Chem.* **64**, 833 (1960).
69. F. J. Bergersen, *Biochim. Biophys. Acta* **50**, 576 (1961).
70. L. E. Mortensen, R. C. Valentine, and J. E. Carnahan, *J. Biol. Chem.* **238**, 794 (1963).
71. L. E. Mortensen, *Annu. Rev. Microbiol.* **17**, 115 (1963).
72. W. A. Bulen, J. R. LeComte, R. C. Burns, and J. Hinkson, *in* "Non-Heme Iron Proteins: Role in Energy Conversion" (A. San Pietro, ed.), p. 261. Antioch Press, Yellow Springs, Ohio, 1965.

73. M. Kelly, *Biochim. Biophys. Acta* **171**, 9 (1969).
74. R. W. F. Hardy and R. C. Burns, *Annu. Rev. Biochem.* **37**, 331 (1968).
75. R. C. Valentine, L. E. Mortensen, and J. E. Carnahan, *J. Biol. Chem.* **238**, 114 (1963).
76. M. Tanaka, T. Nakashima, A. Benson, H. F. Mower, and K. J. Yasunobu, *Biochem. Biophys. Res. Commun.* **16**, 422 (1964).
77. J. S. Hong and J. C. Rabinovitz, *J. Biol. Chem.* **245**, 4982 (1970).
78. W. H. Orme Johnson and H. Beinert, *Biochem. Biophys. Res. Commun.* **36**, 337 (1969).
79. L. C. Sieker, E. Adman, and L. H. Jensen, *Nature (London)* **235**, 40 (1972).
80. C. W. Carter, Jr., S. T. Freer, Na. H. Xuong, R. A. Alden, and J. Kraut, *Cold Spring Harbor Symp. Quant. Biol.* **36**, 381 (1972).
81. J. C. Tsibris and R. W. Woody, *Coord. Chem. Rev.* **5**, 417–458 (1970).
82. C. H. Wei, G. R. Wilkes, P. M. Treichel, and L. F. Dahl, *Inorg. Chem.* **5**, 900 (1966).
83. J. A. Ibers, T. J. Marks, W. R. Robinson, J. P. Collman, L. F. Dahl, R. Poilblanc, and M. D. Grillone, *Proc. Int. Conf. Coord. Chem., 14th*, p. 1 (1972).
84. D. Leibfritz, *Angew. Chem., Int. Ed. Engl.* **11**, 232 (1972).
85. R. W. F. Hardy and E. Knight, Jr., *in* "Progress in Phytochemistry" (L. Reinhold, ed.), p. 387. Wiley, London, 1967.
86. J. P. Collman, *Accounts Chem. Res.* **1**, 136 (1968).
87. R. C. Burns, R. D. Holsten, and R. W. F. Hardy, *Biochem. Biophys. Res. Commun.* **39**, 90 (1970).
88. R. W. F. Hardy and E. Knight, Jr., *Biochem. Biophys. Res. Commun.* **23**, 409 (1966).
89. A. J. D'Eustachio and R. W. F. Hardy, *Biochem. Biophys. Res. Commun.* **15**, 319 (1964).
90. R. W. F. Hardy, E. Knight, Jr., and E. K. Jackson, *Science* **157**, 110 (1967).
91. R. Scholhorn and R. H. Burris, *Proc. Nat. Acad. Sci. U.S.* **58**, 213 (1967).
92. A. Lockshin and R. H. Burris, *Biochim. Biophys. Acta* **111**, 1 (1965).
93. M. Kelley, J. R. Postgate, and R. Richards, *Biochem. J.* **102**, 1 (1967).
94. G. N. Schrauzer and G. Schlesinger, *J. Amer. Chem. Soc.* **92**, 1808 (1970).
95. G. N. Schrauzer, G. Schlesinger, and P. A. Doemeny, *J. Amer. Chem. Soc.* **93**, 1803 (1971).
96. G. N. Schrauzer and P. A. Doemeny, *J. Amer. Chem. Soc.* **93**, 1608 (1971).
97. G. N. Schrauzer, P. A. Doemeny, G. W. Kiefer, and R. H. Frazier, *J. Amer. Chem. Soc.* **94**, 3604 (1972).
98. W. E. Newton, J. L. Corlin, P. W. Schneider, and W. A. Bulen, *J. Amer. Chem. Soc.* **93**, 368 (1971).
99. M. D. Kamen and H. Gest, *Science* **109**, 560 (1949).
100. A. E. Shilov, A. K. Shilova, and Yu. G. Borodko, *Kinet. Catal. (USSR)* **10**, 213 (1969).
101. A. Sacco and M. Aresta, *Chem. Commun.* p. 1223 (1968).
102. G. M. Bancroft, M. J. Mays, and B. E. Prater, *Chem. Commun.* p. 39 (1967).
103. J. Chatt, G. J. Leigh, and R. L. Richards, *Chem. Commun.* p. 515 (1969).
104. G. M. Bancroft, M. J. Mays, B. E. Prater, and E. P. Stefanini, *J. Chem. Soc., A* p. 2146 (1970).
105. P. W. Jolly and K. Jonas, *Angew. Chem.* **80**, 705 (1968).
106. M. Rossi and A. Sacco, *Chem. Commun.* p. 471 (1969).
107. J. Chatt, D. P. Melville, and R. L. Richards, *J. Chem. Soc., A* p. 2841 (1969).
108. L. A. P. Kane-Maguire, P. S. Sheridan, F. Basolo, and R. G. Pearson, *J. Amer. Chem. Soc.* **92**, 5865 (1970).

109. P. G. Douglas, R. D. Feltham, and H. G. Metzger, *Chem. Commun.* p. 889 (1970).
110. P. G. Douglas, R. D. Feltham, and H. G. Metzger, *J. Amer. Chem. Soc.* **93**, 84 (1971).
111. F. Bottomley, *Can. J. Chem.* **48**, 357 (1970).
111a. F. Bottomley, *Quart. Rev. Chem. Soc.* **24**, 617 (1970).
112. A. A. Diamantis and G. J. Sparrow, *Chem. Commun.* p. 469 (1969).
113. J. N. Armor and H. Taube, *J. Amer. Chem. Soc.* **91**, 6874 (1969).
114. S. Pell and J. N. Armor, *J. Amer. Chem. Soc.* **94**, 686 (1972).
115. L. S. P. Kane-Maguire, P. S. Sheridan, F. Basolo, and R. G. Pearson, *J. Amer. Chem. Soc.* **90**, 5295 (1968).
116. R. Murray and D. C. Smith, *Coord. Chem. Rev.* **3**, 429 (1968).
117. K. Kuchyuka, *Catal. Rev.* **3**, 11 (1969).
118. G. Henrici-Olive and S. Olive, *Angew. Chem., Int. Ed. Engl.* **8**, 650 (1969).
119. Yu. G. Borodko and A. E. Shilov, *Russ. Chem. Rev.* **38**, 355 (1969).
120. J. E. Fergusson and J. L. Love, *Rev. Pure Appl. Chem.* **20**, 33 (1970).
121. A. D. Allen, *Advan. Chem. Ser.* **100**, 79–94 (1971).
121a. A. D. Allen, R. O. Harris, B. R. Loescher, J. R. Stevens, and R. N. Whiteley, *Chem. Rev.* **73**, 11 (1973).
122. Yu. G. Borodko, S. M. Vinogradova, Yu. P. Myagkov, and D. O. Mozzhukin, *J. Struct. Chem.* (*USSR*) **11**, 251 (1970).
122a. J. Chatt, R. L. Richards, J. E. Fergusson and J. C. Love, *Chem. Commun.* p. 1522 (1968).
123. A. D. Allen and F. Bottomley, *J. Amer. Chem. Soc.* **91**, 1231 (1969).
124. Yu. G. Borodko, A. K. Shilova, and A. E. Shilov, *Russ. J. Phys. Chem.* **44**, 349 (1970).
125. F. Bottomley and S. C. Nyburg, *Chem. Commun.* p. 897 (1966).
126. D. E. Harrison, E. Weissberger, and H. Taube, *Science* **159**, 320 (1968).
127. I. M. Trietel, M. T. Flood, R. E. Marsh, and H. B. Gray, *J. Amer. Chem. Soc.* **91**, 6512 (1969).
128. J. Chatt, A. B. Mikolsky, R. L. Richards, and J. R. Sanders, *Chem. Commun.* p. 154 (1969).
129. A. D. Allen and F. Bottomley, *Can. J. Chem.* **46**, 469 (1968).
130. A. A. Diamantis and G. J. Sparrow, *Chem. Commun.* p. 819 (1970).
131. F. Bottomley and J. R. Crawford, *Chem. Commun.* p. 200 (1971).
132. J. N. Armor and H. Taube, *Chem. Commun.* p. 287 (1971).
133. B. R. Davis and J. A. Ibers, *Inorg. Chem.* **9**, 2768 (1970).
134. I. J. Itzkovitch and J. A. Page, *Can. J. Chem.* **46**, 2743 (1968).
135. J. N. Armor and H. Taube, *J. Amer. Chem. Soc.* **92**, 6874 (1970).
136. E. L. Farquhar, L. Rusnock, and S. J. Gill, *J. Amer. Chem. Soc.* **92**, 416 (1970).
137. A. K. Shilova and A. E. Shilov, *Kinet. Catal.* **10**, 213 (1969).
138. A. Yamamoto, S. Kitazume, and S. Ikeda, *J. Amer. Chem. Soc.* **90**, 1089 (1968).
139. W. H. Knoth, *J. Amer. Chem. Soc.* **90**, 7172 (1968).
140. T. I. Iliades, R. O. Harris, and M. C. Zie, *Chem. Commun.* p. 1709 (1970).
141. I. Takashi, S. Kitazume, A. Yamamoto, and S. Ikeda, *J. Amer. Chem. Soc.* **92**, 3011 (1970).
142. W. H. Knoth, *J. Amer. Chem. Soc.* **94**, 104 (1972).
143. M. M. Taqui Khan and R. K. Andal, submitted to *J. Chem. Soc.* (Dalton).
144. M. M. Taqui Khan and R. K. Andal, unpublished results.
145. J. E. Fergusson and J. L. Love, *Chem. Commun.* p. 399 (1969).
146. Yu. G. Borodko, G. J. Kozub, and Yu. P. Myagkov, *Russ. J. Phys. Chem.* **44**, 643 (1970).

147. J. E. Fergusson, J. L. Love, and W. T. Robinson, *Inorg. Chem.* **11**, 1662 (1972).
148. P. K. Das, J. M. Pratt, R. G. Smith, G. Swinden, and W. J. V. Woolcock, *Chem. Commun.* p. 1539 (1968).
149. J. Chatt, J. R. Dilworth, H. P. Gunz, G. J. Leigh, and Y. Kitahara, *Chem. Commun.* p. 90 (1970).
150. C. M. Elson, J. Gulens, I. J. Itzkovitch, and J. A. Page, *Chem. Commun.* p. 875 (1970).
151. Yu. P. Myagkov, M. A. Kaplunov, Yu. G. Borodko, and A. E. Shilov, *Kinet. Catal.* (*USSR*) **12**, 1030 (1972).
152. Yu. G. Borodko, A. K. Shilova, and A. B. Shilov, *Dokl. Akad. Nauk SSSR* **176**, 1297 (1967).
153. J. Chatt, G. J. Leigh, and R. L. Richards, *J. Chem. Soc., A* 2243 (1970).
154. J. Chatt, D. P. Melville, and R. L. Richards, *J. Chem. Soc., A* p. 895 (1971).
155. P. K. Maples, F. Basolo, and R. G. Pearson, *Inorg. Chem.* **10**, 765 (1971).
156. H. A. Scheidegger, J. N. Armor, and H. Taube, *J. Amer. Chem. Soc.* **90**, 3263 (1968).
157. M. Clive, I. Elson, J. Itzkovitch, and J. A. Page, *Can. J. Chem.* **48**, 1639 (1970).
158. C. M. Elson, J. Gulens, and J. A. Page, *Can. J. Chem.* **49**, 207 (1971).
159. G. Speier and L. Marko, *Inorg. Chim. Acta* **3**, 126 (1969).
160. G. Speier and L. Marko, *J. Organometal. Chem.* **21**, 46 (1970).
161. J. H. Enemark, B. R. Davis, J. A. McGinnety, and J. A. Ibers, *Chem. Commun.* p. 96 (1968).
162. B. R. Davis, N. C. Payne, and J. A. Ibers, *J. Amer. Chem. Soc.* **91**, 1241 (1969).
163. B. R. Davis, N. C. Payne, and J. A. Ibers, *Inorg. Chem.* **8**, 2719 (1969).
164. A. Misono, Y. Uchida, M. Hidai, and M. Araki, *Chem. Commun.* p. 1044 (1968).
165. L. S. Pu, A. Yamamoto, and S. Ikeda, *J. Amer. Chem. Soc.* **90**, 3896 (1968).
166. A. Misono, Y. Uchida, M. Hidai, and T. Kuse, *Chem. Commun.* p. 981 (1968).
167. G. W. Parshall, *J. Amer. Chem. Soc.* **90**, 1669 (1968).
168. L. S. Pu, A. Yamamoto, and S. Ikeda, *Chem. Commun.* p. 189 (1969).
169. A. Yamamoto, S. Kitazume, L. Sunpu, and S. Ikeda, *J. Amer. Chem. Soc.* **93**, 371 (1971).
169a. A. Yamamoto, L. S. Pu, S. Kitazume, and S. Ikeda, *J. Amer. Chem. Soc.* **89**, 3071 (1967).
170. M. Aresta, C. F. Nobile, M. Rossi, and A. Sacco, *Chem. Commun.* p. 781 (1971).
171. L. S. Volkova, V. M. Volkova, and S. S. Chernikov, *Russ. J. Inorg. Chem.* **16**, 1383 (1971).
172. W. E. Silverthorn, *Chem. Commun.* p. 1310 (1971).
173. J. Chatt, J. R. Dilworth, and G. J. Leigh, *Chem. Commun.* p. 687 (1969).
174. J. Chatt, J. R. Dilworth, G. J. Leigh, and R. L. Richards, *Chem. Commun.* p. 955 (1970).
175. J. Chatt, J. R. Dilworth, G. J. Leigh, and R. L. Richards, *J. Chem. Soc., A* p. 702 (1971).
176. B. R. Davis and J. A. Ibers, *Inorg. Chem.* **10**, 578 (1971).
177. J. Chatt, J. R. Dilworth, and G. J. Leigh, *J. Organometal. Chem.* **21**, 49 (1970).
178. J. J. Molewyn Hughes and A. W. B. Garner, *Chem. Commun.* p. 1309 (1969).
179. D. Sellman, *J. Organometall. Chem.* **36**, C27 (1972).
180. M. Hidai, K. Tominari, Y. Uchida, and M. Misono, *Chem. Commun.* p. 814 (1969).
181. M. Hidai, K. Tominari, and Y. Uchida, *J. Amer. Chem. Soc.* **94**, 110 (1972).
182. L. K. Atkinson, A. H. Mawby, and D. C. Smith, *Chem. Commun.* p. 157 (1971).
183. M. Hidai, K. Tominari, Y. Uchida, and A. Misono, *Chem. Commun.* p. 1392 (1969).

184. B. Bell, J. Chatt, and G. J. Leigh, *Chem. Commun.* p. 576 (1970).
185. B. Bell, J. Chatt, and G. J. Leigh, *Chem. Commun.* p. 843 (1970).
186. P. W. Jolly, K. Jonas, C. Kruger, and Y. H. Tsay, *J. Organometall. Chem.* **33**, 109 (1971).
187. G. L. Johnson and W. D. Beveridge, *Inorg. Nucl. Chem. Lett.* **3**, 323 (1967).
188. M. M. Taqui Khan and Q. Anwaruddin, unpublished results.
189. D. Sellman, *Angew Chem., Int. Ed. Engl.* **10**, 919 (1971).
190. A. Misono, Y. Uchida, T. Saito, M. Hidai, and M. Araki, *Inorg. Chem.* **8**, 159 and 168 (1969).
191. J. H. Teuben and H. J. de Liefde Meijer, *Rec. Trav. Chim. Pays-Bas* **90**, 360 (1971).
192. R. M. Marvich and H. H. Brintzinger, *J. Amer. Chem. Soc.* **93**, 2046 (1971).
193. J. E. Bercaw and H. H. Brintzinger, *J. Amer. Chem. Soc.* **93**, 2046 (1971).
194. E. E. Van Tamelen, *Amer. Chem. Soc., Monogr.* **100**, 95–110 (1971).
195. V. M. Volkova and L. V. Volkova, *Russ. J. Inorg. Chem.* **16**, 1382 (1971).
196. R. B. King, *Amer. Chem. Soc. Activation Small Mol., 1972* (1972).
197. P. J. Emmett, "Catalysis." Van Nostrand-Reinhold, Princeton, New Jersey, 1955.
198. R. P. Eischens and J. Jackson, *Proc. Int. Congr. Catal., 3rd, 1964* p. 627 (1965).
199. S. S. Chernikov, S. G. Kuznin, and Yu. G. Borodko, *Russ. J. Phys. Chem.* **8**, (1969).
200. R. Van Hardeveld and A. Van Montfoort, *Surface Sci.* **4**, 396 (1966).
201. Yu. G. Borodko, A. E. Shilov, and A. A. Shteinman, *Dokl. Akad. Nauk SSSR.* **168**, 581 (1966).
202. A. E. Shilov, A. A. Steinmann, and M. B. Tyantun, *Tetrahedron Lett.* p. 4177 (1968).
203. D. C. Owsley and G. K. Helmkamp, *J. Amer. Chem. Soc.* **89**, 4558 (1967).
204. J. Ellermann, F. Poersch, R. Kuntsmann, and R. Kramolowsky, *Angew. Chem., Int. Ed. Engl.* **8**, 203 (1969).
205. L. Pauling, *Tetrahedron* **17**, 119 (1962).
206. C. W. Scherr, *J. Chem. Phys.* **23**, 569 (1955).
207. B. J. Ransil, *Rev. Mod. Phys.* **32**, 239 (1960).
208. J. W. Richardson, *J. Chem. Phys.* **35**, 1829 (1961).
209. J. Chatt, *Proc. Roy. Soc., Ser. B* **172**, 327 (1969).
210. J. Chatt, *Platinum Metals Rev.* **13**, 9 (1969).
211. W. C. Price, *in* "Molecular Spectroscopy" (P. Hepple, ed.), pp. 221–236. Institute of Petroleum, London, 1968.
212. P. E. Cade and K. D. Sales, *J. Chem. Phys.* **44**, 1973 (1966).
213. L. Orgel, "Transition Metal Chemistry, Ligand Field Theory," p. 137. Methuen, London, 1960.
214. D. J. Darensbourg, *Inorg. Chem.* **11**, 1436 (1972).
215. J. A. Stanko and T. W. Starinshak, *Inorg. Chem.* **8**, 2156 (1969).
216. H. H. Jaffe and M. Orchin, *Tetrahedron* **10**, 212 (1960).
217. K. G. Caulton, R. L. Dekock, and R. F. Fenski, *J. Amer. Chem. Soc.* **92**, 515 (1970).
218. D. J. Darensbourg and C. L. Hyde, *Inorg. Chem.* **10**, 431 (1971).
219. G. M. Bancroft, M. J. Mays, and B. E. Prater, *Chem. Commun.* p. 39 (1967).
220. G. M. Bancroft, R. E. H. Garrod, A. G. Maddock, M. J. Mays, and B. E. Prater, *J. Amer. Chem. Soc.* **94**, 647 (1972).
221. G. Henrici-Olive and S. Olive, *Angew. Chem., Int. Ed. Engl.* **6**, 873 (1967).
222. M. F. Volpin, M. A. Illatovskaya, and V. B. Shur, *Kinet. Catal.* (*USSR*) **11**, 279 (1970).
223. M. E. Volpin, M. A. Illatovskaya, L. V. Kosyakova, and V. B. Shur, *Chem. Commun.* p. 1074 (1968).

224. E. E. Van Tamelen, G. Boche, and R. Greeley, *J. Amer. Chem. Soc.* **90**, 1674 (1968).
225. E. E. Van Tamelen, R. B. Fechter, S. W. Schneller, G. Boche, R. H. Greeley, and B. Akermark, *J. Amer. Chem. Soc.* **91**, 1551 (1969).
226. E. E. Van Tamelen and D. A. Seeley, *J. Amer. Chem. Soc.* **91**, 5194 (1969).
227. D. R. Gray and C. H. Brubaker, *Chem. Commun.* p. 1239 (1969).
228. L. G. Bell and H. Brintzinger, *J. Amer. Chem. Soc.* **92**, 4464 (1970).
229. E. E. Van Tamelen, R. B. Fechter, and S. W. Schneller, *J. Amer. Chem. Soc.* **91**, 7196 (1969).
230. G. N. Nechiporenko, G. M. Tabrina, A. K. Shilova, and A. E. Shilov, *Dokl. Akad. Nauk SSSR* **164**, 1062 (1965); *Chem. Abstr.* **63**, 6686a (1965).
231. R. Maskill and J. M. Pratt, *Chem. Commun.* p. 950 (1967).
232. H. Brintzinger, *J. Amer. Chem. Soc.* **88**, 4305 (1966).
233. H. Brintzinger, *J. Amer. Chem. Soc.* **89**, 687 (1967).
234. H. Brintzinger, *J. Amer. Chem. Soc.* **88**, 4307 (1966).
235. G. Henrici-Olive and S. Olive, *Angew. Chem., Int. Ed. Engl.* **7**, 386 (1968).
236. E. E. Van Tamelen, *Accounts Chem. Res.* **3**, 361 (1970).
237. H. H. Brintzinger and L. S. Bartell, *J. Amer. Chem. Soc.* **92**, 1105 (1970).
238. A. Yamamoto, M. Ookaw, and S. Ikeda, *Chem. Commun.* p. 841 (1969).
239. M. E. Volpin, V. B. Shur, R. V. Kudryavtsev, and L. A. Prodayko, *Chem. Commun.* p. 1038 (1968).
240. M. E. Volpin, A. A. Kelvi, V. B. Shur, N. A. Katkov, I. M. Mekaeva, and R. V. Kudryavtsev, *Chem. Commun.* p. 246 (1971).
241. G. W. Parshall, *J. Amer. Chem. Soc.* **89**, 1822 (1967).
242. A. E. Shilov, A. K. Shilova, E. F. Kavashina, and T. A. Vorontsova, *Chem. Commun.* p. 1590 (1971).
242a. Yu. G. Borodko, I. N. Ivleva, L. M. Kachapina, E. F. Kavashina, A. K. Shilova, and A. E. Shilov, *Chem. Commun.* p. 169 (1973).
243. A. E. Shilov and A. K. Shilova, *Russ. J. Phys. Chem.* **44**, 164 (1970).
244. N. T. Denisov, O. N. Efimov, N. I. Shuvalova, A. K. Shilova, and A. E. Shilov, *Russ. J. Phys. Chem.* **44**, 1693 (1970).
245. A. Shilov, N. T. Denisov, O. N. Efimov, N. I. Shuvalova, N. Shuvalov, and A. K. Shilova, *Nature (London)* **231**, 460 (1971).
246. J. Chatt, *Amer. Chem. Soc. Sym. Activation Small Mol., 1972* (1972).
247. J. Chatt, G. A. Heath, and R. L. Richards, *Chem. Commun.* p. 1010 (1972).

4

Activation of Carbon Monoxide

I. Introduction

Carbon monoxide is one of the most important and useful ligands in synthetic inorganic chemistry, and its complexes with transition metals have been the subject of extensive research because of their interesting structures and chemical bonding, the many and varied reactions which they undergo, and their important applications to synthesis. The extensive investigations of carbonyl complexes now underway are expected to produce an abundance of new compounds and new reaction types in the years to come. The unusual structures, bonding, and properties of these compounds have opened up virtually limitless vistas of research for theoretical chemists, as well as for those interested in the synthesis of new organic and organometallic compounds.

Basic research on carbonyl complexes has been reinforced and given added vigor by the many industrial applications of these compounds. Metal carbonyls are intermediates in the formation of synthetic hydrocarbons by the Fischer-Tropsch process. The synthesis of alcohols and aldehydes in the oxo reaction takes place through the formation of metal carbonyl intermediates. Hundreds of industrial compounds are now synthesized by the incorporation of one or more carbonyl groups in an organic molecule, mainly through the catalytic effects of metal carbonyls.

The present chapter deals with only those reactions in which carbon monoxide is activated in the coordination sphere of a metal ion. Carbon monoxide exchange reactions will be considered first since such reactions provide an insight into the nature and reactivities of metal carbonyls that act as intermediates in the metal-activated insertion of carbon monoxide into organic molecules. This treatment is followed by a description of these insertion reactions and their synthetic applications, for cases where metal ion activation of the carbon monoxide has been demonstrated or seems to be indicated. The final section deals with the metal-catalyzed oxidation of carbon monoxide. The oxo reaction, in which an alkene or alkyne is first activated by π-bond coordination to the metal ion, followed by addition of carbon monoxide and hydrogen, is treated in Volume II.

II. Carbon Monoxide Exchange Reactions

The carbonyls selected as examples for this review of carbon monoxide exchange reactions are $Mn(CO)_5X$ (X = halogen); R—$Co(CO)_4$ (R = alkyl or aryl); $Co_2(CO)_8$, $Ni(CO)_4$, and $Fe(CO)_5$. These carbonyls may be considered as complexes of carbon monoxide where each carbonyl group donates a pair of electrons to the central metal ion with back π bonding from the metal ion to carbon monoxide. These donor groups may be replaced by other donors capable of forming π bonds (as π acceptors) with the central metal ion, such as nitric oxide, tertiary phosphines, arsines, stibines, and cyanide ion.

Uncoordinated (i.e., "free") carbon monoxide has a bond length of 1.131 Å [1], with a bond dissociation energy of 257.3 kcal/mole. The low polarity of free carbon monoxide is indicated by its low dipole moment of 0.14 Debye units [2].

On the basis of infrared measurements halopentacarbonylmanganese, $Mn(CO)_5X$, has been reported [3] to possess a distorted octahedral structure with the axial carbon monoxide ligand more strongly bound to the central metal ion than the planar carbonyl groups. This finding is confirmed by the carbon monoxide exchange studies on $Mn(CO)_5X$ reported by Wojcicki and Basolo [4], which indicated that at 31°–38°C (in toluene), one of the carbon monoxide groups exchanges more slowly than the other four. The rate of exchange of the four equivalent carbon monoxide groups does not depend on the concentration of carbon monoxide, and the exchange proceeds by a dissociative S_N1 path. The relative rates of exchange of carbon monoxide in the $Mn(CO)_5X$ compounds decrease in the order, $Mn(CO)_5Cl$ > $Mn(CO)_5Br$ > $Mn(CO)_5I$, as the "softness" of the halide ligands increases. Substitution of the carbonyl groups in $Mn(CO)_5X$ by tertiary phosphines, arsines, stibines, phosphites, and pyridine also takes place by a similar S_N1 mechanism [5].

It was reported by Angelici and Basolo [5] that the rate of formation of the mono-substituted product, $Mn(CO)_4XL$ (L = tertiary phosphines, arsines, phosphite) from $Mn(CO)_5X$ and the ligand L is independent of the concentration of the ligand, but depends on the first power of $Mn(CO)_5X$

concentration. Since the rate of substitution of carbon monoxide by the ligand in $Mn(CO)_5X$ was the same as the rate of exchange of the four equatorial carbonyl groups, it was concluded [5] that the incoming ligand L substitutes an equatorial carbon monoxide in $Mn(CO)_5X$.

In alkyl and acyl manganese pentacarbonyls, $RMn(CO)_5$ and $RCOMn(CO)_5$, the axial carbon monoxide is again more strongly bound than the four equatorial carbonyl groups. This information is based on studies of substitution reactions of methylmanganese pentacarbonyl, where the incoming ligand goes to an equatorial position [6]. The mechanism of substitution reactions of methylmanganese pentacarbonyl, $CH_3Mn(CO)_5$, will be treated in detail under carbonyl insertion reactions, Section III,A.

In dicobalt octacarbonyl, $Co_2(CO)_8$, all eight carbonyl groups exchange at the same rate [7]. This is also true of the four carbonyl groups in nickel tetracarbonyl, $Ni(CO)_4$, which [7] exchange with ^{14}CO by a dissociative mechanism. Nickel tetracarbonyl and dicobalt octacarbonyl exchange with carbon monoxide faster than do the carbonyls of chromium, manganese, and iron [$Cr(CO)_6$, $Mn_2(CO)_{10}$, and $Fe(CO)_5$]. The driving force for the dissociative exchange of $Ni(CO)_4$ and $Co_2(CO)_8$ has been postulated [7] to be the stabilizing effect of π bonding on the remaining carbonyl groups in the transition state.

Unlike the manganese compounds, alkyl and acyl cobalt tetracarbonyls, $RCo(CO)_4$ and $RCOCo(CO)_4$, are stable only below 0°C because of their high reactivities. These compounds have been characterized in solution by their infrared spectra. Although exchange studies have not been carried out on them, infrared evidence based on reactivities indicate that all four carbonyl groups in $RCo(CO)_4$ are equally reactive, and the alkyl compound has been postulated to be in equilibrium with the acyl derivative [8], in accordance with reaction (1). A band at 1724 cm^{-1} in the infrared spectra of

$$RCo(CO)_4 \rightleftharpoons RCOCo(CO)_3 \tag{1}$$

$RCo(CO)_4$ has been assigned to the acyl group in $RCOCo(CO)_3$ [8,9]. The formation of acylcobalt tricarbonyl as an intermediate in the reactions of alkylcobalt tetracarbonyl is further indicated by the formation of acylcobalt tricarbonyl–triphenylphosphine, $RCOCo(CO)_3(PPh_3)$ by the reaction of alkylcobalt tetracarbonyl with triphenylphosphine without the evolution of carbon monoxide. Heating alkyl or acyl cobalt tetracarbonyl with trimethyl phosphite leads to the formation of acylcobalt dicarbonyl–bis (trimethyl phosphite), $RCOCo(CO)_2[P(OCH_3)_3]_2$ [10]. The alkyl and acyl cobalt carbonyls become more stable as the number of phosphite ligands increases.

Although nickel tetracarbonyl, $Ni(CO)_4$, and dicobalt octacarbonyl, $Co_2(CO)_8$, exchange with ^{14}CO by a dissociative mechanism, the cyclopentadienyl derivatives $(C_5H_5)Ni(CO)_2$ and $(C_5H_5)Co(CO)_2$ exchange through second-order displacement reactions. The rates of exchange of these compounds depend [11] on both carbon monoxide and substrate concentrations. According to Wojcicki and Basolo [11], substitution of carbonyl by cyclopentadienyl groups produces a change in electronic structure because of the difference in donor capacities of carbon monoxide and the cyclopentadienyl radical. This difference is reflected in a different reaction pathway for the exchange reaction.

Iron pentacarbonyl, $Fe(CO)_5$, exchanges very slowly [7] with ^{14}CO. However, it has been reported by Basolo *et al.* [12] that $Fe(CO)_5$, $Fe(CO)_4(PPh_3)$, and $Fe(CO)_3(PPh_3)_2$ become very labile and undergo quite rapid carbon monoxide exchange even at $-30°$ to $-20°C$ in the presence of trifluoroacetic acid or sulfuric acid in a medium of 1,2-dichloroethane. It is interesting to note that under these conditions iron pentacarbonyl exchanges by a dissociative mechanism and that two of the carbonyl groups exchange more rapidly than the remaining three. Similarly, only three of the four carbonyl groups in $Fe(CO)_4PPh_3$ undergo exchange rapidly even at $-30°C$, after an induction period depending on the concentration of the acid catalyst. Infrared measurements show the existence in acid solution of hydrido species involving a metal–hydrogen bond which presumably are intermediates in the exchange reactions.

III. Carbon Monoxide Insertion Reactions

A. Alkyl and Acyl Manganese Pentacarbonyls

Coffield *et al.* [13] have reported the reversible formation of acylmanganese pentacarbonyl by heating alkylmanganese pentacarbonyl in an atmosphere of carbon monoxide [Eq. (2)]. This was the first report of a carbon monoxide insertion reaction in a metal–carbon bond. On the basis of experiments with

$$RMn(CO)_5 \xrightleftharpoons{CO} RCOMn(CO)_5 \quad (2)$$

^{14}CO, it was shown later by Coffield *et al.* [14] that ^{14}CO entering or leaving the organic moiety of the molecule is not derived from the carbon monoxide present in the reaction atmosphere, but comes from a coordinated carbonyl group. In the reaction of alkylmanganese pentacarbonyl with primary and secondary amines [15], no carbon monoxide escaped during the substitution reaction, Eq. (3). The rate of substitution of the carbonyl group of $RMn(CO)_5$

$$RMn(CO)_5 + NHR'R'' \longrightarrow RCOMn(CO)_4NHR'R'' \quad (3)$$

by amines depends on the nature of the organometallic substituent R and on the amines involved. Thus methylmanganese pentacarbonyl, $CH_3Mn(CO)_5$, forms acetylmanganese tetracarbonylamine, $CH_3COMn(CO)_4$(amine), much more readily than do the corresponding phenyl and benzyl derivatives. Primary and secondary aliphatic amines react with alkyl and aryl manganese pentacarbonyls more readily than do the aromatic amines. The extent of carbon monoxide insertion reaction with a particular alkyl or aryl manganese tetracarbonylamine is measured by the reaction of the acyl or aroyl derivative formed with sodium methoxide, whereby fission of the acyl or aroyl manganese bond takes place with the formation of the corresponding methyl ester, as shown in Eq. (4).

$$CH_3COMn(CO)_4NH_2C_6H_{11} + NaOCH_3 \rightleftharpoons CH_3COOCH_3 + NaMn(CO)_4NH_2C_6H_{11} \quad (4)$$

This reaction affords a very useful method of preparation of esters by the carboxylation of alkyl or aryl halides through the carbon monoxide insertion reaction. This reaction will be considered in detail under synthetic applications (Section IV,A,5).

Keblys and Filbey [15] proposed a bimolecular concerted mechanism, Eq. (5), for carbon monoxide insertion in $Mn(CO)_5$. The attack of the

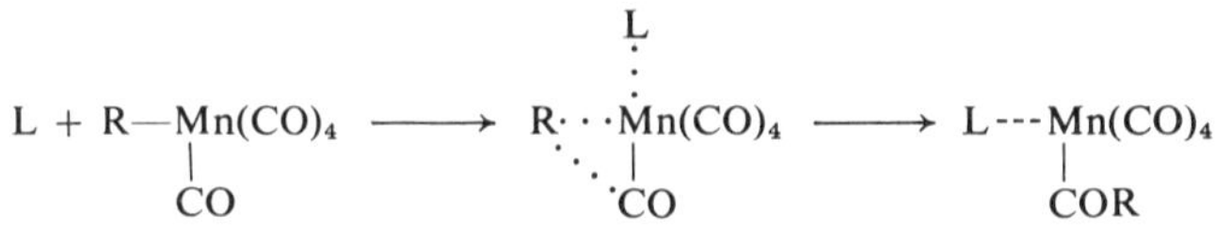

(5)

1

ligand L on $RMn(CO)_5$ results in the formation of a transition state **1** where partial bond formation takes place between the alkyl group and carbon monoxide.

Calderazzo and Cotton [16] measured the equilibrium and the rates of carbonylation and decarbonylation reactions of methylmanganese pentacarbonyl in various solvents at four temperatures. The reaction of methylmanganese pentacarbonyl with carbon monoxide [Eq. (6)] was found to be first order with respect to both reactants. The rate constants k_1 and k_{-1} were

$$\underset{\mathbf{2}}{CH_3Mn(CO)_5} + CO \underset{k_{-1}}{\overset{k_1}{\rightleftharpoons}} \underset{\mathbf{3}}{CH_3COMn(CO)_5} \quad (6)$$

reported to be $8.99 \times 10^{-3} M^{-1}\,sec^{-1}$ and $2.45 \times 10^{-5}\,sec^{-1}$, respectively, at 30°C in 2,2′-diethoxydiethyl ether as solvent. The effect of changing the solvent was considered due mainly to a bulk dielectric constant effect (i.e., a polar solvent helps to stabilize a polar transition state). Solvent participation in the transition state was not observed for the solvents studied. Although the experimental evidence thus favored a concerted mechanism, the possibility of a two-step mechanism involving rapid rearrangement of **2** followed by subsequent addition of carbon monoxide to the coordinatively unsaturated species $CH_3COMn(CO)_4$, illustrated in Eqs. (7) and (8), cannot be completely ruled out. Available data on the various rates and equilibrium constants in the substitution reaction of methyl- and acetylmanganese pentacarbonyl by triphenylphosphine support the possibility that the reaction steps (7) and (8) may be important in the carbonyl insertion reaction of **2**.

$$\underset{\mathbf{2}}{CH_3Mn(CO)_5} \underset{k_{-1}'}{\overset{k_1'}{\rightleftharpoons}} \underset{\mathbf{4}}{CH_3COMn(CO)_4} \quad (7)$$

$$CO + \underset{\mathbf{4}}{CH_3COMn(CO)_4} \underset{k_{-2}}{\overset{k_2}{\rightleftharpoons}} \underset{\mathbf{3}}{CH_3COMn(CO)_5} \quad (8)$$

The carbon monoxide substitution reactions of **2** and **3**, Eqs. (9) and (10), by triphenylphosphine in 2,2′-diethoxydiethyl ether proceed at rates first order in the carbonyl compounds and independent of triphenylphosphine concentration.

$$\underset{\mathbf{2}}{CH_3Mn(CO)_5} + PPh_3 \underset{k_{-3}}{\overset{k_3}{\rightleftharpoons}} \underset{\mathbf{5}}{CH_3COMn(CO)_4PPh_3} \quad (9)$$

$$\underset{\mathbf{3}}{CH_3COMn(CO)_5} + PPh_3 \underset{k_{-4}}{\overset{k_4}{\rightleftharpoons}} \underset{\mathbf{5}}{CH_3COMn(CO)_4PPh_3} + CO \quad (10)$$

The values of the rate constants k_3 and k_4 at 30.5°C are 6.6×10^{-5} and 3.6×10^{-5} sec^{-1}, respectively. The zero-order dependence of the rates of reactions (9) and (10) on triphenylphosphine concentration may possibly be due to the inability of the bulky molecule to attack a six-coordinate manganese atom [17]. Reactions (9) and (10) may therefore proceed by a two-step mechanism analogous to the reaction between methylmanganese pentacarbonyl and carbon monoxide illustrated in Eqs. (7) and (8).

Mawby *et al.* [18] studied the reaction of methylmanganese pentacarbonyl, **2**, with cyclohexylamine, triphenylphosphine, and triphenyl phosphite in various solvents—ethers, nitroethane, methanol, *n*-hexane, and mesitylene. It was found that these solvents not only exert a dielectric constant effect, but also participate directly in the transition state by forming a complex with the central metal ion. In nitromethane and ethers, **2** reacts with cyclohexylamine to give the adduct of **4**, $CH_3COMn(CO)_4 \cdot C_6H_{11}NH_2$, at a rate first order in methylmanganese pentacarbonyl concentration, but independent of the amine concentration. In *n*-hexane, the reaction was first order in both the reactants. In methanol, the rate was dependent on amine concentration but approached a limiting value at high concentrations of the amine. Similar results were obtained with triphenylphosphine and triphenyl phosphite as ligands. The results may be explained [18] on the basis of a mechanism involving a solvent-assisted dissociation, S_N1, of methylmanganese pentacarbonyl, followed by addition of the amine, triphenylphosphine, or trimethyl phosphite. Only in the limiting case where the coordinating ability of the solvent becomes very low is there the possibility of a concerted bimolecular mechanism, in which the nucleophilic reagent takes over the role of the solvent. The mechanism proposed for these reactions is indicated by Eq. (11).

$$\underset{\mathbf{2}(\mathrm{M})}{CH_3Mn(CO)_5} \underset{k_{-1}}{\overset{k_1}{\rightleftharpoons}} CH_3COMn(CO)_4(\text{solvent}) + \underset{\mathrm{L}}{\text{amine}} \underset{k_{-2}}{\overset{k_2}{\rightleftharpoons}} \underset{\mathbf{4}\cdot\mathrm{L}(\mathrm{ML})}{CH_3COMn(CO)_4(\text{amine})} \tag{11}$$

$$\frac{d[\mathrm{ML}]}{dt} = k_2[\mathrm{L}]\frac{k_1[\mathrm{M}] + k_{-2}[\mathrm{ML}]}{k_{-1} + k_2[\mathrm{L}]} - k_{-2}[\mathrm{ML}] \tag{12}$$

Taking $k_2[\mathrm{L}] \gg k_{-1}$, and neglecting the back reaction involving k_{-2} (i.e., k_{-2} or [ML], or both, is small), one obtains the expression given in Eq. (13).

$$\frac{d[\mathrm{ML}]}{dt} = k_1[\mathrm{M}] \tag{13}$$

Equation (13) indicates first-order dependence of the rate of reaction on the concentration of methylmanganese pentacarbonyl, **2**. Where the concentration of the ligand L is very small, as in the case of the reaction of carbon monoxide with **2**, $k_2[\mathrm{L}]$ becomes very small compared to k_{-1} and the overall rate becomes first order with respect to ligand and metal carbonyl concentration [M].

$$\frac{d[\mathrm{ML}]}{dt} = \frac{k_1k_2}{k_{-1}}[\mathrm{M}][\mathrm{L}] \tag{14}$$

The proposed participation of the solvent in the coordination sphere of the metal ion is supported by the fact that the rate of the reaction of **2** with the ligand L is greater in methanol than in nitromethane, although the dielectric constant of nitromethane (37.4) is higher than that of methanol (31.2).

There are two possible pathways for the reaction involving the formation of the acylmanganese complex, **4**, from the alkylmanganese pentacarbonyl, **2**: (1) insertion of the carbon monoxide group previously bonded to manganese in the metal–carbon bond of **2**; (2) migration of the methyl group to a coordinated carbonyl group of $CH_3Mn(CO)_5$.

In order to decide between the two pathways, the reverse (dissociation) reaction (i.e., de-insertion) of **5**, *trans*-$CH_3COMn(CO)_4PPh_3$, was followed [18] by the rate of disappearance of the acetyl carbon–oxygen bond. It was argued [18] that if the reaction follows pathway (15) there should be no change in the stereochemistry of the product, which should also be *trans*. Migration of the methyl group, however, would change the geometric arrangement of the groups in the product from *trans* to *cis*. It was reported that **5** gave a dissociation (reverse of carbon monoxide insertion) reaction in which only *cis*-$CH_3Mn(CO)_4PPh_3$, **7**, was formed, thus supporting the migration of a methyl group [pathway (16)].

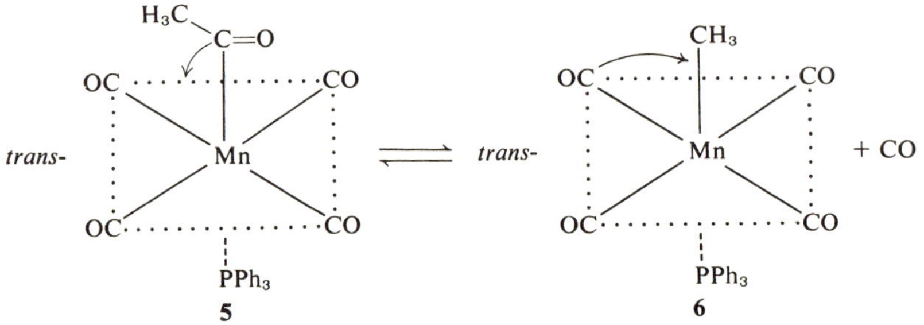

(15)

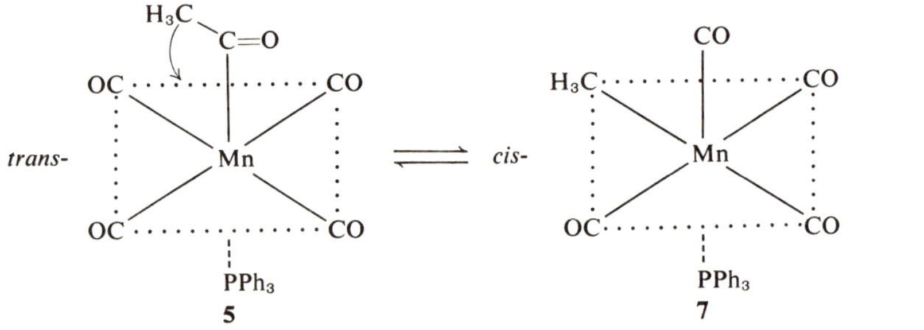

(16)

The evidence for methyl migration in the decarbonylation of $CH_3COMn\cdot(CO)_4PPh_3$ [18] cannot be considered as firmly established because of the isomerizations found to take place during the formation of $CH_3COMn\cdot(CO)_4PPh_3$ from $CH_3Mn(CO)_5$ and triphenylphosphine [19,20] and in the formation of $CH_3COMn(CO)_4[P(OCH_2)_3CR]$ from $CH_3Mn(CO)_5$ and the corresponding phosphite [$P(OCH_2)_3CR$, R = CH_3, C_2H_5]. Examples of such isomerization reactions are illustrated by Eqs. (17)–(19). Methylmanganese pentacarbonyl, $CH_3Mn(CO)_5$, **2**, when treated with the phosphite $P(OCH_2)_3CR$ gave *cis*-$CH_3COMn(CO)_4[P(OCH_2)_3CR]$. Both methyl migration, [Eq. (17)] and CO insertion, [Eq. (18)] are consistent with the observed retention of configuration in the formation of the acetyl complex. In steps (17) and (18) the methyl or carbonyl groups *cis* to the attacking ligand L migrate to a *cis* position to yield the final *cis* product. Insertion of a *trans* carbonyl group in the manganese–methyl bond, as indicated by reaction (19), can be excluded from consideration since it would result in the formation of *trans*-$CH_3COMn(CO)_4[P(OCH_2)_3CR]$. The possibility that the *trans* isomer is actually formed in reaction (19) and then undergoes a fast irreversible

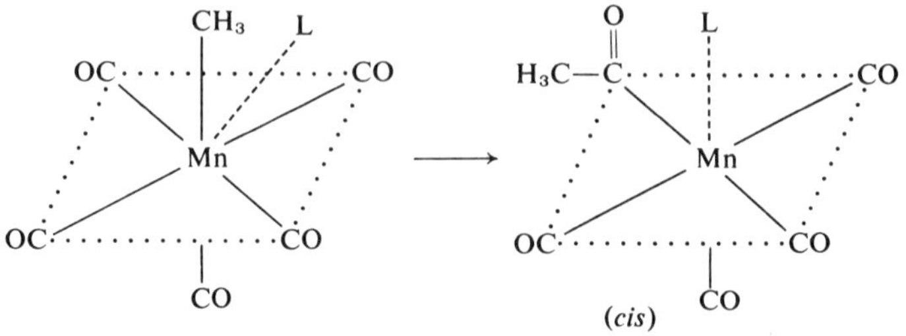

(17)

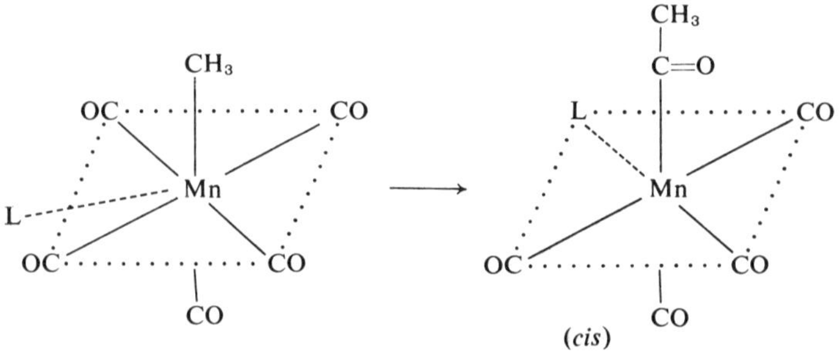

(18)

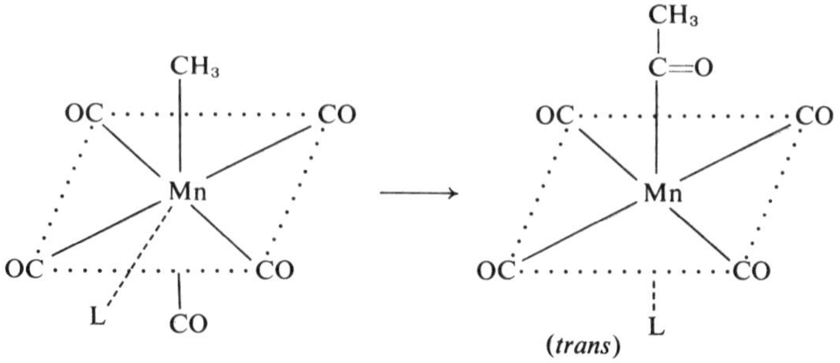

(19)

L = $P(OCH_2)_3CR$, 4-methyl-(or ethyl)- 2,6,7-trioxa-1-phosphabicyclo[2.2.2]octane

rearrangement to the corresponding *cis* isomer is unlikely because the *trans* product predominates in the following equilibrium:

$$cis\text{-}CH_3COMn(CO)_4L \rightleftharpoons trans\text{-}CH_3COMn(CO)_4L \quad (20)$$

The reaction of methylmanganese pentacarbonyl with ^{13}CO afforded acetylmanganese pentacarbonyl **8**, in which labeled carbon monoxide is not inserted in the metal–carbon bond, and the newly formed acetyl and coordinated ^{13}CO groups are in mutually *cis* positions. Formation of the *cis* isomer in the reaction may be explained by attack of the ligand ^{13}CO at a position on the metal ion *cis* to methyl or carbonyl, followed by methyl migration or carbonyl insertion in the manner depicted by Eq. (17) and (18). Decarbonylation of **8**, *cis*-$CH_3COMn(CO)_4{}^{13}CO$, gave both *cis*- and *trans*-$CH_3Mn(CO)_4{}^{13}CO$ complexes **9** and **10** in the ratio 1:2 [22]. Whereas carbon monoxide migration can give only the *cis* product [Eq. (20)], methyl migration explains the formation of both the *cis* and *trans* products [routes (21) and (22)]. Thus methyl migration is favored over carbon monoxide migration in the decarbonylation of *cis*-$CH_3COMn(CO)_4{}^{13}CO$.

CO Migration

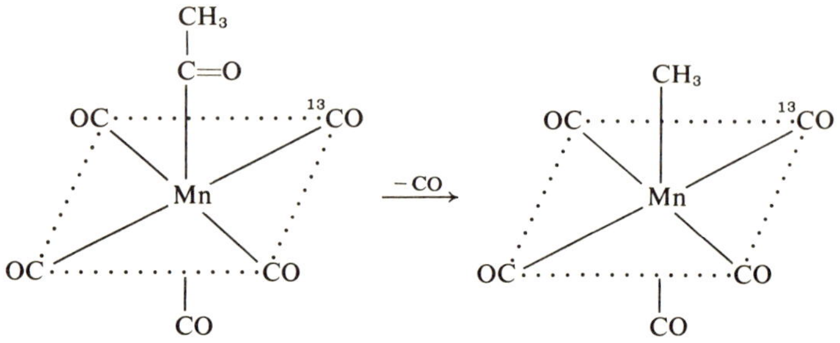

8 (*cis*) 9 (21)

Methyl Migration

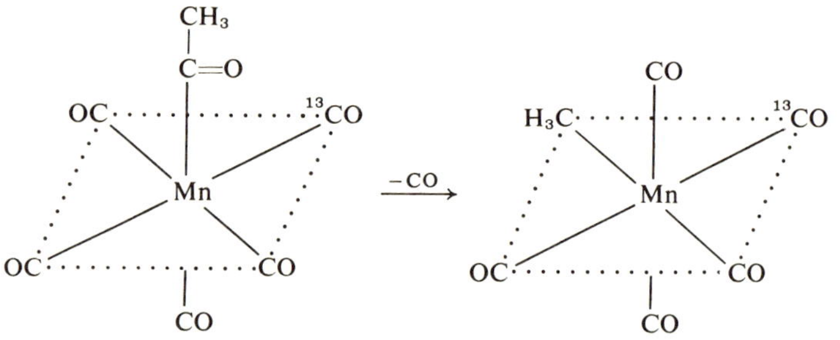

8 (*cis*) 9 (22)

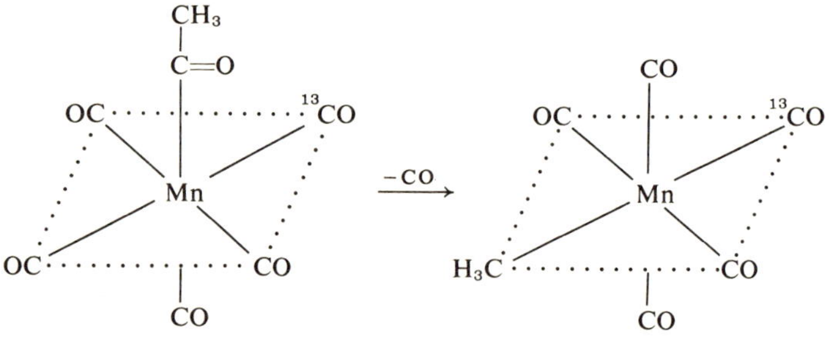

8 (*trans*) 10 (23)

Noack *et al.* [23] have studied the reaction of methylmanganese pentacarbonyl with triphenylphosphine [Eq. (9)] by NMR spectrometry. It was found that reaction (9) initially gave the *cis*-acetyl isomer (appearance of NMR peak at τ 7.68) which was then converted to the *trans* isomer in a slow reaction, until an equilibrium mixture [Eq. (24)] of the *cis* and *trans*-acetyl isomers was obtained. The value of k_1 at 30°C in acetone solution is

$$cis\text{-}CH_3COMn(CO)_4PPh_3 \underset{k_{-1}}{\overset{k_1}{\rightleftharpoons}} trans\text{-}CH_3COMn(CO)_4PPh_3 \qquad (24)$$

2.5×10^{-4} sec^{-1} [23]. The *cis-trans* equilibrium depicted in Eq. (24) is solvent-dependent. The initial formation of *cis*-acetyl product [Eq. (9)] may be explained by both methyl migration and carbonyl insertion mechanisms in the manner suggested by Green and Wood [21].

The conversion of the *cis*-acetyl to the *trans* isomer has been explained [23] on the basis of the attack of a solvent molecule on the *cis*-acetyl species with the possible formation of a hexacoordinated solvated species *cis*-[$CH_3COMn \cdot (CO)_4S$] (S = solvent). Since isomerism follows first-order kinetics it was suggested that the solvated species *cis*-[$CH_3COMn(CO)_4S$] undergoes rearrangement of the ligands to form another solvated species, *trans*-[$CH_3COMn(CO)_4S$], which yields *trans*-$CH_3COMn(CO)_4PPh_3$ by solvent displacement, as shown in Eq. (25).

$$cis\text{-}CH_3COMn(CO)_4L \underset{+L}{\overset{+S}{\rightleftharpoons}} cis\text{-}[CH_3COMn(CO)_4S] \underset{k_{-1}}{\overset{k_1}{\rightleftharpoons}} trans\text{-}[CH_3COMn(CO)_4S] \underset{+S}{\overset{+L}{\rightleftharpoons}} trans\text{-}CH_3COMn(CO)_4L \qquad (25)$$

In the decarbonylation reaction of $CH_3{}^{13}COMn(CO)_5$ with PPh_3 it was found [23] that all the ^{13}CO was retained in the reaction product, $CH_3{}^{13}COMn(CO)_4PPh_3$, when the reaction was conducted in hydrocarbon (heptane) or donor (2,2′-dimethoxydiethyl ether) solvents. The intensity of the IR acetyl bands indicates that both *cis*- and *trans*-acetyl isomers are formed in the reaction. Noack *et al.* [23] suggested dissociation of $CH_3{}^{13}COMn(CO)_5$ to a coordinately unsaturated species $CH_3{}^{13}COMn(CO)_4$ in a slow step. The value of the first-order rate constant at 30°C associated with this dissociative step varies from 2.46×10^{-5} to 1.99×10^{-6} sec^{-1}, depending on the solvent. Solvation of the species $CH_3{}^{13}COMn(CO)_4$ gives the intermediate $CH_3{}^{13}COMn(CO)_4S$, which is the same type as that suggested for reactions (9) and (24), and can either remain unchanged or rearrange to a *trans* species. Subsequent attack by PPh_3 on the *cis*- or *trans*-solvated $CH_3{}^{13}COMn \cdot (CO)_4S$ species would give a mixture of *cis* and *trans* products as indicated in reaction (26).

It may be concluded that the carbonyl insertion reactions that are promoted by a nucleophilic reagent [Eq. (17)–(19)] usually take place by initial formation of intermediates through migration of alkyl or carbonyl groups situated *cis* with respect to the attacking nucleophile. The *cis* product either retains its configuration or rearranges to an equilibrium mixture of the *cis*

$$CH_3{}^{13}COMn(CO)_5 \underset{+CO(fast)}{\overset{-CO(slow)}{\rightleftharpoons}} CH_3{}^{13}COMn(CO)_4$$

$$\underset{}{\overset{S}{\rightleftharpoons}} \left\{ \begin{array}{l} cis\text{-}CH_3{}^{13}COMn(CO)_4S \overset{L}{\rightleftharpoons} cis\text{-}CH_3{}^{13}COMn(CO)_4L + S \\ trans\text{-}CH_3{}^{13}COMn(CO)_4S \overset{L}{\rightleftharpoons} trans\text{-}CH_3{}^{13}COMn(CO)_4L + S \end{array} \right. \quad (26)$$

and *trans* forms in one or more subsequent steps. Solvation of the intermediates plays an important role in the catalysis of these reactions but probably does not greatly influence the ratio of *cis* to *trans* products, provided that the reaction conditions allow a reasonable approach to equilibrium.

B. π-Cyclopentadienyl(alkyl)molybdenum Tricarbonyl

Craig and Green [24] have studied the kinetics of the carbonyl insertion reaction of π-cyclopentadienyl(alkyl)molybdenum tricarbonyl, **11**, π-$(C_5H_5)(R)Mo(CO)_3$, effected through attack by phosphorus-containing ligands in acetonitrile solvent, to give the corresponding acyl complexes, **12,** π-$(C_5H_5)(COR)Mo(CO)_2L$, according to reaction (27). The relative

$$\underset{\mathbf{11}}{\pi\text{-}(C_5H_5)(R)Mo(CO)_3} + L \longrightarrow \underset{\mathbf{12}}{\pi\text{-}(C_5H_5)(RCO)Mo(CO)_2L} \quad (27)$$

(L = PR'_3, where R′ = CH_3, C_2H_5, $C_6H_5CH_2$, $CH_2{=}CH{-}CH_2$, and alkoxy groups)

reactivities of π-$(C_5H_5)(R)Mo(CO)_3$ toward phosphorus-containing nucleophiles decrease with the nature of R in the order, $C_2H_5 > CH_3 > C_6H_5CH_2 > CH_2{=}CH{-}CH_2$. The rate of the reaction is solvent-dependent, proceeding about a thousand times faster in a donor solvent such as acetonitrile, than in tetrahydrofuran. For a particular solvent, the rate of carbonylation of π-$(C_5H_5)(R)Mo(CO)_3$ is independent of the concentration of the nucleophile. This supports a mechanism involving the dissociation of π-$(C_5H_5)(R)Mo(CO)_3$ in a rate-determining step to form a possibly solvated acyl complex, π-$(C_5H_5)(COR)Mo(CO)_2S$. [Eq. (28)]. The solvated acyl complex then reacts with the nucleophile in a fast step, Eq. (29), where $k_2 \gg k_{-1} > k_1$.

$$\pi\text{-}(C_5H_5)(R)Mo(CO)_3 + S \underset{k_{-1}}{\overset{k_1}{\rightleftharpoons}} \pi\text{-}(C_5H_5)(COR)Mo(CO)_2S + CO \quad (28)$$

$$L + \pi\text{-}(C_5H_5)(COR)Mo(CO)_2S \xrightarrow{k_2} \pi\text{-}(C_5H_5)(COR)Mo(CO)_2L \quad (29)$$

Butler *et al.* [25] have found that the rates of the carbonyl insertion reaction of π-$(C_5H_5)CH_3Mo(CO)_3$ with the nucleophiles PPh_3, $P(n\text{-}C_4H_9)_3$, and $P(OC_6H_5)_3$ are independent of the nucleophile concentration in a donor solvent such as acetonitrile. In a hydrocarbon solvent (toluene), however, the rate of carbonylation of π-$(C_5H_5)CH_3Mo(CO)_3$ follows two reaction paths, one that is independent of nucleophile concentration and one path that is first order with respect to the nucleophile, L.

$$\begin{aligned} k_{obs} &= k_1[M] + k_2[M][L] \\ [M] &= \pi\text{-}(C_5H_5)CH_3Mo(CO)_3;\ L = PR_3 \end{aligned} \tag{30}$$

With the nucleophiles $P(n\text{-}C_4H_9)_3$, $P(n\text{-}OC_4H_9)_3$, PPh_3, and $P(OC_6H_5)_3$, k_1 at 50.7°C was almost a constant, $5.8 \times 10^{-6}\ sec^{-1}$, but k_2 showed a decrease in value with decreasing nucleophilicity of the ligand L. Thus values of k_2 decrease in the order $P(n\text{-}C_4H_9)_3 > P(n\text{-}OC_4H_9)_3 > PPh_3 > P(OC_6H_5)_3$. On the basis of these kinetic considerations, Butler *et al.* [25] have suggested a solvent- or nucleophile-assisted mechanism in the carbonyl insertion reactions of **11** [Eq. (31)]. The rate of the reaction depends on nucleophile concentration when either the nucleophilicity of the attacking nucleophile is very high or the coordinating power of the solvent is extremely low, as in the case of a hydrocarbon solvent. In a strongly coordinating solvent such as aceteontrile, the solvent-assisted mechanism prevails and the rate of the carbonylation of π-$(C_5H_5)CH_3Mo(CO)_3$ becomes independent of the concentration of nucleophile.

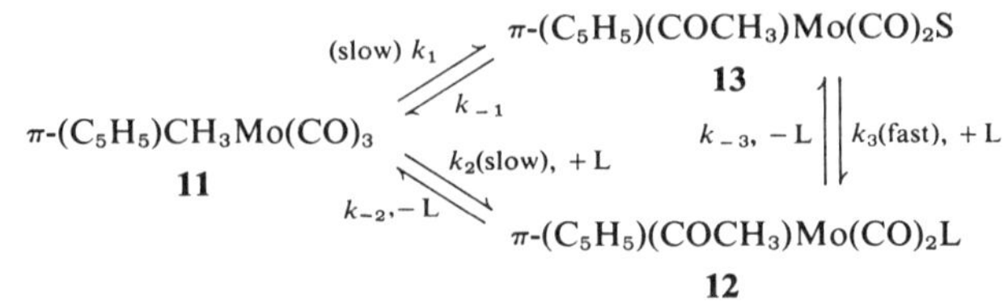

(31)

At a steady-state concentration of the solvated intermediate, **13**, the rate expression is given by Eq. (32).

$$-\frac{d[M]}{dt} = k_1[M] - k_{-1}\frac{k_1[M] + k_{-3}[ML]}{k_{-1} + k_3[L]} + k_2[M][L] - k_{-2}[ML] \tag{32}$$

When the reaction tends to go to completion, k_{-2} and k_{-3} may be neglected and $k_3[L] \gg k_{-1}$. The rate expression then simplifies to Eq. (33).

$$-\frac{d[M]}{dt} = k_1[M] + k_2[M][L] \tag{33}$$

C. π-Cyclopentadienyl(alkyl)iron Dicarbonyl

π-Cyclopentadienylmethyliron dicarbonyl, **14**, reacts with the ligands $P(C_5H_5)_3$, $P(n\text{-}C_4H_9)_3$, $P(OC_6H_5)_3$, and $P(OC_4H_9)_3$ in refluxing tetrahydrofuran to give almost quantitative yields of the acyl complexes $\pi\text{-}(C_5H_5)\cdot(COCH_3)Fe(CO)L$ where L is the phosphine or phosphite ligand [26]. In a hydrocarbon solvent such as hexane, no acyl derivative of **14** was formed. With π-bonding nucleophiles that are weaker ligands than phosphines and phosphites, such as iodide, dialkyl sulfide, and *p*-toluidine, the carbonyl insertion reaction does not take place.

The kinetics of the reactions of π-cyclopentadienylmethyliron dicarbonyl, **14**, with the ligands PPh_3, $P(n\text{-}C_4H_9)_3$, and $P(n\text{-}OC_4H_9)_3$ in tetrahydrofuran has been studied by Butler *et al.* [25]. In tetrahydrofuran the rate of carbonyl insertion depends on the concentration of the ligand. The mechanism given in Eq. (34) was proposed [25]. If k_{-1} becomes comparable in magnitude to

$$\underset{\mathbf{14}}{\pi\text{-}(C_5H_5)CH_3Fe(CO)_2} \underset{k_{-1}}{\overset{k_1}{\rightleftharpoons}} \underset{\mathbf{15}}{\pi\text{-}(C_5H_5)(COCH_3)Fe(CO)S} \quad \xrightleftharpoons[-L,\ k_{-2}]{+L,\ k_2} \quad \underset{\mathbf{16}}{\pi\text{-}(C_5H_5)(COCH_3)Fe(CO)L + S} \tag{34}$$

$k_2[L]$, there will be competition for the solvated intermediate **15** that results in the dependence of the rate on phosphine or phosphite concentration. Using the steady-state approximation for the concentration of the solvated species **15**, the rate of reaction (34) may be written as given in Eq. (35).

$$-\frac{d[M]}{dt} = k_1[M] - k_{-1}\left(\frac{k_1[M] + k_{-2}[ML]}{k_{-1} + k_2[L]}\right) \tag{35}$$

Neglecting $k_{-2}[ML]$, with the assumption that the reaction goes to completion over the range of concentration employed, the rate expression becomes Eq. (36).

$$-\frac{d[M]}{dt} = \frac{k_1k_2[M][L]}{k_{-1} + k_2[L]}$$

or (36)

$$\frac{1}{k_{obsd}} = \frac{k_{-1}}{k_1k_2}\frac{1}{[L]} + \frac{1}{k_1}$$

A plot of $1/k_{obsd}$ against $1/[L]$ gave a straight line with intercept equal to $1/k_1$ and slope of k_{-1}/k_1k_2, from which the ratio of the constants k_{-1}/k_2 was calculated [25]. The ratio k_{-1}/k_2 is a measure of the nucleophilic strength of the attacking ligand. The smaller this ratio (the larger is k_2), the greater is the nucleophilicity of the ligand L and the better it competes with the solvent for the organometallic complex. The decrease is in the order $P(n\text{-}C_4H_9)_3$ > $P(n\text{-}OC_4H_9)_3$ > $P(C_6H_5)_3$, in accord with the observation of Bibler and Wojcicki [26]. A coordinating solvent appears to provide a low-energy path for the reaction through formation of the solvated intermediate, since there was no detectable formation of the acyl derivative **16** in poor or nondonor solvents such as hexane.

Although there is no conclusive evidence in support of a methyl migration in the carbonylation reaction of **14** and the corresponding molybdenum complex, $\pi\text{-}(C_5H_5)CH_3Mo(CO)_3$, the relative rates of acylation, Mo > Fe, suggest the cleavage of the metal–alkyl carbon bond with methyl migration to an adjacent carbonyl group. The rate of acylation of iron and molybdenum methyl carbonyl complexes correlates with the probable order of metal–carbon bond strength, iron > molybdenum [25].

D. Alkyl and Aryl Cobalt Tetracarbonyls

1. *Displacement of Carbonyl Groups by Ligands*

Breslow and Heck [8] have presented infrared evidence that alkylcobalt tetracarbonyls are in equilibrium with acylcobalt tricarbonyls. A band at 5.8 microns in the infrared spectrum of the former compounds prepared from alkyl halide and sodium cobalt tetracarbonyl at 0°C was attributed [8] to the presence of acylcobalt tricarbonyls. The reduction [9] of acylcobalt tetracarbonyls by hydrogen is inhibited by the presence of carbon monoxide. This inhibition may be due to competition between carbon monoxide and the reducing agent [H] for the coordinately unsaturated $RCOCo(CO)_3$ species [27] [Eq. (37)]. Except for the band at 1724 cm^{-1} in the infrared, no direct

$$\underset{\mathbf{17}}{RCOCo(CO)_4} \underset{+CO}{\overset{-CO}{\rightleftharpoons}} \underset{\mathbf{18}}{RCOCo(CO)_3} \xrightarrow{[H]} RCHO + HCo(CO)_3 \tag{37}$$

$$\underset{\mathbf{18}}{RCOCo(CO)_3} \rightleftharpoons \underset{\mathbf{19}}{RCo(CO)_4}$$

evidence is yet available for the presence of acylcobalt tricarbonyls in the acylcobalt tetracarbonyl–alkylcobalt tetracarbonyl systems.

Heck and Breslow [28] have found that triphenylphosphine reacts with alkylcobalt tetracarbonyls to form acylcobalt tricarbonyl–triphenylphosphines without the evolution of carbon monoxide. The same phosphine adducts can be prepared directly from acylcobalt tetracarbonyls and triphenylphosphines with the evolution of carbon monoxide [Eq. (38)]. Based on an analogous reaction with acylmanganese pentacarbonyls, it has been postulated [28] that the carbon monoxide displaced from acylcobalt tetracarbonyl is one of the coordinated carbonyls and does not come from the organic (acyl) part of the molecule. On the basis of the same analogy, it is one of the coordinated carbon monoxide groups that takes part in the formation of acylcobalt tetracarbonyl from alkylcobalt tetracarbonyl.

$$\begin{array}{l} RCOCo(CO)_4 + PPh_3 \xrightarrow{-CO} \\ RCo(CO)_4 + PPh_3 \longrightarrow \end{array} RCOCo(CO)_3PPh_3 \quad (38)$$

Heck [29] has studied the kinetics of the reaction between acylcobalt tetracarbonyl and triphenylphosphine, The variation of reaction rate with varying concentration of reactants indicated the reaction to be first order with respect to acylcobalt tetracarbonyl and independent of the concentration of the ligand used. The mechanism given in Eqs. (39) and (40) was proposed [30]. There is no competition by carbon monoxide for the

$$RCOCo(CO)_4 \underset{k_{-1}}{\overset{k_1}{\rightleftharpoons}} CO + [RCOCo(CO)_3 \rightleftharpoons RCo(CO)_4] \quad (39)$$

$$RCOCo(CO)_3 + PPh_3 \xrightarrow{k_2} RCOCo(CO)_3PPh_3 \quad (40)$$

triphenylphosphine reaction with $RCOCo(CO)_3$ for over 80% of completion, so that $k_2[PPh_3]$ is seen to be much larger than $k_{-1}[CO]$. Under these conditions the rate expression reduced to Eq. (41).

$$-\frac{d[RCOCo(CO)_4]}{dt} = k_1[RCOCo(CO)_4] \quad (41)$$

Heck [30] has reported the effect of varying the acyl group on the rate of the dissociation of acylcobalt tetracarbonyl into acylcobalt tricarbonyl or alkylcobalt tetracarbonyl and carbon monoxide. The results are presented in Table I.

The data in Table I indicate that steric factors are quite significant in determining the rate of dissociation of acylcobalt tetracarbonyl. Two methyl groups in the isobutyryl compound have $\frac{1}{40}$ of the effect of three methyl groups in the *t*-butyl compound (Nos. 3 and 4), while isovalerylcobalt tetracarbonyl is as reactive as *n*-hexanoylcobalt tetracarbonyl (Nos. 2 and 5). Thus it seems that steric strain may be an important factor in the reactivities of the acylcobalt tetracarbonyls. The effect of adding electron-withdrawing groups on the rate of dissociation is shown in the series, 4-chlorobutyryl, methoxyacetyl, and trifluoroacetyl. The rate of dissociation steadily decreases in the series, with the slowest rate observed for trifluoroacetylcobalt tetra-

TABLE I

EFFECT OF ACYL GROUP STRUCTURE ON THE RATE OF DISSOCIATION OF ACYLCOBALT TETRACARBONYL

	Acylcobalt tetracarbonyl	Relative rate (Acetyl compound = 1)
1	$CH_3COCo(CO)_4$	1.0
2	$CH_3(CH_2)_4COCo(CO)_4$	1.1
3	$(CH_3)_2CHCOCo(CO)_4$	2.1
4	$(CH_3)_3CCOCo(CO)_4$	86
5	$(CH_3)_2CHCH_2COCo(CO)_4$	1.2
6	$CH_2Cl(CH_2)_2COCo(CO)_4$	0.8
7	$CH_3OCH_2COCo(CO)_4$	0.3
8	$CF_3COCo(CO)_4$	0.1
9	$CH_3CH{=}CH_2COCo(CO)_4$	64
10	$C_6H_5COCo(CO)_4$	34
11	$C_6H_5CH_2COCo(CO)_4$	1.3

[a] $k = 2.55 \pm 0.18 \times 10^{-2}$ sec^{-1} (25°C) for the acetyl compound.

carbonyl. Benzoylcobalt tetracarbonyl dissociates at a rate about 34 times faster than the acetyl compound. For the phenylacetyl compound the interposition of a methylene group between the phenyl and acyl groups seems to eliminate the effect of the aromatic ring.

The effect of substituents on the rate of dissociation of $RCOCo(CO)_4$ supports a transition state in which R group (alkyl or aryl) is partially bound both to the metal atom and to carbon monoxide, as shown in Eq. (42).

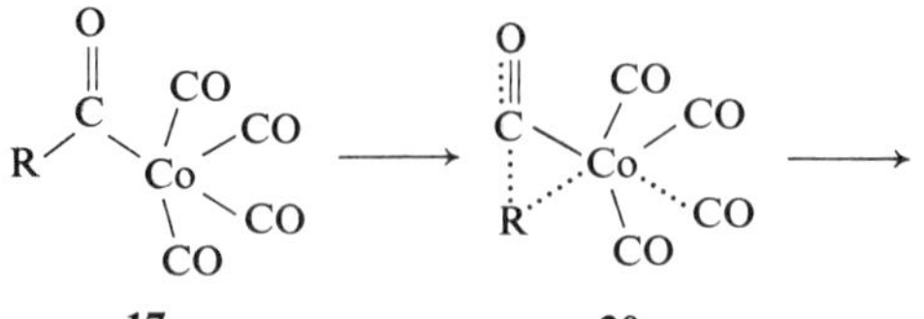

17 **20**

$$CO + [RCo(CO)_4 \rightleftharpoons RCOCo(CO)_3] \quad (42)$$

19 **18**

The direct interaction between the alkyl group R and the cobalt ion visualized in the mechanism given in Eq. (42) is supported by the higher rate of dissociation of the benzoyl derivative and crotonyl derivative, where the aromatic ring and the propenyl group are considered capable of supplying electrons to the metal ion by π bonding. Electron-withdrawing groups substituted on R make dissociation of carbon monoxide more difficult by decreasing the electron density around the metal ion. The strong effect of the *t*-butyl group may be both steric and electronic, since the methyl groups can partially supply electrons by hyperconjugation. The insertion of carbon monoxide between the bulky *t*-butyl group and the metal would tend to relieve the steric repulsions between them. The exact nature of the promotion effect by pivalyl is not clear from the data presented [30].

It is noted that the above mechanism suggested by Heck [30] indicates the formation of an acylcobalt tricarbonyl by the dissociation of an acylcobalt tetracarbonyl through an alkylcobalt tetracarbonyl intermediate. It is also possible that the acylcobalt tricarbonyl is formed directly from the acylcobalt tetracarbonyl by rearrangement and elimination of carbon monoxide from the transition state.

2. *Lower Carbonyls of Cobalt*

Displacement reactions similar to those described above also occur with lower carbonyls. Thus acylcobalt tricarbonyl–triphenylphosphine reacts [10] with trimethyl phosphite, $P(OCH_3)_3$, to form acylcobalt dicarbonyl–trimethyl phosphite–triphenylphosphine by the displacement of a molecule of carbon monoxide, Eq. (43). π-Allylcobalt tricarbonyl [27,29] reacts with triphenyl-

$$RCOCo(CO)_3PPh_3 + P(OCH_3)_3 \longrightarrow RCOCo(CO)_2P(OCH_3)_3PPh_3 + CO \tag{43}$$

phosphine to form π-allylcobalt dicarbonyl–triphenylphosphine by the displacement of one molecule of carbon monoxide, as shown in Eq. (44).

$$\pi\text{-}C_3H_5Co(CO)_3 + PPh_3 \longrightarrow \pi\text{-}C_3H_5Co(CO)_2PPh_3 + CO \tag{44}$$

The rate of the reaction of π-allylcobalt tricarbonyl, $C_3H_5Co(CO)_3$, with triphenylphosphine is first order with respect to π-allylcobalt tricarbonyl up to 0.06 *M* concentration of triphenylphosphine [30]. A dissociative S_N1 mechanism similar to that proposed for the dissociation of acylcobalt tetracarbonyls has been postulated [29]. The rate of the displacement reaction of π-allylcobalt tricarbonyl with triphenylphosphine is $1.14 \pm 0.6 \times 10^{-2}$ sec^{-1} at 25°C. The π-allyl complex is thus less reactive than the acyl complex. The effect of substituents on the rate of dissociation of π-allylcobalt tricarbonyl is, however, not as clear as for the acyl compound.

Nitrosylcobalt tricarbonyl [31] in tetrahydrofuran reacts with triphenylphosphine by a bimolecular mechanism, first order in the carbonyl compound and first order in triphenylphosphine. For the displacement reaction (44) the relative order of reactivities of the carbonyl compounds mentioned above gave the following decreasing sequence: $HCo(CO)_4 > CH_3COCo(CO)_4 > C_3H_5Co(CO)_3 > NOCo(CO)_3$. Comparing the rate of dissociation of the acylcobalt tetracarbonyl with that of acylmanganese pentacarbonyl, the cobalt complex is more labile and dissociates about 2500 times faster than the manganese complex.

3. *Isomerization of Acyl and Alkyl Groups in Cobalt Carbonyls*

The nature of the transition state in the dissociation of acylcobalt tetracarbonyl was elucidated by the work of Takegami *et al.* [32—35] on the isomerization reactions of alkyl or acyl halides with sodium tetracarbonyl cobaltate, $NaCo(CO)_4$, in an atmosphere of either carbon monoxide or nitrogen. The isomerized alkyl groups were identified by the reaction of acylcobalt tetracarbonyl with iodine and alcohol [35], which converts the acyl group to an ester of the alcohol employed. The halides *n*-propyl chloride, α-ethyl butyryl chloride, and isobutyryl chloride, when treated in this manner, were found to yield both normal and chain-isomerized esters. The ratio of the normal to the iso ester was found to depend on (1) temperature of the reaction, (2) solvent [35], (3) reaction atmosphere [32,33,35], and (4) nature of the acyl group [33,35].

Straight-chain isomers were formed from branched-chain isomers at higher temperatures. Thus isobutyryl bromide gave butyrate and isobutyrate, but *n*-butyryl chloride gave only butyrate. Solvents have a pronounced effect on the rate of isomerization and the ratio of isomers obtained. In the isomerization of isobutyrylcobalt tetracarbonyl the ratio of iso/normal ester formed is about zero in benzene and dioxane [35], 1:5 in tetrahydrofuran, 1:3 in ethyl acetate, 1:1 in diethyl ether, and 1.6:1 in isopropyl ether [32]. Although a specific mechanism was not advanced for the participation of the solvent, the authors [35] attributed the effect of the solvent to its tendency to coordinate with the cobalt complexes. This suggestion is supported by a similar solvent-assisted migration of the alkyl group in the substitution reaction of alkylmanganese pentacarbonyls [6] and by the coordination of diethyl ether with sodium cobalt carbonylate [36].

The reaction atmosphere also affects the rate of isomerization to a marked extent. In all cases studied, isomerization proceeds more rapidly under a nitrogen atmosphere than under carbon monoxide. When the pressure of carbon monoxide was increased to 2 atm, isomerization was slower than that observed under 1 atm of carbon monoxide [33], suggesting a dissociative mechanism, or solvent–carbon monoxide competition.

For the isomerization of the isobutyryl complex, the solvent-assisted mechanism in Eqs. (45)–(47) has been suggested. The reaction steps suggested for the isomerization reaction are supported [33] by the observed evolution of carbon monoxide in the second step followed by gas absorption in the subsequent step. The increased rate of reaction under nitrogen atmosphere supports the suggestion of isomerization at the cobalt tricarbonyl stage. The coordination of the solvent in the transition state helps the reaction by avoiding the necessity of forming a high-energy complex of cobalt with a lower coordination number.

The structure of the acyl group has a marked effect on the ratio of isomers formed. The acyl halide *n*-butyryl chloride is not isomerized in either tetrahydrofuran or isopropyl ether. In both these solvents, the straight-chain compound was isolated, indicating the greater stability of the straight-chain compound over a branched isomer or the lack of a suitable pathway to the

$$(CH_3)_2CHCOBr + NaCo(CO)_4 \longrightarrow (CH_3)_2CHCOCo(CO)_4 + NaBr \qquad (45)$$

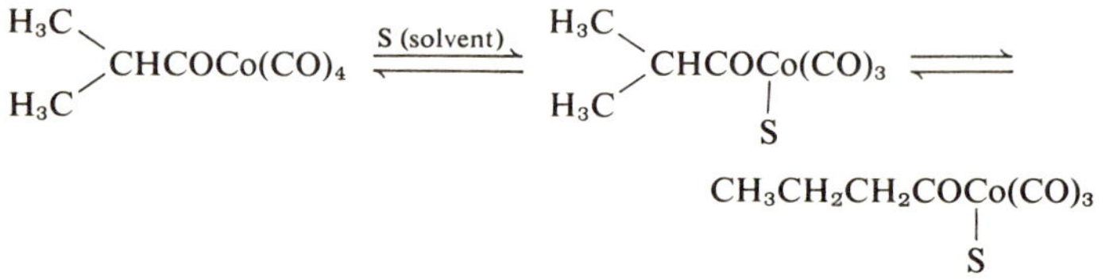
(46)

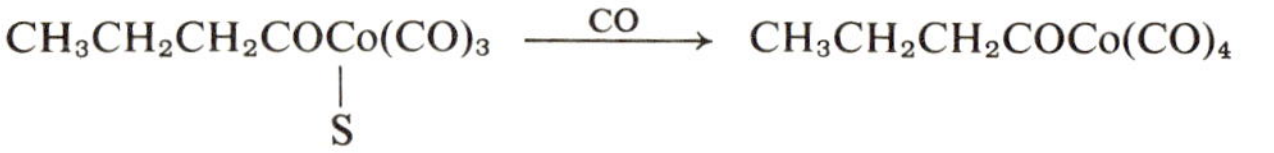
(47)

branched-chain compound. No isomerization was found [37] in the reaction of propylene oxide with sodium tetracarbonyl cobaltate, in view of the fact that ethyl γ-hydroxy-*n*-butyrate was obtained as the main product. The reaction probably proceeds as in Eq. (48). The isomerization of 2-ethyl-

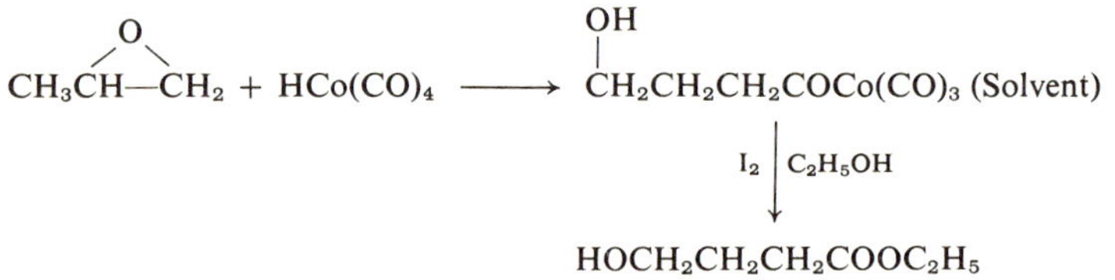
(48)

butyryl chloride gave *n*-caproyl chloride, as indicated in Eq. (49) Whether the isomerization takes place directly from **21** to **23**, or through another branched isomer, α-methylvalerylcobalt tricarbonyl, **22**, is not clear at present.

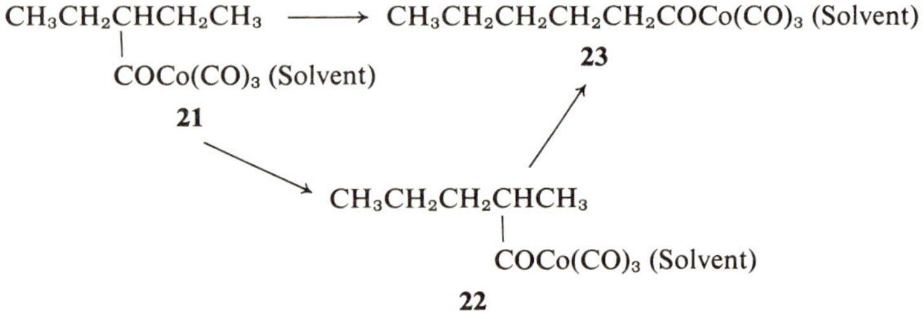
(49)

E. Platinum, Palladium, Rhodium, and Iridium Complexes

Carbonyl insertion reactions of platinum and palladium complexes have been observed [38] in the case of tertiary phosphine complexes of the type, *trans*-$MXR(PEt_3)_2$, where M = Pd or Pt, X = Cl, Br, or I, and R = alkyl or aryl). Carbonylation of palladium compounds, which have been found to be more reactive than those of platinum, proceeds readily at room temperature and at ordinary pressures. Carbonylation of platinum compounds takes place at 90°C and at 50–100 atm of carbon monoxide. All the compounds investigated have the *trans* configuration, as indicated by their dipole moments, and possess very strong carbonyl stretching bands in the 1600–1700 cm^{-1} region. The coordinated halogen of the acyl compounds formed in these carbonyl insertion reactions may be replaced by other anions. The carbonylation of *trans*-$PtI(COCH_3)(PEt_3)_2$ was reversible; the compound reverts back to the original methyl complex.

It was suggested by Booth and Chatt [38] that the carbonyl insertion reaction for these complexes takes place through a five- coordinated intermediate, represented by **25**, where carbon monoxide attacks an axial position of the metal complex. Partial bond formation then takes place between the alkyl group and carbon monoxide, with the ultimate formation of the acyl complex, **26**. Mechanistic details of the reaction are not known. The fact that carbonylation reactions of palladium are faster than those of platinum is considered due to the greater ability of palladium complexes to readily form five-coordinated complexes as intermediates.

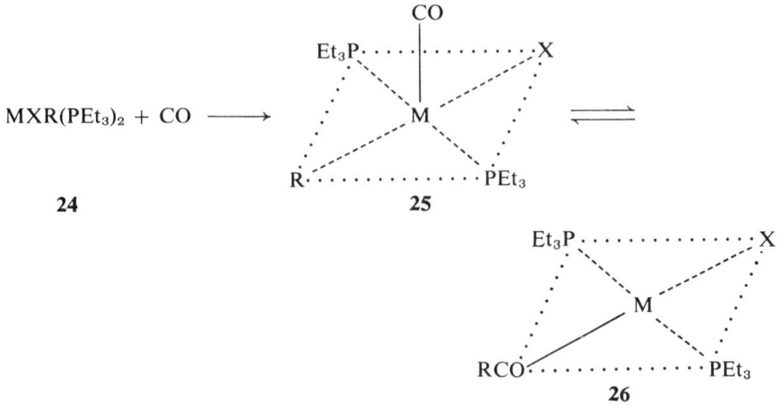

Bis(triphenylphosphine)ethyldichlororhodium(III), $(PPh_3)_2(C_2H_5)RhCl_2$, in chloroform solution at 25°C and 1 atm undergoes a very rapid reaction with carbon monoxide, with the formation of yellow crystals of *trans*-bis-(triphenylphosphine)carbonylchlororhodium(I) [30] and the liberation of ethyl chloride [39]. At −60°C, however, exposure of $(PPh_3)_2(C_2H_5)RhCl_2$ to carbon monoxide results in the formation of a yellow solution of the labile complex, $(PPh_3)_2(C_2H_5)(CO)RhCl_2$. The presence of carbon monoxide coordinated to rhodium(III) was confirmed by the IR band at 2060 cm^{-1}. On warming this carbonyl complex to $-30°C$, the yellow solution becomes orange and orange crystals of the acyl complex $(PPh_3)_2(C_2H_5CO)RhCl_2$ separate. The formation of the acyl complex was confirmed by the appearance of the acyl carbonyl band at 1701 cm^{-1}.

An acyl derivative was not formed by the reaction of carbon monoxide with $(PPh_3)_2(C_2F_4H)RhCl_2$. At $-50°C$ in chloroform this complex absorbs 1 mole of carbon monoxide to form $(PPh_3)_2(C_2F_4H)(CO)RhCl_2$ as yellow crystals.

Bis(triphenylphosphine)vinyldichlororhodium(III) reacts with carbon monoxide in chloroform solution at 25°C to form yellow crystals of bis(triphenylphosphine)acryloyldichlororhodium(III), $(PPh_3)_2(COCH{=}CH_2)RhCl_2$. The acryloyl group was confirmed by the presence of IR bands at 1707 and 1585 cm^{-1}. At $-50°C$, the reaction of bis(diphenylphosphine)vinyldichlororhodium, $(PPh_3)_2(CH{=}CH_2)RhCl_2$, with carbon monoxide gave an intermediate carbonyl, $(PPh_3)_2(CH{=}CH_2)(CO)RhCl_2$, which rapidly rearranged to the acrylolyl derivative at 25°C.

The kinetics and mechanism of carbonylation reactions of rhodium complexes have not yet been elucidated. The scheme in Eq. (50) is an attempt to rationalize the formation of carbonyl and acyl or aroyl octahedral rhodium complexes, and the conversion of these complexes to the planar species characteristic of rhodium(I), with the elimination of alkyl halide.

Displacement of solvent from the coordination sphere of complex **27** forms the carbonyl derivative **28**. Alkyl migration or carbonyl insertion gives rise to the acyl complex **29**. In the formation of the coordinately unsaturated acyl or aroyl complex **29** of rhodium(III), a *cis* addition of carbon monoxide to a metal–alkyl bond or *cis* migration of the alkyl group to a coordinated carbonyl group is suggested. The complexes **28** and **29** are in equilibrium, the position of which depends on the strengths of the metal–ligand bonds in both species and on the migrational ability of the R group (assuming alkyl or aryl migration). The equilibrium constant K for the propionyl complex is considerably less than that of the acetyl analog. For the benzoyl derivative K

$$K = \frac{[(PPh_3)_2R(CO)RhCl_2]}{[(PPh_3)_2(COR)SRhCl_2]}$$

is very large. The stabilities of the acyl and aroyl derivatives decrease in the order $C_2H_5 > CH_3 > C_6H_5$, in accord with the order of increasing stabilities of the corresponding metal–carbon bonds. The alkyl or aryl complex **28** or the acyl or aroyl complex **29** then dissociates to the *trans* rhodium(I) complex **30** by the elimination of an alkyl or aryl halide. The factors governing the stabilities of the alkyl or aryl derivatives, $RhX_2(CO)(R)(PR'_3)_2$, or the acyl or aroyl derivatives, $RhX_2(RCO)(PR'_3)_2S$, toward dissociation of RX depend on the nature of X (halogen), R (alkyl or aryl), and R′ (alkyl, alkoxy, aryl).

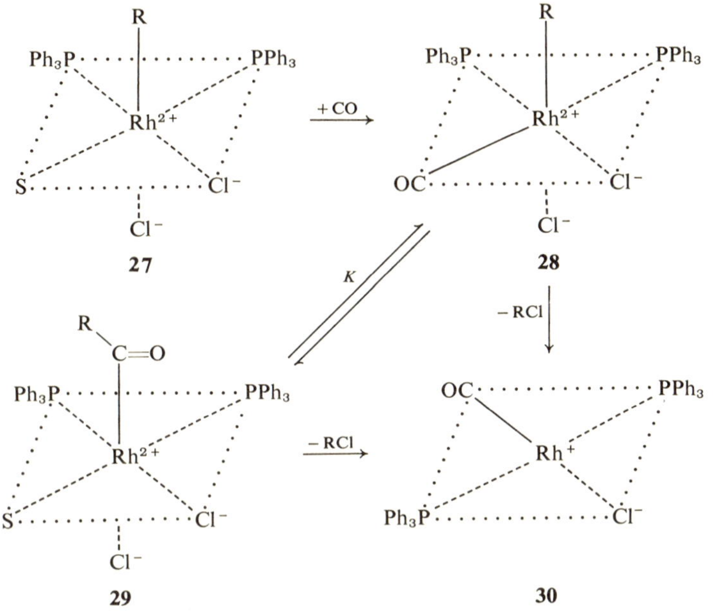

(50)

Glyde and Mawby [40,41] have investigated the kinetics of the carbonyl insertion reactions, Eq. (51).

$$Ir(CO)_2Cl_2RL + L' \longrightarrow Ir(CO)(COR)Cl_2LL' \qquad (51)$$

(R = Me, L = L′ = $AsPh_3$; R = Et, L = L′ = $AsPh_3$; R = Et, L = L′ = $AsMe_2Ph$; R = Et, L = $AsMe_2Ph$, L′ = $AsPh_3$)

The reaction was found to be first order in the iridium(III) complex and independent of concentration of the ligand L′. Change of solvents over a wide range of dielectric constant had little effect on reaction (51). The activation parameters of the reaction were reported as $\Delta H^{\dagger} = 21$–25 kcal/mole and $\Delta S^{\dagger} = 1.8$–6.7 eu. On the basis of the independence of the reaction on the nature of the solvent and the more positive entropy of activation than those systems (−20 to −30 eu) where a solvent-assisted mechanism has been proposed, Glyde and Mawby [40,41] suggested that the solvent plays no major part in the reaction and proposed the mechanism indicated in Eq. (52).

$$Ir(CO)_2Cl_2RL \underset{k_{-1}}{\overset{k_1}{\rightleftharpoons}} Ir(CO)(COR)Cl_2L \xrightarrow[\text{fast}\ (k_2)]{+L'} Ir(CO)(COR)Cl_2LL' \qquad (52)$$

$(k_2 \gg k_{-1} \gg k_1)$

IV. Synthetic Applications of Carbonyl Insertion Reactions

A. Insertion of Carbon Monoxide in Metal–Carbon Bonds

1. *Carbonylation of Alkyl and Aryl Halides through Organometallic Intermediates*

Ryang *et al.* [42–51] have prepared a variety of compounds including aldehydes, symmetrical ketones, and acyloins by the direct interaction of carbon monoxide with alkyl or aryl lithium [45,47,49,50,52], aryl sodium [43], aryl potassium [46], aryl magnesium [44], and aryl chromium, nickel, cobalt, copper, and cadmium [51]. The yield of various carbonylated compounds and their derivatives depend on the nature of the organometallic catalyst, solvent, and temperature.

Phenyllithium [45], phenylsodium [43], phenylpotassium [46], and phenylmagnesium bromide [44] were prepared by the reaction of bromobenzene with the respective metals in dry ether in the absence of air. Carbon monoxide was bubbled through the solution at 1 atm pressure at a rate of 0.5–0.6 liter/minute. The reaction temperature varied from $-10°C$ in the case of phenyllithium reactions to 30°–80°C for phenylsodium. Carbonylation of phenyllithium in the carbonylated fraction gave benzophenone, benzoin, and a trace of benzoic acid, and in the noncarbonylated fraction biphenyl, triphenylmethane, and benzhydrol. Phenylpotassium at $-35°C$ gave 25% benzoin, 2–3% benzhydrol, and 4% benzoic acid. Phenylsodium and phenylmagnesium bromide at 30°–80°C gave more or less the same series of products as did phenyllithium, the yields depending on temperature of the reaction. The variation of the yield of carbonylated products with temperature and solvent for the reaction of organometallic compounds with carbon monoxide is presented in Table II.

The polarity of the solvent and the temperature both seem to affect the nature and yield of the carbonylated products. In nonpolar solvents such as benzene, the yield of biphenyl and of triphenylmethylcarbinol are high, whereas the yield of these products is low in polar solvents such as tetrahydrofuran and ether. The yield of both biphenyl and triphenylmethylcarbinol increases with temperature. In benzene, benzoin is formed only at lower temperatures. The effect of temperature on the formation of benzoin and benzophenone is also apparent in the reaction of phenylcobalt with carbon monoxide (Nos. 5–7, Table II). The yields of benzoin and benzophenone are high at $-25°C$. The highest yield of benzoin is obtained at $-35°C$.

Ryang and Tsutsumi [44,45] suggested reactions (53)–(60) for the formation of various carbonylated products in the reaction of organometallic compounds with carbon monoxide.

TABLE II

Yields of Carbonylated Products in the Reaction of Organometallic Compounds with Carbon Monoxide

	Organometallic compounds	T (°C)	Solvent[a]	Time (hours)	% Yield of reaction products: Benzoin	Benzophenone	Triphenylmethylcarbinol	Triphenylmethane	Biphenyl	Benzoic acid
1	Phenyl Na	30	THF	4	17.4	—	—	—	—	6
2	Phenyl Na	30	C_6H_6	4	15.1	—	15.8	—	13.9	6.6
3	Phenyl Na	80	C_6H_6	4	—	—	23.2	—	13.5	12.0
4	Phenyl Na	30	Ether	6	—	2.5	5.0	5.9	4.8	6.3
5	Phenyl Co	−10	Ether		—	—	—	5.5	85	2.0
6	Phenyl Co	−25	Ether	2	12.0	11.5	—	22	44.2	2.3
7	Phenyl Co	−35	Ether	2	35.0	Trace	5	—	30	—
8	Phenyl Cr halide	−25	THF	5	—	38	—	—	15	—
9	Phenyl Cr halide	−25	Ether	5	—	20	—	—	20	—
10	Phenyl Ni halide	12	Ether	3	23	—	6	—	17	—
11	Phenyl Ni halide	21	Ether	3	—	—	—	—	60	—
12	Phenyl Ni halide	0	THF	5	19	11.5	5	—	19	—
13	Phenyl Cu	−5	Ether	5	—	—	—	—	39	—
14	Phenyl Cu	32	Ether	4	—	—	—	—	42	—
15	Phenyl Cd halide	15	THF	3	—	—	—	—	8	—

[a] THF = tetrahydrofuran; ether = $(C_2H_5)_2O$.

In the reaction of a metal alkyl with carbon monoxide a mixed-metal carbonyl aryl complex of the metal ion may be first formed, followed by a carbonyl insertion reaction probably of the same type as that described for the $RMn(CO)_5$ or $RCo(CO)_4$ complexes. The aroyl compound RCOM then condenses with itself and yields acyloins (e.g., benzoin) on hydrolysis.

$$\mathrm{RM + CO \longrightarrow RMCO \rightleftharpoons RCOM \rightleftharpoons R\overset{\displaystyle :}{C}(OM)} \tag{53}$$

$$\mathrm{RCOM + R\overset{\displaystyle :}{C}(OM) \longrightarrow R{-}C({=}O){-}C(M)(OM){-}R \xrightarrow{\text{hydrolysis}} R{-}C({=}O){-}C(H)(OH){-}R} \tag{54}$$

$$\mathrm{RCOM + RM \longrightarrow R{-}C(M)(OM){-}R \xrightarrow{\text{hydrolysis}} R{-}C(H)(OH){-}R} \tag{55}$$

$$\mathrm{RCOM + RX \longrightarrow R{-}C({=}O){-}R + MX} \tag{56}$$

$$\mathrm{2\,RM + 3\,CO \longrightarrow R{-}C({=}O){-}R + 2\,MCO} \tag{57}$$

$$\mathrm{R{-}M^{n+}\cdots n\,Cl^- \longrightarrow R\cdot + M^{n+}\cdots n\,Cl^-} \tag{58}$$

$$\mathrm{2\,R\cdot \longrightarrow R{-}R} \tag{59}$$

$$\mathrm{RCOM \xrightarrow{\text{hydrolysis}} RCHO \xrightarrow{+O} RCOOH} \tag{60}$$

Formation of ketones takes place by the reaction of RCOM with RX; or alternatively two molecules of RM may react with carbon monoxide to give the ketone and metal carbonyl. Biaryl is obtained by the dissociation of RM followed by combination of free aryl radicals. Triphenylcarbinol is probably formed by a similar free-radical process. Hydrolysis of RCOM (R = phenyl), followed by oxidation of the aldehyde, yields benzoic acid. In most of the reactions where water is carefully removed from the alcohol solvent before the reaction is carried out, the yield of benzoic acid is very low. The postulation of a mixed-metal carbonyl alkyl or aryl complex as the first step in the carbonylation of organometallic compounds is supported by the fact that only alkali metals, plus magnesium, cobalt, chromium, nickel, and iron, which are known to form such mixed-ligand complexes, take part in the formation of the carbonylated products. Copper and cadmium, which cannot form such carbonyls, are inactive in the formation of carbonylated reaction products. In the case of copper and cadmium the only product obtained is biphenyl (entries 13–15, Table II).

Support for the reactions proposed above involving insertion of carbon monoxide into an organic molecule from the coordination sphere of a metal ion is provided [49] by the reaction of phenyllithium with iron pentacarbonyl, in which benzoin is obtained as the main product, along with small quantities of biphenyl and triphenylmethane. Similarly, *o*-methylphenyllithium gave on treatment [49] with iron pentacarbonyl, di-*o*-tolycarbinol in 35% yield. The reaction of *m*-methylphenyllithium with iron pentacarbonyl gave 34.2% *m*-methylbenzaldehyde.

It was postulated [48] that aryllithium reacts with iron pentacarbonyl to form an aroyllithium compound, **33**, Eqs. (61) and (62). The aroyllithium either reacts with hydrogen ions to form the aldehyde or, depending on the reaction conditions, undergoes other reactions [(53)–(56), (60)] with the formation of various products. Thus in the formation of RCOLi by the reaction

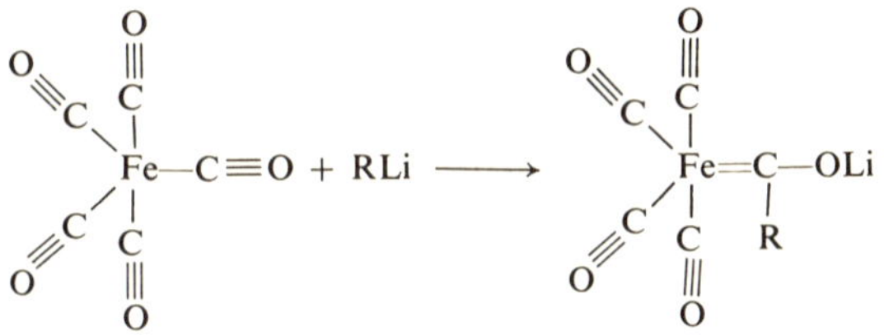

31 **32** (61)

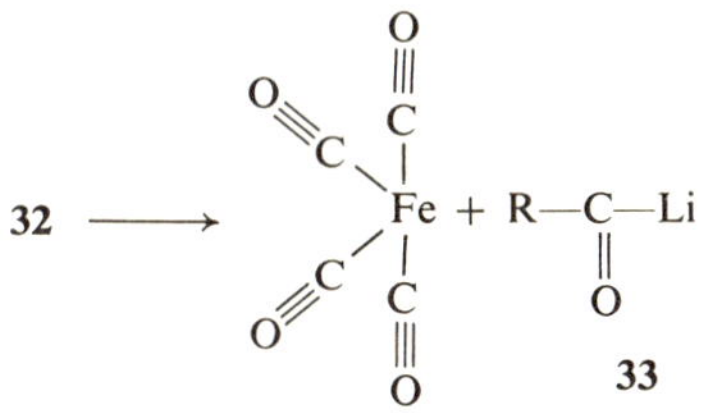

33 (62)

$$R{-}\underset{\underset{O}{\|}}{C}{-}Li + H^+ \longrightarrow RCHO + Li^+$$

33

of aryllithium with iron pentacarbonyl, activated carbon monoxide enters the organic moiety through the coordination sphere of the metal ion. Aryllithium compounds undergo similar reactions with nickel tetracarbonyl with the formation of acyloins [53] as the main product. The carbon monoxide insertion reaction in organometallic compounds thus permits the isolation of carbonyl compounds such as aldehydes, ketones, acyloins, carboxylic acids, and their derivatives by employing appropriate reaction conditions.

2. *Synthesis of Aldehydes*

Ryang *et al.* [48] have reported conditions for the preferential formation of aldehydes in the reaction of aryllithium with iron pentacarbonyl at −50° to −60°C. In order to minimize the possibility of side reactions leading to benzoin, ketones, and benzhydrol derivatives, iron pentacarbonyl was added rapidly within 2 minutes to a diluted solution of aryllithium in ether at −60°C and was then allowed to react for 2 hours at −50° to −60°C. Dilution of the organometallic compound with ether and rapid addition of the carbonyl apparently decreased the concentration of free aryllithium which enters into side reactions with RCOLi. Table III presents the starting materials and the yields of corresponding aldehydes obtained by Ryang *et al.* [48]. It may be seen from the results shown in Table III that high yields of aldehydes are frequently obtained.

Cook [54] and Collman [55] have synthesized aldehydes in 80–90% yield by the reaction of alkyl or aralkyl halides with sodium tetracarbonyl ferrate (2−),$Na_2Fe(CO)_4$, **34**, in tetrahydrofuran under a carbon monoxide atmosphere. The carbonyl **34** was obtained by the reduction of iron pentacarbonyl in tetrahydrofuran with sodium amalgam. The mechanism suggested [55] for the formation of aldehydes is illustrated by Eq. (63)–(66).

$$Fe(CO)_5 \xrightarrow[THF]{Na-Hg} \underset{\mathbf{34}}{Na_2Fe(CO)_4} + CO \tag{63}$$

$$\underset{\mathbf{34}}{Na_2Fe(CO)_4} + RX \longrightarrow \underset{\mathbf{35}}{Na[RFe(CO)_4]} + NaX \tag{64}$$

(R = alkyl or aralkyl)

$$\underset{\mathbf{35}}{Na[RFe(CO)_4]} + CO \longrightarrow \underset{\mathbf{36}}{Na[RCOFe(CO)_4]} \tag{65}$$

$$\underset{\mathbf{36}}{Na[RCOFe(CO)_4]} + HOAc \longrightarrow RCHO + NaOAc + Fe(CO)_4 \tag{66}$$

Oxidative addition of alkyl or aralkyl halide to **34** [reaction (64)] results in formation of the iron(0) alkyl or aralkyl complex **35**. Carbon monoxide insertion in the carbon–iron bond of **35** results in the formation of the acyl complex **36**. Protonation of **36** with acetic acid (or other proton donor) gives the aldehyde and the carbonyl $Fe(CO)_4$, which can be reconverted to the original anion $Fe(CO)_4{}^{2-}$.

3. *Synthesis of Ketones*

Symmetrical ketones have been synthesized [46,47,49] in high yields by the direct reaction of alkyl or aryl lithium derivatives with carbon monoxide in anhydrous ether at −70°C. The organolithium derivative was obtained from metallic lithium and the corresponding alkyl or aryl bromide in dry

TABLE III

SYNTHESIS OF ALDEHYDES BY THE REACTION OF ARYLLITHIUM WITH $Fe(CO)_5$ IN ETHER AT −50° to −60°C

Starting material	Aldehyde obtained	Yield (%)
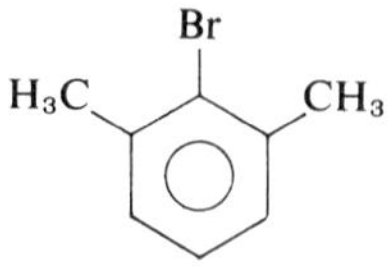	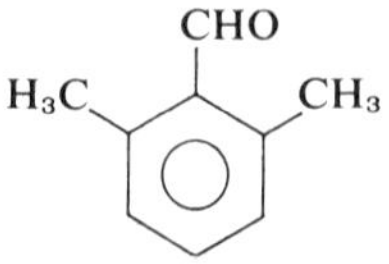	36.4
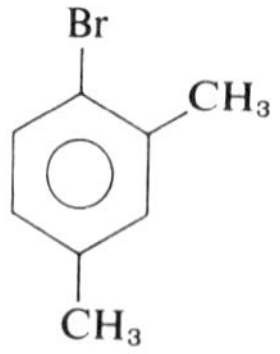	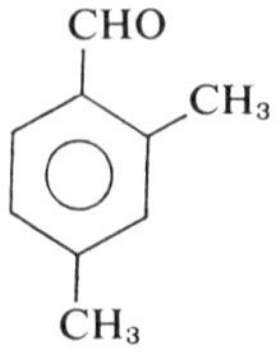	65.4
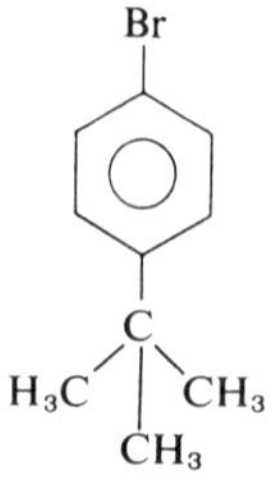	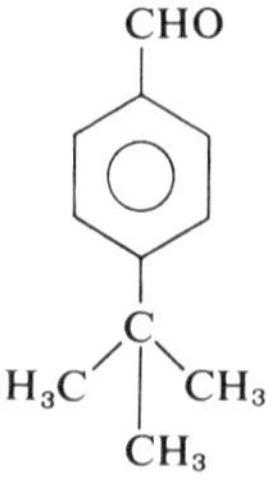	47.0
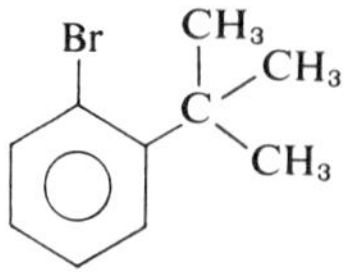	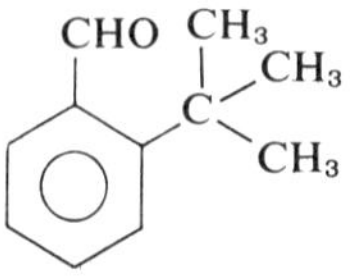	54.3
	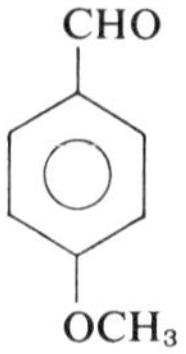	24.2
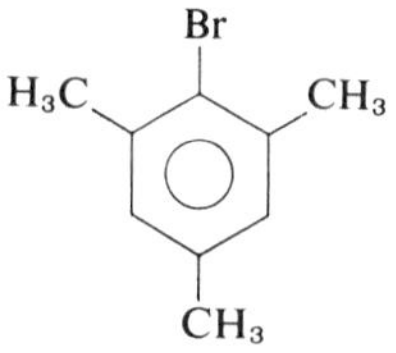	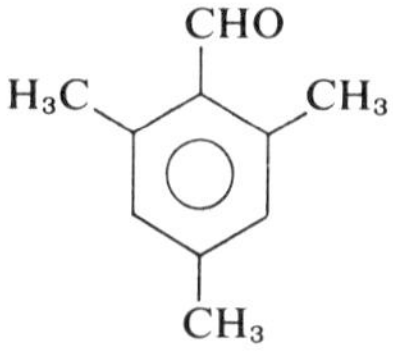	60.5

TABLE IV
Synthesis of Ketones by the Direct Interaction of CO with Alkyl or Aryl Lithium in Ether

T (°C)	Starting material	Ketone	Yield	References
−70	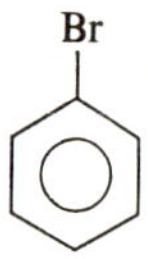	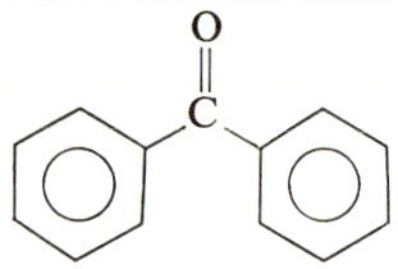	55	46, 47
−70		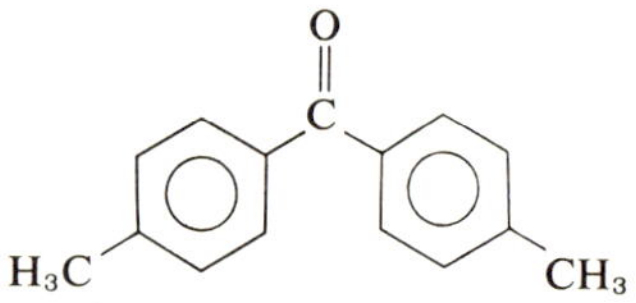	50	46, 47
−60	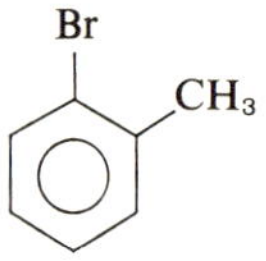	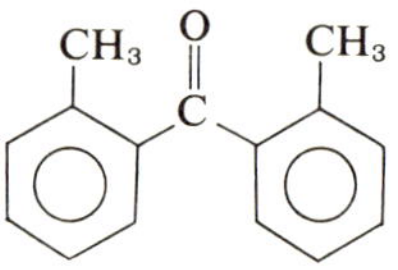	43.8	49
−60	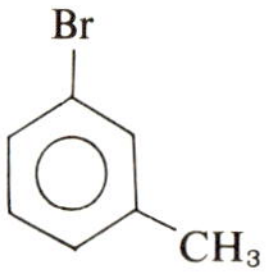	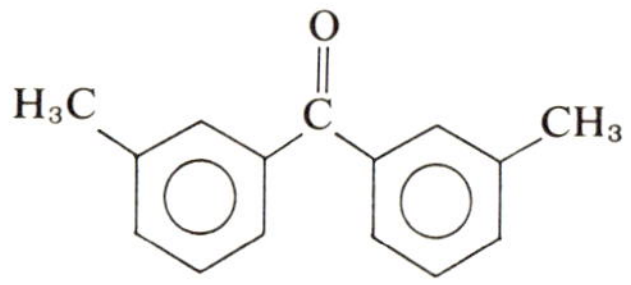	48.1	49
−60	$BrCH{=}CH_2$	Polyvinyl ketone Divinyl ketone	—	52
−5, −10	$(CH_2)_4Br_2$	Cyclopentanone and polymer	—	49
−70	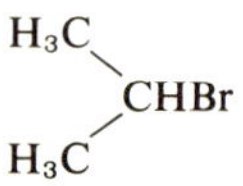	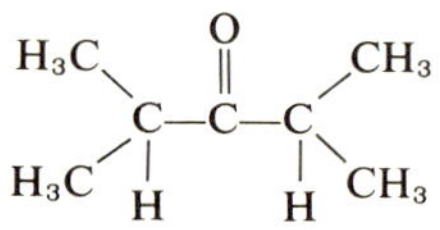	30.0	46, 49
−70	$CH_3(CH_2)_2Br$	$CH_3(CH_2)_2\overset{O}{\overset{\|}{C}}(CH_2)_2CH_3$	40.0	46, 49
−70	$CH_3(CH_2)_3Br$	$CH_3(CH_2)_3\overset{O}{\overset{\|}{C}}(CH_2)_3CH_3$	28.0	44, 49

ether, and carbon monoxide was bubbled through the solution. Table IV lists the ketones prepared by this method.

Ketones were obtained [55] in 80–90% yield by the reaction of **34**, $Na_2Fe(CO)_4$, in tetrahydrofuran with alkyl or aralkyl halides under carbon monoxide. The reaction sequence of Eqs. (64), (65), and (67) for the synthesis of ketones differs from the aldehyde synthesis shown above in Eqs. (63)–(66) principally in the fact that excess alkyl halide is available for reaction with the carbonyl insertion product **36**.

$$\mathbf{34} + RX \longrightarrow \underset{\mathbf{35}}{Na[RFe(CO)_4]} + NaX \tag{64}$$

$$\mathbf{35} + CO \longrightarrow \underset{\mathbf{36}}{Na[RCOFe(CO)_4]} \tag{65}$$

$$\mathbf{36} + R'X \longrightarrow RCOR' + NaX + Fe(CO)_4 \tag{67}$$

4. *Synthesis of Acyloins*

Acyloins have been obtained [53] by the reaction of nickel tetracarbonyl with an organolithium compound, RLi, in ether solution at −70°C, followed by hydrolysis with ethanol and hydrochloric acid. Alkyllithiums usually yield symmetrical ketones and aryllithiums give acyloins by this method. Table V lists the acyloins obtained by the reaction of aryllithium compounds with nickel tetracarbonyl.

TABLE V
SYNTHESIS OF ACYLOINS BY THE INTERACTION OF ARYLLITHIUM WITH $Ni(CO)_4$[a]

Starting material	Product	Yield (%)
Phenyllithium	Benzoin	36
m-Tolyllithium	*m*-Toluoin	61
p-Tolyllithium	*p*-Toluoin	71
p-Methoxyphenyllithium	*p*-Anisoin	15
	p-Anisil	23

[a] Reference [53].

5. *Synthesis of Carboxylic Acids and Derivatives*

Bliss and Southworth [56] studied the carboxylation of aryl halides by the treatment with iron, nickel, or cobalt carbonyls at 200°C under pressure in aqueous hydrochloric acid. In some cases carbon monoxide was also introduced into the reaction mixture. Carbonylation proceeds through the intermediate formation of acyl metal complexes, RCOM, with subsequent hydrolysis to the carboxylic acids. *p*-Chlorotoluene gave *p*-toluic acid by this method. Esters were formed in the presence of alcohols, alkoxyalkanes, and phenols [57]. Yamamoto and Sato [58] reported the carboxylation of iodobenzene by nickel tetracarbonyl in an atmosphere of carbon monoxide under a pressure of 120 atm at 220°C to give benzoic acid in very good yield. Dichlorobenzene (*o*:*p*, 45:55) in the presence of water, nickel tetracarbonyl, and iodine, under a pressure of 200 atm of carbon monoxide at 310°C gave a mixture of the *o*- and *p*-dicarboxylic acids in approximately equivalent amounts.

Cyclic anhydrides of aromatic dicarboxylic acids have been obtained [59,60] by the treatment of monohalogenated aromatic compounds in benzene with an alkali metal carbonate or phosphate and nickel tetracarbonyl at 325°C in the presence of carbon monoxide under pressure. Formation of the aromatic cyclic anhydride is preceded by formation of intermediates of the type C_6H_5COM and $C_6H_4(COM)(COOCOC_6H_5)$. The latter rearranges to the cyclic anhydride and the metal salt of an aromatic carboxylic acid. By this method bromobenzene was converted to phthalic anhydride in 73.3% yield. *o*-Bromotoluene gave 3-methylphthalic acid anhydride.

Aroyl halides have also been obtained [61] from monochloro- or monobromobenzene and sodium fluoride in the presence of nickel tetracarbonyl as a catalyst at 378°C and 100 atm of carbon monoxide. Chlorobenzene gave benzoyl fluoride. *o*-Dichlorobenzene yielded *o*-chlorobenzoyl fluoride. Other nickel compounds such as nickel(II) chloride, nickel(II) bromide, and tetracyanonickelate(II), are equally effective as catalysts, presumably because they are converted to nickel carbonyl under the reaction conditions employed.

Aryl iodides are converted to aroate esters [62] by nickel tetracarbonyl in alcohol solvents. A reaction of the type of Eq. (68) probably takes place.

$$ArX \xrightarrow{Ni(CO)_4} [Ar{-}\overset{\overset{\displaystyle O}{\|}}{C}{-}\overset{\overset{\displaystyle X}{|}}{Ni}(CO)_n] \xrightarrow{ROH} Ar{-}\overset{\overset{\displaystyle O}{\|}}{C}{-}OR + HX + Ni(CO)_4 \quad (68)$$

$$\underset{\mathbf{37}}{Pd(PPh_3)_4} + RX \longrightarrow \underset{\mathbf{38}}{RPdX(PPh_3)_2} + 2\ PPh_3 \quad (69)$$

$$(R = \text{alkyl, aryl};\ X = Cl^-, Br^-, I^-)$$

$$\underset{\mathbf{38}}{RPdX(PPh_3)_2} + CO \rightleftharpoons \underset{\mathbf{39}}{RCOPdX(PPh_3)_2} \quad (70)$$

$$\underset{\mathbf{39}}{RCOPdX(PPh_3)_2} + R'OH + 2\ PPh_3 \longrightarrow \underset{\mathbf{40}}{RCOOR'} + HX + \underset{\mathbf{37}}{Pd(PPh_3)_4} \quad (71)$$

The yields of aroate esters obtained are quite high. In aprotic solvents the final step does not occur, and arils are obtained in good yield.

The high reactivity of triphenylphosphinepalladium(0) complexes with respect to the displacement of phosphine ligands has been utilized [63] for the preparation of esters from alkyl or aryl halides according to reactions (69)–(71). The reaction of an alkyl or aryl halide with $Pd(PPh_3)_4$, **37**, results in the displacement of two triphenylphosphine ligands to give the palladium(II) complex **38**. Carbon monoxide insertion in the R—Pd(II) bond results in the acylpalladium(II) complex **39**, which reacts with an alcohol R′OH to give the ester **40**.

Complexes having labile palladium(II)–carbon bonds such as **38** have been useful in several alkylation or arylation reactions [63], including the alkylation of olefins (Volume II, Chapter 5).

B. Carbonylation of Unsaturated Compounds

1. *The Reppe Carboxylation Reaction*

Reppe *et al.* [64,65] developed an important process for the carboxylation of unsaturated compounds by carbon monoxide under pressure in the presence of nickel tetracarbonyl and suitable hydrogen donors such as alcohols. The overall reaction for alkenes and alkynes is represented by Eqs. (72)–(74). The use of an acid in place of alcohol as the hydrogen donor results in the formation of an unsaturated acid anhydride [66].

$$R{-}CH{=}CH_2 + CO + R'OH \xrightarrow[200^\circ C,\ NiX_2]{Ni(CO)_4} R{-}CH_2{-}CH_2{-}COOR' + R{-}\underset{}{\overset{CH_3}{\overset{|}{C}}}H{-}COOR' \tag{72}$$

$$R{-}C{\equiv}C{-}H + CO + R'OH \xrightarrow[180^\circ C,\ NiX_2]{Ni(CO)_4} R{-}CH{=}CH{-}COOR' \tag{73}$$

$$C_6H_5C{\equiv}CH + CO + H_2O \xrightarrow[65^\circ C,\ CH_3COOH]{Ni(CO)_4} C_6H_5\overset{CH_2}{\overset{\|}{C}}{-}COOCOCH_3 \tag{74}$$

Chiusoli *et al.* [67–69] reported the carboxylation of allyl chloride by carbon monoxide under a pressure of 50 atm at 199°C in methanol with nickel tetracarbonyl as catalyst. Methyl esters of 2- and 3-butenoic acids were obtained, as shown in Eq. (75), with the 2-butenoic acid ester as the major

$$CH_2{=}CH{-}CH_2X + CO + CH_3OH \xrightarrow[100^\circ C]{Ni(CO)_4} \begin{matrix} CH_2{=}CH{-}CH_2COOCH_3 \\ + \\ CH_3{-}CH{=}CH{-}COOCH_3 \end{matrix} \tag{75}$$

product. The unsaturated esters are probably formed by the reaction sequence of Eq. (76) [65]. Carbonylation of 3-chloropropyne by carbon monoxide under pressure with nickel tetracarbonyl as the catalyst yielded [Eq. (77)] itaconic acid [68].

An interesting modification of the nickel tetracarbonyl-catalyzed carbonylation reaction is the double insertion of carbon monoxide and acetylene in chlorallyl derivatives with the formation of dienoic acid esters [68]. Double insertion of carbon monoxide and acetylene in iodobenzene at 120°C at a pressure of 30 atm, Eq. (78), results in the formation of benzoyl propionate [70]. The aroyl group reacting with acetylene and carbon monoxide is

$$R{-}CH{=}CH{-}CH_2X + Ni(CO)_4 \longrightarrow [R{-}CH{=}CH{-}CH_2CO{-}Ni(CO)_3]^+X^-$$

$$[R{-}CH{=}CH{-}CH_2CO{-}Ni(CO)_3]^+X^- \xrightarrow{CO,\ R'OH} R{-}CH{=}CH{-}CH_2{-}COOR' + Ni(CO)_4 + HX \qquad (76)$$

$$[R{-}CH{=}CH{-}CH_2CO{-}Ni(CO)_3]^+X^- \longrightarrow [RCH_2{-}CH{=}CHCONi(CO)_3]^+X^- \xrightarrow{CO,\ R'OH} RCH_2{-}CH{=}CHCOOR' + Ni(CO)_4 + HX$$

$$HC{\equiv}C{-}CH_2Cl + 2\ CO + 2\ H_2O \xrightarrow{Ni(CO)_4} H_2C{=}C(COOH){-}CH_2COOH \qquad (77)$$

$$C_6H_5I + CH{\equiv}CH + Ni(CO)_4 + R'OH + HX \longrightarrow C_6H_5COCH_2CH_2COOR' + NiIX + 2\ CO \qquad (78)$$

$(R' = H, \text{alkyl})$

normally expected to give an unsaturated ester, $C_6H_5COCH{=}CH{-}COOR'$. The double bond becomes hydrogenated in reaction (78) with the formation of only the saturated product.

Methyl *cis*-hexa-2,5-dienoate has been synthesized by Guerrieri and Chiusoli [71] by treatment of π-allylnickel halides of the general composition $[\pi\text{-}C_3H_5Ni(CO)(PR_3)X]$ (X = Cl, Br, I; R = alkyl or aryl) at 0°C and atmospheric pressure with a 70:30 carbon monoxide–acetylene mixture in a solvent consisting of 80 volume % toluene and 20 volume % methanol. The carbonylation reaction of the alkyne may be represented by Eq. (79).

$$C_3H_5Ni(CO)PR_3Br + CH{\equiv}CH + 3\ CO + CH_3OH \longrightarrow CH_2{=}CH{-}CH_2{-}CH{=}CH{-}COOCH_3 + Ni(CO)_3(PR_3) + HBr \qquad (79)$$

In ether solution the π-allylnickel bromide exists as a dimer, **41**, which absorbs carbon monoxide to give an intermediate butenoylnickel dicarbonyl

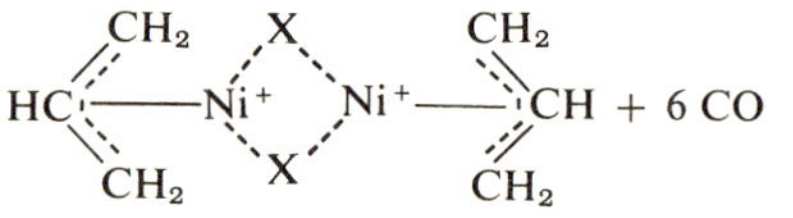

$$\mathbf{41} + 6\ CO \longrightarrow 2\ CH_2{=}CH{-}CH_2CONi(CO)_2X \quad \mathbf{42} \qquad (80)$$

41 **42**

$$\mathbf{42} \xrightarrow{CH_3OH} CH_2{=}CH{-}CH_2COOCH_3 + 2\ Ni(CO)_4$$

halide, **42**. This complex then reacts with methanol and carbon monoxide in a separate step [72] to give methyl 3-butenoate and nickel tetracarbonyl.

2. *Catalysis by Cobalt Carbonyls*

Dicobalt octacarbonyl, $Co_2(CO)_8$, can catalyze the carbonylation of alkenes by carbon monoxide under pressure. When the reaction is conducted in an alcohol medium, saturated esters are obtained in good yield. The hydrogen needed for the reaction is furnished by reaction of the alcohol with $Co_2(CO)_8$, which results in formation of hydridocobalt tetracarbonyl, $HCo(CO)_4$ [73]. Ercoli [74] has prepared aliphatic mono- and dicarboxylic acids and alicyclic acids from olefins in acetone or ether at 100°–190°C, at a pressure of 50–250 atm of carbon monoxide in the presence of dicobalt octacarbonyl. With this synthetic method, cyclohexene is converted to cyclohexanecarboxylic acid in 86% yield.

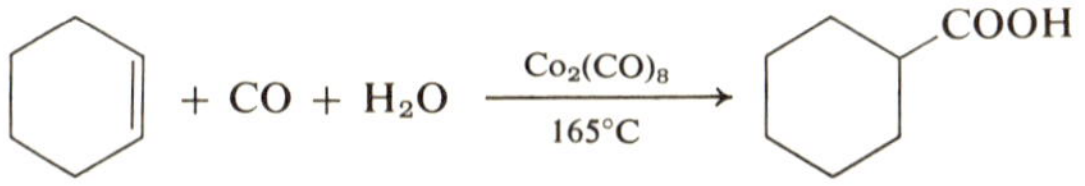

(81)

Carbonylation of 1,2-propylene oxide in methyl alcohol under carbon monoxide pressure of 300 lb/inch² yielded methyl β-hydroxybutyrate [75], as shown in Eq. (82). When the reaction was conducted in benzene under the

$$CH_3\overset{\frown O \frown}{CH{-}{-}CH_2} + CO + CH_3OH \xrightarrow[130°C]{Co_2(CO)_8} CH_3{-}\overset{OH}{\overset{|}{C}H}{-}CH_2{-}COOCH_3 \qquad (82)$$

same conditions in the absence of methyl alcohol, 3-butenoic acid was obtained in 81% yield [76].

π-Allylcobalt tricarbonyl, **43**, reacts [77] with methanol and carbon monoxide in the presence of a base such as dicyclohexylethylamine or sodium alkoxide. One mole of carbon monoxide is absorbed, and methyl allyl ether is obtained. With sodium methoxide the reaction is that given in Eq. (83).

$$HC\begin{matrix} CH_2 \\ \vdots \\ CH_2 \end{matrix}{-}Co(CO)_3 + NaOCH_3 + CO \longrightarrow NaCo(CO)_4 + CH_2{=}CHCH_2OCH_3 \qquad (83)$$

43

Butyrolactones and β-alkylbutyrolactones have been obtained [78] from the reaction of oxacyclobutane or its 3-alkyl derivatives with carbon monoxide under a pressure of 100–700 atm and at 120°–250°C in the presence of metal carbonyls as catalysts. From 3,3-dimethyloxacyclobutane, **44**, β,β-dimethylbutyrolactone, **45**, was obtained as shown in Eq. (84), in 58% yield.

Cyclic ketones were also prepared from nonconjugated polyolefins by reaction with carbon monoxide under a pressure of 1600 lb/inch^2 at 165°C with dicobalt octacarbonyl as catalyst [79]. Similarly, 1,5-hexadiene gave 35% 2,5-dimethylcyclopentanone and 6% 2,5-dimethylcyclopent-2-enone [Eq. (85)].

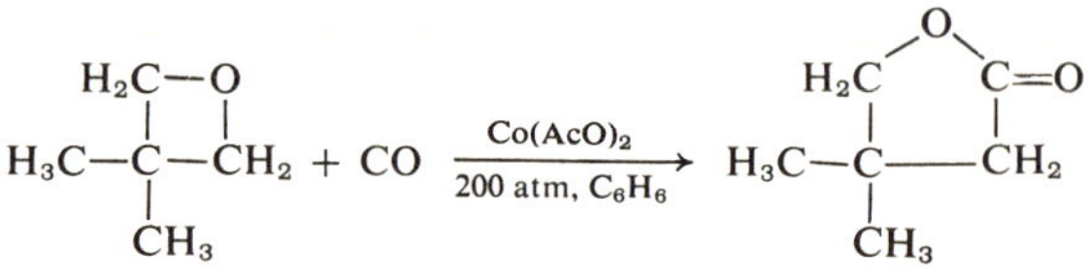

44 45 (84)

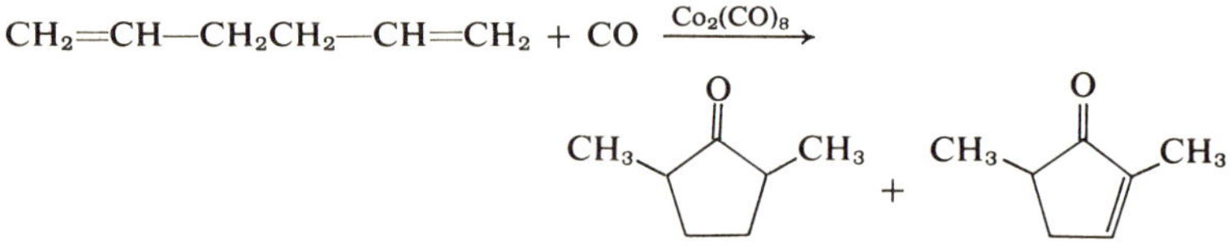

(85)

3. *Catalysis by Palladium(II) Chloride*

Carbonylation of alkenes may also be carried out with palladium(II) chloride as a catalyst [80–89]. The reaction proceeds through the initial formation of an olefin–palladium(II) complex followed by attack of carbon monoxide in the absence or in the presence of an alcohol to produce an acyl chloride or ester, respectively. Although the reaction starts in the homogeneous phase, palladium metal is precipitated at the end of the reaction. β-Chloroacyl chlorides have been obtained by nucleophilic attack of carbon monoxide on an olefin–palladium chloride complex [80,82]. The overall reaction may be expressed as follows:

$$(RCH{=}CH_2)_2Pd_2Cl_4 + 2\,CO \longrightarrow 2\,R{-}\underset{\displaystyle Cl}{\underset{|}{CH}}{-}CH_2COCl + 2\,Pd^0$$

The mechanism given in Eq. (86) has been proposed [80] for the formation of the β-chloroacyl chloride from an olefin–palladium chloride complex, **46**. The reaction of carbon monoxide with the binuclear olefin–Pd(II) chloride complex **46** is indicated as resulting in the formation of a mononuclear carbonyl complex **47**. Further attack by carbon monoxide on this complex results in the transfer of a chloride ion to the more highly substituted carbon atom of the olefin, with the concerted conversion of the π complex **47** to the σ complex **49** through the proposed transition state **48**. Insertion of carbon monoxide then probably takes place in the metal–carbon σ bond with the formation of an acyl complex **50** that collapses finally to the chloroacyl chloride derivative and metallic palladium. α-Olefins up to 1-hexene always yield straight-chain β-chloroacyl chlorides since the formation of a carbon–palladium σ bond takes place at the least-crowded carbon atom. On the other hand, α-olefins higher than 1-hexene yield only small proportions of straight-chain acyl derivatives. A mechanistic interpretation or rationalization of this behavior has not yet been suggested.

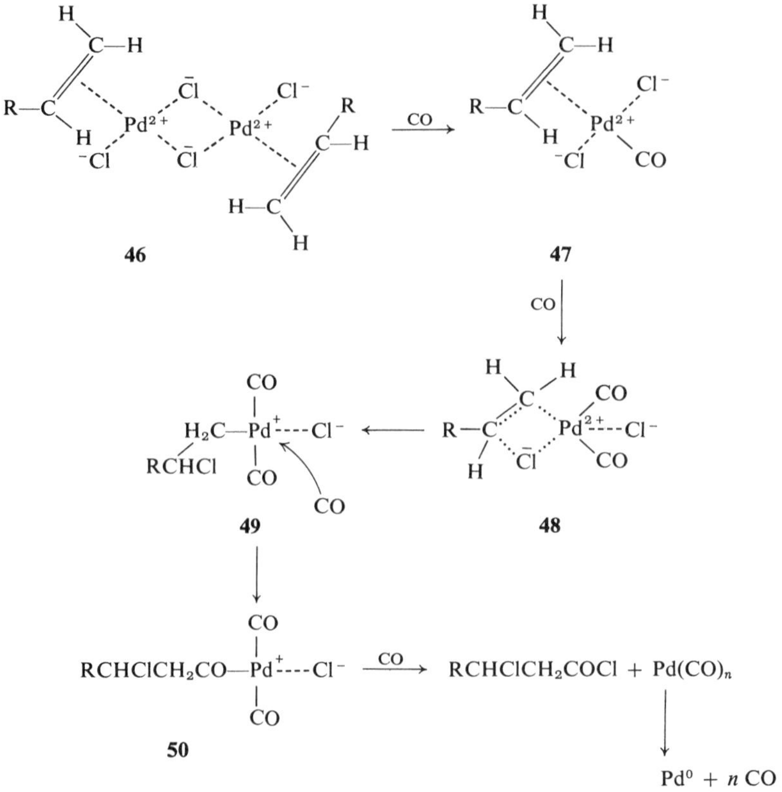

(86)

The π-allylpalladium chloride complex **51**, obtained by the reaction of allyl chloride or allyl alcohol with palladium chloride, carbonylates readily at 70°C under carbon monoxide pressures of about 100 kg/cm^2. In the presence of ethyl alcohol, ethyl 3-butenoate, ethyl 2-butenoate, and ethyl isobutyrate are obtained [83,84]. The overall reaction may be written as Eq. (87). The relative amounts of the products vary with temperature and with the concentration of hydrogen chloride built up in the reaction mixture.

Carbonylation of butadiene and isoprene [85,86] produces, besides the usual carbonylated products, varying amounts of 1,4-dihalides, the yield of the latter depending on the solvent. In the case of the butadiene–palladium dichloride complex **52** carbonylation in benzene yields a 3:2 molar ratio of 1,4-dichloro-2-butene, **54**, and 5-chloro-3-pentenoyl chloride, **56**. In ethanol,

1,4-dichloro-2-butene and ethyl 3-pentenoate, **57**, are obtained in the ratio of 1:2. The formation of 1,4-dichloro-2-butene, **54**, may be explained on the basis of the nucleophilic attack of chloride ion on carbon atom 1 of the

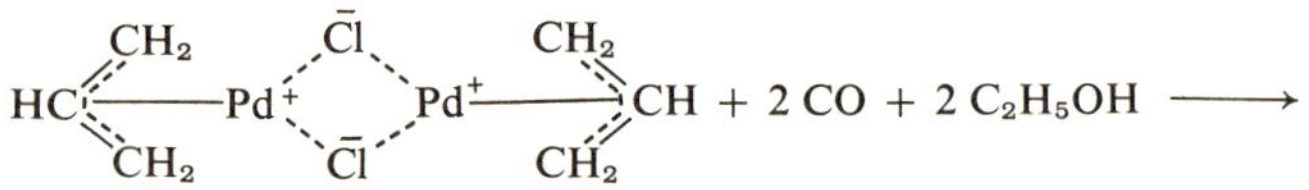

51

$$2\,HCl + \begin{cases} CH_2{=}CH{-}CH_2COOC_2H_5 \\ + \\ CH_3{-}CH{=}CH{-}COOC_2H_5 \\ + \\ (CH_3)_2CH{-}COOC_2H_5 \end{cases} \quad (87)$$

allylpalladium dichloride complex **53**. Nucleophilic attack of carbon monoxide on carbon atom 1 of **53** yields 5-chloro-3-pentenoyl chloride, **56**, through the intermediate complex **55**. Ethyl 3-pentenoate, **57**, is formed in ethanol solution by solvolysis of **55** and dehalohydrogenation at carbon atom

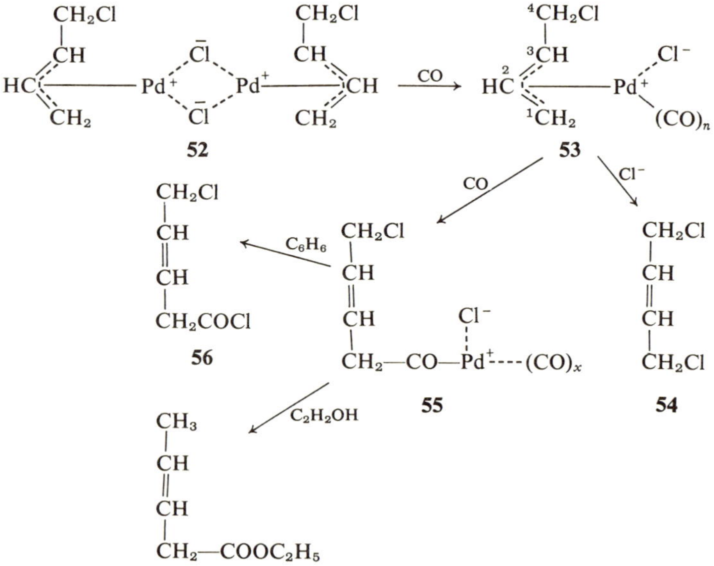

52 **53** **56** **55** **54** **57**

5. Carbonylation of an allene-palladium dichloride complex, **58**, in ethyl alcohol yields ethyl 3-chloro-3-butenoate [87] [Eq. (88)]. Carbonylation of 1,5-cyclooctadiene, **59**, and of 1,3-cyclooctadiene, **61**, at 100 °C, at a carbon monoxide pressure of 100 kg/cm^2, gave respectively, ethyl 4-cyclooctene carboxylate, **60**, and ethyl 3-cyclooctene carboxylate, **62** [88], as shown in Eqs. (89) and (90).

Carbonylation of 1,5,9-cyclododecatriene, **63**, Eq. (91), with palladium chloride as a catalyst gave a mixture of mono- and diesters [89].

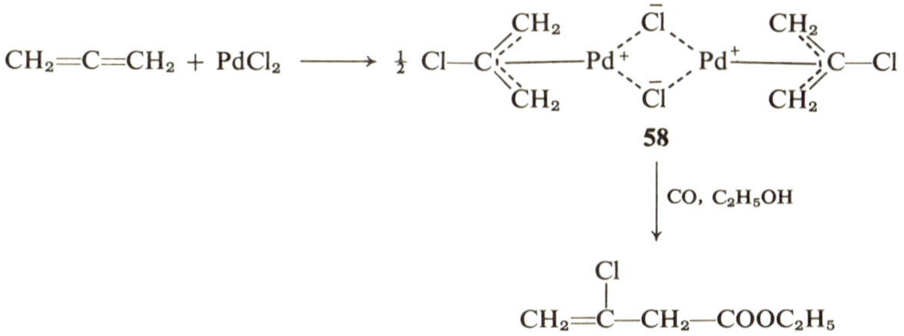

(88)

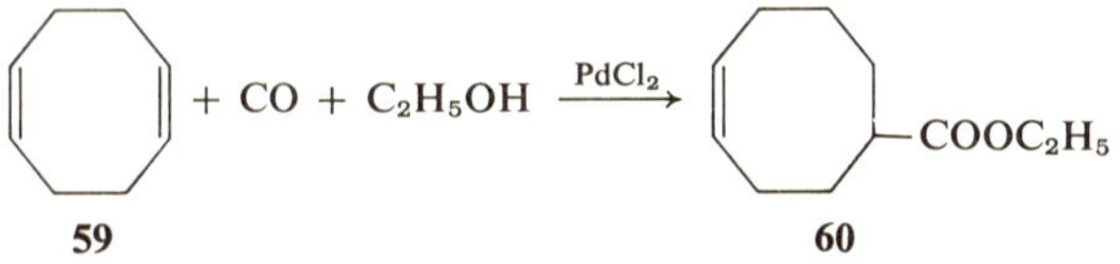

(89)

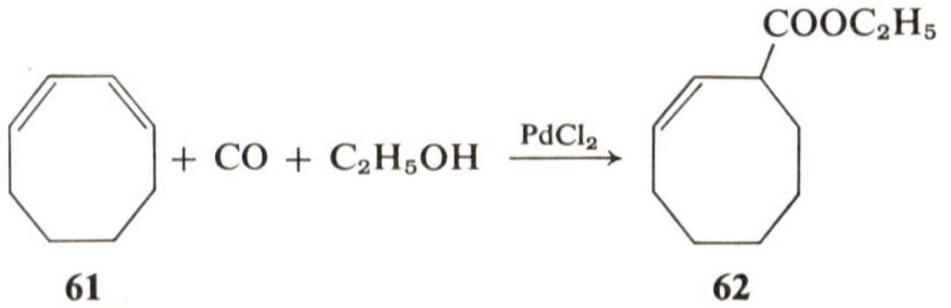

(90)

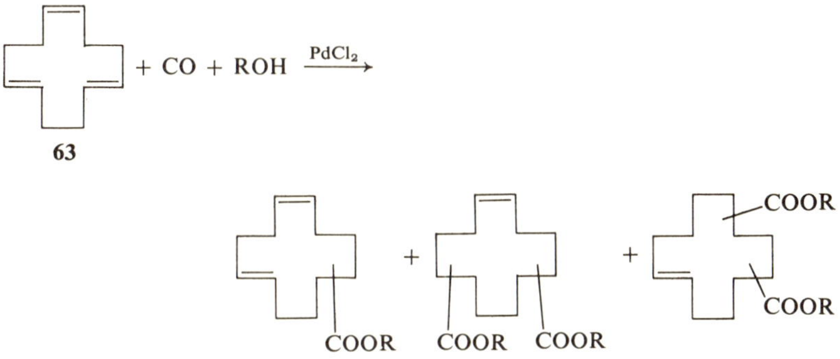

(91)

Muconic acid was obtained along with maleic and fumaric acids by the carbonylation of acetylene under pressure in a benzene solution of dibenzonitriledichloropalladium [90]. The catalyst solution was first stirred with a 1:1 mixture of carbon monoxide and acetylene at room temperature for several hours. Carbon monoxide was then charged into the mixture at pressures up to 100 kg/cm^2, and the mixture was heated to 100°C. The solution was then treated with methanol to convert the acid chlorides to the corresponding methyl esters. Methyl maleate, methyl fumarate, and methyl muconate (38% combined yield) were then separated from the reaction mixture. The overall carbonylation reaction may be represented as shown in Eq. (92).

$$(2x + y + z)\ CH{\equiv}CH + 2(x + y + z)\ CO + (x + y + z)\ PdCl_2 \longrightarrow$$

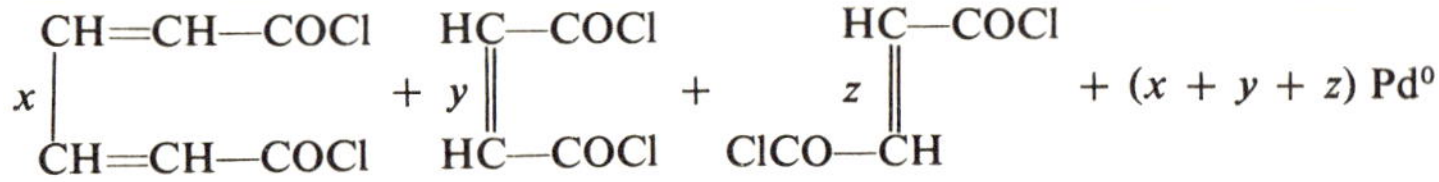

(92)

Brewis and Hughes [91a] carried out the transannular addition of carbon monoxide to cycloocta-1,5-diene, **64**, in tetrahydrofuran at 150°C and 1000 atm pressure in the presence of 1% diiodobis(tributylphosphine)palladium(II) to give bicyclo[3.3.1]nona-2-en-9-one, **68**, in 40–45% yield. The IR spectra of the product indicated the presence of a *cis* double bond and a six-membered ring ketone. Cycloocta-1,3-diene was also obtained in 45–57% yield. The mechanism of Scheme I was suggested for the formation of the cyclic ketone

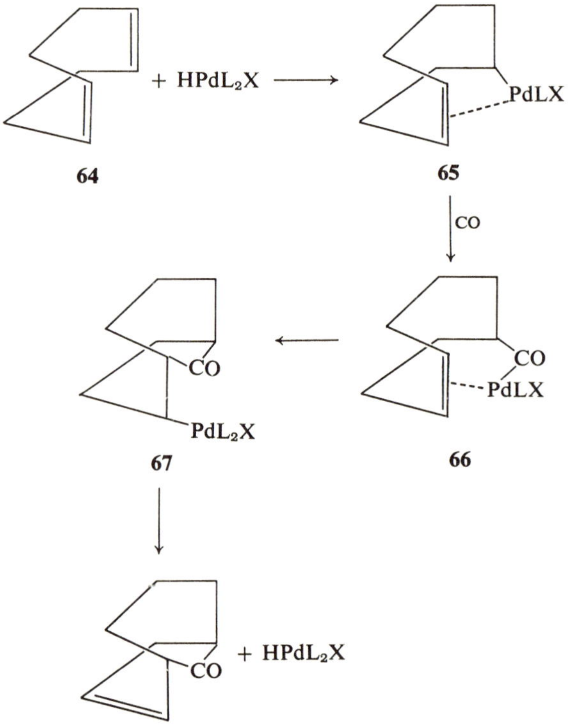

(L = PR_3, X = halogen)

Scheme I. Transannular addition of CO to cycloocta-1,5-diene.

[91a]. Carbonyl insertion in the metal–carbon bond of **65** gives a mixed ligand acyl π-complex **66**. Insertion of the alkene group into the metal–carbon bond of the acyl moiety takes place intramolecularly to give the transannular intermediate **67**, which is converted to the product and catalyst by a standard σ-π shift [70].

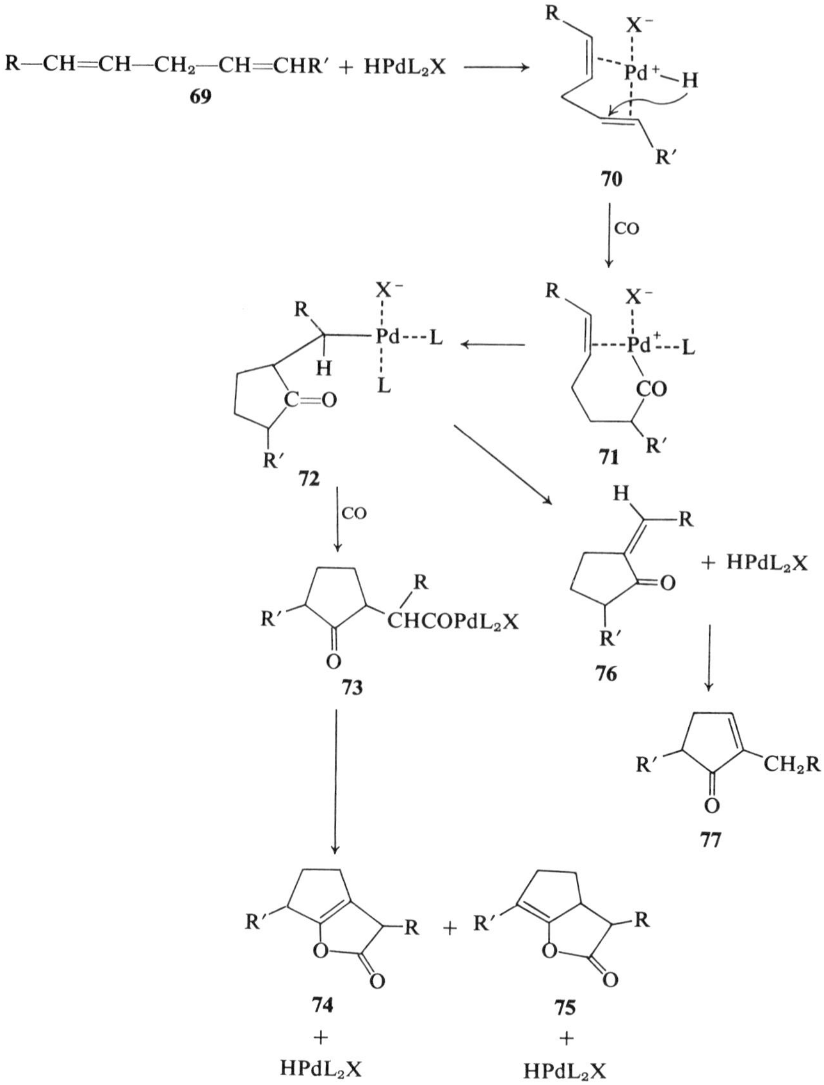

(L = tributylphosphine)

Scheme II. Carbonylation of 1,4-dienes.

Carbonylation of 1,4- or 1,5-dienes [91b] with carbon monoxide at 200°C and 1000 atm in a 2% solution of diiodobis(tributylphosphine)palladium(II) in tetrahydrofuran resulted in the formation of cyclic ketones of the general formula **77** and a mixture of enol-lactones **74** and **75** in an overall yield of less than 10%. Penta-1,4-diene (**69**, R = H, R′ = H) gave 2-methylcyclopent-2-enone (**77**, R = R′ = H) and a mixture of **74** and **75** (R, R′ = H) in an overall yield of 4%. Hexa-1,5-diene produced 2,5-dimethylcyclopent-2-enone (**77**, R = H, R′ = CH_3) in 6% yield and enol-lactones **74** and **75** (R = H, R′ = CH_3) in 10% yield. Hepta-1,6-diene gave a mixture of isomeric 5-ethyl-2-methylcyclopent-2-enone (**77**, R = H, R′ = C_2H_5) and 2-ethyl-5-methylcyclopent-2-enone (**77**, R = CH_3, R′ = CH_3) in an overall yield of 16%. Octa-1,7-diene failed to react. Thus the yield of the cyclic ketones gradually increased from penta-1,4-diene to hepta-1,6-diene. The reason for the failure of octa-1,7-diene to form the cyclic ketone is not yet clear, but must certainly be due in part to lowered probability of cyclization with increased separation of the double bonds. The 1,5- and 1,6-dienes may have reacted by isomerization to the intermediate 1,4-diene, **69**, prior to carbonylation.

The cyclization of the diene **69** to the ketone **77** proceeds through the insertion of one molecule of carbon monoxide into a metal–carbon bond and formation of an alk-5-enoylpalladium complex **71**. The unsaturated ketone **77** is formed by the elimination of a $HPd(L)_2X$ group from **72**, followed by isomerization of the unsaturated ketone **76** to **77**. Insertion of a second molecule of carbon monoxide in **72** gives the intermediate **73** which loses $HPd(L)_2X$ and cyclizes to the enol-lactones **74** and **75**.

A reaction of considerable commercial interest [92] is the palladium(II) and copper(II) acetate-catalyzed one-step synthesis of β-acetoxycarboxylic acids, **78**, from α-olefins, carbon monoxide, and molecular oxygen in acetic acid/acetic anhydride. The overall reaction is represented by Eq. (93). Pyrolysis

$$R{-}CH{=}CH_2 + CO + \tfrac{1}{2}O_2 + AcOH \xrightarrow{Pd(II)/Cu(II)} R{-}\underset{\substack{|\\ OAc}}{CH}{-}CH_2{-}COOH \qquad (93)$$

78

of **78** gives α,β-unsaturated carboxylic acids, **79**. A small amount of lithium chloride is essential for higher yields of **78**. The nature of the participation of lithium chloride in reaction (93) is not understood. The function of the copper(II) is to catalyze the oxidation of Pd(0) to Pd(II) by molecular oxygen, thus assuring a completely catalytic process. Reaction sequence (94) has been proposed as the mechanism of reaction (93).

$$R{-}\underset{\substack{|\\ OAc}}{CH}{-}CH_2{-}COOH \xrightarrow[\text{pyrolysis}]{-HOAc} R{-}CH{=}CH{-}COOH$$

78 **79**

Reaction of the α-olefin, carbon monoxide, palladium(II), and acetate gives a π-bonded intermediate **80**, which reacts with coordinated acetate giving a π–σ rearrangement to form the intermediate **81**. Carbonyl insertion followed by nucleophilic attack of acetate ion on the coordinated carbonyl group gives the ester **82** and Pd^0. Hydrolysis of **82** gives the β-acetoxy-carboxylic acid **78**.

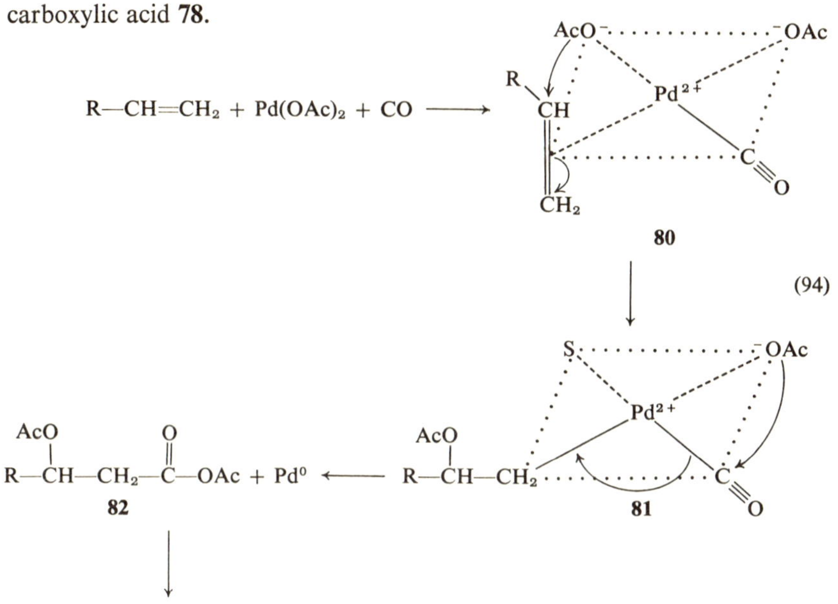

$AcOCHRCH_2COOH + HOAc$

78

$$Pd^0 + \tfrac{1}{2} O_2 + 2\, CH_3COOH \xrightarrow{Cu(II)} Pd(OAc)_2 + H_2O$$

C. Insertion of Carbon Monoxide in Metal–Oxygen Bonds

Synthesis of Carboxylic Acids

Methanol solutions of mercuric acetate readily absorb 1 mole of methanol and 1 mole of carbon monoxide per mole of solute to form an intermediate compound, $HgC_4H_6O_4$ [93]. This substance, with the same formula as mercuric acetate, contains only one residual acetate group, which may be replaced by other anions. Carbon monoxide may be regenerated from the compound by heating or by treatment with concentrated hydrochloric acid. The IR spectrum shows a carbonyl stretching vibration at 1190 cm^{-1}, characteristic of the formate group. The compound may thus be regarded as a derivative of methyl formate, with the arrangement of functional groups indicated in Eq. (95) [93]. In reaction (95) carbon monoxide may be considered to be inserted into a metal–oxygen bond to form the formate derivative.

The reaction of arylmercury(II) ions with carbon monoxide at high pressure yields derivatives of the corresponding carboxylic acids. The basic nitrate of phenylmercury(II) reacts with carbon monoxide at 247 atm pressure at 100°C in benzene solution to produce phenylmercury(II) benzoate as the principal product, with precipitation of mercury [94]. The reaction is inhibited by anions, such as chloride and acetate, that form stable covalent complexes with the phenylmercury(II) ion, thus indicating that the phenylmercury(II) ion is involved in the reaction with carbon monoxide. The stoichiometry of the reaction may be represented as shown in Eq. (96).

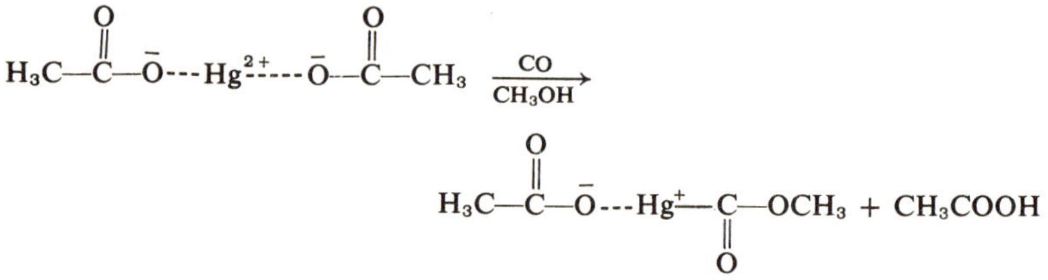
(95)

$$(RHg)_2(OH, NO_3) + CO \longrightarrow RHgOCOR + HNO_3 + Hg \qquad (96)$$

Alkylnitratomercury(II) complexes in benzene, water, or methanol react with carbon monoxide at 25–250 atm. Metallic mercury is precipitated [95] and in water or moist benzene carboxylic acids are formed in 35–50% yield. In methanolic solution the corresponding methyl esters are obtained as in Eqs. (97) and (98). 1-Nitrobutane and 1-methoxy-2-nitroethane are also

$$RHgONO_2 + CO + H_2O \longrightarrow RCOOH + Hg + HNO_3 \qquad (97)$$

$$RHgONO_2 + CO + CH_3OH \longrightarrow RCOOCH_3 + Hg + HNO_3 \qquad (98)$$

(R = *n*-butyl or 2-methoxyethyl)

formed in small quantities. Thus alkylnitratomercury seems to react with carbon monoxide to form an intermediate complex that decomposes either to a nitroalkane or to a carboxylic acid derivative.

D. Insertion of Carbon Monoxide in the Metal–Nitrogen Bond

1. *Carbonylation of Amines*

Formamide and *N*-substituted formamides are formed by the reaction of ammonia [96,97], primary [64,65,96], secondary [54,64,98], and tertiary [64,65] aliphatic amines, respectively, with carbon monoxide at ordinary temperature and at atmospheric pressure in the presence of metal carbonyls, such as mercurycobalt tetracarbonyl, $HgCo(CO)_4$ [90], dicobalt octacarbonyl [98], and nickel tetracarbonyl [64,65]. A complete survey of this type of carbonylation reaction has been published by Bird [99,100]. Arylamines [64] react with carbon monoxide in the presence of $Ni(CO)_4$ to yield formanilide and *N*,*N*-diphenylurea. Hydrazine hydrate [101,102] reacts with carbon monoxide at 45°C under pressure (900 atm) in the presence of catalytic amounts of iron pentacarbonyl to give semicarbazide. At 100°C and 500 atm pressure of carbon monoxide both urea and semicarbazide are formed. Wender *et al.* [103] demonstrated the reversibility of the conversion of substituted hydrazines to the corresponding substituted ureas by studying ^{14}CO exchange with *N*,*N*′-diphenylurea at 230°C with cobalt octacarbonyl as a catalyst.

Sternberg *et al.* [98] proposed the mechanism given in Eq. (99) for the formation of substituted formamides by the reaction of dimethylamine or piperidine with carbon monoxide in the presence of dicobalt octacarbonyl.

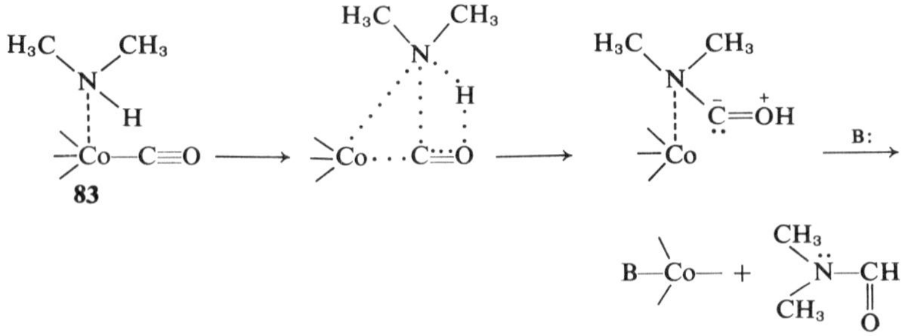

(99)

Intermediate **83** is first formed between cobalt octacarbonyl and the amine, as indicated by infrared measurements. In the subsequent step, carbon monoxide is transferred from the metal to dimethylamine, which then dissociates to dimethylformamide. *N*,*N*,*N*′,*N*′-Tetramethylurea was also obtained in the reaction of dimethylamine with carbon monoxide. Its formation is summarized by overall equation (100). It is seen that the addition of a second

$$2\,(CH_3)_2NH + CO \xrightarrow{Co_2(CO)_8} (CH_3)_2N{-}\overset{\displaystyle O}{\overset{\|}{C}}{-}N(CH_3)_2 + H_2 \qquad (100)$$

mole of the amine to the intermediate suggested above, **83**, may give a second intermediate, **84**, which could lead directly to the product of the reaction, tetramethylurea.

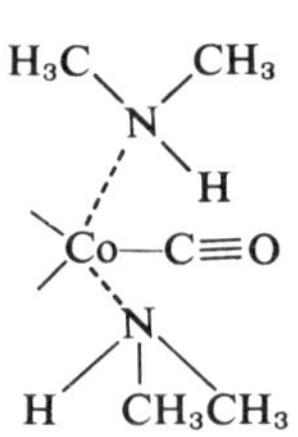

84

The manganese carbonyls, $Mn_2(CO)_{10}$ and $CH_3Mn(CO)_5$, catalyze the carbonylation of primary amines almost exclusively to 1,3-dialkylureas [104] with the evolution of hydrogen. The reactions of primary amines are more rapid than those of secondary amines; aniline reacts very slowly, and ammonia does not react at all. Low yields of *N*-alkylformamides are also formed. The overall reaction may be represented by Eq. (101).

$$2\,RNH_2 + CO \xrightarrow[106\text{–}136\text{ atm, heptane}]{Mn_2(CO)_{10}\text{ or }CH_3Mn(CO)_5} (RNH)_2CO + H_2 \quad (101)$$

No dialkylurea was obtained in the presence of anhydrous manganese(II) acetate. Thus carbonylation of the amine does not take place unless carbon monoxide is present in the coordination sphere of the metal ion. The catalyst is converted to ionic species of the type $[Mn(CO)_{6-n}L_n]^+[Mn(CO)_{5-n}L_n]^-$ (where L may be primary amine, alkylformamide, or dialkylurea). The overall reaction may occur in two steps, Eqs. (102) and (103).

$$RNH_2 + CO \xrightarrow{k_1} RNHCHO \quad (102)$$

$$RNHCHO + H_2NR \xrightarrow{k_2} (RNH)_2CO + H_2 \quad (103)$$

$$(k_2 \gg k_1)$$

Steps (102) and (103) appear to take place within the coordination sphere of the metal ion. The *N*-alkylformamide formed as an intermediate may remain coordinated to the metal ion and then undergo rapid nucleophilic attack by a second molecule of the amine.

Tsuji and Iwamoto [105] carried out the carbonylation of primary amines at 180°C under a carbon monoxide pressure of 100 kg/cm^2 with palladium dichloride catalyst. Dialkylurea and dialkyloxamide were formed and hydrogen was evolved, as shown in Eq. (104).

$$4\,RNH_2 + 3\,CO \xrightarrow{PdCl_2} RNHCONHR + RNHCOCONHR + H_2 \quad (104)$$

$$(R = n\text{-decyl})$$

The use of palladium on carbon as the catalyst reduced the yield of oxamide, and alkylformamide was obtained as the major product [105]. This reaction is an interesting example of the formation of an oxamide by a possible coupling reaction between two coordinated *N*-alkylformamide residues. In the case of lower aliphatic amines, the yield of dialkylurea decreased and that of formamide increased.

Primary amines react with palladium(II) chloride at 65°–85°C and at 1.4 atm of carbon monoxide to yield isocyanates, with subsequent reduction of palladium to elemental palladium (Pd^0). The yields of isocyanate in the case of *n*-butylamine and aniline were found to be 50 and 68%, respectively

$$RNH_2 + CO + PdCl_2 \longrightarrow RNCO + Pd^0 + 2\,HCl \quad (105)$$

[106]. The overall reaction may be expressed as Eq. (105). A detailed mechanism for this reaction has not yet been suggested.

2. *Synthesis of Heterocyclic Nitrogen Compounds*

Falbe and Kort [107,108] reported that the reaction of α,β- or β,γ-unsaturated aliphatic amides with carbon monoxide in the presence of cobalt carbonyl, cobalt salts, or Raney cobalt at 160°–280°C and 100–300 atm yields succinimides or glutarimides in 65–95% yield. Acrylamide, **85** (R = H), gives succinimides, **86** (R = H), in 81% yield [Eq. (106)]. Partial isomerization takes place with longer chain carbonamides. Thus crotonamide, **87** (R = H R′ = CH_3), yields 68% methylsuccinimide and 19% glutarimide, **88** (R = H, R″ = H). With cyclohexene-1-carbonamide, **89**, hexahydrophthalimide, **90**, is obtained in 91% yield. Under the same reaction conditions, the use of

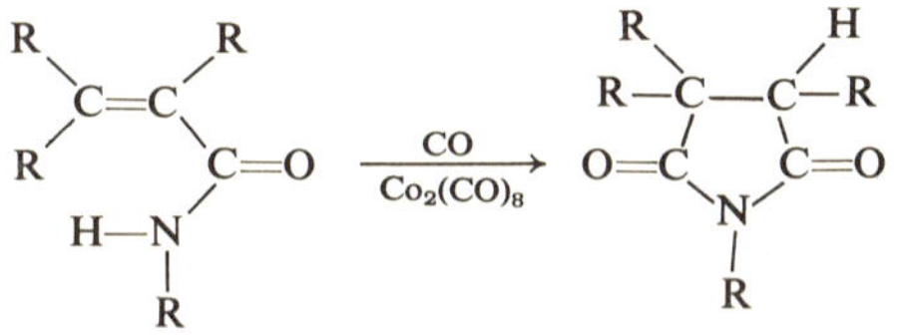

85 **86** (106)

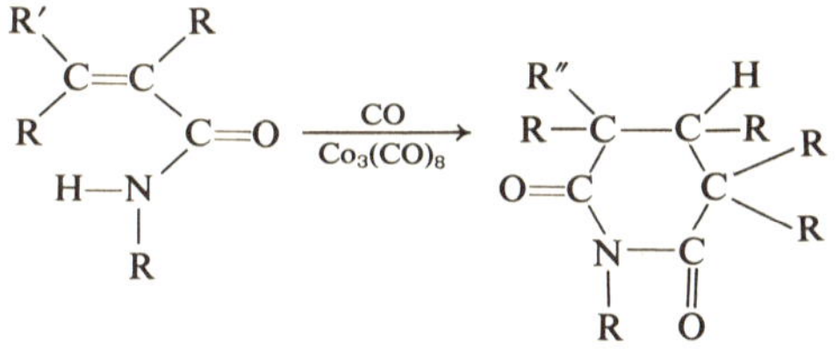

87 **88** (107)

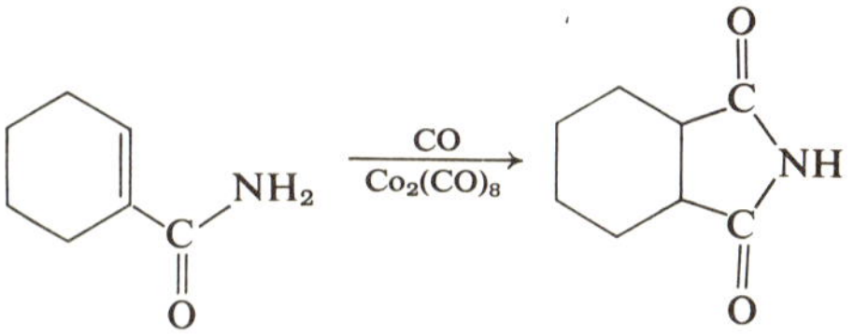

89 **90** (108)

unsaturated alcohols in place of unsaturated amides results in the formation of lactones [109], as indicated in Eq. (109).

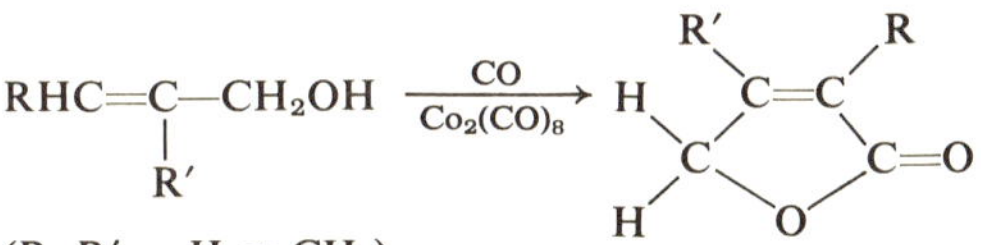

(R, R′ = H or CH_3) (109)

The compounds that take part in the cyclization reactions depicted above have functional groups in a position that favors ring closure. The best catalysts for the reaction are rhodium and iron carbonyls and cobalt octacarbonyl. Nickel tetracarbonyl is inactive as a catalyst in these cyclization reactions of unsaturated amides and alcohols. The mechanism given in Eq. (110) has been suggested for unsaturated amides. The unsaturated alcohols react in an analogous manner. Under the reaction conditions

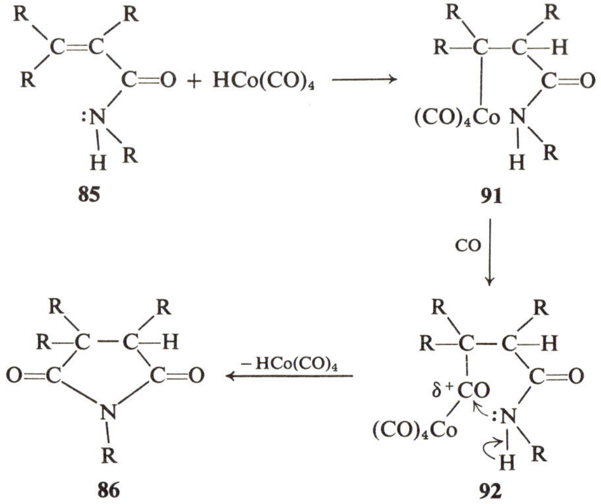

(110)

employed the carbonyl is converted to a hydrocarbonyl by interaction with the active hydrogen of the amine or alcohol. Cobalt hydrocarbonyl adds to the double bond in the unsaturated amide **85** to form the alkylcobalt carbonyl derivative **91**. Insertion of a molecule of carbon monoxide in the metal–carbon bond of **91** forms the reactive intermediate complex **92**. The $Co(CO)_4^-$ moiety and active hydrogen attached to nitrogen then dissociate in a ring closure reaction to give the succinimide **86**. The ring closure and the displacement of the metal carbonyl moiety may be a concerted reaction, as indicated above.

The reaction of benzaldehyde anil, **93**, in benzene with carbon monoxide under 100–200 atm pressure at 220°–230°C, in the presence of dicobalt octacarbonyl as catalyst, gives an 80% yield of 2-phenylphthalimidine [110], [Eq. (111)]. An analogous compound was obtained from 1-naphthaldehyde

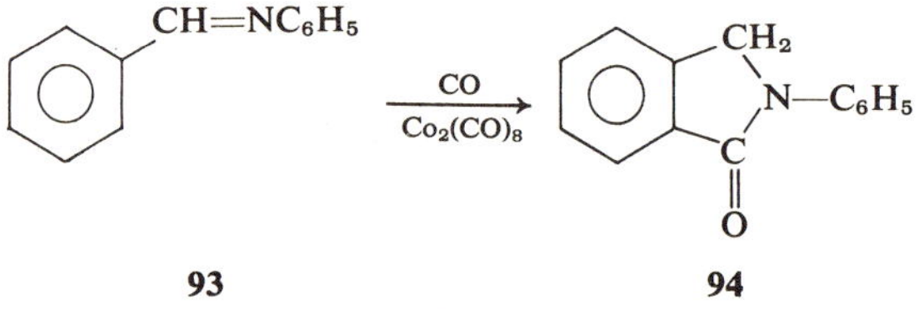

(111)

anil. The reaction takes place smoothly in nonpolar solvents, while in tetrahydrofuran and ether no phthalimidines were obtained. Other cobalt complexes and compounds, such as cobalt acetylacetonate and cobaltous carbonate, are also effective as catalysts, but the reactions presumably occur through the formation of cobalt carbonyl intermediates. Iron pentacarbonyl is not as effective a catalyst as is dicobalt octacarbonyl, while nickel tetracarbonyl is inactive. In the absence of the catalyst no formation of phthalimidines takes place. Derivatives with substituents on the *para* position of the aniline moiety (OCH_3, H, OH, Cl) and in the *para* position of the aldehyde ring [OH, $(CH_3)_2N$] also gave good yields of phthalimidines [111]. Nitro substituents on both the aldehyde and aniline rings of the anil produced complications and lower yields. Phthalimidines were also obtained from Schiff's bases derived from aldehydes and aliphatic amines and also from Schiff's bases obtained from condensation of acetophenone and benzophenone with aniline. All attempts to prepare 2-hydroxyphthalimidines by the reaction of aldoximes and ketoximes with carbon monoxide failed; substituted ureas were obtained.

Under similar reaction conditions [112] substituted benzaldehyde anils were converted to the 6-chloro, 3-methyl, and 6-dimethylamino derivatives of 2-phenylphthalimidine. The same investigator [112] also reported the preparation of 2-methylphthalimidine, 2-(1-naphthyl)phthalimidine, 2-(dimethylaminophenyl)phthalimidine, and *N,N'*-ethylenediphthalimidine.

The cobalt carbonyl-catalyzed carbonylation of benzophenone phenylhydrazone, **95** (R = NHC_6H_5), is illustrated in Eqs. (112a) and (112b). Reaction with carbon monoxide under pressure at 230°–240°C in the presence of dicobalt octacarbonyl as catalyst gave a 70% yield of 3-phenylphthalimidine-*N*-carboxyanilide, **96** [113]. At 210°–220°C, however, 3-phenylphthalimidine, **97**, was obtained in 50% yield with 12% formation of the carboxyanilide derivative **96**. At 190°–200°C only 3-phenylphthalimidine was reported [113] in 25% yield. Thus the yield of *N*-substituted phthalimidines seems to increase with temperature.

In the reaction scheme indicated by Eqs. (112c)–(112e), it is seen that benzophenone semicarbazone, **95**, R = $NHCONH_2$, reacts [114] with carbon monoxide under pressure at 235°–245°C in the presence of dicobalt octacarbonyl to yield a variety of products, 3-phenylphthalimidine–*N*-carboxyanilide (**96**), 3-phenylphthalimidine (**97**), 3-phenyl-2-(carboxybenzhydrylamido)phthalimidine (**98**), and *N,N'*-bis(benzhydryl)urea (**99**). At 200°–220°C, however, the reaction of **95** with carbon monoxide and $Co_2(CO)_8$ produced bis(benzophenone) azine, **100** and benzophenone 4-benzhydrylsemicarbazone, **101**. Carbonylation of both **100** and **101** at 230°–240°C yielded **97**, **98**, and **99**. Thus **100** and **101** seem to be intermediates in the formation of *N*-substituted phthalimidines.

Rosenthal and Millard [115] described the direct high-pressure carbonylation of aromatic nitriles at 230°–240°C with dicobalt octacarbonyl as catalyst. Thus benzonitrile gave a 22% yield of *N*-benzylphthalimidines, **102**, along with 13% benzamide. The solvents used were a mixture of pyridine and benzene. *N*-Formylpiperidine was also detected in the reaction products. In the absence of pyridine, the yield of *N*-benzylphthalimidine amounted to only 2% of the theoretical value.

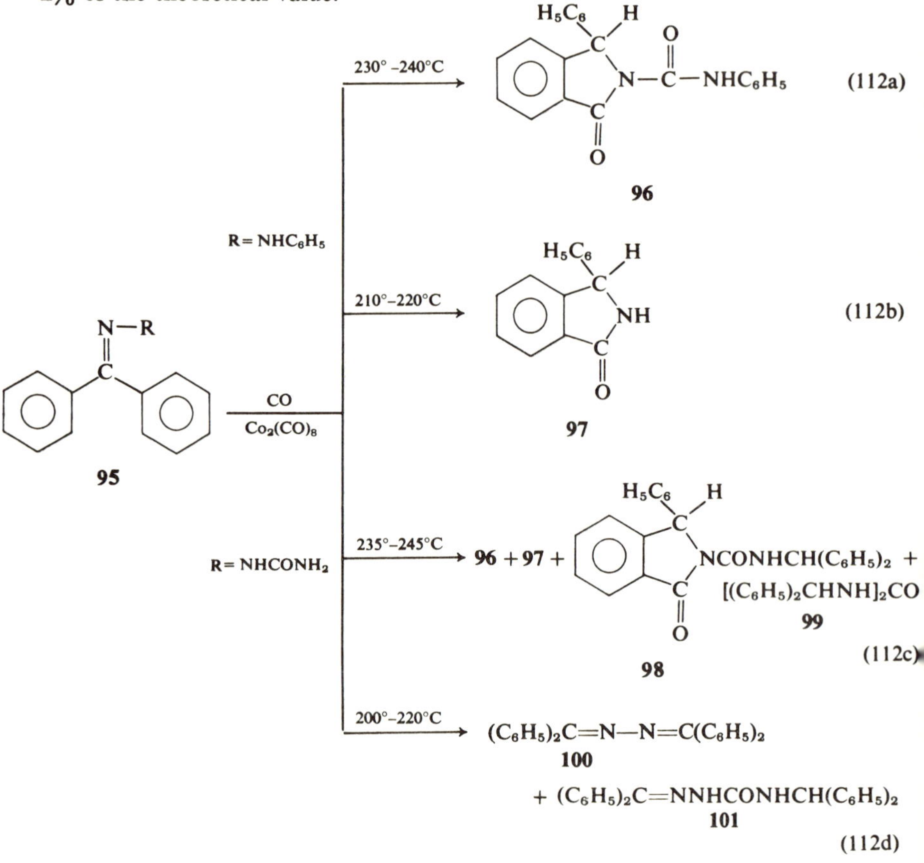

$$\mathbf{100, 101} \xrightarrow[230°-240°C]{CO + Co_2(CO)_8} \mathbf{97} + \mathbf{98} + \mathbf{99} \quad (112e)$$

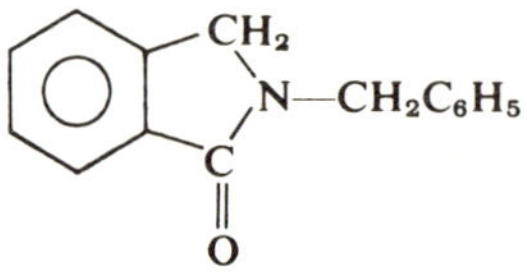

102

Murahashi *et al.* [116–118] and Horiie and Murahashi [119] obtained a 49% yield of 2-phenylindazolone, **104**, from azobenzene, **103**, by carbonylation under 150 atm pressure at 190°C in aromatic hydrocarbon solvents with dicobalt octacarbonyl as catalyst. When the reaction temperature was raised to 220°–230°C two molecules of carbon monoxide were incorporated into the azobenzene molecule with the formation of 3-phenyl-2,4-dioxo-1,2,3,4-tetrahydroquinazoline (**105**) as the main reaction product. Heating **104** at 230°C gave a quantitative yield of **105**. These reactions are summarized in Eq. (113). Thus 2-phenylindazolone may be considered as an intermediate in the formation of the quinazolone. Effective catalysts for the reaction are iron pentacarbonyl, dicobalt octacarbonyl and cobalt(II) compounds. Under the reaction conditions employed, nickel tetracarbonyl was found to be ineffective as a catalyst. The reaction is inhibited by polar solvents, the best solvents being aromatic hydrocarbons. Since quinazolones yield anthranilic acids on hydrolysis, the reaction affords a new synthetic route for the formation of anthranilic acid. In the case of substituted azobenzenes, ring closure occurred on the ring containing the *ortho–para*-directing substituents. Azobenzenes containing *meta*-directing (electron-withdrawing) substituents do not cyclize.

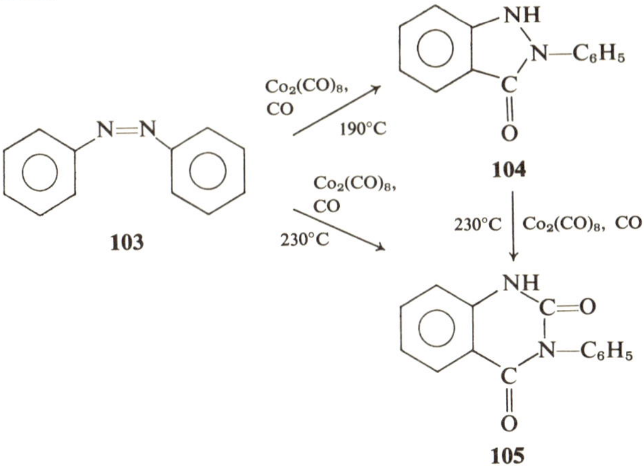

(113)

Prichard [120] studied the carbonylation of a series of azo compounds, RN=NR′ (R,R′ = phenyl, hydroxyphenyl, alkoxyphenyl, naphthyl, aminotolyl, aminophenyl, hydroxynaphthyl, acetamidophenyl, and halophenyl), in cyclohexane under pressure at 250°C in the presence of catalytic amounts of $Ni(CO)_4$. In this manner 2-(2-carboxyphenyl)indazolone and 3-phenylquinazoline-2,3-dione were synthesized.

3. *Mechanism of Pthalimidine Formation*

Horiie and Morahashi [119] have proposed a mechanism for phthalimidine formation on the basis of their observations of the steric effects of methyl and ethyl substituents on the aromatic rings of the anil. It was proposed that a bond is first formed between nitrogen and cobalt, and partial bond formation between nitrogen and a coordinated carbon monoxide group then occurs, as indicated in Eq. (114a) by the transition state **106**. The formation of the N–CO bond is then stabilized by simultaneous formation of five-membered ring **107** by substitution at the aromatic carbon atom *ortho* to the aldehyde group and transfer of the *ortho* hydrogen to the unsaturated carbon atom of the anil. A similar reaction occurs with azo compounds.

In cases where the unsaturated bond is absent, as in *N*-benzylcyclohexylamine, carbonylation results in the substitution of the amine nitrogen with a formyl group, as indicated by the conversion of **108** to **109**, Eq. (114b). The reactions of unsaturated carbonamides with carbon monoxide in the presence of dicobalt octacarbonyl, which result in the formation of succinimides or glutarimides [107], probably proceed by a similar mechanism, as is indicated (Section IV,D,2) by the transformation **85** to **86** via **91** and **92** [Eq. (110)].

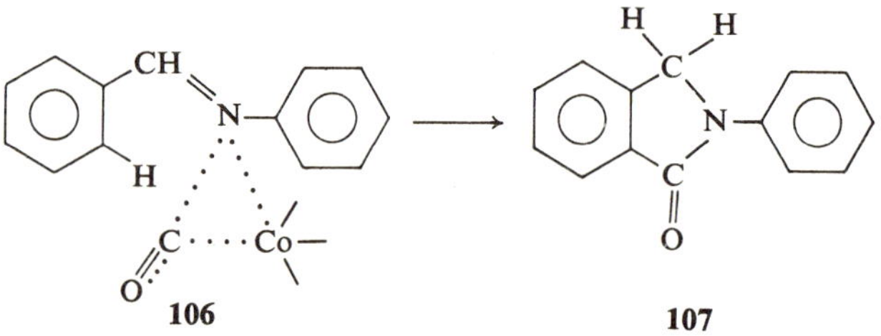
(114a)

106 **107**

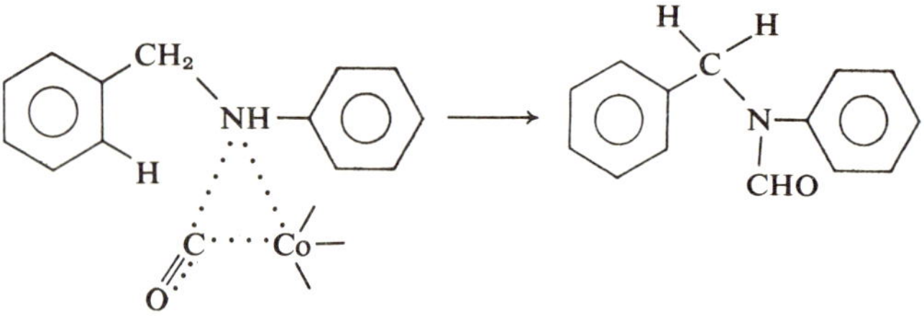
(114b)

108 **109**

V. Catalysis of Oxidation of Carbon Monoxide by Metal Ions and Metal Complexes

Harkness and Halpern [121] have described the catalytic effects of mercury(II) and silver(I) on the oxidation of carbon monoxide by potassium permanganate. The reduction of mercury(II) in aqueous perchloric acid by carbon monoxide proceeds by the stoichiometric equation (115). On the

$$2\,Hg^{2+} + CO + H_2O \longrightarrow Hg_2^{2+} + CO_2 + 2\,H^+ \qquad (115)$$

basis of kinetic measurements in dilute perchloric acid, the rate law, Eq. (116), was given.

$$-\frac{d[CO]}{dt} = k[CO][Hg^{2+}] \tag{116}$$

In the proposed mechanism given in Eqs. (117)–(119), the first step is postulated to be insertion of carbon monoxide between mercury(II) and a coordinated water molecule, followed by rapid electron transfer and the formation of elemental mercury and mercurous ion.

$$Hg^{2+}\text{-----}OH_2 + CO \xrightarrow{k} [\text{---}Hg\text{—}\overset{\overset{\displaystyle O}{\|}}{C}\text{—}OH]^+ + H^+ \tag{117}$$

$$[Hg\text{--}\overset{\overset{\displaystyle O}{\|}}{C}\text{—}OH]^+ \xrightarrow{\text{fast}} Hg + CO_2 + H^+ \tag{118}$$

$$Hg + Hg^{2+} \xrightarrow{\text{fast}} Hg_2{}^{2+} \tag{119}$$

Support for carbonyl insertion as the rate-determing step is provided by the fact that under similar conditions a methyl formate derivative, $CH_3COO\text{---}Hg\text{---}COOCH_3$, is obtained when carbon monoxide is passed into a methanolic solution of mercuric acetate [93]. The reduction of silver(I) by carbon monoxide in aqueous $HClO_4$ was found [121] to be very slow in the temperature range of 26°–54°C, which is the operable range for the mercuric ion reduction. The reaction with silver(I) ion attained measurable rates only above 100°C.

The reaction of silver(I) ion with carbon monoxide in aqueous $HClO_4$ is enhanced and proceeds readily at room temperature and pressure when silver(I) is complexed with ammonia, ethylenediamine, primary or secondary amines. A twenty-fold ratio of ligand to metal was maintained so as to ensure complete formation of the complex AgL_2 as the predominant species in solution. The overall stoichiometry of the reaction is given by Eq. (120).

$$2\,AgL_2{}^+ + CO + 2\,H_2O \longrightarrow 2\,Ag + CO_3{}^{2-} + 4\,HL^+ \tag{120}$$

The rate law may be expressed as Eqs. (121) or (122).

$$-\frac{d[CO]}{dt} = k_{exp}[CO][AgL_2{}^+][HL^+]^{-1} \tag{121}$$

$$-\frac{d[CO]}{dt} = k[CO][LAgOH] \tag{122}$$

The rate constant k_{exp} increases by a factor of 140 when ammonia is replaced by diethylamine as a ligand. The reaction was not influenced by precipitation of silver or by other surface effects. Nakamura and Halpern [122] proposed the stepwise mechanism given in Eqs. (123)–(125) for the oxidation of carbon monoxide by silver–amine complexes. The rate-determining step in the

$$AgL_2^+ + H_2O \overset{K}{\rightleftharpoons} [L\text{---}Ag\text{---}OH] + HL^+ \tag{123}$$

$$[L\text{---}Ag\text{---}OH] + CO \xrightarrow{k} L\text{---}Ag\text{—}\overset{\overset{\displaystyle O}{\|}}{C}\text{—}OH \tag{124}$$

$$L\text{---}Ag\text{—}\overset{\overset{\displaystyle O}{\|}}{C}\text{—}OH + [L\text{---}Ag\text{---}OH] \xrightarrow{\text{fast}} 2\,Ag + CO_3^{2-} + 2\,HL^+ \tag{125}$$

$(k_{exp} = kK)$

sequence involves carbon monoxide addition to a hydroxo silver(I) complex. Since the reactions were carried out in the presence of a large excess of ligand, which acts as a buffer in each case, the rate constant was pseudo-first-order, but varied considerably with pH. In general, the rate increased with pH, as the concentration of the hydroxo species AgLOH, also increases.

McAndrew and Peters [123] have studied the oxidation of carbon monoxide by silver(I) in acetate and perchlorate solutions between 60° and 110°C under pressures of up to 30 atm of carbon monoxide. In perchlorate solution the rate law may be described as in Eq. (126).

$$-\frac{d[CO]}{dt} = \frac{k_2[Ag^+]^2[CO]}{[H^+]} \tag{126}$$

The mechanism shown in Eqs. (127) and (128) has been advanced [123].

$$Ag^+ + CO + H_2O \overset{K}{\rightleftharpoons} Ag\text{—}\overset{\overset{\displaystyle O}{\|}}{C}\text{—}OH + H^+ \tag{127}$$

$$Ag\text{—}\overset{\overset{\displaystyle O}{\|}}{C}\text{—}OH + Ag^+ \xrightarrow{k} 2\,Ag + CO_2 + H^+ \tag{128}$$

In the presence of acetate, the rate law takes the form indicated by Eq. (129).

$$-\frac{d[CO]}{dt} = k_1[AgOAc][CO] + \frac{k_2[Ag^+]^2[CO]}{[H^+]} + \frac{k_3[Ag^+][AgOAc][CO]}{[H^+]} \tag{129}$$

Under these conditions, the suggested mechanism is indicated by Eq. (130)–(135).

$$Ag^+ + {}^-OAc \rightleftharpoons AgOAc \tag{130}$$

$$AgOAc + CO \xrightarrow{k_1} Ag\text{—}\overset{\overset{\displaystyle O}{\|}}{C}\text{—}OAc \tag{131}$$

$$\mathrm{Ag{-}\overset{\displaystyle O}{\overset{\|}{C}}{-}OAc + Ag^+ + H_2O \xrightarrow{\text{fast}} 2\,Ag^0 + CO_2 + HOAc + H^+} \tag{132}$$

$$\mathrm{Ag^+ + CO + H_2O \overset{K_1}{\rightleftharpoons} Ag{-}\overset{\displaystyle O}{\overset{\|}{C}}{-}OH + H^+} \tag{133}$$

$$\mathrm{Ag{-}\overset{\displaystyle O}{\overset{\|}{C}}{-}OH + Ag^+ \xrightarrow{k_2} 2\,Ag^0 + CO_2 + H^+} \tag{134}$$

$$\mathrm{Ag{-}\overset{\displaystyle O}{\overset{\|}{C}}{-}OH + AgOAc \xrightarrow{k_3} 2\,Ag^0 + CO_2 + HOAc} \tag{135}$$

where k_1, k_2, and k_3 are, respectively, 0.04, 3.7×10^{-6} and $1.1 \times 10^{-5}\ M^{-1}\ \mathrm{sec}^{-1}$ at 90°C.

The intermediates in reactions (131) and (133) are formed, as in Nakamura and Halpern's mechanism [122], by the insertion of carbon monoxide in the silver–oxygen bond of the ligand. The resulting formyl group is then oxidized to carbon dioxide by silver(I) in steps (132), (134), and (135). Reaction (131) is acid-independent and the formation of Ag—COOAc is rate-determining. In the acid-dependent reactions, (134) and (135), the oxidation of the intermediate Ag—COOH is rate-determining, while its formation is a pre-equilibrium process.

In the presence of nickel(II) ammine or 1,2-diaminopropane complexes, the oxidation of carbon monoxide proceeds according to the overall stoichiometry [124] of Eq. (136). In aqueous ammonia the rate law for the

$$\mathrm{NiL_n{}^{2+} + 5\,CO + H_2O \longrightarrow Ni(CO)_4 + CO_2 + \mathit{n}\,L + 2\,H^+} \tag{136}$$

reaction has the form given by Eq. (137).

$$-\frac{d[\text{complex}]}{dt} = \frac{k'[\mathrm{Ni^{2+}}][\mathrm{CO}][\mathrm{OH^-}]}{[\mathrm{NH_3}]^{0.4}} \tag{137}$$

$$[k' = 8.5 \times 10^{12} \exp(-18{,}600/RT)\ M^{-2}\ \text{second}^{-1}]$$

The mechanism shown by Eqs. (138) and (139) was suggested [124].

$$\mathrm{[Ni(NH_3)_n]^{2+} + CO + OH^- \xrightarrow{k} [(NH_3)_nNi{-}\overset{\displaystyle O}{\overset{\|}{C}}{-}OH]^+} \tag{138}$$

$$\mathrm{[(NH_3)_nNi{-}\overset{\displaystyle O}{\overset{\|}{C}}{-}OH]^+ + 4\,CO \xrightarrow{\text{fast}} Ni(CO)_4 + CO_2 + NH_4{}^+ + (\mathit{n} - 1)\,NH_3}$$

$$(k = k'[\mathrm{NH_3}]^{-0.4}) \tag{139}$$

Carbon monoxide may be oxidized to CO_2 by copper(II) in aqueous solution according to the stoichiometry [125] of Eq. (140).

$$\mathrm{2\,Cu^{2+} + 3\,CO + H_2O \longrightarrow 2\,Cu(CO)^+ + CO_2 + 2\,H^+} \tag{140}$$

Reaction (140) proceeds only at very high carbon monoxide pressures (1360 atm). At pressures lower than 1360 atm, the reaction becomes heterogeneous because of the precipitation of metallic copper. Under homogeneous conditions, the reaction proceeds according to the rate law given in Eq. (141).

$$-\frac{d[Cu^{2+}]}{dt} = \frac{k_1'[Cu^{2+}][Cu(CO)^+]}{[H^+]} + \frac{k_2'[Cu^{2+}]^2}{[H^+]} P_{CO} + k_3'[Cu^{2+}]^2 P_{CO} \qquad (141)$$

The reaction was suggested to proceed [125] by two paths, one of which (path A) is independent of carbon monoxide pressure due to the formation of a stable cuprous carbonyl complex $Cu(CO)^+$. It was suggested [125] that the rate-determining steps in both reactions (paths A and B) are the decomposition of a carbon monoxide insertion complex $(Cu—C—OH)^+$ by cupric ion in accordance with Eq. (142)–(148).

Path A

$$Cu^+ + CO \xrightarrow{\text{fast}} Cu(CO)^+ \qquad (142)$$

$$Cu(CO)^+ + H_2O \underset{}{\overset{K_3}{\rightleftharpoons}} (Cu—\overset{\overset{\text{O}}{\|}}{C}—OH) + H^+ \qquad (143)$$

$$(Cu—\overset{\overset{\text{O}}{\|}}{C}—OH) + Cu(II) \xrightarrow{k_3} CO_2 + CuH^+ + Cu(I) \qquad (144)$$

$$CuH^+ + Cu(II) \xrightarrow{\text{fast}} 2\,Cu(I) + H^+ \qquad (145)$$

Path B

$$Cu(II) + CO + H_2O \overset{K_1}{\rightleftharpoons} [Cu—\overset{\overset{\text{OH}}{|}}{C}—OH]^{2+} \qquad (146)$$

$$[Cu—\overset{\overset{\text{OH}}{|}}{C}—OH]^{2+} + Cu(II) \xrightarrow{k_1} CO_2 + 2\,Cu(I) + 2\,H^+ \qquad (147)$$

$$Cu(II) + CO + H_2O \overset{K_2}{\rightleftharpoons} (Cu—\overset{\overset{\text{O}}{\|}}{C}—OH)^+ + H^+ \qquad (148)$$

$$(Cu—\overset{\overset{\text{O}}{\|}}{C}—OH)^+ + Cu^{2+} \xrightarrow{k_2} CO_2 + 2\,Cu^+ + H^+ \qquad (148a)$$

The equilibrium and rate constants in Eqs. (142)–(148) are related to the constants in the rate law (141) by the following relationships: $k_1' = K_3k_3$, $k_2' = K_2k_2$, and $k_3' = K_1k_1$.

Reduction of permanganate to manganese dioxide by carbon monoxide in acid and neutral solution and to manganate ion in basic solution was found [121] to proceed rapidly in the temperature range 28°–50°C. The rate law for the reduction of permanganate in the absence of catalysts was found to be that given in Eq. (149).

$$-\frac{d[CO]}{dt} = k[CO][MnO_4^-] \qquad (149)$$

Permanganate reduction is catalyzed by silver(I) and mercury(II) ions but not by copper(II), iron(III), cadmium(II), or thallium(III). The rate law observed [121] for the silver(I)- and mercury(II)-catalyzed oxidation of carbon monoxide by permanganate is represented by Eq. (150).

$$-\frac{d[CO]}{dt} = k[CO][MnO_4^-][M^{n+}] \quad (150)$$

A reactive intermediate involved in the catalytic oxidation may be a permanganate complex of the catalytic metal ion. The enthalpy and entropy of activation for the uncatalyzed oxidation of carbon monoxide by permanganate are 13 kcal/mole and −17 eu, respectively. The enthalpies of activation are significantly lower for the silver(I) and mercury(II)-catalyzed reactions, the values being 1.2 kcal/mole for the former and 6.4 kcal/mole for the latter. The lowering of the enthalpies of activation provide a measure of the extent of activation of carbon monoxide by the metal ion that occurs in reactions (117) and (124).

Zakumbaeva *et al.* [126] reported 35% oxidation of carbon monoxide at 20°C in a solution containing palladium(II) and copper(II) ions as catalysts in the molar ratio of 1:5 in a medium 0.2 *M* in chloride and bromide ions, the molar ratio of Cl^- to Br^- being maintained at 1:5. In the absence of bromide ions the extent of oxidation was only 12% at 20°C and 6% at 40°C. The authors attributed the catalytic effect of palladium(II) to the presence in solution of the species $[PdCl_3Br]^{2-}$ and $[PdCl_2Br_2]^{2-}$. The addition of acetate ions increased the extent of oxidation of carbon monoxide to 50% at 20°C and to 55% at 40°C. An increase in pH of the solution increased the extent of oxidation to 70% at 20°C, indicating that hydroxo species of the metal ions are probably the active intermediates, an observation similar to that of Nakamura and Halpern [122]. A mechanism for the Cu(II)-catalyzed oxidation of carbon monoxide has not yet been suggested. Copper(II) probably assists in the reoxidation of palladium(0) to palladium(II) when air is bubbled through the solution, thus restoring the reactive palladium(II) complexes.

An interesting example of the oxidation of carbon monoxide within the coordination sphere of a metal ion is the formation of allylic palladium(II) chloride **113**, from sodium chloropalladite, allylic chlorides, and carbon monoxide. It was found by Dent *et al.* [127] that passage of carbon monoxide through an aqueous methanolic solution of sodium chloropalladite containing allyl chloride gave an almost quantitative yield of diallyldichlorodipalladium, $Pd_2Cl_2(C_3H_5)_2$. The mechanism shown in Eq. (151) has been proposed by Nicholson *et al.* [128] for this reaction.

Allyl chloride, carbon monoxide, and water are coordinated to the palladium(II) salt to form a complex, pictured as **110**. Hydroxide ion migration then gives the carbonylate complex **111**, which breaks down to give hydrogen chloride, carbon monoxide, and the allylpalladium chloride monomer, **112**, which then dimerizes.

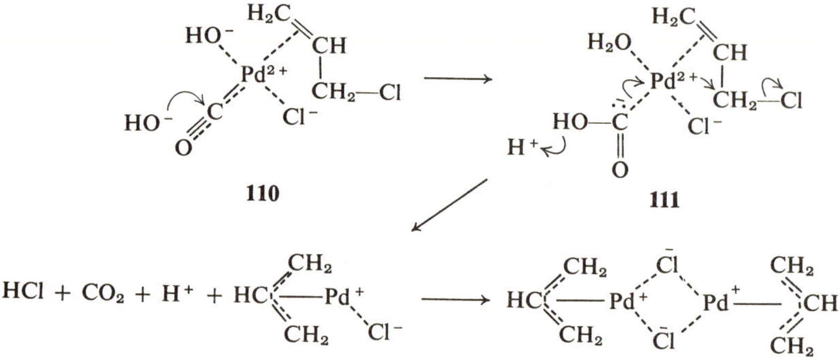

110 **111** (151)

112 **113**

James and Rempel [129] and James [130] have studied the oxidation of carbon monoxide to carbon dioxide by rhodium(III), with the formation of an anionic species of rhodium(I), $[Rh(CO)_2Cl_2]^-$. The overall stoichiometry of the reaction is given by Eq. (152). The rhodium(I) complex may be

$$Rh^{3+} + 3\,CO + H_2O \longrightarrow Rh(CO)_2^+ + CO_2 + 2\,H^+ \qquad (152)$$

precipitated from the yellow solution of the anion $[Rh(CO)_2Cl_2]^-$ by the addition of a large cation such as tetraphenylarsonium. The reduction and carbonylation of rhodium(III) exhibits autocatalysis and fits the rate law Eq. (153).

$$\frac{d[Rh(CO)_2^+]}{dt} = [Rh^{3+}](k_1' + k_2'[Rh(CO)_2^+]) \qquad (153)$$

In this equation, k_1' and k_2' are pseudo first- and second-order constants, respectively, corresponding to $k_1[CO]$ and $k_2[CO]$, respectively. At 80°C in 3 M hydrochloric acid, $k_1' = 2.5 \times 10^{-4}$ sec^{-1} and $k_2' = 4.5 \times 10^{-2}$ M^{-1} sec^{-1}. The mechanism given in Eqs. (154)–(157) is proposed for the oxidation of carbon monoxide by rhodium(III).

$$Rh(H_2O)^{3+} + CO \xrightarrow{k_1} Rh{-}\overset{\overset{\displaystyle O}{\|}}{C}{-}OH^{2+} + H^+ \qquad (154)$$

$$Rh{-}\overset{\overset{\displaystyle O}{\|}}{C}{-}OH^{2+} \xrightarrow{\text{fast}} Rh^+ + CO_2 + H^+ \qquad (155)$$

$$Rh^+ + 2\,CO \xrightarrow{\text{fast}} Rh(CO)_2^+ \qquad (156)$$

$$CO + Rh(H_2O)^{3+} \xrightarrow[Rh(CO)_2^+]{k_2} Rh^+ + CO_2 + 2\,H^+ \qquad (157)$$

The first step in the above mechanism is carbon monoxide insertion in a Rh(III)—OH_2 bond, followed by dissociation of a proton, and is similar to the rate-determining step proposed by Nakamura and Halpern [122] in the oxidation of carbon monoxide by silver amine complexes [Eq. (124)]. The unstable Rh(III)—CO_2H species then dissociates in a fast step to carbon dioxide and rhodium(I). The coordinately unsaturated rhodium(I) species combines in a fast step with carbon monoxide to yield the carbonyl complex, $Rh(CO)_2{}^+$. It must be emphasized that all the rhodium species are in 3 *M* hydrochloric acid and are therefore coordinated to chloride ions and water molecules to various extents, depending on the number of other coordinated groups and on the relative affinities of the metal ion for water and chloride ligands in the medium employed. The details of the last autocatalytic reaction of rhodium(III) have not been elucidated [129]. The apparent termolecular step may perhaps proceed by a more probable reaction path involving a mixed-valence, bridged transition state of the type Rh(I)$\cdots$X$\cdots$Rh(III) (X = CO or halogen), in which the carbon monoxide inserted between rhodium(III) and H_2O may be a carbonyl coordinated to the rhodium(I). The lowering of activation energy by about 10 kcal/mole for the termolecular path compared to the first path (k_1) supports such a transition state, $\Delta H^{\dagger}(k_2) = 18$ kcal/mole, $\Delta H^{\dagger}(k_1) = 29$ kcal/mole.

The reaction of rhodium(III) chloride with carbon monoxide provides a convenient method for the preparation of rhodium(I) complexes. The reaction product $[Rh(CO)_2Cl_2]^-$, when treated with triphenylphosphine, evolves 1 mole of carbon monoxide and gives an excellent yield of *trans*-$[Rh(CO)Cl(PPh_3)_2]$.

VI. Reactions of Coordinated Carbon Monoxide

Coordinated carbon monoxide is susceptible to nucleophilic attack by many nucleophiles. Examples of such reactions including the attack by R (alkyl, aryl) and L (phosphine) have been discussed in Section III. The present section deals with topics not covered previously. Attack of an alkoxy group OR^- on $M(CO)_n$ leads to products of the type $M(CO)_{n-1}(COOR)$. Formation of metal–carbene derivatives takes place by the attack of nucleophiles, e.g.,

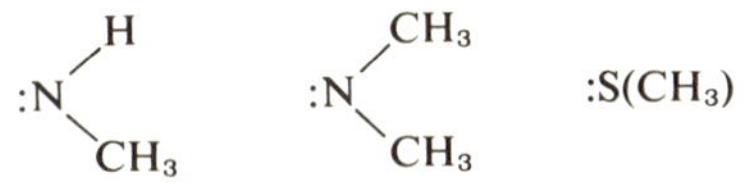

on the coordinated carbonyl group.

Reactions of cationic metal carbonyls with alcohols to form the corresponding alkoxy carbonyl complexes have been investigated with $PtClH(PPh_3)_2$

[131–134], $PtCl(PPh_3)_2(CO)^+$ [135], $Ir(CO)_3(PPh_3)_2$ [136,137], and $Mn(CO)_4 \cdot (PPh_3)_2$ [138]. The formation of alkoxy carbonyls from cationic metal carbonyls proceeds according to the mechanism given in Eq. (158), where M indicates a metal ion coordinated with additional ligands such as Cl^-, H^-, PPh_3, and CO.

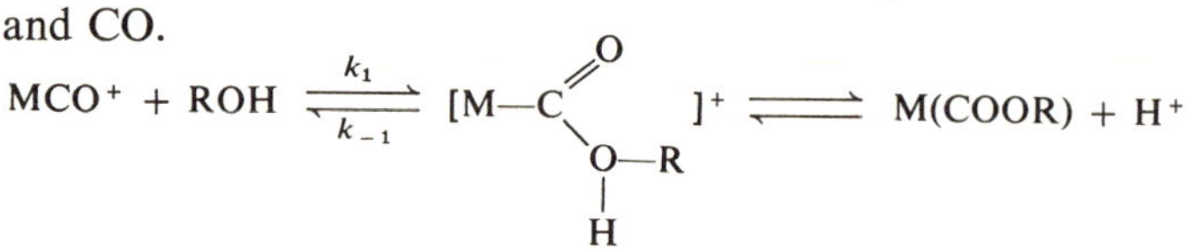

(158)

Byrd and Halpern [139] have investigated the kinetics of the reaction of $PtCl(PPh_3)_2(CO)^+$ with a series of alcohols, ROH (R = alkyl, aralkyl), and reported that the reaction follows the rate law of Eq. (159).

$$\frac{d[PtCl(PPh_3)_2(COOR)]}{dt} = k_1[PtCl(PPh_3)_2(CO)^+][ROH] - k_{-1}[PtCl(PPh_3)_2(COOR)][H^+] \quad (159)$$

With various alcohols, the rate constant k_1 decreased in the order $CH_3OH > CH_3OCH_2CH_2OH \gtrsim CH_3CH_2OH > CH_3CH_2CH_2OH \gtrsim C_6H_5CH_2OH > (CH_3)_2CHOH > (CH_3)_3COH$, as would be expected if there is considerable dependence on steric factors. The values of $K_{eq} = k_1/k_{-1}$ also follow the rate constant k_1 in regard to variation of the alcohol.

Metal complexes that contain the grouping M=C(OR)(R′) are called metal–carbene derivatives [140,141]. These compounds have been prepared from neutral metal carbonyls as shown in Eq. (160).

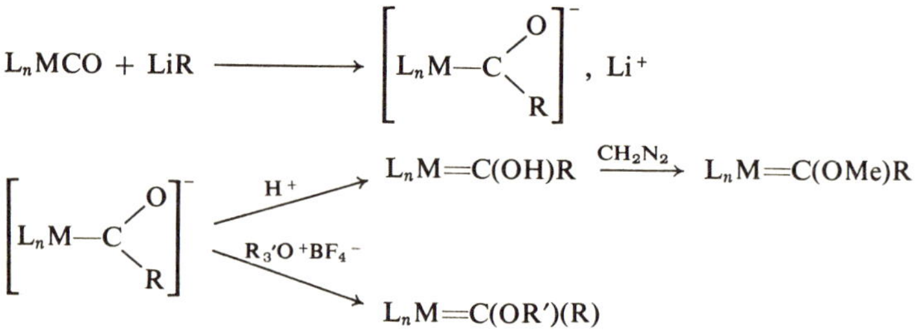

(160)

[R = alkyl or H, R′ = CH_3 or C_6H_5, M = $M'(CO)_5$ or $M'(PPh_3)(CO)_4$,
M′ = Cr^0, Mo^0, W^0
or M = $M''(\pi\text{-}C_5H_5)(CO)_2$, M″ = Mn, Re

When there is a choice of carbonyl groups within a molecule, nucleophilic attack always occurs at the carbonyl group with the greater stretching force constant [142], indicating that there is a direct correlation between the apparent carbonyl stretching force constant and the charge residing on the carbonyl carbon atom. In other words, the less the electron drift toward the metal ion, the lower is the back π bonding. With respect to the opposite flow of charge resulting from σ and π bonding, the latter seems to be more important in determining reactivity with a nucleophilic reagent.

At present metal–carbene complexes are of interest because of their unique chemical bonds, and no important synthetic applications have been reported thus far.

References

1. G. Glockler, *J. Phys. Chem.* **62**, 1049 (1958).
2. J. E. Boggs, C. M. Crain, and J. E. Whiteford, *J. Phys. Chem.* **61**, 481 (1957).
3. H. B. Gray, E. Billig, A. Wojcicki, and M. Farona, *Can. J. Chem.* **41**, 1281 (1963).
4. A. Wojcicki and F. Basolo, *J. Amer. Chem. Soc.* **83**, 525 (1961).
5. R. J. Angelici and F. Basolo, *J. Amer. Chem. Soc.* **84**, 2495 (1962).
6. R. J. Mawby, F. Basolo, and R. G. Pearson, *J. Amer. Chem. Soc.* **86**, 3994 (1964).
7. F. Basolo and A. Wojcicki, *J. Amer. Chem. Soc.* **83**, 520 (1961).
8. D. S. Breslow and R. F. Heck, *Chem Ind.* (*London*) p. 467 (1960).
9. R. F. Heck and D. S. Breslow, *J. Amer. Chem. Soc.* **83**, 1097 (1961).
10. R. F. Heck, *J. Amer. Chem. Soc.* **85**, 1220 (1963).
11. A. Wojcicki and F. Basolo, *J. Inorg. Nucl. Chem.* **17**, 77 (1961).
12. F. Basolo, A. T. Brault, and A. J. Poe, *J. Chem. Soc., London* p. 676 (1964).
13. T. H. Coffield, R. D. Closson, and J. Kozikowski, *J. Org. Chem.* **22**, 598 (1957).
14. T. H. Coffield, J. Kozikowski, and R. D. Closson, *Proc. Int. Conf. Coord. Chem.*, (*London*) April 6–11, 1959, Paper 26.
15. K. A. Keblys and A. H. Filbey, *J. Amer. Chem. Soc.* **82**, 4204 (1960).
16. F. Calderazzo and F. A. Cotton, *Inorg. Chem.* **1**, 30 (1962).
17. F. Calderazzo and F. A. Cotton, *Chim. Ind.* (*Milan*) **46**, 1165 (1964).
18. R. J. Mawby, F. Basolo, and R. G. Pearson, *J. Amer. Chem. Soc.* **86**, 5043 (1964).
19. F. Calderazzo and K. Noack, *J. Organometal. Chem.* **4**, 280 (1965).
20. C. S. Kraihanzel and P. K Maples, *J. Amer. Chem. Soc.* **87**, 5267 (1965).
21. M. Green and D. C. Wood, *J. Amer. Chem. Soc.* **88**, 4106 (1966).
22. K. Noack and F. Calderazzo, *J. Organometal. Chem.* **10**, 101 (1967).
23. K. Noack, M. Ruch, and F. Calderazzo, *Inorg. Chem.* **7**, 345 (1968).
24. R. J. Craig and M. Green, *Chem. Commun.* p. 1246 (1967).
25. I. S. Butler, F. Basolo, and R. G. Pearson, *Inorg. Chem.* **6**, 2074 (1967).
26. J. P. Bibler and A. Wojcicki, *Inorg. Chem.* **5**, 889 (1966).
27. R. F. Heck and D. S. Breslow, *J. Amer. Chem. Soc.* **83**, 4023 (1961).
28. R. F. Heck and D. S. Breslow, *J. Amer. Chem. Soc.* **84**, 2499 (1962).
29. R. F. Heck, *J. Amer. Chem. Soc.* **85**, 655 (1963).
30. R. F. Heck, *J. Amer. Chem. Soc.* **85**, 651 (1963).
31. R. F. Heck, *J. Amer. Chem. Soc.* **85**, 657 (1963).
32. Y. Takegami, C. Yokokawa, Y. Watanabe, and Y. Okuda, *Bull. Chem. Soc. Jap.* **37**, 181 (1964).
33. Y. Takegami, C. Yokokawa, Y. Watanabe, H. Masada, and Y. Okuda, *Bull. Chem. Soc. Jap.* **37**, 1190 (1964).
34. Y. Takegami, C. Yokokawa, Y. Watanabe, and H. Masada, *Bull. Chem. Soc. Jap.* **37**, 672 (1964).
35. Y. Takegami, C. Yokokawa, Y. Watanabe, H. Masada, and Y. Okuda, *Bull. Chem. Soc. Jap.* **38**, 787 (1965).

36. W. Hieber, O. Vohler, and G. Braun, *Z. Naturforsch. B* **13**, 192 (1958).
37. R. F. Heck, *J. Amer. Chem. Soc.* **85**, 1460 (1963).
38. G. Booth and J. Chatt, *J. Chem. Soc., A* p. 634 (1966).
39. M. C. Baird, J. T. Mague, J. Osborn, and G. Wilkinson, *J. Chem. Soc., A* p. 1347 (1967).
40. R. W. Glyde and R. J. Mawby, *Inorg. Chim. Acta* **4**, 331 (1970).
41. R. W. Glyde and R. J. Mawby, *Inorg. Chem.* **10**, 854 (1971).
42. M. Ryang, H. Miyamolo, and S. Tsutsumi, *Nippon Kagaku Zasshi* **82**, 1276 (1961); *Chem. Abstr.* **58**, 11387a (1963).
43. M. Ryang and S. Tsutsumi, *Nippon Kagaku Zasshi* **82**, 875 (1961); *Chem Abstr.* **58**, 11265c (1963).
44. M. Ryang and S. Tsutsumi, *Nippon Kagaku Zasshi* **82**, 878 (1961); *Chem. Abstr.* **58**, 11387f (1963).
45. M. Ryang and S. Tsutsumi, *Bull. Chem. Soc. Jap.* **34**, 1341 (1961).
46. M. Ryang and S. Tsutsumi, *Bull. Chem. Soc. Jap.* **35**, 1121 (1962).
47. M. Ryang, Y. Sawa, T. Miki, and S. Tsutsumi, *Technol. Rep. Osaka Univ.* **12**, 187 (1962); *Chem. Abstr.* **58**, 5558c (1963).
48. M. Ryang, I. Rhee, and S. Tsutsumi, *Bull. Chem. Soc. Jap.* **37**, 341 (1964).
49. M. Ryang, Y. Sawa, H. Masada, and S. Tsutsumi, *Kogyo Kazaku Zasshi* **66**, 1086 (1963); *Chem. Abstr.* **62**, 7670n (1965).
50. M. Ryang, Y. Sawa, T. Hashimoto, and S. Tsutsumi, *Bull. Chem. Soc. Jap.* **37**, 1704 (1964).
51. M. Ryang, K. Yoshida, H. Yokoo, and S. Tsutsumi, *Bull. Chem. Soc. Jap.* **38**, 636 (1965).
52. Y. Sawa, T. Miki, M. Ryang, and S. Tsutsumi, *Technol. Rep. Osaka Univ.* **13**, **(537–62) (229–36)** (1963); *Chem. Abstr.* **59**, 10097f (1963).
53. S. K. Myeong, Y. Sawa, M. Ryang, and S. Tsutsumi, *Bull. Chem. Soc. Jap.* **38**, 330 (1965).
54. M. P. Cook, Jr., *J. Amer. Chem. Soc.* **92**, 6080 (1970).
55. J. P. Collman, *Proc. Int. Conf. Coord. Chem. 14th*, p. 9 (1972).
55a. J. P. Collman and N. W. Hoffman, *J. Amer. Chem. Soc.* **95**, 2690 (1973).
56. H. Bliss and R. W. Southworth, U.S. Patent 2,565,461 (1951); *Chem. Abstr.* **46**, 2577h (1952).
57. G. E. Tabet, U.S. Patent 2,565,463 (1951); *Chem. Abstr.* **46**, 2578a (1952).
58. K. Yamamoto and K. Sato, Japanese Patent 2424 (1952); *Chem. Abstr.* **48**, 2105f (1954).
59. W. W. Prichard, U.S. Patent 2,689,750 (1954); *Chem. Abstr.* **49**, 6308d (1955).
60. W. W. Prichard, U.S. Patent 2,680,751 (1954); *Chem. Abstr.* **49**, 6308f (1955).
61. W. W. Prichard, U.S. Patent 2,696,503 (1954); *Chem. Abstr.* **49**, 15966i (1955).
62. N. L. Bauld, *Tetrahedron Lett.* **27**, 1841 (1963).
63. O. N. Temkin, O. L. Kaliya, G. K. Shestakov, S. M. Brailovskii, R. M. Flid, and A. P. Aseeva, *Kinet. Catal.* (*USSR*) **11**, 1333 (1970).
64. W. Reppe and W. J. Schweckendieck, *Ann.*, **560** 104 (1948).
65. W. Reppe, H. Kroper, H. J. Pistor, and H. Schlenk, *Justus Liebigs Ann. Chem.* **582**, 38 (1953).
66. E. R. H. Jones, T. Y. Shen, and M. C. Whiting, *J. Chem. Soc., London* p. 766 (1951).
67. G. P. Chiusoli, *Atti. Accad. Naz. Lincei Cl. Sci. Fis. Mat. Natur., Rend.* **26**, 790 (1959). *Chem. Abstr.* **54**, 8709g (1960).
68. G. P. Chiusoli, *Angew. Chem.* **72**, 74 (1960).
69. G. P. Chiusoli and S. Merzoni, *Z. Naturforsch. B* **17**, 850 (1962).

70. G. P. Chiusoli, S. Merzoni, and G. Mondelli, *Tetrahedron Lett.* **38**, 2777 (1964).
71. F. Guerrieri and G. P. Chiusoli, *Chem. Commun.* p. 781 (1967).
72. R. J. Heck, *J. Amer. Chem. Soc.* **86**, 2013 (1963).
73. G. Natta, P. Pino, and R. Ercoli, *J. Amer. Chem. Soc.* **74**, 4496 (1952).
74. R. Ercoli, U.S. Patent 2,911,422 (1959); *Chem. Abstr.* **54**, 7590d (1960).
75. J. L. Eisenmann, R. L. Yamortino, and J. F. Havard, Jr., *J. Org. Chem.* **26**, 2102 (1961).
76. W. A. McRae and J. L. Eisenmann, U.S. Patent 3,024,275 (1962); *Chem. Abstr.* **57**, 2077i (1963).
77. R. F. Heck and D. S. Breslow, *J. Amer. Chem. Soc.* **85**, 2779 (1963).
78. H. Nienburg and G. Elsehnig, German Patent 1,066,572 (1959); *Chem. Abstr.* **55**, 10323h (1961).
79. P. R. Klemchuk, U.S. Patent 2,995,607 (1962); *Chem. Abstr.* **56**, 1363e (1962).
80. J. Tsuji, M. Morikawa, and J. Kiji, *Tetrahedron Lett.* **16**, 1061 (1963).
81. J. Tsuji, M. Morikawa, and J. Kiji, *Tetrahedron Lett.* **22**, 1437 (1963).
82. J. Tsuji, M. Morikawa, and J. Kiji, *J. Amer. Chem. Soc.* **86**, 4851 (1964).
83. J. Tsuji, J. Kiji, and M. Morikawa, *Tetrahedron Lett.* **26**, 1811 (1963).
84. J. Tsuji, J. Kiji, S. Imamura, and M. Morikawa, *J. Amer. Chem. Soc.* **86**, 4350 (1964).
85. J. Tsuji, J. Kiji, and S. Hosaka, *Tetrahedron Lett.* **12**, 605 (1964).
86. J. Tsuji and S. Hosaka, *J. Amer. Chem. Soc.* **87**, 4075 (1965).
87. J. Tsuji and T. Susuki, *Tetrahedron Lett.* **34**, 3027 (1965).
88. J. Tsuji, S. Hosaka, J. Kiji, and T. Susuki, *Bull, Chem. Soc. Jap.* **39**, 141 (1966).
89. J. Tsuji and T. Nogi, *Bull. Chem. Soc. Jap.* **39**, 146 (1966).
90. J. Tsuji, M. Morikawa, and N. Iwamoto, *J. Amer. Chem. Soc.* **86**, 2095 (1964).
91a. S. Brewis and P. R. Hughes, *Chem. Commun.* p. 6 (1966).
91b. S. Brewis and P. R. Hughes, *Chem. Commun.* p. 71 (1967).
92. D. Medema, R. Van Helden, and C. F. Kohl, *Inorg. Chim. Acta.* **3**, 255 (1969).
93. J. Halpern and S. F. A. Kettle, *Chem. Ind. (London)* p. 668 (1961).
94. L. R. Barlow and J. M. Davidson, *Chem. Ind. (London)* p. 1656 (1965).
95. J. M. Davidson, *Chem. Commun.* p. 126 (1966).
96. W. Hieber and L. Schuster, *Z. Anorg. Chem.* **287**, 214 (1956).
97. W. Hieber and H. Hensinger, *J. Inorg. Nucl. Chem.* **4**, 179 (1957).
98. H. W. Sternberg, I. Wender, R. A. Friedel, and M. Orchin, *J. Amer. Chem. Soc.* **75**, 3148 (1953).
99. C. W. Bird, *Chem. Rev.* **62**, 283 (1962).
100. C. W. Bird, "Transition Metal Intermediates in Organic Synthesis." Academic Press, New York, 1967.
101. H. J. Sampson, U.S. Patent 2,589,289 (1952); *Chem. Abstr.* **46**, 11234b (1952).
102. H. J. Sampson, U.S. Patent 2,589,290 (1952); *Chem. Abstr.* **46**, 11234c (1952).
103. I. Wender, S. Friedman, W. A. Steiner, and R. S. Anderson, *Chem. Ind. (London)* p. 6194 (1958).
104. F. Calderazzo, *Inorg. Chem.* **4**, 293 (1965).
105. J. Tsuji and N. Iwamoto, *Chem. Commun.* p. 380 (1966).
106. E. W. Stern and M. L. Spector, *J. Org. Chem.* **37**, 596 (1966).
107. J. Falbe and F. Korte, *Angew. Chem. Int. Ed. Engl.* **1**, 266 (1962).
108. J. Falbe and F. Korte, *Chem. Ber.* **98**, 1928 (1965).
109. J. Falbe, *Angew. Chem., Int. Ed. Engl.* **5**, 435 (1966).
110. S. Murahashi, *J. Amer. Chem. Soc.* **77**, 6403 (1955).
111. S. Murahashi and S. Horiie, *J. Amer. Chem. Soc.* **78**, 4816 (1956).

112. W. W. Prichard, U.S. Patent 2,841,591 (1958); *Chem. Abstr.* **52**, 20197f (1958).
113. A. Rosenthal and M. R. S. Wier, *Can. J. Chem.* **40**, 610 (1962).
114. A. Rosenthal and J. Gevvay, *Chem. Ind.* (*London*) p. 1623 (1963).
115. A. Rosenthal and S. M. Millard, *Can. J. Chem.* **41**, 2504 (1963).
116. S. Murahashi, S. Horiie, and T. Jo, *Nippon Kagaku Zasshi* **79**, 72 (1958).
117. S. Murahashi, S. Horiie, and T. Jo, *Nippon Kagaku Zasshi* **79**, 75 (1958).
118. S. Murahashi, S. Horiie, and T. Jo, *Bull. Chem. Soc. Jap.* **33**, 81 (1960).
119. S. Horiie and S. Murahashi, *Bull. Chem. Soc. Jap.* **33**, 247 (1960).
120. W. W. Prichard, U.S. Patent, 2,769,003 (1957); *Chem. Abstr.* **51**, 7412d (1957).
121. A. C. Harkness and J. Halpern, *J. Amer. Chem. Soc.* **83**, 1258 (1961).
122. S. Nakamura and J. Halpern, *J. Amer. Chem. Soc.* **83**, 4102 (1961).
123. R. T. McAndrew and E. Peters, *Can. Met. Quart.* **3**, 153 (1964).
124. E. Hirsch and E. Peters, *Can. Met. Quart.* **3**, 137 (1964).
125. J. J. Byerley and E. Peters, *Can. J. Chem.* **47**, 313 (1969).
126. G. D. Zakumbaeva, N. F. Noskova, E. N. Konaev, and D. V. Sokolvikii, *Dokl. Akad. Nauk SSSR* **159**, 1323 (1964).
127. W. T. Dent, R. Long, and A. J. Wilkinson, *J. Chem. Soc., London* p. 1585 (1964).
128. J. K. Nicholson, J. Powell, and B. L. Shaw, *Chem. Commun.* p. 174 (1966).
129. B. R. James and G. L. Rempel, *Chem. Commun.* p. 158 (1967).
130. B. R. James, "Symposium on the Activation of Small Molecules." 1972.
131. H. C. Clark, P. W. R. Corfield, K. R. Dixon, and J. A. Ibers, *J. Amer. Chem. Soc.* **89**, 3360 (1967).
132. H. C. Clark, K. R. Dixon, and W. J. Jacobs, *J. Amer. Chem. Soc.* **91**, 1346 (1969).
133. H. C. Clark, K. R. Dixon, and W. J. Jacobs, *Chem. Commun.* p. 93 (1968).
134. H. C. Clark and W. J. Jacobs, *Inorg. Chem.* **9**, 1229 (1970).
135. J. E. Byrd and J. Halpern, *J. Amer. Chem. Soc.* **93**, 1634 (1971).
136. L. Malatesta, G. Caglio, and M. Angoletta, *J. Chem. Soc., London* p. 6974 (1965).
137. L. Malatesta, G. Caglio, and M. Angoletta, *J. Chem. Soc., A* p. 1836 (1970).
138. T. Kruck and M. Noack, *Chem. Ber.* **97**, 1693 (1964).
139. J. E. Byrd and J. Halpern, *J. Amer. Chem. Soc.* **93**, 1634 (1971).
140. M. L. H. Green, L. C. Mitchard, and M. G. Swanwick, *J. Chem. Soc., A* p. 794 (1971), and references therein.
141. F. A. Cotton and C. H. Lukehart, *in* "Progress In Inorganic Chemistry" (S. J. Lippard, ed.), Vol. 16, p. 487–613. John Wiley, New York.
142. D. J. Darensbourgh and M. Y. Darensbourgh, *Inorg. Chem.* **9**, 1691 (1970).

5

Activation of Nitric Oxide

I. Introduction

It has been seen that the activation of hydrogen, nitrogen, carbon monoxide, and oxygen by metal ions involves the formation of reactive species from molecules that are relatively unreactive at room temperature. Nitric oxide, on the other hand, is a fairly reactive molecule, as evidenced by its oxidation and metal complex formation reactions. Therefore the influence of metal ions is generally one of modifying the course of the reactions of nitric oxide, although some increases in reaction rates are also observed. As will be shown below, the activation (and modification) of reactions generally occurs through electron transfer with a metal ion.

Nitric oxide is an unusual and interesting diatomic molecule because of the fact that it exists as a free radical in the gas phase and has the ability to combine with other free radicals to give stable molecules with paired spins. It has a low dipole moment [1] of 0.5 Debye with the negative charge centered on the nitrogen, the less electronegative element. The nitric oxide molecule has a low ionization potential [2] of 9.5 eV and can readily lose an electron to give the nitrosyl ion, NO^+, which is isoelectronic with carbon monoxide and molecular nitrogen. The removal of an electron from the nitric oxide molecule takes place from an antibonding orbital, increasing the bond order from $2\frac{1}{2}$ to 3 and increasing the bond dissociation energy [3] from 6.49 to 11.6 eV, the latter being comparable to the dissociation energy of carbon monoxide, 11.1 eV.

II. Transition Metal Nitrosyls

Nitric oxide combines with a number of transition metal ions to form nitrosyls, in which the nitric oxide molecule is generally considered to be bonded through the nitrogen atom. Although substantial evidence has accumulated for metal–nitrogen coordination in metal nitrosyls, there are still a few cases where bonding between metal and oxygen is considered a possibility [2]. In forming a bond to a metal ion, nitric oxide can either accept an electron from the metal ion to form "bent" nitrosyls of the type $M^{(n+1)+}\ {}^{-}\ddot{\underset{\cdot\cdot}{N}}{=}\ddot{O}{:}$, or it can transfer an electron to the metal ion to form linear compounds of the type $M^{(n-1)+}\ \ddot{N}{\equiv}\overset{+}{O}{:}$. From a structural viewpoint the bent and linear arrangement of the metal–nitrosyl bond has been viewed in terms of negative NO^-[4–15] and positive NO^+ [16–28] ligands, respectively. Although some principles and generalizations have been derived from structural [4–28] and spectral [29–31,40] studies of nitrosyl complexes, a complete assessment of the factors that determine one bonding mode over the other have not been fully described [30]. The preparation and properties of metal nitrosyl complexes have been covered in a review by Johnson and McCleverty [32] to which the reader is referred for further details. The present section will deal mostly with the structure and bonding in transition metal nitrosyls and a few reactions of binary nitrosyls that are of importance in nitrosation reactions.

A. Binary Nitrosyl Complexes

There are relatively few examples of stable binary nitrosyls [2]. On the other hand, carbonyl nitrosyls, cyano nitrosyls, and mixed tertiary phosphine and arsine nitrosyls are quite stable. Nitrosyl halo compounds are relatively unstable and most of them cannot be isolated from solution. A study of the formation and dissociation of cobalt(II) halo nitrosyls was conducted by Fraser *et al.* [33–35]. A blue compound, $CuX_2\cdot NO$, is formed when nitric oxide is absorbed by a nonaqueous solution of cupric bromide or cupric chloride. The nitrosyl complex dissociates in solution [33] according to Eq. (1).

$$CuX_2\cdot NO \rightleftharpoons CuX_2^- + NO^+ \tag{1}$$

The dissociation constant corresponding to Eq. (1) decreases as the solvent is varied from methanol to butanol (i.e., as the dielectric constant decreases). It was possible to prepare solid $CuF_2\cdot NO$ only in butanol and in higher alcohols [33]. The equilibrium in Eq. (2) is complete only in alcoholic solutions and is shifted to the right as the dielectric constant of the alcohol

$$CuX_2 + NO \rightleftharpoons CuX_2\cdot NO \tag{2}$$

decreases. The formation in solution of copper(II) nitrosyls containing iodide, cyanide, and thiocyanato groups has also been reported [35]. Cupric salts in ethanol solution absorb nitric oxide to give violet complexes [36]. Triethanol–nitrosyl–copper(I) ion, $[Cu(C_2H_5OH)_3NO]^{2+}$, is diamagnetic in ethanolic solution and the IR spectrum (1854 cm^{-1} for the dichloride) shows that the complex should be considered as containing the NO^+ group coordinated to copper(I). On decomposition of this nitrosyl complex, the cupric ion is regenerated. This fact, together with the IR and the magnetic data, leads to the formula $[Cu(C_2H_5OH)_3NO]^{2+}$. Although copper(I) nitrosyls are stable in alcohols, they are very labile in aqueous solution. The aqueous complexes may serve as catalysts for nitrosation reactions, as indicated below in Sections III,A and III,B.

Aqueous solutions of cobaltous sulfate containing ammonia absorb [37] nitric oxide to form the complex $[Co(NH_3)_5NO]^{2+}$. In the formation of this complex compound, the metal ion donates one electron to nitric oxide to form a bond of the type $M{-}{-}{-}:\ddot{N}{=}\ddot{O}\bar{:}$, a reaction similar to the formation of $[Co(CN)_5H]^{3-}$ from $[Co(CN)_5]^{3-}$ and hydrogen. Silvestroni and Ceciarelli [38,39] have studied polarographically the reaction of nitric oxide with the cobalt(II)–histidine chelate containing a 2:1 ratio of ligand to metal and have shown that the metal chelate system exhibits with nitric oxide the same reversible carrier property that it shows with molecular oxygen. It was found [39] that the absorption of nitric oxide by the dihistidine–cobalt(II) system is complete in 15–20 minutes at pH 7.5 and at a nitric oxide pressure of 490 mm. The equilibrium constant for reaction (3) was reported [39] to be 10^4 at 20°C. Nitric oxide is very rapidly displaced from the complex by acidification with

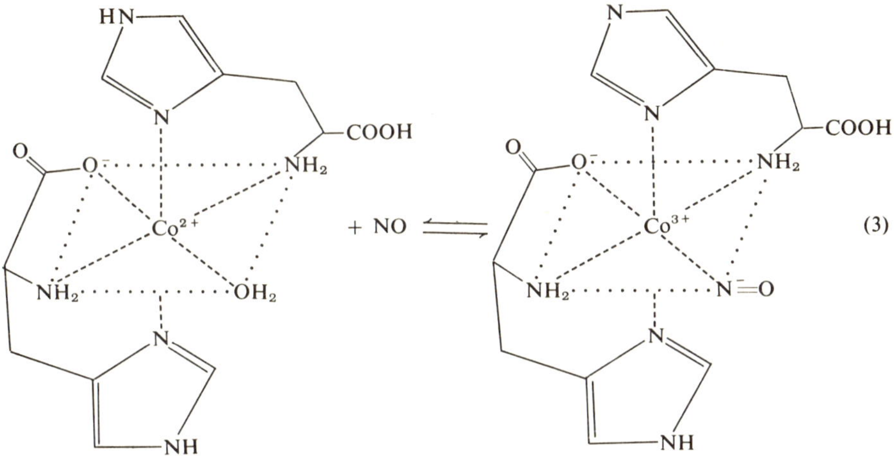

1 2 (3)

perchloric acid and may be dissociated more slowly by bubbling nitrogen through the solution. The bis(histidino)cobalt(II) complex is oxidized rapidly by molecular oxygen in the presence of activated charcoal as a catalyst to a cobalt(III) complex which no longer absorbs nitric oxide.

A number of paramagnetic iron nitric oxide complexes were prepared by Griffith *et al.* [36]. The magnetic susceptibilities and the N—O stretching frequencies given in Table I indicate the presence of the NO^+ ligand in these

TABLE I

PROPERTIES OF SOME IRON–NITRIC OXIDE COMPLEXES

Formula	Magnetic susceptibility	IR frequency (cm^{-1})	Formal valence state of Fe
$Fe(H_2O)_5NO^{2+}$	3 e	1730–1880	Fe(I)
$Fe(NH_3)_5NO^{2+}$	—	1730–1880	Fe(I)
$Fe(C_2H_5OH)_5NO^{3+}$	4 e	1775	Fe(II)
$Fe(CN)_5NO^{3-}$	1 e	—	Fe(III)

complexes. It was concluded that the first three complexes are of the high-spin type, in which an electron had been donated by nitric oxide to the metal ion. The pentacyanonitrosyl complex is believed to have a low-spin electronic configuration, with the single unpaired electron probably localized mainly on the coordinated nitric oxide ligand.

The applications of metal nitrosyls have not yet been fully exploited commercially and most compounds now known seem to be mainly of academic interest. Copper(II), iron(III), and cobalt(II) nitrosyls containing suitable ligands such as water, ammonia, and amines provide convenient catalytic routes for nitrosation and related reactions. In a few cases copper(II) compounds have been found to be effective catalysts in the nitrosation of secondary amines. Interest has very recently developed in the elucidation of the mechanisms of nitrosation reactions, which are now thought to be more complex than the carbonyl insertion reactions described in Chapter 4. The catalysis of nitrosation reactions at nitrogen and oxygen centers will be discussed in Section III.

B. STRUCTURE AND BONDING IN METAL NITROSYLS

Coordinate bonding in metal nitrosyls involves both σ and π bonds. A σ bond is formed by the overlap of an *sp* orbital of nitric oxide with a suitable metal orbital, and formation of a d_π–p_π bond involves the back donation of electrons from filled metal t_{2g} orbitals to empty antibonding p_π orbitals of nitric oxide. In linear nitrosyls, nitric oxide may be regarded as having a formal positive charge and the extent of backbonding to form a d_π–p_π bond should be stronger than the analogous effect in metal carbonyls, although a quantitative comparison cannot be made. In bent nitrosyls, there is ligand-to-metal donation of the p_π–d_π type from π^* nitric oxide molecular orbital to suitable empty metal d orbitals [29].

On the basis of observed infrared spectra low-frequency ν(N—O) vibrations have been assigned [40] to M^{n+1}—N^-=O and higher frequency vibrations to the $M^{(n-1)+}$—N≡O^+ structure. It may be readily seen from Table II that this rule is misleading because there is so much overlap in the ν(N—O) frequencies of linear and bent nitrosyls in the 1620–1720 cm^{-1} range that assignment to a particular configuration is not generally possible. The

TABLE II

SUMMARY OF METAL NITROSYL COMPLEXES

	Complex	M—N—O angle	M—N (Å)	N—O (Å)	ν(NO) (cm^{-1})	References
	Four-Coordinate Bent Nitrosyls					
1	$[Ir(NO)_2(PPh_3)_2]^+$	164°	1.77	1.21	1760–1715	8
2	$Ni(N_3)(NO)(PPh_3)_2$	153	1.69	1.16	—	15
	Four-Coordinate Linear Nitrosyls					
3	$[Co(NO)_2]_2-[\mu(NO_2)_2]$	179	1.67	1.13	1867	22
	Five-Coordinate Bent Nitrosyls					
4	$[IrCl(NO)(CO)(PPh_3)_2]^+$	124	1.97	1.16	1680	4,5
5	$[IrI(NO)(CO)(PPh_3)_2]^+$	125	1.89	1.17	1720	6
6	$Ir(CH_3)I(NO)(PPh_3)_2$	120	1.91	1.23	1525	9
7	$IrCl_2(NO)(PPh_3)_2$	123	1.94	1.03	1560	10
8	$[RuCl(NO)_2(PPh_3)_2]^+$	178	1.74	1.16	1640	12,13
		138	1.85	1.17	1687	
9	$[Os(OH)(NO)_2(PPh_3)_2]^+$	128	1.98	—	1632	14
	Five-Coordinate Linear Nitrosyls					
10	$[Ru(NO)(diphos)_2]^+$	174	1.73	1.20	1673	23
11	$Ru(H)(NO)(PPh_3)_3$	176	1.79	1.18	1640	16
12	$[Ir(H)(NO)(PPh_3)_3]^+$	175	1.68	1.21	1715	26
13	$[Os(NO)(CO)_2(PPh_3)_2]^+$	175	1.89	1.12	1750	27
14	$Fe(NO)(S_2CN(CH_3)_2)_2$	170	1.72	1.10	—	25
15	$Mn(NO)(CO)_3(PPh_3)$	178	1.78	1.15	—	18
16	$Mn(NO)(CO)_4$	180	1.80	1.15	—	19
	Six-Coordinate Bent Nitrosyls					
17	$[Co(NO)(en)_2Cl]^+$	121	1.81	1.16	1611	7
18	$[Co(NO)(NH_3)_5]^{2+}$	119	1.87	1.15	1610	11
	Six-Coordinate Linear Nitrosyls					
19	$[Fe(NO)(CN)_5]^{2-}$	178	1.63	1.13	1938	17
20	$[Mn(NO)(CN)_5]^{3-}$	173	1.65	—	1725	21
21	$[Cr(NO)(CN)_5]^{3-}$	176	1.71	1.21	1645	24
22	$[Ru(NO)(OH)(NO_2)_4]^{2-}$	180	1.75	1.13	1907	28
23	$Ru(NO)(S_2CNR_2)_3$ (R = Me, Et)	170	1.72	1.17	1830	20

ν(N—O) stretching frequency also depends on the coordination number and oxidation state of the metal ion in the complex and the geometry of the complex. There is no simple correlation between ν(NO) and the metal–nitrogen or metal–oxygen bond distances. For isoelectronic, isostructural complexes, the metal–nitrogen distances in linear nitrosyls are lower than those of the bent nitrosyls, indicating extensive d_π–p_π interaction in the former complexes.

Finn and Jolly [31] have measured the nitrogen 1s binding energies in a number of linear and bent nitrosyls and have come to the conclusion that there is a definite correlation between the nitrogen 1s binding energy of a coordinated nitrosyl group and the electron density on that group. Bent nitrosyls have lower 1s binding energy (~400.2 eu) than linear nitrosyls (~402.6 eu). Among the linear nitrosyls the binding energy increases with increasing d_π–p_π backbonding from the metal atom. This is in accord with the observation by Manoharan and Gray [29] that the ν(N—O) decreases in the series $Fe(CN)_5NO^{2-}$, $Mn(CN)_5NO^{2-}$, $Mn(CN)_5NO^{3-}$, $Cr(CN)_5NO^{3-}$, $V(CN)_5 \cdot NO^{5-}$ from 1939 cm^{-1} for the iron complex to 1575 cm^{-1} for the analogous vanadium nitrosyl. For this series of metals, the extent of d_π–p_π backbonding decreases in the same order, as indicated by the calculated electron densities in the $d_{xz}d_{yz}$ levels of the metal–NO bonds.

Table II presents a list of bent and linear nitrosyl complexes with various structural parameters including the M—N—O bond angle, the M—N and N—O bond lengths, and ν(N—O), the stretching frequencies of the coordinated nitrosyl groups.

The lowest coordination number encountered in metal nitrosyls is 4, although very few examples of such complexes are known. The structure [8] of the complex $[Ir(NO)_2(PPh_3)_2]ClO_4$, **3**, is illustrated in Fig. 1. Complex **3**

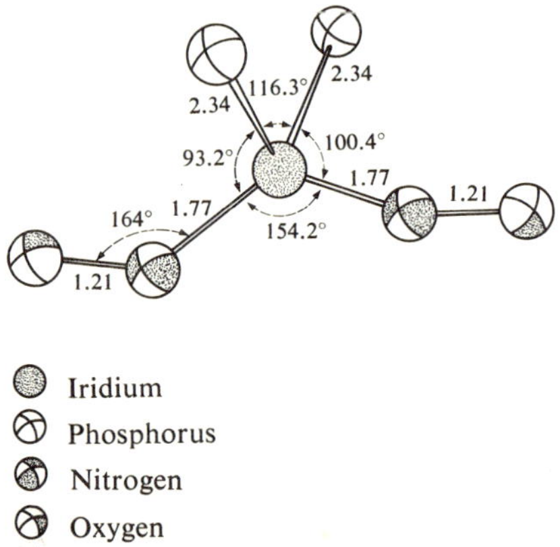

Fig. 1. Structure of $[Ir(NO)_2(PPh_3)_2]$ ClO_4. According to Mingos and Ibers [8].

is regarded as a distorted tetrahedron with the metal ion considered as Ir^- and nitric oxide as NO^+. The M—N—O bond angle of 164° is, however, 16° smaller than expected for linear NO^+ complexes. The complex has a distorted tetrahedral structure (Fig. 1), the distortion being due to the nonbonded interaction of the phenyl rings of the coordinated phosphine groups [8]. Other distorted tetrahedral complexes of metal ions having a formal d^{10} configuration include $Ni(N_3)(NO)(PPh_3)_2$, **4**, recently reported by Enemark [15] and $[Co(NO)_2]_2-\mu(NO_2)_2$, **5**, prepared by Strouse and Swanson [22]. Complex **4** is a bent nitrosyl with the M—N—O angle of 153° (Fig. 2). The oxidation state of the metal ion in **5** may be regarded as (−1) with nitric oxide coordinated as NO^+. The M—N—O angle of 179° confirms this formulation.

Data on five-coordinate metal nitrosyl complexes with bent and linear NO groups are given in Table II. The five-coordinate bent nitrosyls $[IrCl(NO)\cdot(CO)(PPh_3)_2]^+$ [4,5] (Fig. 3), **6**; $[IrI(NO)(CO)(PPh_3)_2]^+$ [6], **7**; $[Ir(CH_3)\cdot I(NO)(PPh_3)_2]^+$ [9], **8**; and $IrCl_2(NO)(PPh_3)_2$ [10], **9** (Fig. 4), may be considered formally to be complexes of d^6 iridium(III) with NO^-. The M—N—O bond angle for these complexes varies from 120° to 124°, as expected of a NO^- group having sp^2-type bonds at the nitrogen atom. Pierpont *et al.* [12,13] have reported a ruthenium complex of the composition $[RuCl(NO)_2(PPh_3)_2]ClO_4$, **10**, with both linear and bent nitrosyl groups. This cationic complex may be described as a distorted tetragonal pyramid. The nitrosyl group in the basal plane (Fig. 5) is bonded to the ruthenium atom in an essentially linear manner and may be considered as NO^+. The second nitrosyl group is bonded in the apical position with a Ru—N—O bond angle of 138° and may be regarded as NO^-. The oxygen atom of the bent nitrosyl group in **10** is oriented towards the basal NO^+ group. Similar

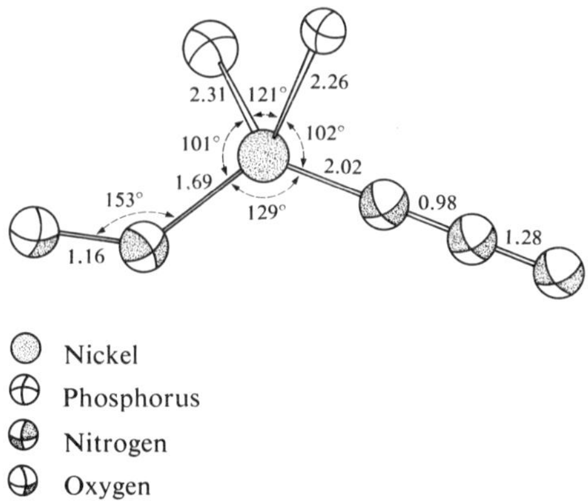

Fig. 2. Structure of $Ni(N_3)(NO)(PPh_3)_2$. After Enemark [15].

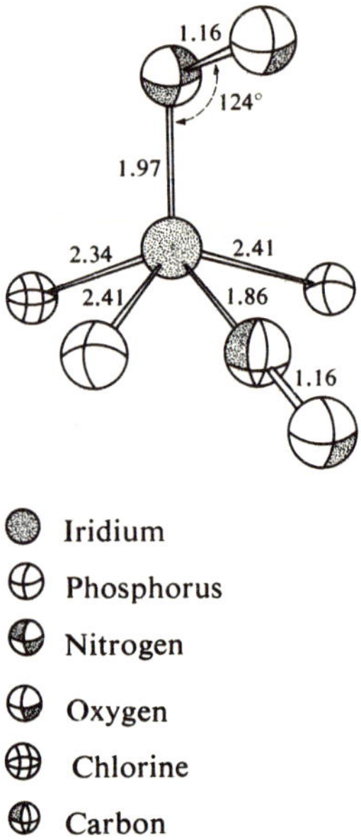

Fig. 3. Structure of $[IrCl(NO)(CO)(PPh_3)_2]\ BF_4$. After Hodgson *et al.* [4,5].

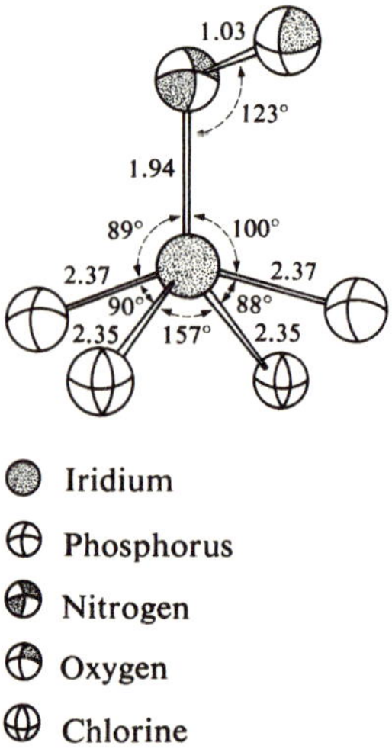

Fig. 4. Structure of $IrCl_2(NO)(PPh_3)_2$. After Mingos and Ibers [10].

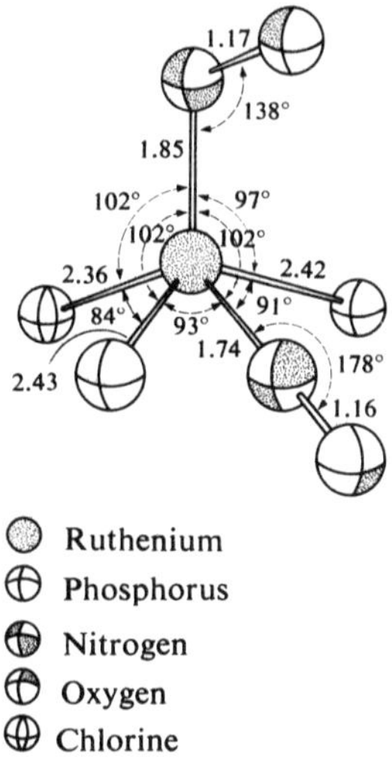

Fig. 5. Structure of $[RuCl(NO)_2(PPh_3)_2]^+$. After Pierpont and Eisenberg [12,13].

orientation of the bent NO group towards the basal phosphine was observed for complexes **6–9** [4,5,6,9,10]. It was suggested by Pierpont and Eisenberg [13] that the lone pair on oxygen of the apical bent nitrosyl may have slight interaction with a basal phosphine in complexes **5–9** and with the basal NO^+ in complex **10**. The apical and basal nitrosyl groups in complex **10** undergo electronic rearrangement in solution whereby the bond angles and bond lengths of the apical and basal nitrosyl groups are interchanged **10** → **12**. This rearrangement was studied [13] in the isotopically labeled complex ion

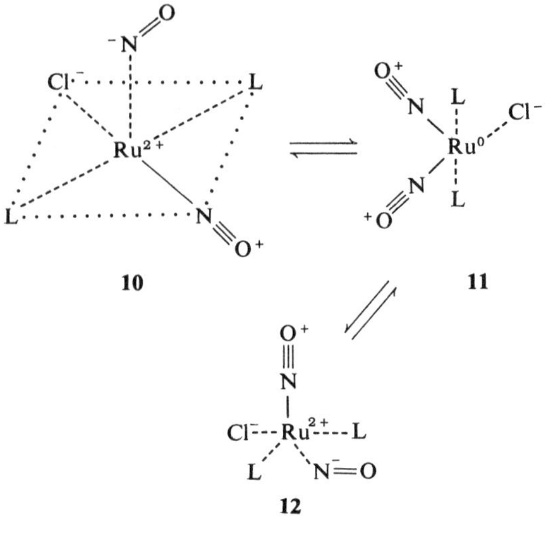

(1)

$[RuCl(^{14}NO)(^{15}NO)(PPh_3)_2]^+$ by monitoring the ν(NO) values in the infrared spectrum. It was suggested that the scrambling of the labeled nitrosyl groups occurs through a trigonal bipyramidal intermediate **11** with equivalent (linear or slightly bent) NO groups as shown in Eq. (1).

Recently Collman *et al.* [41] have demonstrated a rapid equilibrium in nitrobenzene solution of trigonal bipyramidal cobalt(I) complexes, $CoCl_2(NO)L_2$ [L = $P(n\text{-}C_4H_9)_3$, $P(CH_3)(C_6H_5)_2$] having linear nitrosyl groups, **13**, with the corresponding square-pyramidal cobalt(III) complexes having bent nitrosyl groups, **14**. The interconversions **13** $\rightleftharpoons$ **14** and **10** $\rightleftharpoons$ **12**

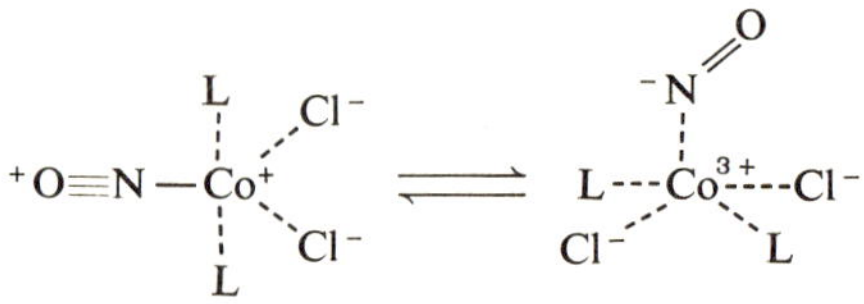

13 **14**

are examples of hybridization tautomerism where the two equilibrated forms undergo intramolecular redox interchange of the nitrosyl groups by appropriate electron shifts.

The configuration [14] of the complex $[Os(OH)(NO)_2(PPh_3)_2]PF_6$ is similar to that of **10** with basal linear and apical bent nitrosyl groups with distorted tetragonal pyramidal geometry of the complex. The apical bent nitrosyl group leans towards the basal nitrosyl in a manner similar to that observed [12] in **10**.

Five-coordinate nitrosyl complexes with linear nitrosyl groups have been listed in Table II (10–16). These complexes are all trigonal bipyramidal, with the nitrosyl in the axial position in complexes 11 and 12 and in the equatorial position in 10 and 13–16. For axially bound nitrosyl complexes the metal ion is displaced from the equatorial plane by about 0.5 Å towards the axial nitrosyl group. This displacement is due to the different coordinating affinities of the nitrosyl and hydride ligands in compounds 11 and 12 (Table II). As a consequence of metal displacement, the N—M—P bond angles are greater than 90°. There are also variations in the P—M—P (M = Ir, Ru) angles from the expected 120° in complexes 11 and 21, Table II. The structures of the complexes $[Ru(NO)(diphos)_2]ClO_4$ [23], Ru(H)(NO)·$(PPh_3)_3$ [16], $IrH(NO)(PPh_3)_3]^+$ [28], and $[Os(NO)(CO)_2(PPh_3)_2]ClO_4$ [27] are shown in Figs. 6–9.

The five-coordinate nitrosyl complexes 10–13, 15, and 16 (Table II) may be considered as complexes of d^8 metal ions [ruthenium(0), osmium(0), manganese(−1), and iridium(I)], coordinated to NO^+. Complex 14 (Table II), $Fe(NO)(S_2CNMe_2)_2$, may be considered [25] to be a complex of a d^7 metal ion, iron(I) with NO^+. The M—N—O bond angles range from

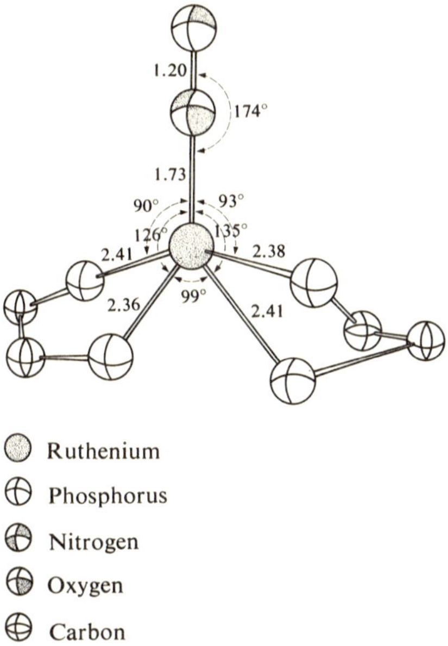

Fig. 6. Structure of $[Ru(NO)(diphos)_2]ClO_4$. After Pierpont *et al.* [23].

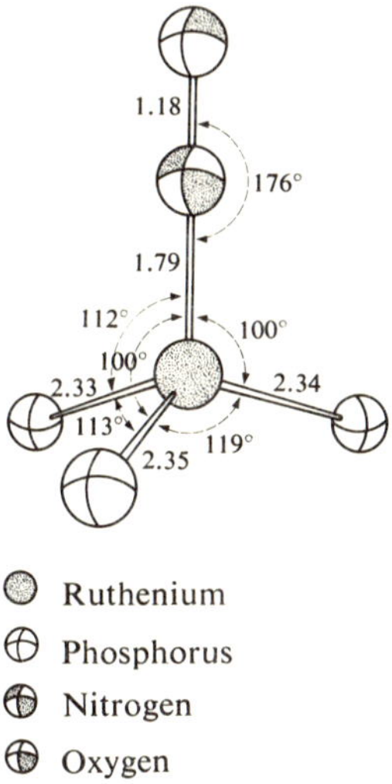

Fig. 7. Structure of $Ru(H)(NO)(PPh_3)_3$. After Pierpont and Eisenberg [16].

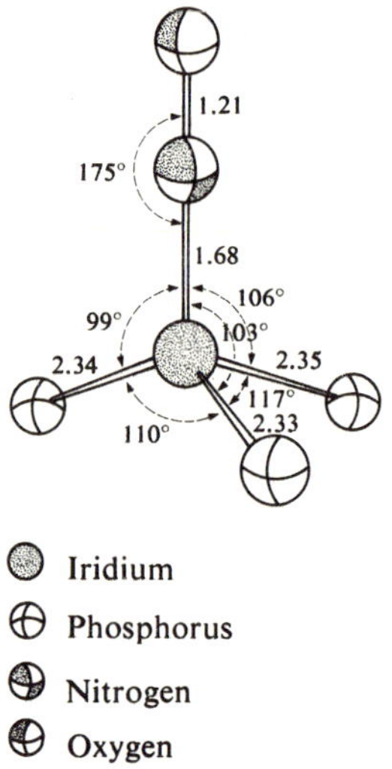

Fig. 8. Structure of $[Ir(H)(NO)(PPh_3)_3]^+$. After Mingos and Ibers [8].

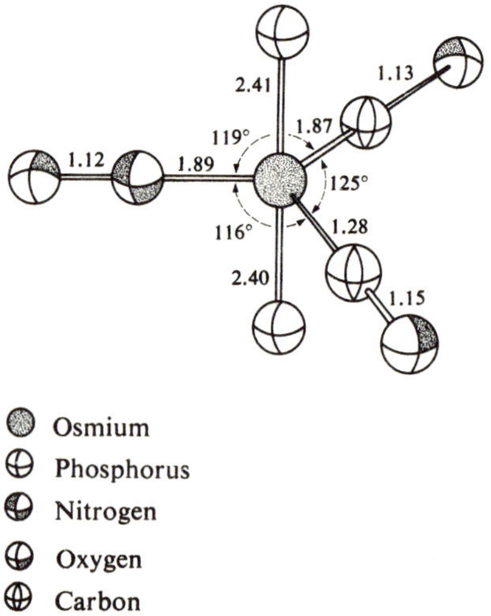

Fig. 9. Structure of $[Os(NO)(CO)_2(PPh_3)_2]ClO_4$. After Clark *et al.* [27].

175° to 180° except for $Fe(NO)(S_2CNMe_2)_2$, which has an M—N—O bond angle of 170°.

Examples of six-coordinate bent nitrosyl complexes are the ions $[Co(NO)(en)_2Cl]^+$, **15**, and $[Co(NO)(NH_3)_5]^{2+}$, **16**, in which a d^6 cobalt(III) ion may be considered to be coordinated to a bent NO^- group. The structure of complex **16** is illustrated in Fig. 10.

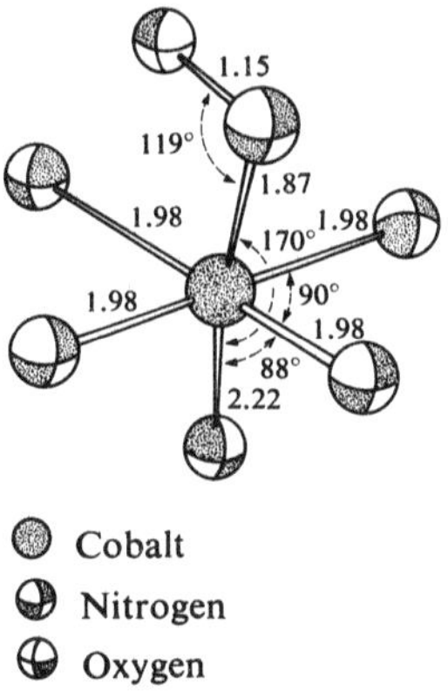

Fig. 10. Structure of $[Co(NH_3)_5(NO)]Cl_2$. After Pratt *et al.* [11].

Six-coordinate linear nitrosyls are octahedral complexes 19, 20, 22, and 23 (Table II), where a d^6 metal ion—iron(II), manganese(I), or ruthenium(II)—is coordinated to a linear NO^+ group. The chromium(I) complex $[CrNO\cdot(CN)_5]^{3-}$ has a d^5 ion coordinated to NO^+. The M—N—O bond angles in the octahedral metal nitrosyls vary from 170° to 180°C, indicating a straight coordination of NO^+ group. Figures 11 and 12 illustrate the structure of the complexes $Ru(NO)(S_2CNEt)_3$ [20] and $[Cr(CN)_5(NO)]^{3-}$ [24], respectively.

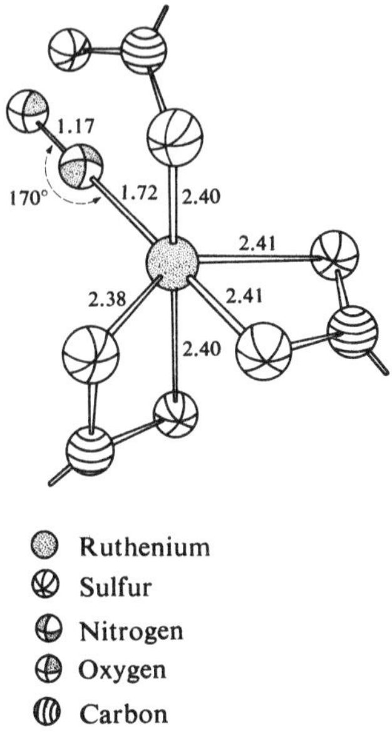

Fig. 11. Structure of $Ru(NO)(S_2CNEt)_3$. After Domenicano *et al.* [20].

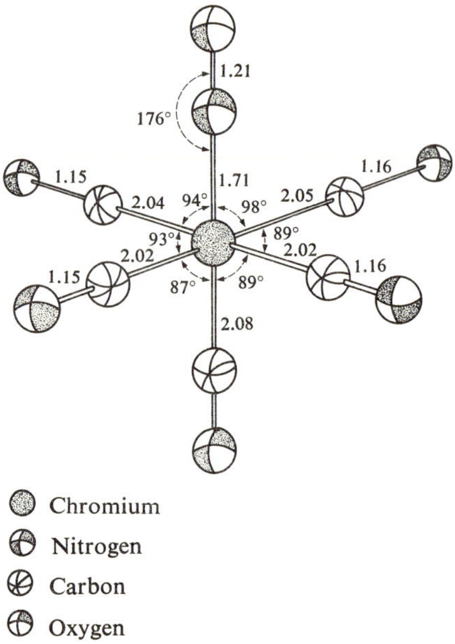

Fig. 12. Structure of $[Cr(CN)_5(NO)]^{3-}$. After Enemark *et al.* [24].

The bonding of the nitrosyl group in five- and six-coordinate complexes and interpretation of the bending of the nitrosyl group in tetragonal complexes has been advanced by Pierpont and Eisenberg [16,30] and by Mingos and Ibers [26]. The molecular orbital diagrams for five- and six-coordinated nitrosyls based on the suggestion of Pierpont and Eisenberg [30] and of Manoharan and Gray [29] are illustrated in Fig. 13. Electron distributions II and IV in Fig. 13 apply to a d^6 metal ion coordinated to NO^+. The arrangement in III is for a d^8 metal ion coordinated to NO^+ and in I and V, for a d^6 metal ion coordinated to NO^-.

The molecular orbital energy levels for the square-pyramid [30] and octahedral [29] complexes are reasonably similar with respect to the molecular orbital energies shown, with differences mainly in the disposition of the antibonding level $e\pi^*(NO)$, $a_1d_{z^2}$ and $b_1d_{x^2-y^2}$. For the square-pyramid structure (II) and for the octahedral complexes of C_{4v} symmetry, the lowest σ-bonding level contains 10 and 12 electrons, respectively, corresponding to electron pairs from five and six ligands, respectively. The π^b nitrosyl orbital has four electrons in the degenerate $(\pi_{2p_x,2p_y})^4$ levels. The next higher level is a metal-based degenerate orbital d_{xz}, d_{yz} with some $\pi^*(NO)$ character. The next level, b_2, which is occupied by two electrons, is also metal-based. Thus the total number of electrons in II and IV are 20 and 22, respectively, neglecting the localized σ-bonding pair between nitrogen and

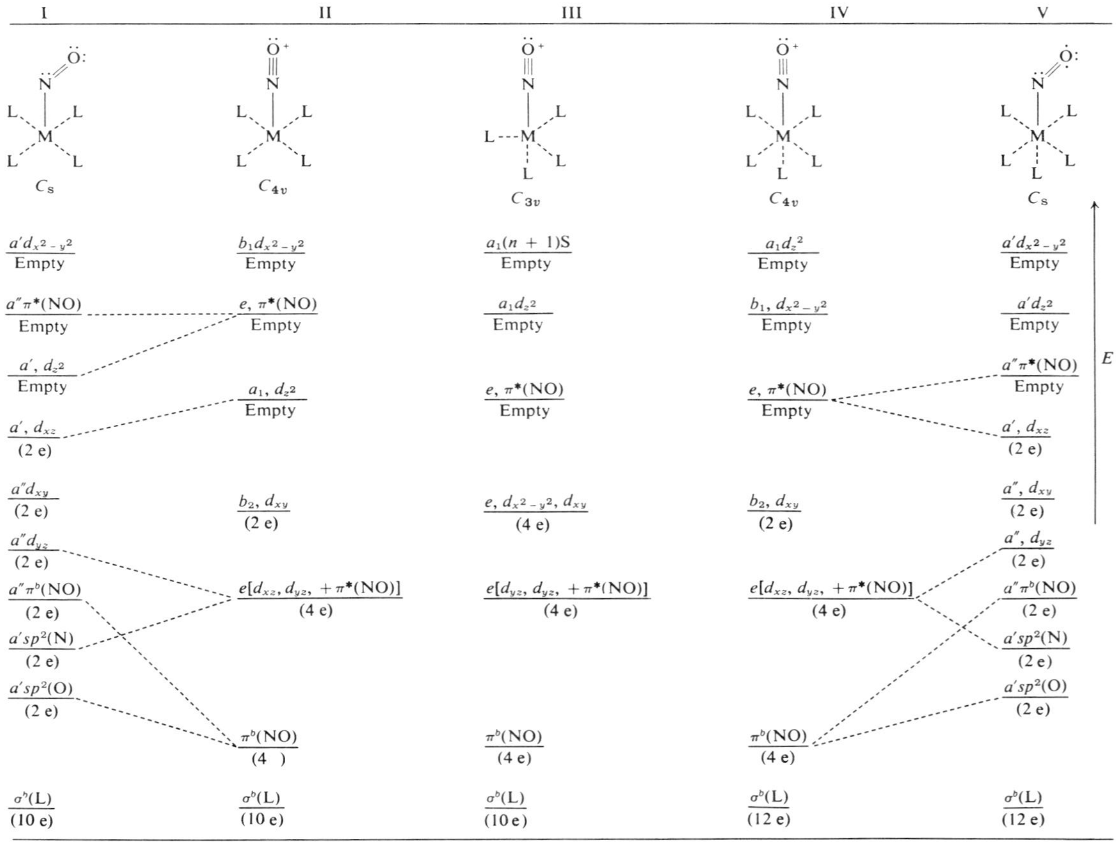

Fig. 13. An orbital correlation diagram for linear and bent nitrosyls in five- and six-coordinate systems. The σ-bonding pair between N—O and one nonbonding pair on oxygen are localized in all the structures.

oxygen, and one nonbonding pair on the oxygen. The addition of two electrons to II or IV to form a NO^- complex requires addition of electrons in the antibonding a_1 level in II or in the antibonding e level in IV. This is avoided by distortion of the linear nitrosyl group in complexes II and IV (C_{4v}) to bent nitrosyl groups in I and V (C_S), respectively. This distortion removes the degeneracy of the levels π^b(NO), $e[d_{xz}, d_{yz}, + \pi^*(NO)]$ and $e\pi^*$(NO) and provides additional levels $a'(sp_2)$ to accommodate the nonbonding pairs on nitrogen and oxygen. Thus complexes of d^6 metal ions with NO^- should have a tetragonal C_S symmetry, as indeed is the case (Table II, entries 4–9, 17–18).

The trigonal bipyramidal geometry (C_{3v}) has been observed primarily in d^8 metal complexes having a linear NO^+ group (entries 10–16, Table II). The molecular orbital diagram suggested by Pierpont and Eisenberg is shown in Fig. 13 (III). For a 22-electron system the electronic configuration is $(\sigma^b L)^{10}[\pi^b(NO)]^4[e(d_{xz}, d_{yz}, + \pi^*(NO)]^4$ and $(e,d_{x^2-y^2}, {}_{xy})^4$. The highest filled level for the trigonal bipyramidal case $(e,d_{x^2-y^2}, {}_{xy})^4$ has slight antibonding character [16]. The result is weakening of σ bonding in the trigonal plane of the complex. The complexes nevertheless have strong metal–NO d_π–p_π bonding due to the NO^+ group. In the alternate situation of the bonding of a d^6 metal ion with a bent nitrosyl NO^- (C_S symmetry), the arrangement leads to more effective bonding [16] with the loss of some d_π–p_π bonding due to the bent M—N—O bond. Thus the configurations M^{n+}---NO^+ and $M^{(n+2)+}$---NO^- are equivalent. As explained above, in some cases, such as $CoCl_2(NO)L_2$, **13**, the conversion barrier appears to be so small that one type can be easily converted into another [16].

C. Reactions of Coordinated Nitric Oxide

1. *Nucleophilic Attack on Coordinated Nitric Oxide*

Meyer *et al.* [42,43] have reported the formation of a coordinated nitrite, **19**, in the reaction of 2,2′-bipyridyl nitrosyl complexes of ruthenium(II) $[Ru(bipy)_2NOX]^{2+}$ with hydroxide ion in basic solution, as indicated by reaction sequence (4). The coordinated nitrosyl in **17** behaves as NO^+ in its reaction towards hydroxide ion. Complex **19** on protonation (i.e., in acid solution) is converted back to **17**. The reaction may be considered to proceed through an intermediate, **18**, as the result of nucleophilic attack by the hydroxide ion on the coordinated NO^+ group in **17**. The intermediate **19** then may react with more base, or with acid, to give the nitrito complex or the starting material **17**.

Reactions of [44] nitrosylpentammineruthenium(II), **20**, with hydroxylamine, hydrazine, or ammonia produce (dinitrogen oxide)pentammineruthenium(II), **21**. In the case of hydrazine and ammonia, dinitrogenpentammineruthenium(II), **22**, has also been isolated [44] along with **21**. These reactions are indicated by Eq. (5) and (6). The major product of reaction with hydroxylamine and with hydrazine is complex **21**. With

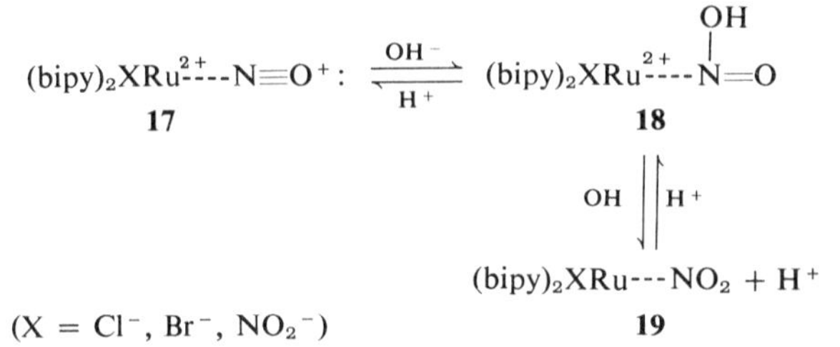

(4)

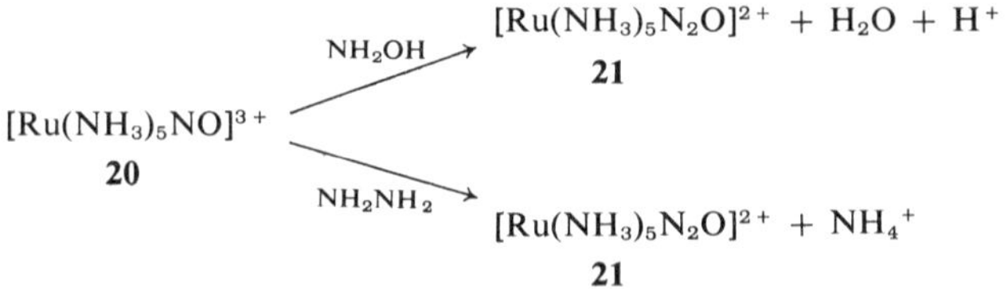

(5)

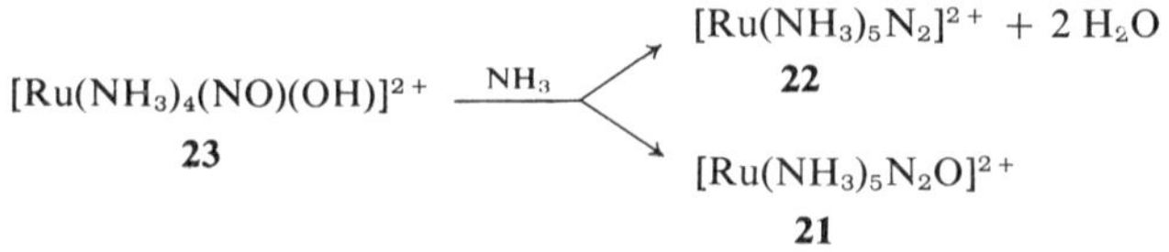

(6)

ammonia, however, intermediate **23** is formed, which may react with more ammonia to give **21** and **22**. The dinitrogen complex **22** is not directly obtained by the reaction of **21** with N_2H_4. Hydrazine promotes the decomposition of **21** to $[Ru(NH_3)_5(OH)]^{2+}$, which reacts with hydrazine to give **22**.

Reaction of complex **20** with hydroxide gave [45] $[Ru(NH_3)_5NO_2]^+$, **24**, as the major product along with $[Ru(NH_3)_4(NH_2)(NO)]^{2+}$, **25**. The nitrito complex **24** may be explained to have been formed by a reaction similar to (4) whereas **25** is obtained by the direct deprotonation of a coordinated ammonia.

$$\underset{\mathbf{20}}{Ru(NH_3)_5NO^{3+}} + 2\,OH^- \longrightarrow \underset{\mathbf{24}}{Ru(NH_3)_5NO_2^{+}} + H_2O \quad (7)$$

$$\underset{\mathbf{20}}{Ru(NH_3)_5NO^{3+}} + OH^- \longrightarrow \underset{\mathbf{25}}{Ru(NH_3)_4(NH_2)(NO)^{2+}} + H_2O \quad (8)$$

Grundy *et al.* [46] have described the protonation of coordinated nitrosyl groups in $OsCl(CO)(NO)(PPh_3)_2$, **26**; $Os(NO)_2(PPh_3)_2$, **27**; and $Ir(NO)\cdot(PPh_3)_3$, **28**, with 1, 2, and 3 moles of hydrogen chloride, respectively, to form $OsCl_2(CO)(HNO)(PPh_3)_2$, **29**; $OsCl_2(NHOH)(NO)(PPh_3)_2$, **30**; and

$IrCl_3(NH_2OH)(PPh_3)_2$, **31**. The protonation steps are reversible, as indicated by Eq. (9), (10), and (11), and generally take the form of reversible oxidative addition reactions. [Eq. (9) seems to be an exception.]

Clarkson and Basolo [47] have investigated the addition of molecular oxygen to $[Co(en)_2(NO)]^{2+}$ and Schiff base complexes $[L_4Co(NO)]$ in the presence of a base, B, to form nitrato complexes $[Co(en)_2(NO_3)]^{2+}$ and $[L_4Co(NO_3)]$ [where L_4 represents a tetradentate Schiff base such as bis-(acetylacetone)ethylenediimine]. The reaction at room temperature of a solution of $[Co(en)_2NO](ClO_4)_2$ in acetonitrile with oxygen results in the

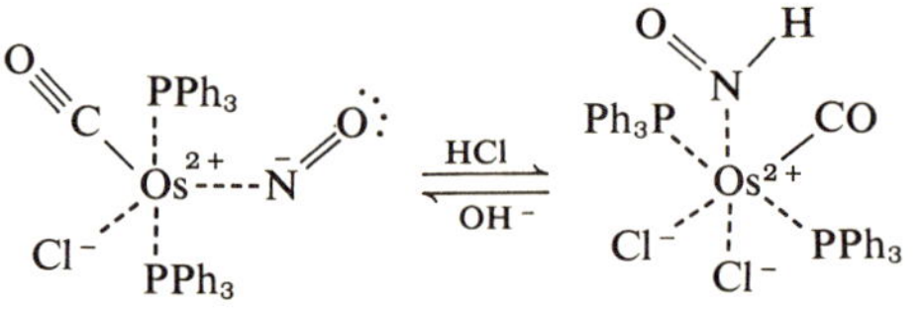

26 **29** (9)

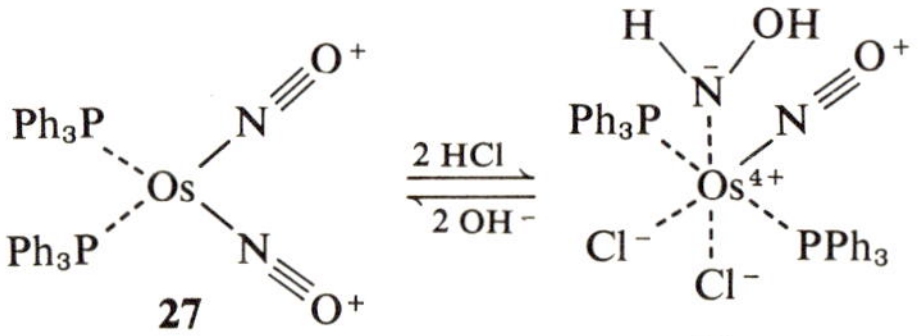

27 **30** (10)

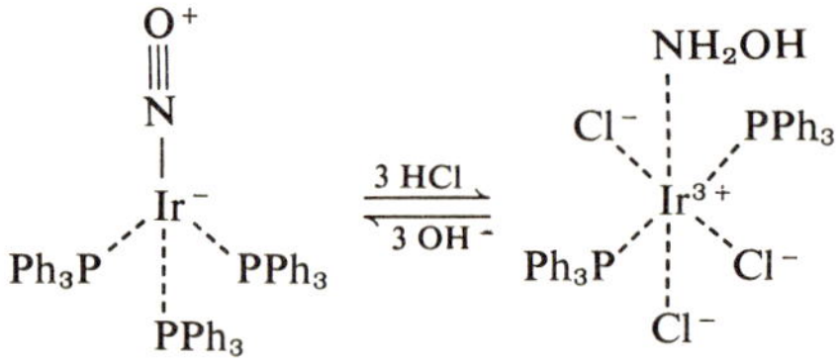

28 **31** (11)

$$L_4Co(NO) + B \rightleftharpoons BL_4Co(NO) \qquad (12)$$

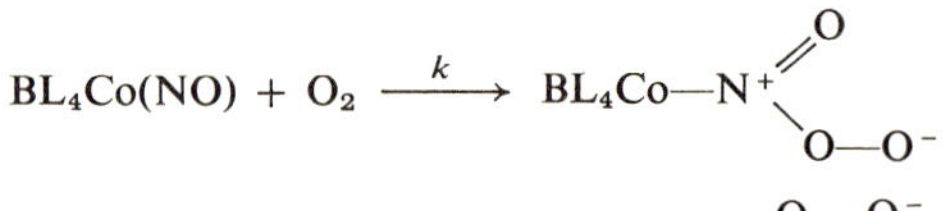

(13)

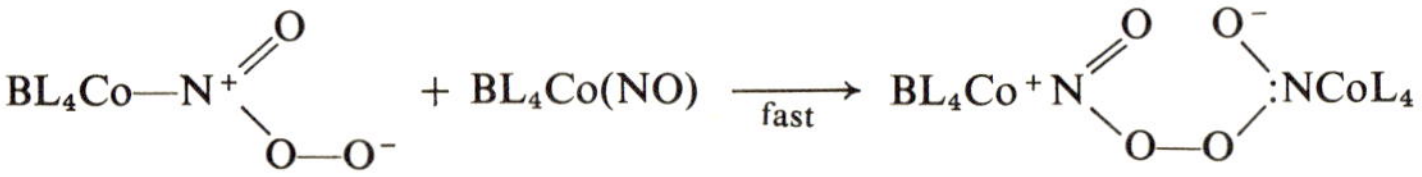

(14)

(15)

[L_4 = N,N-ethylenebis(1-acetonylethylideneiminato)
= N,N′-ethylenebis(benzoylisopropylideneiminato)]

formation of the nitro complex $[Co(en)_2(CH_3CN)(NO_2)](ClO_4)_2$. In the case of Schiff's base complexes, $[L_4Co(NO)]$, the mechanism given in Eqs. (12)–(15) has been advanced [47] for the oxidation of coordinated nitric oxide.

III. Nitrosation of Organic Compounds

Nitrosation may be defined as the introduction of an —N=O group in an organic molecule with or without displacement of a hydrogen atom. Nitrosation reactions may be carried out at carbon, nitrogen, or oxygen centers and are referred to in the literature [48] as *C*-nitrosation, *N*-nitrosation, and *O*-nitrosation, respectively. The nitroso compounds formed in nitrosation reactions are useful in organic chemistry as (a) a source of carbonium ions in the deamination of primary amines, (b) a source of diazo compounds, (c) a convenient route for the conversion of oximes to the corresponding carbonyl compounds under mild conditions, (d) in the alteration of ring size, and (e) as a convenient source of unsymmetric hydrazines by the reduction of dialkylnitrosamines. Reactions (c)–(e) may be metal-catalyzed and are discussed at some length under applications of nitrosation in this chapter (Section IV).

The reagents usually employed in nitrosation reactions are the following: acidic aqueous nitrous acid, alkyl nitrites, nitrosyl halides, and nitrosyl sulfuric acid. The active nitrosating species formed with these reagents are $NO \cdot NO_2$, $NO \cdot Cl$, $NO \cdot OH_2^+$, or NO^+, the most active being the NO^+ species. The reaction involves an electrophilic attack by NO^+ on the nonbonding or polarizable p or π electrons of the organic molecule at carbon, nitrogen, or oxygen centers [49].

Nitrosation at aliphatic carbon atoms requires the presence of an electron-attracting group (usually unsaturated) on the atom adjacent to that being nitrosated. The nitrosyl ion then attacks the enolized π bond to form a nitrosyl which gives the oxime by a prototropic shift as shown in the Eq. (16).

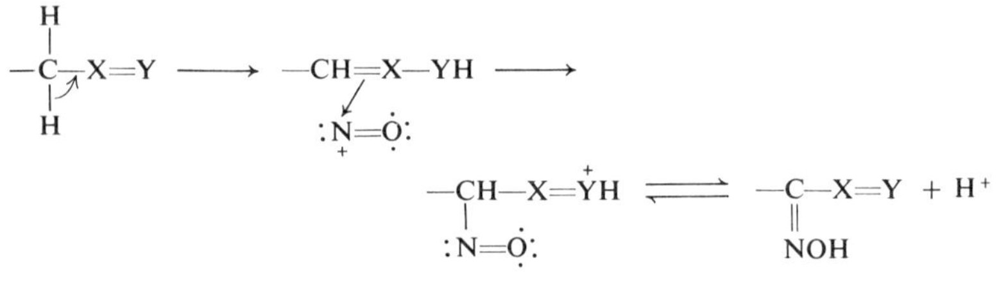
(16)

The general mechanism of *N*-nitrosation may be summarized by reaction sequence (17). In the case of primary amines, deamination takes place with the formation of an intermediate carbonium ion, **35**, and the various products formed by its subsequent reactions. In the case of secondary amines the nitrosamine **32** can be isolated. If R is aryl, **34** is stable and diazonium salts may be obtained.

O-Nitrosation results in the formation of nitrite esters from alcohols. The reaction of the hydroxyl group with NO^+ takes place either in acidic nitrous acid or with any of the nitrosating reagents described above. The mechanism of this reaction may be summarized as indicated by Eq. (18). The reaction

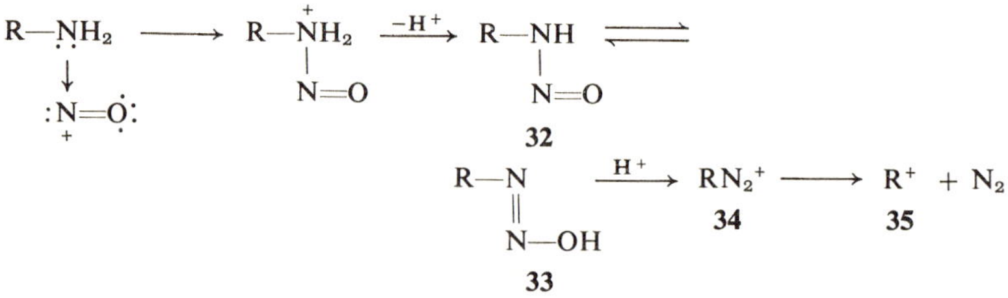

(17)

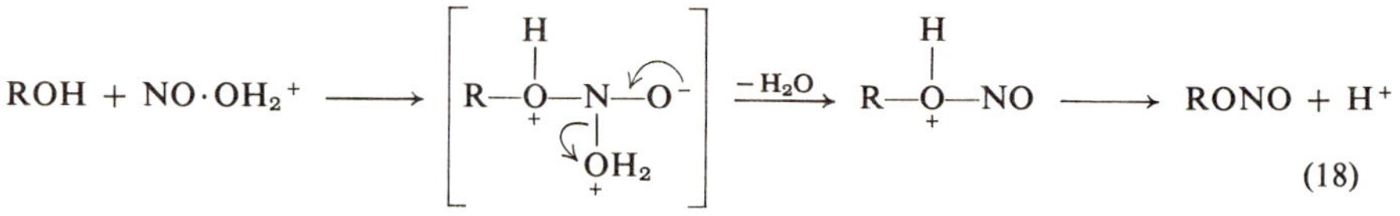

(18)

36

involves an attack of the nitrous-acidium ion on a water molecule, resulting in a replacement of the H_2O group by the ROH molecule, followed by fast dissociation of a proton. According to Austin [48] the transition state involved in the attack of $O{=}N{-}O^+H_2$ on the alcohol is probably an addition complex, **36**.

A. Metal Ion-Catalyzed Nitrosation

Nitrosation reactions involving nitrosyl chloride or nitrosyl sulfonic acid, $NOSO_3H$, as reagents are usually catalyzed by aluminum halides. In such reactions the metal ion acts as a Friedel Crafts-type catalyst, by the polarization of the ON—Cl or ON—SO_3H bonds, and the subsequent generation of the active electrophile NO^+. Such reactions do not come under the direct activation of nitric oxide by metal ions, and as such are not considered in the present chapter.

In direct nitrosation reaction with nitric oxide, the molecule may be activated by metal ions in a manner analogous to activation of carbon monoxide in carbonylation reactions. As in the case of the activation of carbon monoxide, nitric oxide is activated without the rupture of the nitrogen–oxygen bond and is inserted in an organic molecule through reaction with a coordinated metal nitrosyl. Metal ions studied as catalysts for homogeneous nitrosation involving nitric oxide are copper (II) [50–53], iron(II) [51,54], manganese(II) [51,53], and nickel(II) [51,53]. Heterogeneous catalysts [51] such as palladium on carbon, nickel on carbon, and rhodium on carbon are also effective in the nitrosation of secondary amines, but only at higher temperatures (which must still be below the decomposition temperatures of the corresponding nitrosamine). Secondary amines that may be employed for catalytic nitrosation are aliphatic amines such as dimethylamine, dipropylamine, and their derivatives containing the substituents hydroxyl, halo, cyano, carbonyl, carboxyl, and nitro. The metal ions employed as homogeneous catalysts were used in the form of their sulfates, chlorides, and pyridinium chlorides. Nitric oxide was introduced either as the pure gaseous compound, or diluted with a stream or nitrous oxide. The pressure employed falls in the range of 7–28 atm depending on the substrate used. The temperature of the reaction may range from 30° to 96°C and the yields of the nitrosamine obtained vary from 70 to 95%, depending on the catalyst and the substrate. Because of the variation of experimental conditions in the patent literature, the relative catalytic activities of the metal ions used cannot be established with certainty, but in cases where copper(II) was used as the catalytic metal, the yields seem to be higher than when other homogeneous or heterogeneous catalysts were employed.

Recently Minisci and Galli [54] described the use of iron(II) as a catalyst in the nitrosation of aliphatic secondary amines in methanol solution. A solution of ferrous sulfate in methanol was saturated with nitric oxide and the secondary alkylamine chloride was added with stirring to the resulting solution over a period of 20 minutes. Water was then added to the solution and the nitrosamine was extracted with ether. With *sec*-butylamine as the starting material, the yield of dibutylnitrosamine was 70%. The compounds *N*-nitrosodiethylamine, *N*-nitrosopiperidine, and *N*-nitrosomorpholine were prepared in a similar manner.

Cobalt(III) compounds have not yet been used as catalysts in nitrosation reactions, but there is no reason to believe that cobalt nitrosyls should not be effective intermediates for this purpose.

In cases of homogeneous catalytic nitrosation in the presence of metal ions, the mechanism of nitrosation is not very clearly understood. On the basis of the close analogy between nitric oxide and carbon monoxide, it is suggested that the metal ion-catalyzed nitrosation reaction may involve the formation of a mixed-ligand metal nitrosyl complex with the organic compound, as the secondary ligand, followed by insertion of nitric oxide into the organic molecule.

B. Kinetics and Mechanism of Metal Ion-Catalyzed Nitrosation

Brackman and Smit [55,56] have studied the homogeneous catalytic nitrosation of diethylamine (*N*-nitrosation) [55] and of primary alcohols (*O*-nitrosation) [56] by nitric oxide with copper(II) ion as catalyst. The stoichiometries of these reactions are given by Eqs. (19) and (20). In the

$$4\,NO + 2\,R_2NH \longrightarrow N_2O + 2\,R_2N\cdot NO + H_2O \tag{19}$$

$$4\,NO + 2\,ROH \longrightarrow N_2O + 2\,RONO + H_2O \tag{20}$$

nitrosation of secondary amines the solvent used was methanol. At high concentration of the secondary amines and at temperatures in the range 0°–30°C, reaction with the amine is favored over reaction with the solvent, and the nitrosamine is the main product. At lower temperatures, however, reaction with the alcohol predominates, and the alkyl nitrite is the major product. There seems to be competition for NO^+ between the two nucleophiles, secondary amine, and alcohol, the product being controlled by the reaction conditions, as indicated.

The copper(II)-catalyzed nitrosation of diethylamine was found [55] to be half-order with respect to the amine, second order with respect to nitric oxide concentration, and approximately first order with respect to cupric ion concentration. The observed kinetics was explained by the mechanism [55] given in Eqs. (21)–(27).

Any nitrosation reaction involving NO may be described as the disproportionation of the dimeric nitric oxide into NO^- and NO^+. The species NO^- is converted into N_2O, and NO^+ nitrosates the available nucleophile. Copper(II) enters as a catalyst in the reduction of nitric oxide to NO^- and the oxidation of nitric oxide to NO^+, passing through the 1+ and 2+ oxidation states during each cycle of the catalytic process. The free metal ions illustrated in Eqs. (21)–(27) as catalytic species may be 1:1 or 2:1 complexes of the amine or the solvent with Cu(I) and Cu(II). In the rapidly established preequilibrium reaction nitric oxide forms the dimer, which reacts [Eq. (22)] with the primary amine to yield an adduct, $R_2N_3O_2$, **37**, the nature of which was demonstrated by the work of Drago *et al.* [57–60], in which the compound was isolated from an ether solution at low temperature. The participation of Drago's adduct in the Cu(II)-catalyzed nitrosation of secondary amines has been supported [55] by its accelerating effect on the rate of disappearance of nitric oxide in the reaction and the observed conversion by the adduct of the Cu(I) complex to the Cu(II) complex. It was not possible, however, to directly demonstrate the formation of the adduct in the nitrosation reaction because of its instability in the absence of free amine and the inability of the adduct to form insoluble metal salts.

Equation (23) in the above mechanistic sequence involves the catalytic nitrosation of R_2NH by nitric oxide. This reaction may take place by a combination of fast preequilibrium and rate-determining steps, two possible sequences being indicated by Eqs. (28) and (29), and Eqs. (30)–(32). The rate

$$2\,NO \xrightleftharpoons{K} N_2O_2 \qquad (21)$$

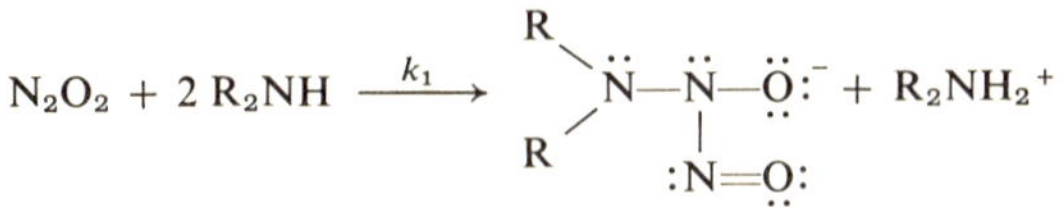

(22)

37

$$Cu^{2+} + NO + R_2NH \xrightarrow{k_5} Cu^+ + H^+ + R_2\ddot{N}{-}\ddot{N}{=}\ddot{O} \qquad (23)$$

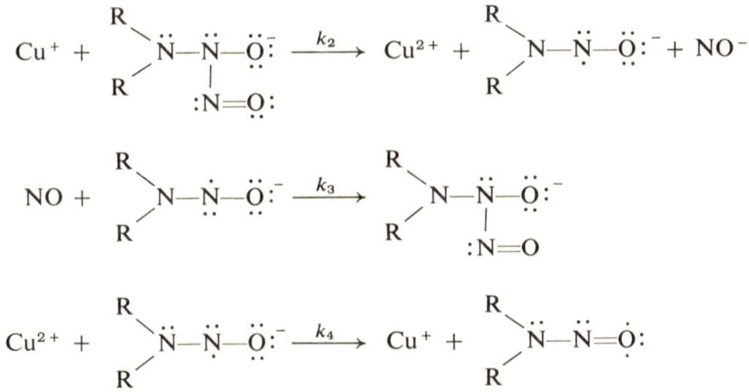

(24)

(25)

(26)

$$2\,NO^- \xrightarrow{2\,H^+} 2\,HNO \longrightarrow H_2N_2O_2 \longrightarrow H_2O + N_2O \qquad (27)$$

$$Cu^{2+} + NO \rightleftharpoons Cu^{\pm}{-}{-}N{\equiv}O^+ \qquad (28)$$

$$Cu^{\pm}{-}{-}N{\equiv}O^+ + R_2NH \xrightarrow{\text{fast}} Cu^+ + H^+ + R_2N\cdot NO \qquad (29)$$

$$Cu^{2+} + R_2NH \rightleftharpoons [Cu{-}{-}{-}NR_2H]^{2+} \qquad (30)$$

$$[Cu{-}{-}{-}NR_2H]^{2+} + NO \rightleftharpoons [{:}O^+{\equiv}N{-}{-}{-}Cu^{\pm}{-}{-}NR_2H]^{2+} \qquad (31)$$

38

$$[{:}O^+{\equiv}N{-}{-}{-}Cu^{\pm}{-}{-}NR_2H]^{2+} \longrightarrow Cu^+ + H^+ + R_2\ddot{N}{-}\ddot{N}{=}\ddot{O}{:} \qquad (32)$$

of nitric oxide disappearance in the presence of copper(II) before the addition of amine suggests that about 20–30% of the copper(II) present occurs as the copper(I) nitrosyl. Formation of copper(I) nitrosyl from copper(II) and nitric oxide in alcoholic solution is substantiated by the work of Fraser and Dasent [33,34]. According to the second alternative scheme, nitrosation of 1:1 or 2:1 amine complex of copper(II) may take place by forming a mixed-ligand complex, **38**, and subsequent nitrosation of the amine occurs by insertion of nitric oxide in the Cu—N bond. Kinetically it is not possible to distinguish between the two schemes shown. The reoxidation of copper(I) to copper(II) by Drago's complex in Eq. (24) was substantiated [55] by independent studies with copper(I). The free radical $R_2N\text{-}N\dot{O}^-$, either reacts with copper(II) to yield copper(I) or with nitric oxide to regenerate the Drago adduct, as is indicated by Eqs. (25) and (26). Assuming steady-state concentrations of $R_2N{-}\dot{N}{-}O^-$, $R_2N_3O_2^-$, and copper(I), expression (33) for the overall rate of nitric oxide disappearance was obtained [55].

$$-\frac{d[\mathrm{NO}]}{dt} = 2\left(\frac{Kk_1k_3k_5}{k_4}\right)^{1/2}[\mathrm{NO}]^2[\mathrm{R_2NH}]^{1/2} \qquad (33)$$

Rate equation (33) is in accord with the observed dependence of the reaction rate on nitric oxide and amine concentrations.

Alkyl nitrites are obtained in the nitrosation with nitric oxide either at high pressure or at low temperatures, according to Eq. (34). The nitrosation of alcohols can take place according to Eqs. (28) and (35). This mechanism is

$$\mathrm{Cu^{2+} + ROH + NO \longrightarrow Cu^+ + RONO + H^+} \qquad (34)$$

$$\mathrm{Cu^{2+} + NO \rightleftharpoons Cu^{\pm}{-}{-}N{\equiv}O^+} \qquad (28)$$

$$\mathrm{Cu^{\pm}{-}{-}N{\equiv}O^+ + RO^- \longrightarrow RONO + Cu^+} \qquad (35)$$

similar to the reaction sequence described above [Eqs. (28) and (29)] for the nitrosation of a secondary amine. Alternatively, the reaction may occur via reaction of nitric oxide with a copper(II) alkoxide, Eq. (36).

$$\mathrm{Cu^{2\pm}{-}{-}\bar{O}R + NO \longrightarrow Cu^+ + RONO} \qquad (36)$$

In the presence of amines, copper(II) alkoxide may be formed by reaction of the solvent with the copper(I) complex, **39**, of the Drago adduct, in accordance with reaction scheme (37). The NO^- formed is stabilized by

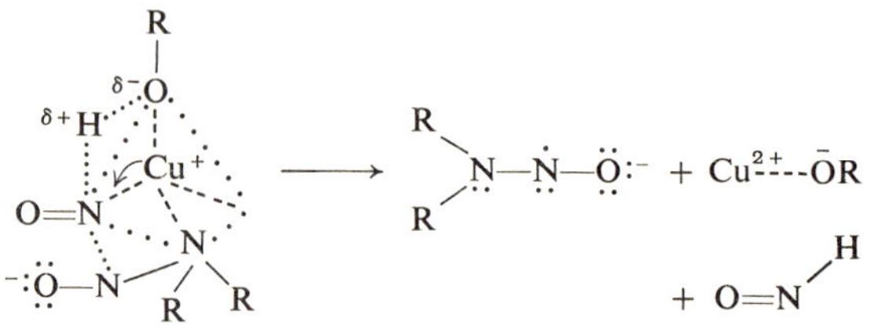

(37)

39

combination with a hydrogen ion, and the copper(II) is coordinated by the methoxide ion, which is displaced by excess amine in accordance with Eq. (38). The amine complex then reacts with nitric oxide to give $R_2N \cdot NO$

$$\mathrm{Cu^{2\pm}{-}{-}OR^- + R_2NH \longrightarrow Cu^{2\pm}{-}{-}NHR_2 + RO^-} \qquad (38)$$

and copper(I) [Eqs. (31) and (32)]. In the presence of excess nitric oxide and copper(II), the alkoxide complex may react directly with nitric oxide before the displacement of alkoxide by amine, in accordance with Eq. (36), and form copper(I) and the alkyl nitrite directly.

According to the reaction schemes described above, copper(II)-catalyzed *N*- or *O*-nitrosation seems to involve formation of mixed amine-nitric oxide or alkoxy–nitric oxide complexes of copper(II) as intermediates, followed by electron transfer from nitric oxide to copper(II) and migration of amine or alkoxide to $—N{\equiv}O^+$ resulting in the formation of nitrosamine or alkyl nitrite. If these nitrosation reactions are compared with carbonyl insertion reactions, in which alkyl groups migrate to coordinated carbon monoxide to form acyl groups, it appears that nitrosation may be formally regarded as a nitric oxide insertion reaction. At present, the mechanisms of these reactions are not known in detail. Since there are many differences between reactions of nitric oxide and carbon monoxide, and the products of nitric oxide insertion do not generally remain coordinated to the metal ion, use of the term "nitric oxide insertion" is primarily a convenience and does not imply parallel reaction mechanisms of these two types of catalysis. Considerable further work is needed to elucidate the mechanisms and determine the intermediates involved in metal ion-catalyzed nitrosation.

C. Catalysis of Nitrosation of Organometallic Compounds

Very interesting classical examples of the nitrosation of organometallic compounds are the reactions of dimethyl- or diethylzinc with nitric oxide, discovered by Frankland [61,62]. Dimethyl- and diethylzinc react with nitric oxide to give products which Frankland formulated as $R_2Zn{\cdot}Zn{\cdot}[ON(NO)R]_2$. On hydrolysis, these compounds give salts of the type $Zn[ON(NO)R]_2$.

Abraham *et al.* [63] have recently reported that dipropylzinc and Grignard reagents such as RMgCl, where R is butyl or isopropyl, absorb nitric oxide when shaken with the gas in ether or hydrocarbon solution. Depending on the organometallic compound employed, either 1 or 2 moles of nitric oxide are absorbed per mole of metal ion. In the case of the Grignard reagents, 1 mole of nitric oxide is absorbed per mole of magnesium compound, and in the case of diisopropylzinc or diethylaluminum chloride, 2 moles of nitric oxide are absorbed per mole of zinc or aluminum alkyl. The mechanism given in Eq. (39) is proposed for the reaction of nitric oxide with the metal dialkyl. Coordination of nitric oxide through oxygen with the formation of the unstable complex **40** was proposed, followed by a 1,3-shift of the alkyl group to form the isomeric complex **41**. Reaction of a second mole of nitric oxide then takes place to give the final product **42**, which on hydrolysis gives alkyl derivatives of *N*-nitrosohydroxylamine in good yields. The product from dipropylzinc was reported [63] to be RZnON(NO)R. It is oxidized slowly to give mainly ROZnON(NO)R and Zn[ON(NO)R]$_2$.

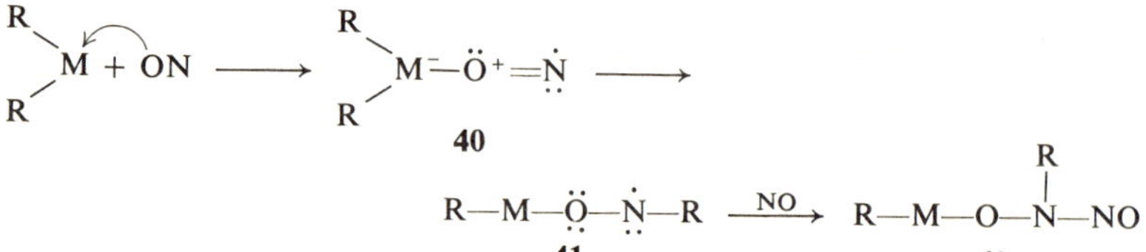

40 **41** **42** (39)

The structures of some of the intermediate zinc complexes postulated in reaction (39) were confirmed by their PMR spectra. The proposal of oxygen coordination to the metal ion is based on the absence of the infrared band characteristic of the nitrosyl (—NO$^+$) group. Oxygen coordination for certain cobalt nitrosyl complexes is indicated by the work of Yamada *et al.* [64]. Nitrosyl insertion reactions of organometallic compounds which give *N*-nitrosoalkyl hydroxylamine derivatives are relatively complex and have not been extensively explored. The mechanisms proposed above are mainly speculative and further work is required to provide a better understanding of these reactions.

Reactions of nitric oxide with aryl Grignards and diaryl metals are very complex, one of the products in such reactions being diphenyl. Diphenylcadmium, for example, was reported [65] to give phenyldiazonium nitrate, $C_6H_5N{=}N\cdot NO_3$, on treatment with nitric oxide. The mechanisms of the reactions of nitric oxide with metal aryls and aryl Grignard reagents have not yet been proposed.

IV. Applications of Metal-Catalyzed Nitrosation Reactions

Metal-catalyzed nitrosation reactions offer considerable promise for synthesis and industrial applications. The most important nitrosation reaction from a synthetic viewpoint is that of the tertiary amines, which produce nitrosamines. The lability of the N—N bond in nitrosamines may effect rearrangement of the NO group to the corresponding *C*-nitroso derivatives [66], which may then take part in a variety of cyclization reactions that result in the formation of new heterocyclic ring systems. Another possible application of nitrosamines is their facile reduction to the corresponding unsymmetric dialkylhydrazines that are otherwise very difficult to prepare. Ring expansion by nitrosation (Demjanov reaction) is also the basis for a promising synthetic method in which the *C*-nitrosation reaction may be advantageously exploited. Careful study of metal ion catalysis of some of these reactions should prove useful in improving yields and obtaining higher specificities. Some representative applications of metal-catalyzed nitrosation reactions are discussed in the following sections.

A. Base-Catalyzed Rearrangement of Nitrosamines

Daeniker [66] has reported the base-catalyzed isomerization of a substituted nitrosamine to the oxime derivative, as indicated by reaction (40).

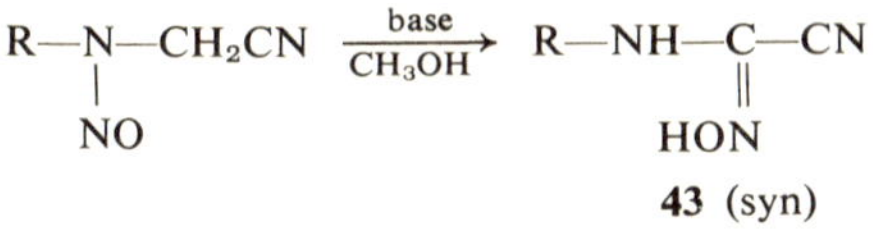

(40)

(R = phenyl, *p*-chlorophenyl, *p*-methoxyphenyl, benzyl, methyl, isopropyl, and *n*-butyl)

The *C*-isonitroso derivative **43** was identified by the products obtained when it was subjected to acid hydrolysis. In the case where R = benzyl (*N*-benzylnitrosoacetonitrile), the reaction gave *N*-benzyl-α-isonitrosoaminoacetonitrile, **44**, which on acid hydrolysis with hydrogen bromide gave the amide **45**. On reduction with hydrogen, the α-imine **46** was formed and was converted by hydrolysis with HCl to the *N*-benzyloxamide, **47**.

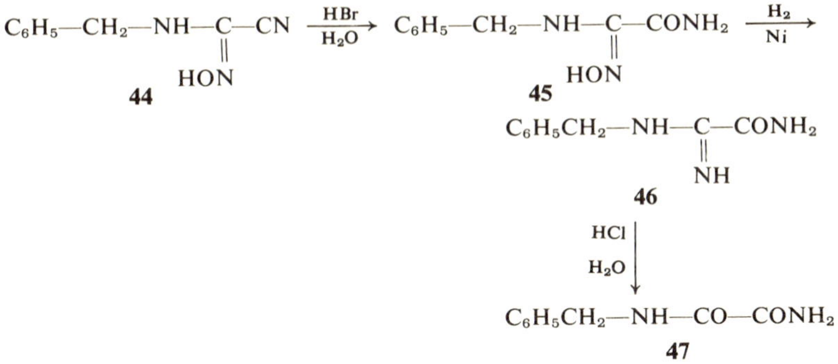

(41)

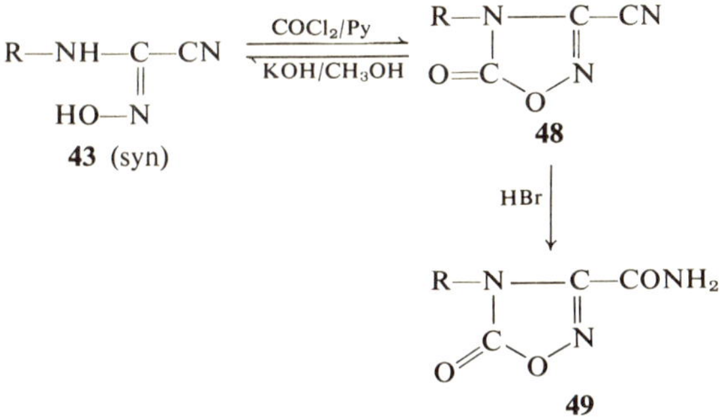

(42)

On treatment with phosgene in pyridine, **43** gave 4-substituted-3-cyano-1,2,4-oxadiazolinone, **48**, which was converted to the corresponding amide,

49, by hydrolysis with hydrobromic acid. Methanolic potassium hydroxide converted **49** to an oxime (a *C*-nitroso compound by a decarboxylation reaction.

Base-catalyzed rearrangements such as the one described above [Eq. (37)] require the presence of a strong electron-withdrawing group in the nitrosamine. The mechanism of Eq. (43) was suggested by Daeniker [66]. This rearrange-

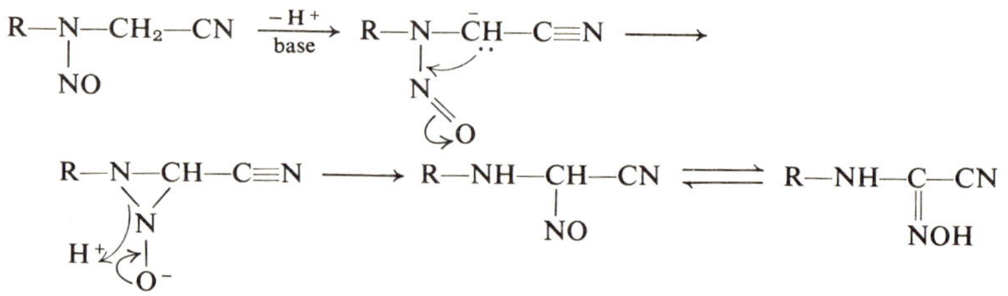

(43)

ment affords a useful pathway for the conversion of *N*-nitroso derivatives into *C*-nitroso compounds (oximes), which can then be converted to ketones by further reaction with nitric oxide. The mechanism given in Eq. (44), suggested by Austin [48], involves electrophilic attack by NO^+ on the π orbital of the oxime. After combination attachment of the nitroso group with the nitrogen atom of the oxime, the resulting carbonium ion, **50**, reacts with the solvent, and the hydroxy nitroso compound thus formed, **51**, undergoes rapid decomposition into the carbonyl compound and $H_2N_2O_2$, which in turn decomposes into nitrous oxide and water.

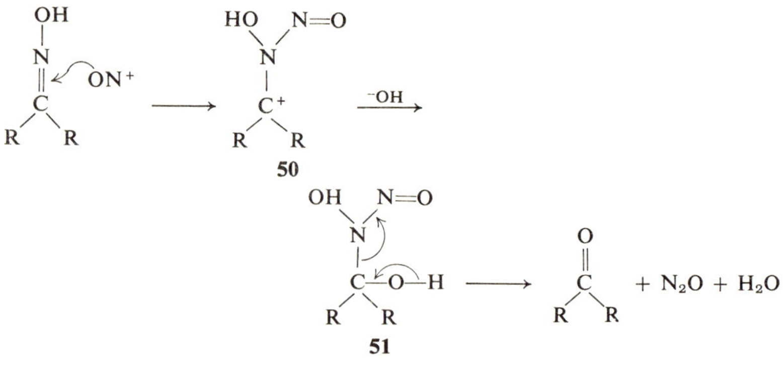

(44)

B. Synthesis of Unsymmetric Hydrazines

One of the most useful synthetic applications of dialkylnitrosamines is reduction to the corresponding unsymmetric hydrazines. The nitrosamines are hydrogenated for this purpose by a variety of heterogeneous catalysts, either with molecular hydrogen under pressure [54,67–73] or with hydrogen produced *in situ* by reaction of zinc or aluminum with sulfuric acid [74,75] or of lithium aluminum hydride in dry ether [76]. The use of small quantities of ferrous ion improves the yield [40,73] of dialkylhydrazine, probably by homogeneous catalysis of the hydrogenation reaction. Table III presents a summary of the catalysts used, conditions of the reactions, and the yields of the unsymmetric hydrazines obtained.

C. Ring Expansion Reactions

Moriconi *et al.* [78] have reported interesting synthetic reactions involving ring expansion by nitrosation. As indicated by reaction scheme (45), addition of *n*-butyl nitrite to a solution of 2-methyl-1-indanone or 2-ethyl-1-indanone, **52**, in toluene, followed by treatment with 3 *N* hydrochloric acid in ethyl acetate at 0°C, produced the cyclic hydroxamic acid 2-hydroxy-3-alkyl isocarbostyril, **54**, in 65–68% yield. At lower acid concentrations or by addition of acid to a mixture of *n*-butyl nitrite and **52**, the intermediate 2-alkyl-2-nitroso-1-indanone, **53**, was obtained in good yields. This compound is the precursor that gives rise to the ring-expanded cyclic hydroxamic acid, **54**, both by acid- and base-catalyzed reactions. The identity of **54** was established by an independent synthesis of the compound involving the

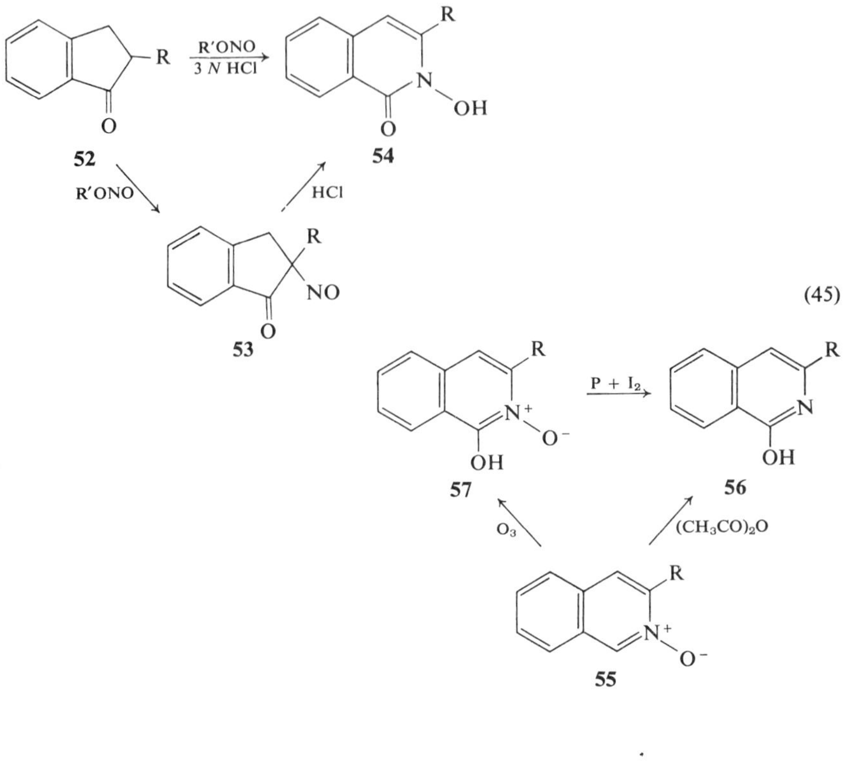

(45)

TABLE III

REDUCTION OF DIALKYLNITROSAMINES TO UNSYMMETRIC HYDRAZINES BY METAL COMPLEXES

Reactant	Catalyst	Solvent	H_2 pressure	T (°C)	Yield (%)	References
$(CH_3)_2N_2O$	Zn, Al + H^+ + $HgCl_2$, H_2PtCl_4 or $HgSO_4$	H_2O	1 atm	101	86	74
$(CH_3)_2N_2O$	$LiAlH_4$	Ether	1 atm	—	—	76
$(CH_3)_2N_2O$	5% Pd + carbon, 0.5 mmole Fe^{2+}	H_2O	40–50 lb/inch^2	45	—	73
$(CH_3)_2N_2O$	5% Pd + carbon + urea	H_2O	1000 lb/inch^2	40	85.5	72
$(CH_3)_2N_2O$	Supported Pt or Pd catalyst on alumina	H_2O	200 lb/inch^2	31	94	54,67,68
C_1–C_3 Nitrosamines	5% Pd + carbon, 0.1–0.2 mmoles Fe^{2+}	C_2H_5OH + OH^-	1000 lb/inch^2	—	99	77
C_1–C_4 Nitrosamines	Rhodium + carbon, palladium + carbon, Ca(II), La(III), or Mg(II)	95% C_2H_5OH 5% H_2O	1000	30–80	20–80	69,70
C_1–C_4 Nitrosamines	5% Pd + carbon	H_2O	450–500 lb/inch^2	40	92–95	71
$(C_3H_7)_2N_2O$	Al + I_2 + H^+ + H_2	C_2H_5OH	1 atm	60–65	83	75
$(C_3H_7)_2N_2O$	5% Pd + carbon + urea	H_2O	1000 lb/inch^2	40	86	72
$(C_6H_5CH_2)_2N_2O$	Al + I_2 + H^+ + H_2	C_2H_5OH	1 atm	60–65	80	75
$(C_3H_7)_2N_2O$	5% Pd—C + urea	H_2O	1000 lb/inch^2	40	100	54,67,68
C_1–C_3 Nitrosamines	5% Pd—C + 0.1–0.2 mmoles Fe^{2+}	Ethanol + OH^-	—	—	99	40
$(CH_3)_2N_2O$	Supported Pt or Pd catalyst on alumina	H_2O	200 lb/inch^2	31	94	54,67,68
C_1–C_4 Nitrosamines $(C_8H_{12})_2N_2O$	Rh—C, Pd—C + Ca(II), La(III), or Mg(II)	95% Ethanol 5% water	100 lb/inch^2	70–80	~99	69,70

ozonization of 3-methylisoquinoline-2-oxide, **55**. Reduction of **54** in glacial acetic acid with iodine and red phosphorus led to 3-methylisocarboxystyril, **56**, which was identical to an authentic sample of **56** obtained by refluxing **55** with acetic anhydride. Spectral evidence suggests that the cyclic hydroxamic acid exists in form **54** rather than its tautomer **57**.

The mechanism of ring expansion of 2-alkyl-1-indanone on nitrosation is presented in the reaction sequence (46), formulas **52–61**. The 2-alkyl-1-indanone on nitrosation gives 2-alkyl-2-nitroso-1-indanone, **53**. Protonation of **53** gives successively the carbonium ion **58** and the oxime intermediate **59**, which then undergoes ring expansion by nitric oxide insertion to give the cyclic hydroxamic acid **54**. The nitrosoindanone can also be converted to the final product by a base-catalyzed rearrangement through intermediates **60** and **61**. Ring expansion by nitric oxide insertion provides a convenient

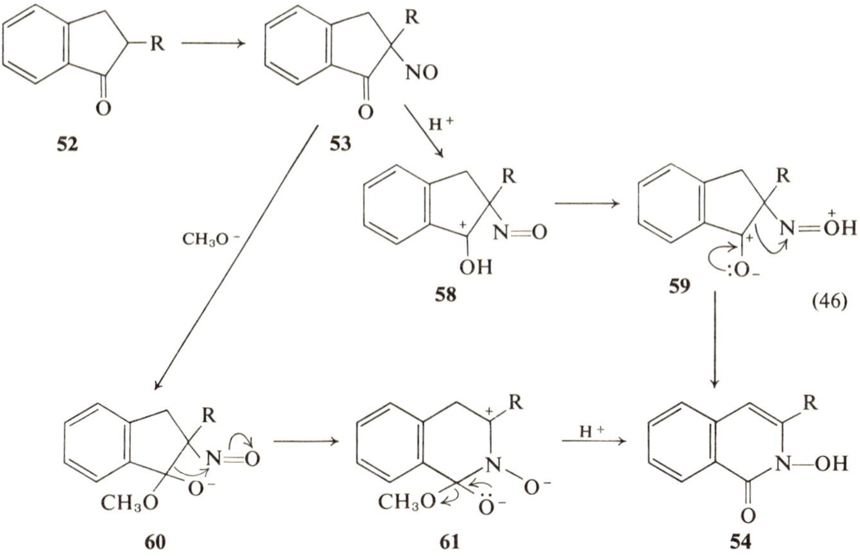

route for the preparation of heterocyclic compounds. Though metal ion catalysis has not been reported thus far, it may be expected that metal nitrosyls can expand the scope and perhaps the specificities of these reactions and can provide interesting opportunities for the development of new synthetic methods.

References

1. H. Brion, C. Moser, and M. Yamazaki, *J. Chem. Phys.* **30**, 673 (1959).
2. R. J. Irving, *Rec. Chem. Prog.* **26**, 115 (1965).
3. J. Berkowitz, *J. Chem. Phys.* **30**, 858 (1959).

4. D. J. Hodgson, N. C. Payne, J. A. McGinnety, R. D. Pearson, and J. A. Ibers, *J. Amer. Chem. Soc.* **90**, 4486 (1968).
5. D. J. Hodgson and J. A. Ibers, *Inorg. Chem.* **7**, 2345 (1968).
6. D. J. Hodgson and J. A. Ibers, *Inorg. Chem.* **8**, 1282 (1969).
7. D. A. Snyder and D. L. Weaver, *Inorg. Chem.* **9**, 2760 (1970).
8. D. M. P. Mingos and J. A. Ibers, *Inorg. Chem.* **9**, 1105 (1970).
9. D. M. P. Mingos, W. T. Robinson, and J. A. Ibers, *Inorg. Chem.* **10**, 1043 (1971).
10. D. M. P. Mingos and J. A. Ibers, *Inorg. Chem.* **10**, 1035 (1971).
11. C. S. Pratt, B. A. Coyle, and J. A. Ibers, *J. Chem. Soc., A* p. 2146 (1971).
12. C. G. Pierpont, D. G. Van der Veer, W. Durland, and R. Eisenberg, *J. Amer. Chem. Soc.* **92**, 4760 (1970).
13. C. G. Pierpont and R. Eisenberg, *Inorg. Chem.* **11**, 1088 (1972).
14. J. M. Waters and K. W. Whittle, *Chem. Commun.* p. 518 (1971).
15. J. H. Enemark, *Inorg. Chem.* **9**, 1952 (1971).
16. C. G. Pierpont and R. Eisenberg, *Inorg. Chem.* **11**, 1094 (1972).
17. P. T. Manoharan and W. C. Hamilton, *Inorg. Chem.* **2**, 1043 (1963).
18. J. H. Enemark and J. A. Ibers, *Inorg. Chem.* **7**, 2339 (1968).
19. B. A. Frenz, J. H. Enemark, and J. A. Ibers, *Inorg. Chem.* **8**, 1288 (1969).
20. A. Domenicano, A. Vaciago, L. Zambonelli, P. L. Loader, and L. M. Venanzi, *Chem. Commun.* p. 476 (1966).
21. A. Tullberg and N. G. Vannerberg, *Acta Chem. Scand.* **20**, 1180 (1966).
22. C. Strouse and B. I. Swanson, *Chem. Commun.* p. 55 (1971).
23. C. G. Pierpont, A. Pucci, and R. Eisenberg, *J. Amer. Chem. Soc.* **93**, 3050 (1971).
24. J. H. Enemark, M. S. Quinby, L. L. Reed, M. J. Stenck, and K. K. Walters, *Inorg. Chem.* **9**, 2397 (1970).
25. G. R. Davies, J. A. J. Jarvis, B. T. Kilbourn, R. H. B. Mais, and P. G. Owston, *J. Chem. Soc., A* p. 1275 (1970).
26. D. M. P. Mingos and J. A. Ibers, *Inorg. Chem.* **70**, 1479 (1971).
27. G. R. Clark, K. R. Grundy, W. R. Roper, J. M. Waters, and K. R. Whittle, *Chem. Commun.* p. 119 (1972).
28. S. H. Simonsen and M. H. Mueller, *J. Inorg. Nucl. Chem.* **27**, 309 (1965).
29. P. T. Manoharan and H. B. Gray, *Inorg. Chem.* **5**, 823 (1966).
30. C. G. Pierpont and R. Eisenberg, *J. Amer. Chem. Soc.* **93**, 4905 (1971).
31. P. Finn and W. L. Jolly, *Inorg. Chem.* **11**, 893 (1972).
32. B. F. G. Johnson and J. A. McCleverty, *Progr. Inorg. Chem.* **7**, 277 (1966).
33. R. T. M. Fraser and W. E. Dasent, *J. Amer. Chem. Soc.* **82**, 348 (1960).
34. R. T. M. Fraser and W. E. Dasent, *J. Inorg. Nucl. Chem.* **17**, 265 (1961).
35. M. Mercer and R. T. M. Fraser, *J. Inorg. Nucl. Chem.* **25**, 325 (1963).
36. W. P. Griffith, J. Lewis, and G. Wilkinson, *J. Chem. Soc., London* p. 3993 (1958).
37. G. D. Sirotkin, *Zh. Neorg. Khim.* **1**, 1750 (1956).
38. P. Silverstroni and L. Ceciarelli, *J. Amer. Chem. Soc.* **83**, 3905 (1961).
39. P. Silverstroni and L. Ceciarelli, *Ric. Sci. Rend. Sez. A*[2] **2**, 121–29 (1962).
40. J. Lewis, R. J. Irving, and G. Wilkinson, *J. Inorg. Nucl. Chem.* **7**, 32 (1958).
41. J. P. Collman, P. Farnham, and G. Dolcetti, *J. Amer. Chem. Soc.* **93**, 1788 (1971).
42. T. J. Meyer, J. B. Godwin, and N. Winterton, *Chem. Commun.* p. 872 (1970).
43. J. B. Godwin and T. J. Meyer, *Inorg. Chem.* **70**, 2150 (1971).
44. F. Bottomley and J. R. Crawford, *Chem. Commun.* p. 200 (1971).
45. F. Bottomley and J. R. Crawford, *Proc. Int. Conf. Coord. Chem. 14th* p. 277 (1972).
46. K. R. Grundy, C. A. Reed, and W. R. Roper, *Chem. Commun.* p. 1501 (1970).
47. S. G. Clarkson and F. Basolo, *Chem. Commun.* p. 670 (1972).

48. A. T. Austin, *Sci. Progr. (London)* **49**, 619 (1961).
49. C. C. Addison and J. Lewis, *Quart. Rev. Chem. Soc.* **9**, 115 (1965).
50. E. L. Reilley, U.S. Patent 2,749,358 (1956).
51. E. L. Reilley, German Patent 1,085,166 (1960).
52. E. L. Reilley, British Patent 867,992 (1961).
53. E. L. Reilley, British Patent 867,993 (1961).
54. F. F. Minisci and R. Galli, *Chim. Ind. (Milan)* **46**, 173 (1964).
55. W. Brackman and P. J. Smit, *Rec. Trav. Chim. Pays-Bas* **84**, 357 (1965).
56. W. Brackman and P. J. Smit, *Rec. Trav. Chim. Pays-Bas* **84**, 372 (1965).
57. R. S. Drago and F. E. Paulik, *J. Amer. Chem. Soc.* **82**, 96 (1960).
58. R. S. Drago and B. R. Karsteller, *J. Amer. Chem. Soc.* **83**, 1819 (1961).
59. R. S. Drago, R. O. Ragsdale, and D. P. Eyman, *J. Amer. Chem. Soc.* **83**, 4337 (1961).
60. R. S. Drago, "Advances in Chemistry Series," No. 36, pp. 143–149. American Chemical Society, Washington, D.C., 1962.
61. H. Frankland, *Justus Liebigs Ann. Chem.* **99**, 345 and 369 (1856).
62. H. Frankland, *Phil. Trans. Roy. Soc. London* **147** (1957).
63. M. H. Abraham, J. H. N. Garland, J. A. Hill, and L. F. Larksworthy, *Chem. Ind. (London)* p. 615 (1962).
64. S. Yamada, N. Nichikawa, and R. Tsuchida, *Bull. Chem. Soc. Jap.* **33**, 930 (1960).
65. A. N. Nesmeyanov and L. G. Markarova, *J. Gen. Chem. USSR* **7**, 2649 (1937); *Chem. Abstr.* **32**, 2095 (1938).
66. H. V. Daeniker, *Helv. Chim. Acta* **47**, 33 (1964).
67. W. P. Moore, Jr. and D. Pickens, U.S. Patent 316,993 (1965).
68. W. P. Moore, Jr. and W. C. Sierichs, U.S. Patent 3,167,588 (1965).
69. G. W. Smith and D. N. Thatcher, *Ind. Eng. Chem., Prod. Res. Develop.* **1**, 117 (1962).
70. D. N. Thatcher, U.S. Patent 3,102,887 (1963).
71. J. B. Tindall, U.S. Patent 3,143,095 (1964).
72. J. B. Tindall, U.S. Patent 3,178,479 (1965).
73. W. B. Tuemmler and H. J. S. Winkler, U.S. Patent 2,979,505 (1961).
74. P. F. Derr, U.S. Patent 2,961,467 (1960); *Chem. Abstr.* **55**, 9280 (1961).
75. H. G. Kazmirowski, H. Goldhann, and E. Carstens, German Patent 1,155,138 (1963).
76. G. Neurath, B. Pirmann, and M. Dunger, *Chem. Ber.* **97**, 1631 (1964).
77. D. A. Lima, U.S. Patent 3,154,538 (1964).
78. E. J. Moriconi, F. J. Creegan, C. K. Donovan, and F. A. Spano, *J. Org. Chem.* **28** 2215 (1963).

Appendix I

Formulas, Bonding, and Formal Charges

The conventions used in representing structural formulas of metal–organic complexes are designed to show the distribution of bonding electrons in a reasonably consistent fashion, as well as to make possible completely unambiguous "electron-bookkeeping" of bonding electrons. An ideal method of representation is impossible for the wide variety of metals and ligands encountered in this book, since seeming inconsistencies will develop for metals or donors at the extremes of the reactivity scale. Thus a consistent way of representing a donor–acceptor (coordinate) bond for electropositive metal ions (e.g., Al^{3+}), with the electrons localized on the electronegative atom, would be misleading for the complexes of highly electronegative metal ions (e.g., Hg^{2+}, Pt^{2+}) for which linkages with the same donor would be highly covalent, with sometimes very little polarity. Thus a consistent way of writing coordinate bonds will lead to inconsistencies in representation of formal charge distribution, which the reader must continually keep in mind, and make allowances for, if this (or any similar) book is to be interpreted correctly.

The objective of accounting for all bonding electrons by the way that bonds and formal charges are represented also provides an automatic method of obtaining the charges of all ionic species. Thus the total charge on a complex ion is equal to the sum of the formal charges indicated at various positions in the formulas. In many instances the total charge is not indicated, but may be deduced by summation of the indicated formal charges. For covalent (homopolar) bond representations, the bonding electrons are (arbitrarily) assumed to be shared equally between the two atoms involved. Similarly, for coordinate (donor–acceptor) linkages, the formal charges are determined by assuming that the bonding electrons are localized on the electronegative donor atom.

The conventions employed for homopolar and coordinate bonding in structural and graphic formulas are shown in the following tabulations. While the isoelectronic NO^+ and CO groups are shown with solid bonds to metal, the negative NO^- group is considered a donor ligand and is represented with dashed lines for coordinated bonding to metal ions.

COVALENT BONDS (SOLID LINE)

Type	Example	
Carbon–carbon	$H_3C—CH_3$	Solid line
Metal hydride	M—H	Solid line
Metal alkyl (σ-bond)	$M—CH_3$	Solid line
Metal carbonyl	M—C≡O	Solid line
Metal nitrosyl	$M—N{\equiv}O^+$	Solid line

DONOR-ACCEPTOR BONDS

Type	Example	
Metal complex (with electronegative atom of ligand)	M^{n+}----Cl^- M^{n+}----NH_3	Broken line
Metal alkene or alkyne complex (π bond)	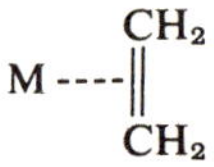	Broken line
Metal–carbon bond formed by an electronegative ion or group	^-NC----Ag^+----CN^-	Broken line
Transition state (partial bonds)	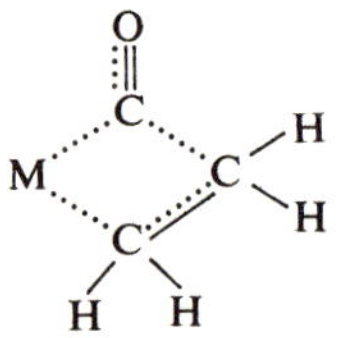	Heavy dotted line
Outline of figures (no bonds)	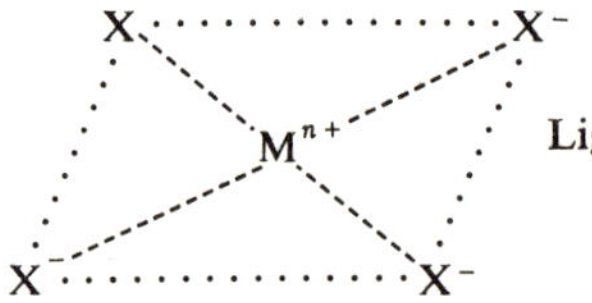	Light dotted line

More complex types of compounds are frequently represented by a simple designation, such as a single solid line to represent a π-allyl complex, as represented below.

π-Allyl complex   Resonance forms

Similarly a sandwich-type complex or a π-complex with an organic aromatic ring is usually represented by a single bond as indicated below.

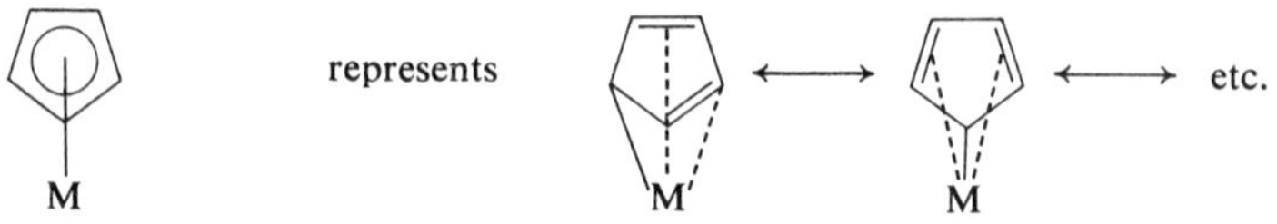

Cyclopentadienyl metal complex   Five resonance forms

Appendix II

Glossary of Terms and Abbreviations

Abbreviation	Full name or formula
Ac	Acetyl
acac	2,4-Pentanedione
AcO	Acetate
ADP	Adenosine diphosphate
$AsPh_3$	Triphenylarsine
ATP	Adenosine triphosphate
B	Base
BPh_4	Tetraphenylboron anion
$BiPh_3$	Triphenylbismuth
CDTA	Cyclohexanediaminetetraacetic acid
Cobaloxime-II	Bis(dimethylglyoximato)cobalt(II)
Cobalamine	Vitamin B_{12}
cp	π-Cyclopentadienyl
das	*o*-Phenylenebis(dimethylarsine)
depe	1,2-Bis(diethylphosphino)ethane
dias	1,2-Bis(diphenylarsino)ethane
diars	1,2-Bis(methylphenylarsino)ethane
dien	Diethylenetriamine
diphos	1,2-Bis(diphenylphosphino)ethane
diphos-2	*cis*-1,2-Bis(triphenylphosphino)ethylene
dipy	2,2-Dipyridyl
dmpe	1,2-Bis(dimethylphosphino)ethane
DMF	Dimethylformamide
DMSO	Dimethyl sulfoxide
DPN	Diphosphopyridine nucleotide
DPNH	Reduced DPN
DTPA	Diethylenetriaminepentaacetic acid
EDTA	Ethylenediaminetetraacetic acid

Glossary (*continued*)

Abbreviation	Full name or formula
en	Ethylenediamine
epr	Electron paramagnetic resonance
fac	Facial
Fenton's reagent	Hydrogen peroxide and an iron(II) salt
glygly(GG)	Glycylglycine
HEDTA	*N*-Hydroxyethylethylenediamine-*N*,*N*′,*N*′-triacetic acid
HIMDA	*N*-Hydroxyethyliminodiacetic acid
HPIP	High potential iron protein
IMDA	Iminodiacetic acid
i-pr, pr^i	Isopropyl
L	Ligand
mer	Meridional
NAD	Pyridineadenine dinucleotide
NADH	Reduced pyridineadenine dinucleotide
NMR	Nuclear magnetic resonance
n-pr, pr^n	Normal propyl
ox	Oxidized
pm	Primary
PMR	Proton magnetic resonance
PPh_3	Triphenylphosphine
PTS	Phthalocyaninetetrasulfonic acid
Py	Pyridine
rac	Racemic
red	Reduced
S	Solvent
salen	*N*,*N*′-Ethylenebis(salicylideneiminato)
$SbPh_3$	Triphenylantimony
sec	Secondary
synthesis gas	Equimolar carbon monoxide and hydrogen
t-Bu or Bu^t	Tertiary butyl
tert	Tertiary
THF	Tetrahydrofuran
tolan	Diphenylacetylene
TPN	Triphosphopyridine nucleotide
$TPNH_2$	Reduced TPN
trien	Triethylenetetraamine
Vaska's complex	Chlorocarbonylbis(triphenylphosphine)iridium(I)
Wilkinson's complex	Chlorotris(triphenylphosphine)rhodium(I)
Zeise's salt	Potassium trichloro(ethylene)platinate(II) monohydrate

Author Index

Numbers in parentheses are reference numbers and indicate that an author's work is referred to although his name is not cited in the text. Numbers in italics show the page on which the complete reference is listed.

A

Abel, E., 123(116), 124(116), *177*
Abel, K., 187, 189, *287*
Abraham, M. H., 381, 382(63), *389*
Acheson, R. H., 161(206), 163(206), 165 (206), *179*
Addison, C. C., 375(49), *389*
Adman, E., 190(79), 191, 192(79), *288*
Ainscough, E. W., 10(64, 65), 23(64, 65), 30(64), 31(64), 33(65), *73*
Akermark, B., 272(225), *292*
Aldren, R. A., 191(80), 192(80), 193(80), *288*
Allen, A. D., 181, 201(1, 2, 7), 202(1, 2, 4, 121, 121a), 203, 207, 217(7), 220(121), 221, 243(1, 2, 4, 121), 244(7, 121), 258 (2, 7), 260(7), *285*, *286*, *289*
Allison, R. M., 185, *287*
Altman, J., 70, *77*
Amma, E. L., 96(53), 97(62), 98(53, 62), 100(53), 101(62), 102(62), *175*
Andal, R. K., 97(63, 64), 98(63, 64), 103 (63, 64), 104(64, 75), *175*, *176*, 216, 244 (143, 144), *289*
Anderson, R. S., 337(103), *355*
Angelici, R. J., 294, 295(5), *353*
Angoletta, M., 20(109), *74*, 352(136, 137), *356*
Antonini, E., 86(11), 88(11), *174*
Anwaruddin, Q., 240, 247(188), *291*
Arai, H., 47, *76*
Araki, M., 226(164), 246(164, 190), 258 (190), *290*, *291*
Ardon, M., 123(111), 124(111), *177*
Aresta, M., 201(101), 229, 230(101), 231 (101), 245(170), 246(170), 247(101), *288*, *290*
Armor, J. N., 201(113, 114), 204, 207, 208, 209, 210, 211(135), 215(135), 220(156), 244(156), 262(156), *289*, *290*
Arnon, D. I., 183, 187, 190, 200, 201, *286*, *287*
Aseeva, A. P., 324(63), 325(63), *354*
Atkinson, L. K., 238, 248(182), *290*
Austin, A. T., 375(48), 376, 384, *389*
Axelrod, J., 154(186, 187, 188), 161(202), 162(202), 163(202), 169(202), *179*
Azim, M. B., 186, *287*

B

Baasbeek, C. J., 131(137), *178*
Bach, M. K., 185, *287*
Bailar, J. C., Jr., 58, 59, 60, 61, 62(189), 67(193), *76*
Baird, M. C., 96(57), 98(57), *175*, 313(39), *354*
Baker, F. B., 124(118), *177*
Ballhausen, C. J., 7(21), *72*
Bancroft, G. M., 201(102, 104), 216, 217 (102), 219(104), 232, 244(104), 245(104), 247(102), 261(104), 262(219, 220), *288*, *291*
Banks, R. G. S., 34, *75*
Barb, W. G., 143(154, 155), 144, 145(155), 146, 148, 149, *178*
Barlow, L. R., 336(94), *355*
Bartell, L. S., 277, 278(237), *292*
Basolo, F., 19(94), *74*, 201(108, 115), 209 (108), 212(108), 216(115), 219(155), 220 (155), 243(115), 244(108), 262(115), *288*, *289*, *290*, 294, 295(5, 6, 7), 296, 298(18), 299(18), 300(18), 305(25), 306(25), 307 (25), 311(6), 374, 375(47), *353*, *388*
Bath, S. S., 19, 20(100), *74*
Battersby, A. R., 136(143), *178*
Bauer, N., 185(47), 187(60), 188, 189, *287*
Bauld, N. L., 324(62), *354*
Baxendale, J. H., 143, 144, 145, 146, *178*
Bayston, J., 34(130), *75*
Beck, M. T., 15, 16, *74*, 113, *176*
Beck, W., 31(121), 33, *75*
Beijerinck, M. W., 182(29), *286*
Beinert, H., 190(78), *288*
Bell, B., 238, 248(185), 253(185), *291*
Bell, L. G., 272(228), *292*
Bennett, M. A., 10(40, 63), 23(40, 63), 25 (40), 28(40), 30(63), 31(63), *73*
Bennett, M. J., 97(60), 98(60), 101, *175*
Benson, A., 190(76), *288*
Bercaw, J. E., 241(193), 249(193), 278, *291*
Bergersen, F. J., 188(69), *287*
Berhauer, K., 112(88), *176*
Berkowitz, J., 357(3), *387*
Beveridge, W. D., 239, 248(187), *291*
Bibler, J. P., 306(26), 307, *353*
Bichin, L. P., 181(17), 266(17), 267(17), 268(17), *286*
Bickel, A., 157(198), *179*
Biellmann, J. F., 20(108), *74*
Biji, P., 94(37), *175*
Billig, E., 294(3), *353*
Birch, A. J., 52(174, 175), 54, *76*
Bird, C. W., 336, *355*
Blake, D. M., 10(38), 23(38), 25(38), 28 (38), *73*
Bliss, H., 323, *354*
Block, K., 155(197), 157(197), 172(197), *179*
Blumberg, W. E., 127(128), *177*
Bobst, A., 155(194), 157(194), *179*
Boche, G., 181(22), 266(22), 269(22), 271 (22, 224), 272(224, 225), *286*, *292*
Boggs, J. E., 294(2), *353*
Bollyky, L., 13, *74*
Bond, G., 183, *286*
Booth, G., 312(38), 313, *354*
Borodko, Yu, G., 181(8), 201(100), 202 (119), 203, 211(100), 217(146), 218, 220 (151), 242(199), 243(100), 244(124), 250, 253, 254(119), 258(119, 152), 260(119), 282(242a), 283(242a), *286*, *288*, *289*, *290*, *291*, *292*
Bottomley, F., 181(2, 4), 201(2, 111, 111a), 202(2, 4), 203, 204(125), 205, 207, 208, 212, 224(125), 243(2, 4, 111), 258(2), *285*, *286*, *289*, 372(44), 373(44, 45), *388*
Brackman, W., 131(137), *178*, 378, 380(55), *389*
Bradbeer, C., 183(31), 184, *286*
Brailovskii, S. M., 324(63), 325(63), *354*
Brault, A. T., 296(12), *353*
Braun, G., 311(36), *354*
Braunstein, A. E., 125(122), *177*
Braus, R. J., 10(35), 23(35), 24(35), 29(35), *73*
Bray, W. C., 145, 147, *178*
Breslow, D. S., 41, *75*, 295(8, 9), 307, 308 (28), 310(27), 327(77), *353*, *355*
Breslow, R., 143(153), 149, 161(153), *178*
Brewis, S., 332, 333(91a), 334(91b), *355*
Bril, W. E., 153, *179*
Brintzinger, H. H., 241(192, 193), 249(192, 193), 272(228), 274, 275, 277, 278, *291*, *292*
Brion, H., 357(1), *387*

Brodie, B. B., 154(186, 187, 188), 161(202), 162(202), 163(202), 169(202), *179*
Brooks, D. W., 122(103), *177*
Brosset, C., 89, *175*
Brown, C. K., 53(177), *76*
Brown, D. A., 70, *77*
Brubaker, C. H., 272(227), *292*
Bufoni, F., 125(123), *177*
Bukreev, V. S., 181(8), 218(8), *286*
Bulen, W. A., 187(52, 53), 188, 190(72), 195, 198(98), 283(72), *287*, *288*
Burmeister, J. L., 10(55), 23(55), 27(55), *73*
Burnett, M. G., 35, *75*
Burns, R. C., 187(53), 188(53), 190(72, 74), 191(74), 193, 195, 196, 283(72), *287*, *288*
Burris, R. H., 183(31), 184, 185, 186, 188 (67), 189(67), 190(67), 193, 195, 196(91), *286*, *287*, *288*
Busch, D. H., 112, *176*
Butler, I. S., 305, 306, 307(25), *353*
Butler, S. A., 10(30), 20(30), 21(30), *73*, 96 (56), 98(56), 103(56), *175*
Butterfield, R. O., 68(205), 69(205), *77*
Byerley, J. J., 347(125), 348(125), *356*
Byrd, J. E., 352(135), 352, *356*

C

Cade, P. E., 253(212), *291*
Caglio, G., 20(109), *74*, 352(136, 137), *356*
Cahill, A. E., 147, *178*
Calcutt, G., 161(210), 167(210), *179*
Calderazzo, F., 20(110, 111), 51(110), *74*, 297, 298(17), 300(16, 19), 301(22), 302 (23), 303(23), 338(104), *353*, *355*
Calligaris, M., 95, 98(41, 42), *175*
Calvin, M., 10(23, 24, 25), 11, 15(23, 24, 25), *73*, 90(28), 112(87a), *175*, *176*
Campbell, P. M., 139(149), 150(149), 161 (149), *178*
Capparella, G., 104(74), *176*
Carnahan, J. E., 184(36, 37), 186(37), 187 (55), 190(55, 70, 75), *287*, *288*
Caroebter, A., 127(132), *177*
Carstens, E., 384(75), 386(75), *389*
Carter, C. W., Jr., 191(80), 192(80), 193, *288*
Castle, J. E., 184(36, 37), 186(37), *287*
Castro, C. E., 64, 66, *77*
Catone, D. L., 3(15), 19, 20(15), 21(15), *72*, 96(46), 98(46), 102(46), *175*
Caulton, K. G., 261(217), *291*
Cavil, B. E., 97(66), 98(66), 104(66), *176*
Ceciarelli, L., 359, *388*
Ceer, A., 154(189), 161(189), 163(189), *179*
Cenini, S., 104, *176*
Chalk, A. J., 12(72), 15, 23(72), *74*
Chance, B., 142, 167(152), *178*
Chapovskaya, N. K., 181(20), 266(20), 267 (20), 269(20), *286*
Charalampous, F. C., 127(129, 130), *177*
Charles, R., 10(40), 23(40), 25(40), 28(40), *73*
Chatt, J., 4, 9, 10(30, 32, 62), 19(94), 20 (30), 21(30), 23(32, 62), 24(32), 28(32), 30(62), 31(62), 62, *72*, *73*, *74*, *76*, 96 (56), 98(56), 103(56), *175*, 181(5), 201(5, 107), 203, 204, 207(128), 218 (149), 219, 233(173, 174), 234(149, 175), 235, 238(184, 185), 243(5), 244(149, 153), 245(153, 154), 247(107, 149, 173, 175, 177), 248(185), 253, 260(107, 149, 153, 173), 261(154), 282(246, 247), 284, 285, *286*, *288*, *289*, *290*, *291*, *292*, 312(38), 313, *354*
Chen, J. Y., 20(107), *74*
Chen, L. S., 96(55), 98(55), 102(55), 107 (55), *175*
Chen, T. T., 127(132), *177*
Cheney, A. J., 10(51, 52, 66), 23(51, 52, 66), 27(51, 52), 28(51, 52), 33(66), *73*
Cher, M., 123(114), 124(113), *177*
Chernikov, S. S., 230(171), 242(199), 247 (171), 250(199), *290*, *291*
Chirelli, F., 84(9), 85(9), 86(9), *174*
Chiriboga, J., 82(4), 83(4), 127(131), *174*, *177*
Chiusoli, G. P., 325, 326, *354*, *355*
Chock, P. B., 21(113), 22, 28, *75*
Chvapil, M., 161(209), 166(209), 167(209), *179*
Ciccone, S., 10(26), 11(26), 17(26), 18(26), *73*
Clark, C. T., 154(187, 188), *179*
Clark, G. R., 96(54), 98(54), 103, *175*
Clark, H. C., 352(131, 132, 133, 134), *356*
Clarkson, S. G., 374, 375(47), *388*
Clive, M., 221(157), *290*

Closson, R. D., 296(13, 14), *353*
Coffey, R. S., 62(197), *76*
Coffield, T. H., 296, *353*
Collman, J. P., 10(33, 47), 21(112), 23(33, 47, 112), 24(33), 26(47), *73*, *75*, 96(45, 50, 51, 52), 98(45, 50, 51, 52), 109(50, 51, 80), *175*, *176*, 181(9, 10), 191(83), 192 (83), 194(86), 201(9), 223(86), 230, 247(9, 10), 260(9), *286*, *288*, 320, 323(55), *354*, 366, *388*
Connick, R. E., 16(88), *74*
Conocchioli, T. J., 147, *178*
Conolly, P. J., 35(142), *75*
Constantopoulos, G., 127(132), *177*
Cook, C. D., 97(69), 98(69), 105(69), *176*
Cook, M. P., Jr., 320, 336(54), *354*
Corfield, P. W. R., 352(131), *356*
Corlin, J. L., 198(98), *288*
Cotton, F. A., 297, 298(17), 300(16), 352 (141), *353*, *356*
Coyle, B. A., 358(11), 361(11), 369(11), *388*
Craig, R. J., 304, *353*
Crain, C. M., 294(2), *353*
Cramer, R. D., 42(152), 56, 57(152, 185), *75*, *76*
Crandall, D. I., 88(16), 112, *174*
Crawford, J. R., 208, *289*, 372(44), 373(44, 45), *388*
Creegan, F. J., 385(78), *389*
Cripps, H. N., 40(147), *75*
Cuppers, H. G. A. N., 55(184), 56(184), *76*
Czapski, C., 11(68), 23(68), *74*

D

Daeniker, H. V., 382(66), 383, 384, *389*
Dahl, L. F., 191(82, 83), 192(82, 83), 194 (82), *288*
Dalgleish, C. E., 161(204, 205), 163(204, 205), 165(205), 168(204, 205), *179*
Daly, J. W., 159(199), *179*
Danon, J., 34(139), *75*
Darensbourg, D. J., 260(214), 261(218), *291*, 352(142), *356*
Darensbourgh, M. Y., 352(142), *356*
Das, P. K., 218(148), 253(148), *290*
Dasent, W. E., 358(33, 34), 379, *388*
Davidson, J. M., 10(62), 23(62), 30(62), 31 (62), *73*, 336(94, 95), *355*
Davidson, N., 123(114), 124(114, 117), *177*
Davies, A. G., 56, *76*, 171(216), *180*
Davies, G. R., 358(25), 361(25), 366(25), *388*
Davis, B. R., 209, 210, 224(161), 225, 234, 235, 253(133, 162, 163, 176), 265(162, 163), *289*, *290*
Davison, V. L., 68(205), 69(205), *77*
Dawson, C. R., 120(98), 122(103), 127(98), *176*, *177*
Dayton, J. C., 13(74), *74*
Deeming, A. J., 10(36), 20(106), 23(36, 106), 25(36, 106), 28(106), *73*, *74*
Dekock, R. L., 261(217), *291*
de Liefde Meijer, H. J., 240, 249(191), *291*
Denisov, N. T., 282(244, 245), 283, 284 (244, 245), *292*
Dent, W. T., 349, *356*
Derr, P. F., 384(74), 386(74), *389*
D'Eustachio, A. J., 186(50), 196(50, 89), *287*, *288*
De Vorkin, H., 123(109), 124(109), 170 (109), *177*
De Vries, G., 35, *75*, 94(37), *175*
Dewhurst, F., 161(210), 167(210), *179*
DeWitt, E. J., 13(75), *74*
Diamantis, A. A., 201(112), 207, 225(112), *289*
Diluzio, J. W., 19, 20(98, 99), 21(98, 99), *74*, *75*
Dilworth, J. R., 218(149), 233(173, 174), 234(149, 175), 235(177), 244(149), 247 (149, 173, 175, 177), 260(149, 173), *290*
Dixon, K. R., 352(131, 132, 133), *356*
Doedens, R. J., 107(77), *176*
Doemeny, P. A., 198, *288*
Dolcetti, G., 366(41), *388*
Dolphin, D., 10(35), 23(35), 24(35), 29(35), *73*
Domenicano, A., 358(20), 361(20), 369, *388*
Donaldson, P. B., 97(60), 98(60), 101, *175*
Donovan, C. K., 385(78), *389*
Douglas, P. G., 201(109, 110), 212, 215, 244(109, 110), *289*
Drago, R. S., 378, *389*
Drinkard, W. C., 132, *178*
Dunger, M., 385(76), 386(76), *389*

Durland, W., 358(12), 361(12), 363(12), 365(12), *388*
Dutton, H. J., 68(206), 70(206), *77*

E

Ebsworth, E., 90(27), *175*
Edwards, L. E., 10(55), 23(55), 27(55), *73*
Efimov, O. N., 282(244, 245), 283(244), 284(244, 245), *292*
Eischens, R. P., 242, *291*
Eisenberg, R., 358(12, 13, 16, 23, 30), 361 (12, 13, 16), 363(12, 13), 365, 366(16, 23), 367, 370, 372(16), *388*
Eisenmann, J. L., 327(75, 76), *355*
Ela, S. W., 181(22), 266(22), 269(22), 271 (22), *286*
Elder, L. W., 123(115), 124(115), *177*
Ellermann, J., 251, *291*
Ellfolk, N., 187, *287*
Elsehnig, G., 327(78), *355*
Elson, C. M., 218(150), 220(158), 221(158), 243(158), 244(150), 260(150), *290*
Elson, I., 221(157), *290*
Elvidge, J. A., 111(86), 112, *176*
Emken, E. A., 68(205), 69(205), *77*
Emmett, P. J., 242, *291*
Enemark, J. H., 224, 225, *290*, 358(15, 18, 19, 24), 361(15, 18, 19, 24), 363, 369(21), 370, *388*
Ercoli, R., 327, *355*
Esposito, R. G., 187(61), *287*
Evans, D., 44(163), 53(176), *76*
Evans, M. G., 143(166), 144, 146(166), *178*
Eyman, D. P., 378(59), *389*

F

Faigenbaum, H. M., 89(22), *175*
Falbe, J., 339, 344(107), *355*
Fallab, S., 90, 112(88), 123, *175*, *176*
Farkas, J., 16(89), *74*
Farnham, P., 366(41), *388*
Farona, M., 294(3), *353*
Farquhar, E. L., 211(136), *289*
Fechter, R. B., 181(22), 266(22), 269(22), 271(22), 272(225), 273(229), *286*, *292*
Feltham, R. D., 201(109, 110), 212(110), 215(109, 110), 244(109, 110), *289*
Fenski, R. F., 261(217), *291*
Ferguson, J. E., 181(5), 201(5), 202(120), 203(122a), 204, 217(145), 218, 243(5), 244(145), 262(145), *286*, *289*, *290*
Filbey, A. H., 296(15), 297, *353*
Finn, P., 358(31), 362, *388*
Fischer, D. J., 184(39), 187(39), 194(39), *287*
Flid, R. M., 324(63), 325(63), *354*
Flood, M. T., 206(127), 253(127), 262(127), 263(127), 265(127), *289*
Fluoronoy, J. H., 13(74), *74*
Flynn, J. H., 42(153), 56, 58, *75*
Foster, D., 10(41), 23(41), 25(41), *73*
Frankel, E. N., 68(205, 206), 69, 70, *77*
Frankland, H., 381, *389*
Fraser, R. T. M., 358, 359(35), 379, *388*
Frazier, R. H., 198(97), *288*
Freer, S. T., 191(80), 192(80), 193(80), *288*
Frenz, B. A., 358(19), 361(19), *388*
Friedel, R. A., 336(98), 337(98), *355*
Frieden, E., 88(18), *174*
Friedman, J. P., 139(149, 150), 150(149, 150, 157), 154(157), 161, 163(157), *178*
Friedman, S., 337(103), *355*
Fusi, A., 104(74), *176*

G

Galli, R., 377, 384(54), 386(54), *389*
Garland, J. H. N., 381(63), 382(63), *389*
Garner, A. W. B., 235(178), 247(178), *290*
Garrod, R. E. H., 262(220), *291*
George, P., 87(15), 88(17), 123(108), 139, 143(154, 155), 144(154, 155, 170), 145 (155), 146(15, 154, 155), 148(154, 155), 149(155), 150(151), 152(151), 161(151), 163(151), *174*, *177*, *178*
Gest, H., 183, 188(32, 33), 193, 200, *286*, *288*
Gevvay, J., 341(114), *356*
Gill, S. J., 211(136), *289*
Gillard, R. D., 5, 18, 19(20), 57(187), *72*, *74*, *76*
Gimesi, I., 15, 16(89), *74*
Ginsberg, A. P., 2(12), *72*
Glockler, G., 294(1), *353*
Glyde, R. W., 315, *354*
Godwin, J. B., 372(42, 43), *388*

Gogan, N., 70(213), *77*
Goldhann, H., 384(75), 386(75), *389*
Gorin, G., 135, *178*
Gorin, M. H., 145, 147, *178*
Gorog, S., 113, *176*
Goryachenkova, E. V., 125(126), 127(126), *177*
Gough, A., 62(197), *76*
Gould, E. S., 82(7), 83(7), 173(7), *174*
Graham, B. W., 97(65), 98(65), 104, *176*
Graham, W. A. G., 10(42, 43), 23(42, 43), 25(42), 26(42, 43), 28(42, 43), *73*
Gray, D. R., 272(227), *292*
Gray, H. B., 7(21), 19(94), 34(139), *72*, *74*, *75*, 206(127), 253(127), 262(127), 263 (127), 265(127), *289*, 294(3), *353*, 358 (29), 361(29), 362, 370, *388*
Greaves, E. O., 10(56), 23(56), 27(56), *73*
Greeley, R., 271(224), 272(224, 225), *292*
Green, D. E., 188(65), *287*
Green, H. J., 161(203), 162(203), *179*
Green, M. L. H., 2(13), 43, 47(162), *72*, *76*, *353*, 352(140), 303, 304, *356*
Griffith, J. S., 87, *174*
Griffith, W. P., 2, 19(7, 93), 34, *72*, *74*, *75*, 359(36), 360, *388*
Grillone, M. D., 191(83), 192(83), *288*
Grinstead, R. R., 87(13), 116, 120, 123(94), 124(94), 149, 154, 161(181), 168(181), *174*, *176*, *179*
Grundy, K. R., 97(66), 98(66), 104(66), *176*, 358(27), 361(27), 366(27), 368(27), 373, *388*
Guerrieri, F., 326, *355*
Gulens, J., 218(150), 220(158), 221(158), 243(158), 244(150), 260(150), *290*
Gunter, B., 13(77), *74*
Gunz, H. P., 218(149), 234(149), 244(149), 247(149), 260(149), *290*
Guroff, G., 159(199), *179*

H

Haber, F., 143, 144, 146, *178*
Hagihara, N., 67, *77*, 97(67), 98(67), 105 (67), *176*
Haight, G. P., Jr., 182, 266, 267(23), 268 (23), *286*
Haines, L. M., 20(103), 23(103), 26(103), 28(103), *74*, 96(58), 98(58), *175*
Hakala, M., 184(41), 185, *287*
Hall, M. C., 102, *176*
Halpern, J., 1, 2(4), 3, 10(26, 27, 28), 11, 12(27, 28, 70, 72, 73), 13, 14, 15, 16(80, 85), 17, 18, 20(107), 21(113), 22, 23(68, 69, 70, 71, 72), 28, 29(116), 30, 35, 36, 37 (145), 41, 42, 43, 44, 47, 58, 66, *72*, *73*, *74*, *75*, *76*, 108, 123(119), 124(119), *176*, *177*, 335(93), 344, 345, 347, 348(121), 349(121), 352, 352, *355*, *356*
Hamilton, E. J., Jr., 147(177), *178*
Hamilton, G. A., 123, 136(144), 139, 142, 143(157, 158), 150, 151, 154(157, 158), 157(144, 158), 158, 159(144), 161, 163 (157), 173, *177*, *178*
Hamilton, W. C., 2(9), 19(9), *72*, 358(17), 361(17), *388*
Hardy, R. W. F., 186, 190(74), 191(74), 193, 195(87), 196, *287*, *288*
Hare, D. G., 171(216), *180*
Hargaden, J. P., 70(213), *77*
Hargrave, K. R., 143(154, 155), 144(154, 155), 145(155), 146(154, 155), 148(154, 155), 149(155), *178*
Harkness, A. C., 344, 345(121), 348(121), 349(121), *356*
Harrifin, J. W., Jr., 139(150), 150(150), *178*
Harris, R. O., 181(2), 201(2), 202(2, 121a), 214(140), 243(2), 244(140), 258(2), *285*, *289*
Harrison, D. E., 181(3), 201(3), 203, 205, 206(126), 243(3, 126), *286*, *289*
Harrod, J. F., 10(26, 27, 28, 59), 11(26), 12 (27, 28), 17, 18, 23(59), 29(59), 42(154), 43(154), 56(154), *73*, *75*
Hart-Davis, A. J., 10(42, 43), 23(42), 25 (42), 26(42, 43), 28(42, 43), *73*
Hartwell, G. E., 10(67), 23(67), 32(67), 33 (67), *74*
Hashimoto, K., 172, *180*
Hashimoto, T., 69, 72(210), *77*, 316(50), *354*
Hatta, G., 10(61), 23(61), 31(61), *73*
Havard, J. F., Jr., 327(75), *355*
Hayaishi, O., 155(197), 157(197), 172, *179*, *180*

Hazelwood, C. M., 161(206), 163(206), 165(206), *179*
Heath, G. A., 282(247), *292*
Heck, R. F., 10(31, 45), 23(31, 45), 24(31), 26(45), 28(31), 41, *73*, *75*, 295(8, 9, 10), 307, 308, 310, 312(37), 313(30), 327(72, 77), *353*, *354*, *355*
Helmkamp, G. K., 250(203), *291*
Henbest, H. B., 55(183), *76*
Henrici-Olive, G., 202(118), 266(221), 269 (221), 270(221), 275, 277, *289*, *291*, *292*
Henry, P. M., 123(112), 124(112), *177*
Hensinger, H., 336(97), *355*
Hessett, B., 10(54), 23(54), 27(54), *73*
Hidai, M., 10(53), 23(53), 27(53), *73*, 226 (164, 166), 236(180), 237(180), 238, 246 (164, 190), 247(180), 248(181, 182, 183), 258(190), *290*, *291*
Hieber, W., 311(36), 336(96, 97), *354*, *355*
Hiembach, P., 97(68), 98(68), 105(68), 106 (68), 108(68), *176*
Hill, J. A., 381(63), 382(63), *389*
Hill, J. M., 123, *177*
Hinkson, J., 190(72), 283(72), *287*
Hirsch, E., 347(124), *356*
Hoch, G. E., 188, 189, 190, *287*
Hodges, R. J., 32(122), 33, *75*
Hodgson, D. J., 358(4, 5, 6,) 361(4, 5, 6), 363(4, 5, 6), 364, 365(4, 5, 6), *388*
Hoffman, N. W., 10(47), 23(47), 26(47), *73*, *354*
Holhn, H. N., 40(147), *75*
Holsten, R. D., 195(87), *288*
Hong, J. S., 190(77), *288*
Horecker, B. L., 127(128), *177*
Horiie, S., 341(111), 343, *355*, *356*
Horn, R. W., 96(50), 98(50), 109, *175*
Hortner, J., 11(68), 23(68), *74*
Hosaka, S., 328(85, 86, 88), 329(85, 86), 330(88), *355*
Howk, B. W., 40(147), *75*
Huffmann, R. E., 124(117), *177*
Huggins, D., 132, *178*
Hughes, P. R., 332, 333(91a), 334(91b), *355*
Hulbust, H. M., 42(153), 56, 58, *75*
Hurych, J., 161(209), 166(209), 167(209), *179*
Hyde, C. L., 261(218), *291*

I

Ibers, J. A., 2(9, 11), 19(9), 23, *72*, 96(44, 47), 98, 99, 100(47), 102(47), 107(77), 113(44), *175*, *176*, 191(83), 192(83), 209, 210, 224(161), 225(161, 162, 163), 234, 235, 253(133, 162, 163, 176), 265(162, 163), *288*, *289*, *290*, 352(131), *356*, 358 (4, 5, 6, 8, 9, 10, 11, 18, 19, 26), 361(4, 5, 6, 8, 9, 10, 11, 18, 19, 26), 362, 363(4, 5, 6, 8, 9, 10), 364, 365(4, 5, 6, 10), 368, 369(11), 370, *387*, *388*
Iguchi, M., 34, *75*
Ikeda, S., 181(11), 213(138), 214(141), 215 (138), 222(11), 224(11), 226(11, 165), 228(168, 169, 169a), 244(138, 141), 245 (11), 246(141), 260(11), 279(238), *286*, *289*, *290*, *292*
Iliades, T. I., 214(140), 244(140), *289*
Illatsovskaya, M. A., 181(16, 18), 266(16, 18), 267(16, 18), 268(16, 18), 269(18), 278(222), *286*, *291*
Imamura, S., 328(84), 329(84), *355*
Indictor, N., 153, *179*
Ingold, K. U., 170(213, 214), *179*
Irving, R. J., 357(9), 358(2, 40), 361(40), 385(40), 386(40), *387*, *388*
Ismail, S. M., 42(159), 63(159), *76*
Itatani, H., 58, 59, 60, 61(189, 193, 195), 62(189), 67(193), *76*
Itzkovitch, I. J., 210, 211, 218(150), 221 (157), 244(150), 260(150), *289*, *290*
Ivleva, I. N., 282(242a), 283(242a), *292*
Iwamoto, N., 331(90), 336(90), 338, *355*

J

Jackson, E. K., 196(90), *288*
Jackson, J., 242, *291*
Jacobs, W. J., 352(132, 133, 134), *356*
Jaff, H. H., 260, *291*
James, B. R., 17(155), 42(154, 155, 156), 43, 44, 53, 56(154), 58, 66, *75*, *76*, 110(81, 82), *176*, 350, 351(129), *356*
Jardine, F. H., 10(29), 20(29), 42(160, 161), 44(163), 46, 47(29, 168), 49(29), 50(29), 51, 52(173), *73*, *76*
Jardine, I., 44, 45(165), 63, *76*
Jarnigan, R. C., 139(147), *178*

Jarvis, J. A. J., 358(25), 361(25), 366(25), *388*
Jauhal, G. S., 97(69), 98(69), 105(69), *176*
Jenner, E. L., 42(152), 56(152), 57(152), *75*
Jensen, L. H., 190(79), 191, 192(79), *288*
Jerina, D. M., 159(199), *179*
Jesson, J. P., 47(171), *76*
Jo, T., 343(116, 117, 118), *356*
Johnson, B. F. G., 358, *388*
Johnson, G. L., 239, 248(187), *291*
Johnson, G. R. A., 143(159), 161(159), 164 (159), *178*
Johnson, N. P., 10(32), 23(32), 24(32), 28 (32), *73*
Jolly, P. W., 201(105), 238, 239, 249(105, 186), *288*, *291*
Jolly, W. L., 358(31), 362, *388*
Jonas, K., 201(105), 238, 239(186), 249 (105, 186), *288*, *291*
Jonassen, H. B., 10(57), 23(57), 27(57), *73*
Jones, C. J., 2(13), *72*
Jones, E. P., 68(206), 70(206), *77*
Jones, E. R. H., 325(66), *354*
Judis, J., 183(33), 188(33), 193(33), 200(33), *286*
Jung, M. J., 20(108), *74*

K

Kachapina, L. M., 282(242a), 283(242a), *292*
Kaeding, W. W., 130(136), *177*
Kaesz, H. D., 2(5), 10(5), *72*
Kakudo, M., 98(71), 102(71), *176*
Kaliya, O. L., 324(63), 325(63), *354*
Kamen, M. D., 183, 188(32), 190, 200, *286*, *287*, *288*
Kane-Maguire, L. A. P., 201(108, 115), 209, 212, 216(115), 243(115), 244(108), 262(115), *288*, *289*
Kang, J. W., 34(131, 132), 37(131, 132), 41, *75*, 96(52), 98(52), *175*, 181(9), 201(9), 247(9), 260(9), *286*
Kaplunov, M. A., 218(151), 220(151), *290*
Karsteller, B. R., 378(58), *389*
Kasai, N., 98(71), 102(71), *176*
Kashiwagi, T., 98(71), 102, *176*
Katlov, N. A., 281(240), *292*
Kauffmann, S., 149(180), 157(180), *179*
Kavashina, E. F., 282(242, 242a), 283(242, 242a), *292*
Kazmirowski, H. G., 384(75), 386(75), *389*
Keblys, K. A., 296(15), 297, *353*
Keeler, R. F., 187, 188(52), *287*
Kelley, M., 196(93), *288*
Kelly, M., 190(73), 195(73), 196(73), *288*
Kelly-Falcoz, F., 127(128), *177*
Kelvi, A. A., 281(240), *292*
Kemball, C., 35(142), *75*
Kemp, A. L. W., 17(155), 42(155), *75*
Kemppi, A., 184(42), 185(42), *287*
Kerekjarko, B., 155(192), *179*
Kern, F., 31(121), 33, *75*
Kertesz, D., 122(101, 104, 106), *176*, *177*
Kettle, S. F. A., 335(93), 345(93), *355*
Khidekel, M. L., 181(6, 8, 18), 201(6), 218 (8), 229(6), 230(6), 246(6), 266(18), 267 (18), 268(18), 269(18), *286*
Khrusch, A. P., 58(192), *76*
Kiefer, G. W., 198(97), *288*
Kiji, J., 328(80, 81, 82, 83, 84, 85, 88), 329 (83, 84, 85), 330(88), *355*
Kilbourne, B. T., 102(72), *176*, 358(25), 361(25), 366(25), *388*
King, N. K., 34, 35, 36, *75*, 147, *178*
King, R. B., 241, 249(196), *291*
King, T. E., 155(196), 157(196), 172(196), *179*
Kinnaird, J. K., 95(39), *175*
Kitahara, Y., 218(149), 234(149), 244(149), 247(149), 260(149), *290*
Kitazume, S., 181(11), 213(138), 214(141), 215(138), 222(11), 224(11), 226(11), 228 (169, 169a), 244(138, 141), 245(11), 246 (141), 260(11), *286*, *289*, *290*
Klemchuk, P. R., 328(79), *355*
Knight, E., Jr., 186(50), 193(85), 196(50, 88, 90), *287*, *288*
Knoth, W. H., 214, 215(142), 244(139), *289*
Knox, L., 2(12), *72*
Kobyashi, H., 88(18), *174*
Kohl, C. F., 334(92), *355*
Kolthoff, I. M., 143(156), 144, 145, 148, *178*, *179*
Konaev, E. N., 349(126), *356*
Kondo, H., 10(61), 23(61), 31(61), *73*
Kooyman, E. C., 157(198), 160(201), 173 (201), *179*

Korinek, G. J., 11(71), 23(71), *74*
Korte, F., 339, 344(107), *355*
Kostner, R., 13(76, 77), *74*
Kosyakova, L. V., 270(223), *291*
Kozikowski, J., 296(13, 14), *353*
Kozub, G. I., 181(8), 217(146), 218(8, 146), *286*, *289*
Kraihanzel, C. S., 300(20), *353*
Kramolowsky, R., 251(204), *291*
Kraut, J., 191(80), 192(80), 193(80), *288*
Kroper, H., 325(65), 336(65), *354*
Kropf, H., 170, *180*
Kruck, T., 352(138), *356*
Kruger, C., 238(186), 239(186), 249(186), *291*
Kubo, H., 187, *287*
Kubota, M., 10(38), 23(38), 25(38), 28(38), *73*, 181(10), 200(10), 247(10), *286*
Kuchyuka, K., 202(117), 203(117), *289*
Kudo, K., 10(53), 23(53), 27(53), *73*
Kudryavtsev, R. V., 280(239), 281(240), *292*
Kuntsmann, R., 251(204), *291*
Kuse, T., 226(166), *290*
Kuznin, S. G., 242(199), 250(199), *291*
Kwiatek, K., 33(123, 124), 34(123, 127, 128), 37, 38(128), 39, 40, 41, 42, *75*

L

Labinger, J. A., 10(35), 23(35), 24(35), 29, *73*
Laing, K. R., 97(65), 98(65), 104(65), *176*
Lamb, A. B., 123(115), 124(115), *177*
Lane, J. R., 34(138), *75*
Laplaca, S. J., 2(9, 11), 19(9), 23, *72*, 96(44), 98, 99, 113(44), *175*
Lapporte, S. J., 109(80), *176*
Larikov, E. I., 181(18), 266(18), 267(18), 268(18), 269(18), *286*
Larksworthy, L. F., 381(63), 382(63), *389*
Lavison, J. J., 96(48), 98(48), 109(48), *175*
Lawrence, R. V., 10(67), 23(67), 32(67), 33(67), *74*
Lawson, D. N., 10(44), 23(44), 26(44), *73*, 96(57), 98(57), *175*
LeComte, J. R., 187(53), 188(53), 190(72), 195(53), 283(72), *287*
Lefier, A., 154(189), 161(189), 163(189), *179*
Legge, J. W., 88(21), *174*
Leibfritz, D., 191(84), 192(84), *288*
Leigh, G. J., 218(149), 219(153), 233(173, 174), 234(149, 175), 235, 238(184, 185), 244(149, 153), 245(153), 247(149, 173, 175, 177), 248(185), 253(185), 260(149, 153, 173), *290*, *291*
Lemberg, R., 88, *174*
L'Eplattenier, F., 10(50, 58), 20(110, 111), 23(50, 58), 26(50), 27(58), 28(50), 51(110), *73*, *74*
Leussing, D. L., 135, *178*
Lever, A. B. P., 111(86), 112, *176*
Levine, W. G., 128(134), *177*
Lewis, J., 358(40), 359(36), 360(36), 361(40), 375(49), 385(40), 368(40), *388*, *389*
Lima, D. A., 386(77), *389*
Lindblom, R. O., 130(136), *177*
Lindsey, R. V., Jr., 42(152), 56(152), 57(152), *75*, *76*
Linsen, B. G., *76*
Little, H. N., 188(66), 189(66), *287*
Loader, P. L., 358(20), 361(20), 369(20), *388*
Lock, C. J. T., 10(56), 23(56), 27(56), *73*
Lockshin A., *288*
Lockwood, W. H., 88(21), *174*
Lodney, F. S., 64(201), 69(201), *77*
Loebl, H., 143(160, 161, 162), 161(160, 161, 162), 163(160, 161, 162), 165(160, 161, 162), *178*
Loescher, B. R., 202(121a), *289*
Long, R., 349(127), *356*
Losada, M., 183(34), 200(34), 201(34), *286*
Love, J. L., 202(120), 203(122a), 217(145, 147), 218(147), 244(145), 262(145), *289*, *290*
Lukehart, C. H., 352(141), *356*
Lukens, L. N., 143(153), 149, 161(153), *178*
Luvalle, J. E., 116(91), *176*

M

McAndrew, R. T., 346, *356*
McClellan, W. R., 40(147), *75*
McCleverty, J. A., 358, *388*

McCormick, B. J., 135, *178*
McGinnety, J. A., 96(47), 98(47), 100, 102 (47), 107, *175*, *176*, 224(161), 225(161), *290*, 358(4), 361(4), 363(4), 364(4), 365 (4), *388*
MacGregor, E. R., 11(69), 23(69), *74*
McKee, H. S., 182(26), 183(26), 185, *286*
McMullen, C. M., 70(213), *77*
McQuillin, F. J., 44, 45(165), 63, *76*
McRae, W. A., 327(76), *355*
Maddock, A. G., 262(220), *291*
Mador, I. L., 33(123), 34(123, 128), 37 (123, 128), 38(128), 41(123), 42(123), *75*
Magee, W. E., 185(46), 186, *287*
Mague, J. T., 10(48), 23(48), 26(48), 51, *73*, *76*, 96(57), 98(57), *175*, 313(39), *354*
Maher, J. P., 41, *74*, *75*
Mahler, H. R., 127(127), *177*
Mahore, H. R., 188(65), *287*
Mais, R. H. B., 358(25), 361(25), 366(25), *388*
Maitlis, P. M., 10(56), 23(56), 27(56), *73*
Malatesta, L., 20(109), *74*, 352(136, 137), *356*
Mann, B. E., 10(66), 23(66), 33(66), *73*
Mann, P. J. G., 123, *177*
Manoharan, P. T., 97(63), 98(63), 103(63), *175*, 358(17, 29), 361(17, 29), 362, 370, *388*
Maples, P. K., 219, 220(155), *290*, 300(20), *353*
Markarova, L. G., 382(65), *389*
Marko, L., 222, 223(160), 245(159), 246 (159), *290*
Marks, T. J., 191(83), 192(83), *288*
Marsh, R. E., 89, 90, *175*, 206(127), 253 (127), 262(127), 263(127), 265(127), *289*
Martell, A. E., 90(28), 94, 116(95), 118(96, 97), 121(100), 123(35, 95, 121a), 124(95), 173(219), *175*, *176*, *177*, *180*, 182, 266 (25), 269(25), 273, *286*
Maruta, M., 123(121a), *177*
Marvich, R. M., 241(192), 249(192), 277 (192), *291*
Marzilli, L. G., 23(117), 30, *75*
Marzilli, P. A., 23(117), 30, *75*
Masada, H., 311(33, 34, 35), 316(49), 319 (49), 320(49), 322(49), *353*, *354*
Maskill, R., 274, *292*
Mason, H. S., 122(102, 104, 107), 125(124), 127(124), 128(133), 155, 157(195, 196), 161(212), 168(212), 172(195, 196), 173, *176*, *177*, *179*
Mawby, A. H., 238(182), 248(182), *290*
Mawby, R. J., 295(6), 298, 299(18), 300 (18), 311(6), 315, *353*, *354*
Maxsted, E. B., 42(159), 63(159), *76*
Mays, M. J., 20(105), *74*, 201(102, 104), 216(202), 217(102), 219(104), 232(102, 104), 244(104), 245(104), 247(102), 261 (104), 262(219, 220), *288*, *291*
Meakin, P., 47(171), *76*
Medalia, A. I., 143(156), 144, 145, 148, *178*, *179*
Medema, D., 334(92), *355*
Meek, D. W., 97(61), 98(61), *175*
Mekaeva, I. M., 281(240), *292*
Melville, D. P., 201(107), 219(154), 245 (154), 247(107), 260(107), 261(154), *288*, *290*
Mercer, M., 358(35), 359(35), *388*
Merz, J. H., 161(207, 208), 166(207, 208), *179*
Merzoni, S., 325(69), 326(70), 333(70), *354*, *355*
Metzger, H. G., 201(109, 110), 212(110), 215(109, 110), 244(109, 110), *289*
Meyer, T. J., 372, *388*
Miki, M., 108(78), 109(78), *176*
Miki, T., 316(47, 52), 320(47), 322(47, 52), *354*
Mikolsky, A. B., 207(128), *289*
Millard, S. M., 342, *356*
Miller, F., 90, *175*
Miller, M. V., 96(55), 98(55), 102(55), 107 (55), *175*
Mills, G. A., 14(82), 34, 35, *74*, *75*
Milne, J. B., 15, 16(85), *74*
Milner, D. L., 10(63), 23(63), 30(63), 31 (63), 55(63), *73*
Mingos, D. M. P., 358(8, 9, 10, 26), 361(8, 9, 10, 26), 362, 363(8, 9, 10), 364, 365 (10), 368, 370, *388*
Minisci, F. F., 377, 384(54), 386(54), *389*
Misono, A., 69(207, 208, 209), *77*, 181(13, 14), 201(14), 222, 223(14), 226, 236(180), 237(180), 238(183), 245(13), 246(14, 164, 190), 247(180), 248(183), 258, 260(14), *286*, *290*, *291*

Mitchard, L. C., 352(140), *356*
Mitchener, J. P., 10(48), 23(48), 26(48), *73*
Miyake, A., 10(61), 23(61), 29, *73*
Miyamolo, H., 316(42), *354*
Molewyn Hughes, J. J., 235(178), 247(178), *290*
Mondelli, 326(70), 333(70), *355*
Montalatici, S., 47(169), 52, *76*
Moore, W. P., Jr., 384(67, 68), 386(67, 68), *389*
Moriconi, E. J., 385, *389*
Morikawa, M., 328(80, 81, 82, 83, 84), 329 (83, 84), 331(90), 336(90), *355*
Moriyama, T., 161(211), 168(211), *179*
Morris, D. E., 10(47), 23(47), 26(47), *73*
Morris, J. H., 10(54), 23(54), 27(54), *73*
Morrison, M., 155(196), 172(196), *179*
Mortensen, L. E., 184(36), 187, 190, *287*, *288*
Mose, S., 66(202), *77*
Moser, C., 357(1), *387*
Mower, H. F., 184(36), 190(76), *287*, *288*
Mozzhukin, D. O., 203(122), *289*
Mueller, M. H., 358(28), 361(28), 366(28), *388*
Muetterties, E. L., 40(147), *75*
Mugdan, M., 153(184), *179*
Muir, W. R., 10(34), 23(34), 24(34), 29, *73*
Murahashi, S., 340(110), 341(111), 343, *355*, *356*
Murakami, M., 34(129, 131, 132), 37(129, 131, 132), 38(129), 41, *75*
Murray, R., 202(116), 250, *289*
Myagkov, Yu. P., 203(122), 217(146), 218, 220(51), *289*, *290*
Myeong, S. K., 319(53), 323(53), *354*

N

Nagihara, N., 98(71), 102(71), *176*
Nagy, F., 13, 14(81), 35, 36(143, 144), 38 (146), 41, *74*, *75*
Nagy, P. L. K., 43, 47(162), *76*
Nakamura, A., 96(59), 97(59, 70), 98(59, 70), 106(70, 76), 107, 108(70, 78), 109(78), *175*, *176*
Nakamura, S., 345, 347, 351, *356*
Nakamura, T., 128(135), *177*
Nakashima, T., 190(76), *288*
Nakon, R., 94, 113, 114(90), 123(35), *175*, *176*
Nancollas, G. H., 94, *175*
Nappier, T. E., Jr., 97(61), 98(61), *175*
Nardin, G., 95(41, 42), 98(41, 42), *175*
Natta, G., 327(73), *355*
Nechiporenko, G. N., 274, *292*
Nelson, J. H., 10(57), 23(57), 27(57), *73*
Nesmeyanov, A. N., 382(65), *389*
Neurath, G., 385(76), 386(76), *389*
Newton, T. W., 124(118), *177*
Newton, W. E., 198, *288*
Ng, F. T. T., 53(178, 180), 110(82), *76*
Nichikawa, N., 382(64), *389*
Nicholas, D. J. D., 184, 186, 187, 194, *287*
Nichols, P., 139(148), *178*
Nicholson, J. K., 349, *356*
Nienburg, H., 327(78), *355*
Ninchilli, D., 95(42), 98(42), *175*
Noack, K., 300(19), 301(22), 302, 303, *353*
Noack, M., 352(138), *356*
Nobile, C. F., 20(101), 21(101), *74*, 229 (170), 245(170), 246(170), *290*
Nofre, C., 154(189), 161, 163(189), *179*
Nogi, T., 328(89), 331(89), *355*
Nord, H., 116, *176*
Norman, R. O. C., 155, 161, 163(190), 164 (190), 165(190), *179*
Noskova, N. F., 349(126), *356*
Nozaki, M., 183(34), 200(34), 201(34), *286*
Nyburg, S. C., 204(125), 205, 224(125), *289*
Nyholm, R. S., 63, *77*

O

Ochiai, E., 110(81), 135, *176*, *178*
O'Connor, C., 53, *76*
O'Connor, C. J., 97, (65), 98(65), 104(65), *176*
Ogata, I., 69(207, 208, 209), *77*
Ogura, Y., 128(135), *177*
Okuda, Y., 311(32, 33, 35), *353*
Olive, S., 202(118), 266(221), 269(221), 270 (221), 275, 277, *289*, *291*, *292*
Ookaw, M., 279(238), *292*
Orchin, M., 260, *291*, 336(98), 337(98), *355*
Orgel, L., 5(19), *72*, 253, *291*
Orioli, P. L., 2(10), 23, *72*

Orme Johnson, W. H., 190(78), *288*
Osaki, S., 88(18), *174*
Osborn, J. A., 10(29, 35, 44, 49, 60), 18(92), 20(29), 23(35, 44, 49, 60), 24(35), 26(44, 49), 29, 42(160, 161), 44(163), 46, 47(29, 167, 168, 169), 49(29), 50(29, 168), 51 (160), 52(169), *73*, *74*, *76*, 96(57), 98(57), *175*, 313(39), *354*
Otsuka, S., 96(59), 97(59, 70), 98(59, 70), 106, 107(76), 108(70, 78), 109(78), *175*, *176*
Owsley, D. C., 250(203), *291*
Owston, P. G., 2(8), *72*, 358(25), 361(25), 366(25), *388*

P

Page, J. A., 210, 211, 218(150), 220(158), 221(157, 158), 243(158), 244(150), 260 (150), *289*, *290*
Parish, R. V., 10(46), 23(46), 26(46), *73*
Park, G. S., 143(166), 146(166), *178*
Parker, C. A., 188(63, 64), *287*
Parshall, G. W., 30(120), *75*, 226, 227(167), 281, 282(241), *290*, *292*
Partridge, J. M., 2(8), *72*
Paul, A. D., 16(88), *74*
Paul, B., 13(77), *74*
Paul, K. G., 152(182), *179*
Paulik, F. E., 378(57), *389*
Pauling, L., 82, 83(1), *174*, 251, *291*
Pavlova, V. K., 182, 270, *286*
Payne, N. C., 96(47), 98(47), 100(47), 102 (47), *175*, 225, (162, 163), 253(162, 163), 265(162, 163), *290*, 358(4), 361(4), 363 (4), 364(4), 365(4), *388*
Pearson, R. G., 10(34), 19(94), 23(34), 24 (34), 29, *73*, *74*, 201(108, 115), 209(108), 212(108), 216(115), 219(155), 220(155), 243(115), 244(108), 262(115), *288*, *289*, *290*, 295(6), 298(18), 299(18), 300(18), 305(25), 306(25), 307(25), 311(6), 358(4), 361(4), 363(4), 364(4), 365(4), *388*, *353*
Peck, H. D., Jr., 183(33), 188(33), 193(33), 200(33), *286*
Peisack, J., 127(128), *177*
Peive, J., 187(62), *287*
Pelichet, M. C., 10(58), 23(58), 27(58), *73*
Pell, S., 201(114), 204, *289*
Perkins, P. G., 10(54), 23(54), 27(54), *73*
Peters, E., 11(69), 23(69), *74*, 346, 347(124, 125), 348(125), *356*
Peters, H. M., 68(205, 206), 70(205, 206), *77*
Phelan, P. F., 29(116), 30, *75*
Pickard, A. L., 108, *176*
Pickens, D., 384(67), 386(67), *389*
Pierpont, C. G., 358(12, 13, 16, 23, 30), 361 (12, 13, 16, 23), 363, 365, 366(16, 23), 367, 370, 372(16), *388*
Piette, L. J., 120(99), 127(99), *176*
Pin Lee, L., 95(40), 98(40), *175*
Pino, P., 327(73), *355*
Pirmann, B., 385(76), 386(76), *389*
Pistor, H. J., 325(65), 336(65), *354*
Pitzer, K. S., 82, 83(2), *174*
Poe, A. J., 296(12), *353*
Poersch, F., 251(204), *291*
Poilblanc, R., 191(83), 192(83), *288*
Posner, A. M., 123(113), 124(113), *177*
Postgate, J. R., 196(93), *288*
Potter, P. E., 10(28), 12(28), 18(28), *73*
Powell, H. K. J., 94, *175*
Powell, J., 349(128), *356*
Prater, B. E., 201(102, 104), 216(102), 217 (102), 219(104), 232(102, 104), 244(104), 245(104), 247(102), 261(104), 262(219, 220), *288*, *291*
Pratt, C. S., 358(11), 361(11), 369, *388*
Pratt, J. M., 34, *75*, 218(148), 253(148), 274, *290*, *292*
Prewitt, C. T., 56(185), 57(185), *76*
Pribanic, M., 36, 37(145), *75*
Price, W. C., 253(211), *291*
Prichard, W. W., 324(59, 60, 61), 341(112), 343, *354*, *356*
Prodayko, L. A., 280(239), *292*
Przywarska-Boniecka, P., 111, 112(87), 127(87), *176*
Pu, L. S., 181(11), 222(11), 224(11), 226(11, 165), 228(168, 169a), 245(11), 260(11), *286*, *290*
Pucci, A., 358(23), 361(23), 366(23), 367 (23), *388*

Q

Quinby, M. S., 358(24), 361(24), 369(24), 370(24), *388*

R

Rabinovitch, B. S., 64(201), 69(201), *77*
Rabinovitz, J. C., 190(77), *288*
Ragsdale, R. O., 378(59), *389*
Ralph, B. J., 161(203), 162(203), *179*
Ramp, R. L., 13(75), *74*
Rampel, G. L., 53(178, 179), *76*
Ramsey, J. B., 123(109), 124(109), 170 (109), *177*
Randaccio, L., 95(41, 42), 98(41, 42), *175*
Ransil, B. J., 253(207), *291*
Redmond, W. J., 184(39), 187(39), 194(39), *287*
Reed, C. A., 10(39), 23(39), 25(39), 28(39), *73*, 96(54), 98(54), 103(54), *175*, 373, *388*
Reed, L. L., 358(24), 361(24), 369(24), 370 (24), *388*
Reilley, E. L., 377(50, 51, 52, 53), *389*
Reinsalu, V. P., 181(2), 201(2), 202(2), 243 (2), 258(2), *285*
Rempel, G. L., 350, 351(129), *356*
Renson, J., 159(199), *179*
Reppe, W., 325, 336(64, 65), *354*
Revesz, A., 123, *177*
Rhee, I., 316(48), 319(48), 320(48), *354*
Rhodes, R. E., 10(22), *73*
Richards, R. E., 196(93), *288*
Richards, R. L., 201(107), 203(122a), 207 (128), 219(153, 154), 233(174), 234(175), 244(153), 245(153, 154), 247(107, 175), 260(107, 153), 261(154), 282(247), *288*, *289*, *290*, *292*
Richardson, J. W., 253(208), 261(208), *291*
Richardson, W. H., 82(6), 83(6), *174*
Ripamonti, A., 95(41), 98(41), *175*
Roberts, E. R., 186, *287*
Robinson, S. D., 10(64, 65), 23(64, 65), 30 (64), 31(64), 33(65), *73*, 96(48), 98(48), 109(48), *175*
Robinson, W. R., 191(83), 192(83), *288*
Robinson, W. T., 217(147), 218(147), *290*, 358(9), 361(9), 362, 363(9), *388*
Rodda, G. K., 155, 161, 163(190), 164 (190), 165(190), *179*
Roper, W. R., 10(39), 21(112), 23(39, 112), 25(39), 28(39), *73*, *75*, 96(45, 54), 97(65, 66), 98(45, 54, 66), 103(54). 104(66), *175*, *176*, 358(27), 361(27), 366(27), 368(27), 373, *388*
Rosenblum, E. D., 184(44), *287*
Rosenthal, A., 341(113, 114), 342, *356*
Rossi, M., 20(101, 102), 21(101), *74*, 181 (12), 201(12, 106), 223, 224, 225, 229 (170), 245(170), 246(12, 106, 170), 259 (106), 260(106), *286*, *288*, *290*
Roughton, F. J. W., 86(12), 88(12, 19), *174*
Rowe, J. M., 2(8), *72*
Ruch, M., 302(23), 303(23), *353*
Rusnock, L., 211(136), *289*
Ryang, M., 316, 319(48, 49), 320, 322(44, 46, 47, 49, 52), 323(53), *354*

S

Sacco, A., 20(101), 102), 21(101), *74*, 181 (12), 201(12, 101, 106), 223, 224, 225, 229(170), 230(101), 231(101), 245(170), 246(12, 106, 170), 247(101), 259(106), 260(106), *286*, *288*, *290*
Saillant, R. B., 2(5), 10(5), *72*
Saito, T., 181(13, 14), 201(14), 222(13), 223 (14), 245(13), 246(14, 190), 258(190), 260 (14), *286*, *291*
Sales, K. D., 253(212), *291*
Salmenoja, E. L., 184(42), 185(42), *287*
Sampson, H. J., 337(101, 102), *355*
Sanders, J. R., 207(128), *289*
Satch, P. A., 127(132), *177*
Sato, K., 324, *354*
Saunders, B. C., 152(183), *179*
Sawa, Y., 316(47, 49, 50, 52), 319(49, 53), 320(47, 49, 53), 322(47, 49, 52), 323(53), *354*
Schaefer, W. P., 89, 90, *175*
Scheidegger, H. A., 220(156), 244(156), 262(156), *290*
Scherr, C. W., 252(206), *291*
Schlenk, H., 325(65), 336(65), *354*
Schlesinger, G., 198(94, 95), *288*
Schneider, K. C., 183(31), 184(31), *286*
Schneider, P. W., 29, *75*, 198(98), *288*
Schneller, S. W., 272(225), 272(229), *292*
Schofield, P., 161(203), 162(203), *179*
Scholhorn, R., 196(91), *288*
Schonbaum, G. R., 139(148), *178*
Schott, H., 97(68), 98(68), 105(68), 106(68), 108(68), *176*
Schorph, K., 31(121), 33, *75*
Schrauzer, G. N., 95(40), 98(40), *175*, 198, *288*

Schrock, R. R., 10(49), 23(49), 26(49), *73*
Schuster, L., 336(96), *355*
Schweckendieck, W. J., 325(64), 336(64), *354*
Scott, R., 182, 266, 267(23), 268(23), *286*
Scutt, P. B., 188(63, 64), *287*
Sears, C. T., Jr., 10(33), 23(33), 24(33), *73*
Seeley, D. A., 272(226), 279(226), *292*
Seigel, W. O., 109(80), *176*
Sellman, D., 236, 240(189), 247(179, 189), *290*, *291*
Senoff, C. V., 181, 201(1, 2), 202(1, 2), 203, 243(1, 2), 258(2), *285*
Seylor, J. K., 33(123), 34(123, 128), 37(123, 128), 38(128), 41(123), 42(123), *75*
Shapeley, J. R., 10(49), 23(49), 26(49), *73*
Shaw, B. L., 4, 9, 10(32, 36, 37, 51, 52, 66), 19(94), 20(106), 23(32, 36, 37, 51, 52, 66, 106), 24(32), 25(36, 37, 106), 27(51, 52), 28(32, 37, 51, 52, 106), 33(66), 62(198), *72*, *73*, *74*, *76*, 96(49), 98(49), 128, *175*, 349, *356*
Shen, T. Y., 325(66), *354*
Sheridan, P. S., 201(108, 115), 209(108), 212(108), 216(115), 243(115), 244(108), 262(115), *288*, *289*
Shestakov, G. K., 324(63), 325(63), *354*
Shiina, 69, 72(210), *77*
Shilov, A. E., 58(192), *76*, 181(8), 201(100), 202(119), 203(100, 124), 211(100), 218(8, 151), 220(151), 243(100), 244(124), 250, 253, 254(119), 258(119, 152), 260 (119), 262(137), 274(230), 282(242, 241a, 243, 244), 283, 284(244), *286*, *288*, *289*, *290*, *291*, *292*
Shilova, A. K., 201(100), 203(100, 124), 211(100), 218(152), 243(100), 244(124), 258(152), 262(137), 274(230), 282(242, 242a, 243, 244), 283(242, 242a, 243, 244), 284(244, 245), 284(245), *288*, *289*, *290*, *292*
Shore, P. S., 161(202), 162(202), 163(202), 169(202), *179*
Shteinman, A. A., 250(201), *291*
Shug, A. L., 188(65), *287*
Shur, V. B., 181(15, 16, 17, 18, 19, 20, 21), 266(15, 16, 17, 18, 19, 20), 267, 268(15, 16, 17, 18, 19), 269(18, 20), 270(21, 222, 223), 278(222), 280(239), 281(240), *286*, *291*, *292*
Shuvalov, N., 282(245), 284(245), *292*
Shuvalova, N. I., 282(244, 245), 283(244), 284(244, 245), *292*
Shvetsov, Yu, A., 181(6, 18), 201(6), 229 (6), 230(6), 246(6), 266(18), 267(18), 268 (18), 269(18), *286*
Siegel, H., 136, *178*
Sieker, L. C., 190(79), 191, 192(79), *288*
Sierichs, W. C., 384(68), 386(68), *389*
Silverstroni, P., 359, *388*
Silverthorn, W. E., 230(172), 233, 247(172), *290*
Simandi, L., 13, 14(81), 35, 36(143, 144), 38(146), 41, *74*, *75*
Simms, R. G., 10(46), 23(46), 26(46), *73*
Simonsen, S. H., 358(28), 361(28), 366(28), *388*
Simplicio, J., 90(32), *175*
Simpson, R. N. F., 20(105), *74*
Singh, R. N., 183(31), 184(31), *286*
Sirotkin, G. D., 359(37), *388*
Skelton, B. W., 96(54), 98(54), 103(54), *175*
Slade, R. M., 10(66), 23(66), 33(66), *73*
Sloan, M., 70(213), *77*
Sloane, N. H., 159(200), *179*
Smas, M. J., 10(67), 23(67), 32(67), 33(67), *74*
Smit, P. J., 378, 380(55), *389*
Smith, C. A., 10(59), 23(59), 29(59), *73*
Smith, D. C., 202(116), 238(182), 248(182), 250, *289*, *290*
Smith, G. W., 384(69), 386(69), *389*
Smith, J. G., 123(119), 124(119), *177*
Smith, J. R. L., 155(193), *179*
Smith, R. G., 218(148), 253(148), *290*
Snyder, D. A., 358(7), 361(7), *388*
Sokolvikii, D. V., 349(126), *356*
Song, K. M., 181(13), 222(13), 245(13), *286*
Sonogashira, K., 67, *77*, 97(67), 98(67), 105 (67), *176*
Southworth, R. W., 323, *354*
Spano, F. A., 385(78), *389*
Sparrow, G. J., 201(112), 207, 225(112), *289*
Spector, M. L., 339(106), *355*
Speier, G., 222, 223(160), 245(159), 246 (159), *290*
Spence, T., 187(60), 189(60), *287*
Spencer, E., 122(107), 128(133), *177*

Stainbank, R. E., 10(37), 23(37), 25(37), 28 (37), *73*, 96(49), 98(49), *175*
Standinger, H., 155(191, 192), *179*
Stanko, J. A., 259(215), 260(215), *291*
Starinshak, T. W., 259(215), 260(215), *291*
Stcherbatiuk, L. J., 125(126), 127(126), *177*
Stefanini, F. P., 20(105), *74*, 201(104), 219 (104), 232(104), 244(104), 245(104), 261 (104), *288*
Stein, G., 123(111), 124(111), 143(159, 160, 161, 162, 163, 164, 165), 161(159, 160, 161, 162, 165), 164(159), 165(160, 161, 162), 166(165), *177*, *178*
Steiner, W. A., 337(103), *355*
Steinmann, A. A., 250(202), *291*
Stenck, M. J., 358(24), 361(24), 369(24), 370(24), *388*
Stephens, R. D., 64, 66(200, 202), *77*
Stern, E. W., 339(106), *355*
Stern, G., 11(68), 23(68), *74*
Sternberg, H. W., 70, *77*, 336(98), 337, *355*
Stevens, J. R., 181(7), 201(7), 202(121a), 217(7), 244(7), 258(7), 260(7), *286*, *289*
Stockwell, P. B., 18(92), *74*
Stolberg, U. G., 42(152), 56(152, 185), 57 (152, 185), *75*, *76*
Strouse, C., 358(22), 361(22), 363, *388*
Sugimoto, R., 123(109), 124(109), 170 (109), *177*
Sun, J. Y., 181(10), 230(10), 247(10), *286*
Sunpu, L., 228(169), *290*
Susuki, T., 328(87, 88), 330(87, 88), *355*
Sutin, N., 147(177), *178*
Swanson, B. I., 358(22), 361(22), 363, *388*
Swanwick, M. G., 352(140), *356*
Swinden, G., 218(148), 253(148), *290*
Swinehart, J. W., 123(110), 124(110), *177*
Sykes, A. G., 94, 95, 111, *175*, *176*

T

Tabet, G. E., 324(57), *354*
Tabrina, G. M., 274(230), *292*
Tagawa, K., 183(34), 187, 190, 200(34), 201 (34), *286*, *287*
Takahashi, M., 41, *75*
Takahashi, S., 97(67), 98(67, 71), 102(71), 105(67), *176*
Takashi, I., 214, 244(141), 246(141), *289*
Takegami, Y., 311, *353*
Tanaka, M., 190(76), *288*
Taqui Khan, M. M., 45, *76*, 97(63, 64), 98 (63, 64), 103, 104(64, 75), 116(95), 118 (96, 97), 123(95), 124(95), *175*, *176*, 182, 216, 240, 244(143, 144), 247(188), 266 (25), 269(25), 273, *286*, *289*, *291*
Tatsuno, Y., 96(59), 97(59, 70), 98(59, 70), 106(70, 76), 107(76), 108(70, 78), 109 (78), *175*, *176*
Taube, H., 147, *178*, 181(3), 201(3, 113), 203, 205, 206(126), 207, 208, 209, 210, 211(135), 215(135), 220(156), 243(3, 126), 244(156), 262(156), *286*, *289*, *290*
Tayim, H. A., 61, *76*
Taylor, I. F., Jr., 96(53), 98(53), 100(53), *175*
Taylor, J. E., 134, *178*
Taylor, K. A., 102(72), *176*
Taylor, R. C., 20(104), 21(104), 57(188), *74*, *76*
Taylor, W. I., 136(143), *178*
Temkin, O. N., 324(63), 325(63), *354*
Temple, R. G., 130(136), *177*
Terry, N. W., III, 97(62), 98(62), 101, 102, *175*
Teuben, J. H., 240, 249(191), *291*
Thatcher, D. N., 384(69, 70), 386(69, 70), *389*
Thomas, D. S., 116(91), *176*
Thompson, D. T., 62(197), *76*
Tindall, J. B., 384(71, 72), 386(71, 72), *389*
Tischer, T. N., 135, *178*
Tokina, A. A., 58(192), *76*
Tolman, C. A., 47(171), *76*
Tominari, K., 236(180), 237(180, 181), 238 (183), 247(180), 248(181, 183), *290*
Torri, K., 161(211), 168(211), *179*
Trapasso, L. E., 13(75), *74*
Treichel, P. M., 191(82), 192(82), 194(82), *288*
Trietel, I. M., 206, 253(127), 262(127), 263, 265(127), *289*
Trocha-Grimshaw, J., 55(183), *76*
Tsay, F. D., 34, *75*
Tsay, Y. H., 238(186), 239(186), 249(186), *291*
Tsibris, J. C., 191(81), *288*
Tsuchida, R., 382(64), *389*

Tsuji, J., 328(80, 81, 82, 83, 84, 85, 86, 87, 88, 89), 329(83, 84, 85, 86), 330(87, 88), 331(89, 90), 336(90), 338, *355*
Tsutsumi, S., 316, 319(48, 49, 53), 320(46, 47, 48, 49), 322(44, 46, 47, 49, 52), 323 (53), *354*
Tuemmler, W. B., 384(73), 385(73), 386 (73), *389*
Tullberg, A., 358(21), 361(21), *388*
Tulupov, V. A., 66, 67, *77*
Tyantun, M. B., 250(202), *291*
Tyson, C. A., 121(100), *176*

U

Uchida, Y., 10(53), 23(53), 27(53), *73*, 181 (13, 14), 201(14), 222(13), 223(14), 226 (164, 166), 236(180), 237(180, 181), 238 (183), 245(13), 246(14, 164, 190), 248 (181, 183), 258(190), 260(14), *286*, *290*, *291*
Udenfriend, S., 154, 159(199), 161(202), 162(202), 163(202), 169(202), *179*
Ukhin, L. Yu, 181(6), 201(6), 229(6), 230 (6), 246(6), *286*
Ullrich, V., 155(191, 192), *179*
Untch, K. G., 159(200), *179*
Uri, N., 144(169, 170, 171), *178*

V

Vaciago, A., 358(20), 361(20), 369(20), *388*
Valentine, D., 96(51), 98(51), 109(51), *175*
Valentine, J., 96(51), 98(51), 109(51), *175*
Valentine, R. C., 187(55), 190(55, 70, 75), *287*, *288*
Van Bekkum, H., 58, *76*
Vanchesan, S., 103(73), *176*
Van der Ent, A., 47(169), 52(169), 55(184), 56(184), *76*
Van der Veer, D. G., 358(12), 361(12), 363 (12), 365(12), *388*
Van Gaal, H., 55(184), 56(184), *76*
Van Gogh, J., 58(191), *76*
Van Hardeveld, R., 250, *291*
Van Helden, R., 157(198), 160(201), 173 (201), *179*, 334(92), *355*
Van Heuvelen, A., 125(125), 127(125), *177*
Van Minnenpathuis, G., 58(191), *76*
Van Montfoort, A., 250, *291*
Vannerberg, N. G., 89, *175*, 358(21), 361 (21), *388*
Van Tamelen, E. E., 181, 241(194), 249 (194), 266(22), 269(22), 271, 272(224, 225, 226), 273, 277, 278, 279, 281, *286*, *291*, *292*
Vant'Hof, L. P., *76*
Varner, J. E., 187(52, 54), 188(52), *287*
Vaska, L., 2(10), 3(14, 15), 10(22), 19, 20 (15, 95, 97, 98, 99, 100), 21(15, 97, 98, 99), 22, 23, 45(115), 49(14), 54(14, 182), 55(182), *72*, *73*, *74*, *75*, *76*, 96(43, 46, 55), 97(62), 98(43, 46, 55), 101(62), 102, 107, *175*
Vastine, E., 181(10), 230(10), 247(10), *286*
Vaucheson, S., 45, *76*
Venanzi, L. M., 358(20), 361(20), 369(20), *388*
Vinogradova, S. M., 203(122), *289*
Virtanen, A. I., 184, 185, 187, *287*
Vleck, A., 89, *175*
Vogt, H., Jr., 89(22), *175*
Vohler, O., 311(36), *354*
Volkov, V. M., 230(171), 241(195), 247 (171), 249(195), *290*, *291*
Volkova, L. S., 230, 241(195), 247(171), 249(195), *290*, *291*
Volpin, M. E., 181, 266(15, 16, 17, 18, 19, 20), 267, 268(15, 16, 17, 18, 19), 269(18, 20), 270(222), 278, 280, 281, *286*, *291*, *292*
Volter, J., 42(151), *75*
Vonderschmitt, D., 112, *176*
Vorontsova, T. A., 282(242), 283(242), *292*

W

Walker, K. A. M., 52(174, 175), 54, *76*
Walling, C., 13, *74*
Walsh, A. D., 82(3), 83(3), *174*
Walters, K. K., 358(24), 361(24), 369(24), 370(24), *388*
Wang, J. H., 84(8), 86(8), 87, 88(8), 139, *174*, *178*
Wang, J. L., 134(139), *178*
Wang, L. C., 183(31), 184(31), *286*
Watanabe, Y., 311(32, 33, 34, 35), *353*
Waters, J. M., 358(14, 27), 361(14, 27), 366 (14, 27), 368(27), *388*

Waters, T. N., 96(54), 98(54), 103(54), *175*
Waters, W. A., 161(207, 208), 166(207, 208), *179*
Weaver, D. L., 358(7), 361(7), *388*
Weber, J. H., 112, *176*
Webster, A. H., 13, 14, 16(80), *74*
Webster, D. E., 32(122), 33(122), *75*
Wei, C. H., 191(82), 192(82), 194, *288*
Weil, J. A., 90(27), 94, 95, *175*
Weininger, M. S., 96(53), 98(53), 100, *175*
Weiss, J., 85(10), 86(10), 87, 88(10), 143, 144, 146, 161(159, 160, 161, 162, 165), 164(159), 165(160, 161, 162), 166(165), *174*, *178*
Weissberger, A., 116, *176*
Weissberger, E., 96(50), 98(50), 109(50), *175*, 205(126), 206(126), 243(126), *289*
Weller, S. W., 14(82), *74*
Wells, P. B., 32(122), 33(122), *75*
Wender, I., 70, *77*, 336(98), 337, *355*
Wheeler, A., 34(134), 35(134), *75*
Wheeler, S., 34(134), 35(134), *75*
Wheelock, K. S., 10(57), 23(57), 27(57), *73*
White, R. F. H., 171(216), *180*
Whiteford, J. E., 294(2), *353*
Whiteley, R. N., 202(121a), *289*
Whiting, M. C., 325(66), *354*
Whittle, K. R., 358(14), 361(14), 366(14), *388*
Wiberley, S. W., 89(22), *175*
Wier, M. R. S., 341(113), *356*
Wilke, G., 97(68), 98(68), 105, 106(68), 108 (68), *176*
Wilkes, G. R., 191(82), 192(82), 194(82), *288*
Wilkins, R. G., 90, *175*
Wilkinson, A. J., 349(127), *356*
Wilkinson, G., 2, 5, 10(29, 44), 18(92), 19 (7, 20, 93), 20(29, 104), 21(104), 23(44), 26(44), 30(6), 34, 42(160, 161), 44(163), 46, 47(29, 167, 168, 169), 49(29), 50(29, 168), 51, 52(169, 173), 53(176, 177), 56 (186), 57(187, 188), 70, *72*, *73*, *74*, *76*, *77*, 96(57), 98(57), *175*, 313(39), *354*, 358 (40), 359(36), 360(36), 361(40), 385(40), 386(40), *388*
Wilmarth, M. K., 10(25), 11, 13(74), 15 (25), *73*, *74*
Wilson, P. W., 183(31), 184(31, 43, 44), 187 (61), 188(65), *286*, *287*
Winfield, M. E., 34, 35, 36, *75*, 147, *178*
Winkler, H. J. S., 384(73), 385(73), 386 (73), *389*
Winogradsky, S., 182(27, 28), *286*
Winterton, N., 372(42), *388*
Witkop, B., 159(199), *179*
Wittenberg, J. B., 88(20), *174*
Wojcicki, A., 294, 295(7), 296, 306(26), 307, *353*
Woo, L., 143(158), 154(158), 155(158), 157(158), 161(158), 164(158), *178*
Wood, D. C., 303, *353*
Woods, M., 95, *175*
Woody, R. W., 191(81), *288*
Woolcock, W. J. V., 218(148), 253(148), *290*
Workman, R. J., 143(158), 154(158), 155 (158), 157(158), 161(158), 164(158), *178*
Wright, W. A., 184(39), 187(39), 194(39), *287*

X

Xuong, Na. H., 191(80), 192(80), 193(80), *288*

Y

Yagupsky, G., 53(176, 177), *76*
Yagupsky, M., 53(177), *76*
Yamada, H., 125(124), 127(124), *177*
Yamada, S., 382, *389*
Yamamoto, A., 181(11), 213, 214(141), 215 (138), 222, 224(11), 226(11, 165), 228 (168, 169, 169a), 244(138, 141), 245(11), 246(141), 279(238), *286*, *289*, *290*, *292*
Yamamoto, K., 324, *354*
Yamamoto, M., 96(59), 97(59), 98(59), 106 (76), 107(76), *175*, *176*
Yamanaka, T., 183, 190, *287*
Yamano, Y., 125(124), 127(124), *177*
Yamasaki, I., 120(99), 122(107), 127(99), 128(133), *176*, *177*
Yamazaki, M., 357(1), *387*
Yamiya, Y., 170(213, 214), *179*
Yamortino, R. L., 327(75), *355*
Yan, J. F., 134(139), *178*

Yasunobu, K. J., 125(124), 127(124), *177*, 190(76), *288*
Yasuoka, N., 98(71), 102(71), *176*
Yatsimirskii, K. B., 82, 83(5), *174*, 182, 270, *286*
Yokokawa, C., 311(32, 33, 34, 35), *353*
Yokoo, H., 316(51), *354*
Yoshida, K., 316(51), *354*
Young, D. P., 153(184), *179*
Young, F. J., 42(160), 46(160), 51(160), *76*
Young, J. F., 10(29), 20(29, 104), 21(104), 42(160, 161), 46(160), 47(29, 167), 49(29), 50(29), 51(160), 56(186), 57(187, 188), *73*, *74*, *76*

Z

Zakumbaeva, G. D., 349, *356*
Zamaraev, C. I., 125(126), 127(126), *177*
Zambonelli, L., 358(20), 361(20), 369(20), *388*
Zelitch, I., 184(44), *287*
Zie, M. C., 214(140), 244(140), *289*
Zito, R., 122(101, 106), *176*, *177*
Zubrzychi, Z., 155(192), *179*

Subject Index

A

Acylmanganese(0) carbonyls, decarbonylation of, 297–304
Acyloins, synthesis of, with nickel carbonyl, 323
Alcohol oxidation, by hydrogen peroxide, 141
Aldehydes, synthesis of, with metal carbonyls, 320, 321
Algae
  blue-green, 183
  mechanism of nitrogen fixation by, 200
Alkene complexes, representation of, 392
Alkenes
  catalytic hydrogenation of
    by complexes of various metal ions, 66
    by iridium(I) complexes, 54
    by palladium(II) complexes, 60
    by platinum(II) complexes, 56
    by rhodium(I) complexes, 46
    by ruthenium(II) complexes, 43
  carbonylation of, 325–335
Alkyl halides
  addition of, to platinum group complexes, 23–27
  mechanism of addition to metal complexes, 28, 29
Alkyl metal complexes, carbon monoxide insertion reactions in, 296
Alkyne complexes, representation of, 392
Alkynes
  carbonylation of, 325
  hydrogenation of, by chromium(II) complexes, 64
π-Allyl complexes, representation of, 393
Amines, carbonylation of, 336
Amino acid oxidase
  action, 125
  model, 125
Ascorbic acid oxidase action, 120
Ascorbic acid oxidation
  catalysis by
    ascorbic acid oxidase, 120
    copper(II) and iron(III) complexes, 118
    copper(II) and iron(III) ions, 116
    vanadyl ion, 118
  by molecular oxygen, 114

B

Bacteria
  free-living
    in aerobic systems, 194, 197
    in anaerobic systems, 190
    mechanism of nitrogen fixation by 190, 194
      model systems for, 198

Bacteria (*continued*)
symbiotic, 182, 183
mechanism of nitrogen fixation by, 189
Benzoic acid oxidation, catalysis of, by copper(II), 130
Bonding orbitals
for binuclear dinitrogen complexes, 256
for edge-on dinitrogen complexes, 257
for $[Ru(NH_3)_5—N_2]^{2+}$, 255
Bonds, covalent, coordinate, representation of, 392

C

Cadmium, diphenyl, nitrosation of, 382
Carbenes, metal, 352
Carbon–hydrogen bond, activation of, by platinum group complexes, 30
Carbon monoxide
complexes
as catalysts in carbonylation reactions, 316–344
exchange reactions of, 294
insertion reactions, 296, 304, 306, 307
mechanism of insertion reactions, 299, 304, 305
mechanism of substitution reactions, 302
substitution reactions of, 298, 294
insertion
of aryl groups in metal complexes, 317
in cobalt complexes, 307
in iridium(III) complexes, 315
in iron(0) complexes, 306
in manganese(0) complexes, 296
mechanism of, 299–304
in molybdenum(0) complexes, 304
mechanism of, 305
in palladium(II) complexes, 313
in platinum(II) complexes, 313
in rhodium(III) complexes, 314
metal-catalyzed oxidation of, 344
oxidation of, by
copper(II), 348
mercury(II), 344
nickel(II), 347
palladium(II) complexes, 350
permanganate, 348
silver(I), 345
Carbonyl complexes, *see also* Carbon monoxide
Carbonyl metal complexes
as hydrogenation catalysts, 68–72
mechanism of hydrogenation catalysis by, 71
Carboxylic acids, synthesis of
with cobalt carbonyls, 325
with mercury(II) complexes, 335
Catalase enzymes
mechanism of action of, 139
oxidation of alcohols by, 141
prosthetic group in, 141
Catalase model systems, 136–139
Catecholase action, 122
Catechol–iron(III) hydrogen peroxide complex, 151
Chromium(III) complexes, nitrogen fixation by, 267
Chromium nitric oxide complex, 361, 370
Cobalt carbonyls
carbonyl replacement in, 307
as catalyst in carbonylation of amines, 337
as catalyst in carbonylation reactions, 327
in cyclicization of
amides, 341
azo compounds, 343
imines, 342, 344
decarbonylation of acyl groups in, 308
dissociation of acyl derivatives, 309
insertion in alkyl and aryl derivatives, 307
isomerization of alkyl and acyl groups, 311
mechanism of carbonyl insertion reactions, 309
in synthesis of heterocyclic nitrogen compounds, 339
in synthesis of phthalamidines, 340
Cobalt(0) dinitrogen complexes
exchange reactions of, 227
hydridodinitrogentris(triphenylphosphine)cobalt(I), 226
structure of, 225
synthesis of, 222

Cobalt–nitric oxide complexes, 361, 369
reaction of, with dioxygen, 374
Cobalt(II)
catalysis, of oxidation of coordinated ligand, 132
complexes, nitrogen fixation by, 267
Copper(II)
catalysis by, of
nitrosation of alcohols, 380
nitrosation of amines, 379
as catalyst for carbon monoxide oxidation, 348
complex catalysis of ascorbic acid oxidation, 118
complexes, catalase action of, 137
ion catalysis of
ascorbic acid oxidation, 116
oxidation of
benzoic acid, 130
methanol, 131
oxidative coupling reactions, 135
Coordinate bonds, representation of, by chemical formulas, 392
Coupling reactions, oxidative, catalysis of, by copper(II), 135
Cyclopentadienyl metal complex, representation of, 393
Cysteine oxidation, catalysis of, by iron(III), 134

D

Deuterium exchange, of cobalt(0) dinitrogen complexes, 227
Dienes
carboxylation of, 330–333
hydrogenation of
catalysis by metal carbonyls, 68–72
with pentacyanocobaltate(II), 38
Dinitrogen
bonding orbitals of binuclear complex, 256
bonding orbitals of ruthenium(II) complex, 255
cobalt(0) complexes of, 222
fixation of, 266–285
iridium(I) complexes of, 230
iron(II) complexes of, 230
manganese(I) complexes of, 240
manganese(II) complexes of, 239
metal complexes of, 200–266
metal complexes of, summary, 243
metal surface complexes of, 242
molecular orbitals of, 18
molecular orbital diagram of
binuclear ruthenium(II) complex, 264
cobalt(I) complex, 265
ruthenium(II) complex, 263
molybdenum(0) complexes of, 236
nonmetallic complexes of, 251
osmium(II) complexes of, 217
platinum(III) complexes of, 241
platinum(IV) complexes of, 241
rhenium(I) complexes of, 233
rhodium(I) complexes of, 229
ruthenium(II) complexes of, 202
stability of, 181
stabilities of metal complexes of, 258–285
structure and bonding of metal complexes of, 251
titanium(III) complexes of, 240
tungsten(0) complexes of, 238
vibrational frequencies in metal complexes of, 258, 260
Dioxygen
amino acid oxidation by, 123
ascorbic acid oxidation by, 114–118
carriers, synthetic
binuclear complexes, 89
mononuclear complexes, 96–110
properties of, 96–97
catalysis of reactions by
ascorbic acid oxidase, 120
catecholase enzymes, 122
tyrosinase, 122
catechol oxidation by, 120
complexes
of amino acid and peptide complexes, 92, 93
of cobalt(II) complexes, 89–98
coordinate bonding in, 106
of iridium(I) complexes, 96, 98
of iron(II)-phthalocyaninetetrasulfonate, 113
of manganese(II)-phthalocyanine, 111
mechanism of
formation of, 102
reaction of, 110–114

Dioxygen, complexes (*continued*)
mononuclear, of cobalt(II) complexes, 96, 101
oxidation of ascorbic acid by, 114
of polyamine metal complexes, 91
reactions of, 108
ruthenium(II) complexes, 103
thermodynamics of formation of, 94
of various metal complexes, 97, 105
insertion reactions, definition of, 80
non-insertion reactions, definition of, 79, 84
oxidation of metal ions by partial insertion reactions, definition of, 81
reactions
catalysis of, by
amino acid oxidase, 125
dioxygenase enzymes, 172
laccase, 128
monooxygenase enzymes, 155
monooxygenase models, 154
tyrosinase, 128
Udenfriend's system, 154
classification of, 79
energetics of, 82
mechanism of action of monooxygenase enzymes, 158
with coordinated nitric oxide, 374

E

Ester synthesis by the Reppe reaction, 325
Exchange reactions of carbon monoxide complexes, 294

F

Fenton's reagent
hydroxylation reactions by, 148
reaction mechanisms of, 143–149
Ferredoxin
clostridial, 191
mechanism of action, 193
model systems, 192
role of, in nitrogen fixation, 195, 197, 201
Formamide derivatives, synthesis of, by metal-catalyzed carbonylation, 336
Formulas of metal complexes, conventions employed, 391

H

Hemoglobin
prosthetic group of, 85
reactions of, 87, 88
Heterocyclic nitrogen compounds, synthesis of, by carbonyl insertion, 339
Hydrazines, unsymmetric
metal catalysis, in synthesis of, 386
synthesis of, by nitrosation, 384
Hydride complexes
molecular orbitals of, 5–9
octahedral, 5, 6
square planar, 7, 8
stability of and lability of, 4
structure of, 3
tetrahedral, 9
Hydrogen
activation
by chloro complexes of various metals, 12, 17
by copper(II) ions, 11
factors involved in, 1, 68
by mercury(I), 11
by pentacyanocobalt(II), 33
by silver(I), 13
by silver(I) complexes, 15
transition states for, 12, 15, 16
molecular
addition to iridium(I) complexes, 21
catalysis of ligand exchange by, 18
reactions of with group VIII metal complexes, 20
Hydrogen peroxide
as a two-electron oxidant, 136
iron(II) catalysis of, 143
complexes, with iron(III) and catechol, 151
Hydrogenation
of alkenes
by complexes of various metal ions, 66
catalysis by
palladium(II) complexes, 60
platinum(II) complexes, 56
mechanism of
palladium(II) catalysis, 62, 63
platinum(II) catalysis, 57, 60, 62
of alkynes
catalysis by
chromium(II) complexes, 64
titanium(II) complexes, 67

mechanism of chromium(II) catalysis, 65
of dienes
catalysis by metal carbonyls, 68
mechanism of, with metal carbonyls, 71, 72
by pentacyanocobaltate(II), 37
of unsaturated compounds
catalysis of, by
iridium(I) complexes, 54
iridium(II) complexes, 55
pentacyanocobaltate(II), 37
rhodium(I) complexes, 46
ruthenium(II) complexes, 42
Hydroxylation, *see* Oxygen

I

Infrared spectra, of metal dinitrogen complexes, 258, 260
Insertion reactions, metal-catalyzed
applications of carbon monoxide insertion, 316
of carbon monoxide, 296
of oxygen, 154, 170
synthesis of
acyloins by, 323
aldehydes by, 320, 321
carboxylic acids, 335
heterocyclic nitrogen compounds, 339–344
ketones by, 320, 322
substituted formamides, 336
substituted ureas, 337
Iridium hydride complexes, deuterium exchange of, 22
Iridium(I)
complexes
addition of various compounds to, 21
catalysis of hydrogenation of alkenes by, 54
mechanism of hydrogenation of alkenes by, 55
dioxygen complexes, 96, 98
dinitrogen complexes, 230
Iridium(III) complexes, carbonyl insertion in, 315
Iridium–nitric oxide complexes, 362, 364, 368
protonation of, 374
Iron carbonyls, alkyl, carbon monoxide insertion in, 306
Iron–nitric oxide complex, 361
Iron(II)
catalysis
of oxidation of hydrogen peroxide, 143
of oxidation of mercaptoacetic acid, 135
complex catalysis, of ascorbic acid oxidation, 118
as hydrogenation catalyst, 6
dinitrogen complexes, 230
Iron(III)
catalysis
of ascorbic acid oxidation, 116
of oxidation of cysteine, 134
catechol–hydrogen peroxide complex, 151
complexes
as catalase models, 137
nitrogen fixation by, 267, 273
Isomerization, of acyl and alkyl cobalt carbonyls, 311
mechanism of, 313

K

Ketones, synthesis of, with metal carbonyls, 320, 322

L

Laccase, oxidation of catechols by, 128
Ligand oxidation, catalysis of by cobalt(II), 132
Lithium alkyls, carbonylation of, 320–323

M

Manganese carbonyls, as catalysts in carbonylation of amines, 338
Manganese–iron oxide complex, 361
Manganesepentacarbonyls
acyl, decarbonylation of, 297
alkyl, carbon monoxide insertion in, 296
Manganese(I)–dinitrogen complexes, 240
Manganese(II)
catalysis of
amino acids by dioxygen, 125
catechol oxidation by dioxygen, 120

Manganese(II) (*continued*)
complexes
as catalase models, 137
as hydrogenation catalysts, 67
dinitrogen complexes, 239
Mercaptoacetic acid oxidation, catalysis of, by iron(II), 135
Mercury(II)
catalysis of carbon monoxide oxidation, 344
complexes, carbonylation of, 336
as oxidant for carbon monoxide, 344
Metal carbenes, 352
Metal–carbon bonds, carbon monoxide insertion in, 316
Metal carbonyls, in synthesis of metal carbenes, 352
Metal catalysis, of nitrosamine reduction, 386
Metal complexes, representation of, 392
Metal ion requirement, for nitrogen fixation, 186
Metapryocatechase action, 172
Methanol oxidation, catalysis of, by copper(II), 131
Microorganisms, photosynthetic, 182, 183
Molecular orbital digrams
for hydrido complexes, 5–9
for nitric oxide complexes, 371
of dinitrogen, 252
of $Co(PPh_3)_3H(N_2)$, 265
of $Ru(NH_3)_5(N_2)^{2+}$, 263
of $[Ru(NH_3)_5]_2(N_2)^{4+}$, 264
Molybdenum complexes, in nitrogen fixation, 198
Molybdenum(0) carbonyls, cyclopentadienyl
carbon monoxide insertion, 304
dinitrogen complexes, 236
mechanism of carbon monoxide insertion, 305
Molybdenum(V) complexes, nitrogen fixation by, 266
Molybdenum(VI) complexes, nitrogen fixation by, 267
Monooxygenase models, 154
mechanism of action, 156
Myoglobin
prosthetic group, 85
reactions of, 87

N

Nickel(0) dinitrogen complexes
structure of, 239
synthesis of, 238
Nickel(II) complexes
as hydrogenation catalysts, 67
nitrogen fixation by, 267
Nickel–nitric oxide complex, 361, 363
Nickel tetracarbonyl, as catalyst in the Reppe reaction, 325
Nitric oxide, *see also* Nitrosyl complexes
complexes of
chromium, 361, 370
cobalt, 361, 369
iridium, 361, 362, 364, 368
iron, 361
manganese, 361
nickel, 361, 363
osmium, 361, 368
ruthenium, 361, 365, 367, 369
coordinated, 358
oxidation of, 374
protonation of, 374
reactions of, 372
reactions with nucleophilic reagents, 373
reactions
rearrangement of nitrosamines, 383
ring expansion, 385
synthesis of unsymmetric hydrazines, 384
Nitrogen fixation
biological
mechanism of, 188
by microorganisms, 182, 183
model systems, 198, 199
products of, from azotobacter, 185
requirements for, 186
substrates for natural fixation, 197
chemical, 266–285
catalysts for
chromium(III), 267
cobalt(II), 267
iridium(III), 273
iron(III), 267
molybdenum(V), 266
molybdenum(VI), 267
nickel(II), 267
osmium(II), 273
palladium(II), 267

platinum(II), 267
ruthenium(II), 273
titanium(II), 270, 271, 273, 280
titanium(III), 280
titanium(IV), 267, 271
tungsten(0), 284
tungsten(VI), 266, 267
various reduced metal compounds, 270
formation of azo compounds, 280
formation of hydrazine, 282
mechanism of, 273–282
mechanism of titanium(II) catalysis of, 273–280
reduction of diimine, 282
table of metal-catalyzed nitrogen fixation, 268
molecular, *see* Dinitrogen
Nitrosamines, rearrangements of, 383
Nitrosation
of alcohols, 376
copper(II) catalysis of, 378
mechanism of, 379
definition of, 375
iron(II) catalysis of, 377
of metal alkyls, 381
of primary amines, 376
Nitrosyl complexes, *see also* Nitric oxide
linear nitrosyl complexes, bond orbitals of, 371
mixed ligand nitrosyl complexes, 358
properties of, 360
structures of, 360–370
table of structural parameters, 361

O

Osmium(II) dinitrogen complexes
binuclear complexes, 220
mechanism of formation, 219
nitrogen fixation by, 273
structure of, 218
synthesis of, 217–222
Osmium nitric oxide complex, 361, 368
protonation of, 374
Oxidase enzymes, catalysis of two-electron oxidations by, 127
Oxygen carriers, natural
properties of, 86
structure of prosthetic group, 85
Oxygen
insertion
by dioxygenase enzymes, 172
free radical reactions, 159, 170, 173
mechanism of mixed function oxidase action, 159
by metal organoperoxides, 171
by metal peroxides, 152
by monooxygenase enzymes, 155
by monooxygenase models, 154
by organic hydroperoxides, 154
partial, 154–170
pyrocatechase action, 173
reaction
with Fenton's reagent, 148
by model peroxidase systems, 149
synthetic applications of, 160–169
total, 170–174
Udenfriend's system, 154
molecular, *see* Dioxygen

P

Palladium(II) complexes
in carbonylation of alkenes, 328
carbonyl insertion in, 313
catalysis of hydrogenation of alkenes by, 56
nitrogen fixation by, 267
as oxidant for carbon monoxide, 350
Pentacyanocobaltate(II)
activation of molecular hydrogen by, 33
catalysis of hydrogenation by, 37
mechanism of hydrogenation with, 39, 40
mechanism of reaction of hydrogen with, 35
Permanganate, as oxidant for carbon monoxide, 348
Peroxidase enzymes, mechanism of action of, 142, 149
Peroxidase models, 149
Phthalamidines, synthesis by carbonyl insertion reactions, 340
mechanism of, 343
Platinum(II) complexes
carbonyl insertion in, 313
catalysis of hydrogenation of alkenes by, 56
nitrogen fixation by, 267

Platinum group metal complexes
activation of hydrogen–carbon bond by, 30
reactions of, with acyl halides, 23–27
reactions of, with alkyl halides, 23–27
Platinum(III) dinitrogen complexes, 241
Platinum(IV) dinitrogen complexes, 241
Polyamine complexes, of various metals, as catalysts for catalase action, 137–139
Pyrocatechase action, 172

R

Reppe synthesis, of carboxylic acid esters, 325
Rhenium(I) dinitrogen complexes
structure of, 235
synthesis of, 233
Rhodium(I) complexes, *see* Alkenes
catalysis of hydrogenation of alkenes by, 46
mechanism of hydrogenation of alkenes by, 49–54
Rhodium(I)
dinitrogen complexes, 229
dioxygen complexes, 96, 102
Rhodium(II) complexes, nitrogen fixation by, 273
Rhodium(III) complexes, carbonyl insertion in, 314
Ruthenium nitric oxide complexes, 361, 365, 367, 369
Ruthenium(II)
catalysis of hydrogenation of alkenes mechanism, 42
complexes
bonding orbital of binuclear dinitrogen complex, 256
bonding orbitals of mononuclear dinitrogen complex, 255
molecular orbital diagram of dinitrogen complex, 263
nitrogen fixation by, 273
dinitrogen complexes, 202
binuclear complexes, 220
bisdinitrogen complexes, 216
mechanism of formation of, 210, 212
nitrogen fixation by, 273
structures of, 205, 206, 210
with tertiary phosphines and arsines, 213
dioxygen complexes, 97, 103

S

Sandwich compounds, representation of, 393
Silver(I), as oxidant for carbon monoxide, 345
Superoxide complexes, of cobalt dimethylglyoxime, 95

T

Titanium(II) complexes, nitrogen fixation by, 270, 271, 273, 280
Titanium(III) dinitrogen complexes, 240
nitrogen fixation by, 280
Titanium(IV) complexes, nitrogen fixation by, 261, 267
Transition metal dinitrogen complexes
synthesis and mechanisms, 200–266
table of properties of, 243
Transition state, representation of, 392
Tungsten(0) dinitrogen complexes, 238
nitrogen fixation by, 284
Tungsten(VI) complexes, nitrogen fixation by, 266, 267
Tyrosinase action, 122
Tyrosinase, catecholase action of, 128
Tyrptophanpyrrolase, 172

U

Udenfriend's system, 154
Unsaturated compounds, conjugated, hydrogenation of by pentacyanocobaltate(II), 38
Urea, alkyl derivatives, synthesis of by metal-catalyzed carbonylation, 337

V

Vanadyl ion catalysis, of ascorbic acid oxidation, 118
Vibrational frequencies, of N–N bond in metal complexes, 258, 260

Z

Zinc alkyls, nitrosation of, 381

A
B 4
C 5
D 6
E 7
F 8
G 9
H 0
I 1
J 2